Handbook of

ELECTRON SPIN RESONANCE

Volume 2

Handbook of

ELECTRON SPIN RESONANCE

Volume 2

Edited by

CHARLES P. POOLE, JR.
and
HORACIO A. FARACH

AIP PRESS

Springer

Editors
Charles P. Poole, Jr.
Horacio A. Farach
Department of Physics and Astronomy
University of South Carolina
Columbia, SC 29208
USA

Library of Congress Cataloging-in-Publication Data
Handbook of electron spin resonance / edited by Charles A. Poole, Jr.,
Horacio A. Farach
p. cm.
Includes bibliographic references and index.
ISBN 1-56396-044-3 (v. 1: hc)
ISBN 0-387-98660-X (v. 2: hc: alk. paper)
1. Electron paramagnetic resonance. 2. Electron nuclear double
resonance spectroscopy. 3. Relaxation phenomena.
I. Farach, Horacio A. II. Title
QC762.P5 1999
537′.364—dc20 93-33834

Printed on acid-free paper.

Production coordinated by Brian Howe and managed by Steven Pisano; manufacturing supervised by Joe Quatela.
Typeset by Asco Typesetters, Hong Kong.
Printed and bound by Hamilton Printing Co., Rensselaer, NY.
Printed in the United States of America.

9 8 7 6 5 4 3 2 1

ISBN 0-387-98660-X Springer-Verlag New York Berlin Heidelberg SPIN 10697590

PREFACE

The field of electron spin resonance (ESR) and its sister field of nuclear magnetic resonance (NMR) were founded about the same time 55 years ago. Both fields have contributed a great deal to our knowledge of physics, chemistry, biology, medicine, and the related sciences. As they have developed and matured they have continuously advanced in the sophistication of their instrumentation and experimental techniques, and they have repeatedly impacted new fields in which they have provided important insights necessary for those fields to move forward. Most of us have watched magnetic resonance imaging (MRI) rise during the past few years to become a major diagnostic tool in medicine, furnishing pictures of internal organs that are superior to those obtained with X-rays. Much of the detailed information provided by MRI is not otherwise obtainable.

When major research fields grow they acquire an extensive literature of experimental data, and these data must be collected in review articles, and eventually in handbooks, to make them useful. Over the years the various subfields of ESR have been reviewed from time to time, and many monographs have been published. Data tabulations have appeared in, e.g., *Advances in Magnetic Resonance*, *Annual Reports of the Chemical Society*, *Chemical Reviews*, *Landolt–Bornstein*, *Magnetic Resonance Review*, *Reports on Progress in Physics*, and *Solid State Physics*. The present editors wrote a chapter on ESR for the three-volume work, *Handbook of Spectroscopy*, which appeared in 1974, edited by J. W. Robinson, and a shorter version of this chapter was included in the one-volume abridged *Handbook of Spectroscopy*, which was published in 1991. For a long time there has been a need for a more comprehensive handbook, and this work hopes to fill that need.

The object of this book is to provide a convenient source of information on ESR. Volume 1 accomplished part of this task, and the present volume represents an attempt to complete this work. The first volume surveyed general aspects of the literature, and provided specific chapters on computer techniques, relaxation, and electron-nuclear double resonance (ENDOR). The present volume contains chapters on sensitivity, resonators, lineshapes, electron spin echo envelope modulation, transition series ions, Hamiltonian types and symmetries, imaging, and high-field ESR. Chapter 8 provides a particularly thorough tabulation of transition ion data.

CONTENTS

CHAPTER I
SENSITIVITY

George A. Rinard, Sandra S. Eaton, and Gareth R. Eaton

Department of Engineering and Department of Chemistry and Biochemistry, University of Denver, Denver, Colorado 80208, USA

Charles P. Poole, Jr. and Horacio A. Farach

Department of Physics and Astronomy, University of South Carolina, Columbia, South Carolina 29208, USA

This chapter describes the various factors that contribute to the sensitivity and shows how to optimize the sensitivity.

CONTENTS

A. INTRODUCTION

For many applications of ESR the number of spins required for a measurement is the key to feasibility.[1] In a typical spectrometer the minimum number of detectable spins depends upon the sample, the resonator, the microwave source and detection system, and on the suppression of noise from various sources. Because so many variables are involved, a wide range of expressions has been presented in the literature, that apply to different cases. For a particular type of sample, a user wants to know how to get the best possible spectrometer performance. In addition, we would like to know how sensitivity varies with microwave frequency. Much of the literature has focused on continuous wave (CW) experiments and that information is outlined here. We also present briefly some factors that are involved for time-domain experiments. Much of the early literature was based on rectangular or cylindrical cavities. Since current ESR spectrometers use a variety of structures, we use the term resonator rather than cavity. Equations are in SI (i.e., MKS) units.

In Sections B to D of this chapter we discuss a relatively general expression for the minimum number of detectable spins, $N_{\min}$, with an explicit statement of the underlying assumptions, a discussion of the expected frequency dependence of sensitivity for three cases, and a summary of experimental results for sensitivity at various microwave frequencies. For those who want further details, additional definitions and derivations of equations are given in Section E.

B. MINIMUM NUMBER OF DETECTABLE SPINS

Historically the sensitivity of ESR spectrometers has been described in terms of the minimum number of spins required to give a signal-to-noise ratio of 1. The fundamental detection equation (1) is the starting point for determining the sensitivity[2,3,4,5]

$$|\Delta E| = \chi'' \eta Q_L \sqrt{Z_0 P}. \tag{1}$$

The derivation of this equation is given in Section E.2. ΔE is the signal voltage at the resonator, η is the resonator filling factor, Q_L is the loaded quality factor Q of the resonator, Z_0 is the characteristic impedance of the line, and P is the microwave power incident on the resonator produced by the external microwave source. The magnetic susceptibility of the sample, χ'', is the imaginary component of the effective RF susceptibility, $\chi = \chi' + i\chi''$, where χ'' results in absorption of energy from the RF field. For a Lorentzian line[2] with linewidth $\Delta\omega$ at the resonance frequency ω_0, $\chi'' = \chi_0(\omega_0/\Delta\omega)$ and for $S = \frac{1}{2}$ we have $\chi_0 = Ng^2\beta^2\mu_0/4V_s k_B T_{\text{sample}}$, where N is the number of spins in volume V_s, and T_{sample} is the temperature of the sample. Substitution for χ'' and x_0 in (1) gives

$$|\Delta E| = \frac{Ng^2\beta^2\mu_0}{4V_s k_B T_{\text{sample}}} \left(\frac{\omega_0}{\Delta\omega}\right) \eta Q_L \sqrt{Z_0 P}. \tag{2}$$

To obtain an expression for the minimum number of detectable spins we equate the ESR signal voltage ΔE, given by (2), to the effective noise voltage at the detector, $E_n = \sqrt{Z_0 P_n}$, where P_n is the noise power at the detector due to all sources. The result is given by (3), which is written as the product of terms, representing the effects, respectively, of the sample, the resonance condition, the cavity, and the detection system[3]

$$N_{\min} = \left(\frac{4k_B}{g^2\beta^2\mu_0} V_S T_{\text{sample}}\right) \left(\frac{\Delta\omega}{\omega_0}\right) \left(\frac{1}{Q_L \eta}\right) \left(\frac{P_n}{P}\right)^{1/2}, \tag{3}$$

where $N_{\min}$ is the number of spins in volume V_S. P_n is often given in the literature as $P_n = F k_B T_{\text{det}} \Delta\nu$, were F is the noise factor of the detection system (see Section E.3), T_{det} is the temperature of the detector, and $\Delta\nu$ is the bandwidth of the detection system. This relationship is based on the assumption that the noise is dominated by detector thermal noise, which may not always be the case. Equation (3) is more general, since P_n includes all noise sources in the spectrometer. The factor $\Delta\omega/\omega_0$ can be replaced by $\Delta B/B_0$ where ΔB is the linewidth and B_0 is the resonant field. A sample calculation based on (3) is given in Section D.1.

Inherent in (3) are multiple assumptions that are discussed in the following sections. Equation (3) provides a useful summary of the functional dependence of sensitivity on T, Q_L, η, P, and the performance of the detection system. It is also useful in comparing spectrometer performance when Q_L and η are known. However, to examine the dependence of sensitivity on ω requires analysis of the frequency dependence of Q_L and η. A more detailed consideration of noise is given in Section C.4.

1. Sample and the Resonance Condition

a. Lineshape, Hyperfine Splitting, and $S > \frac{1}{2}$

The expressions for the effect of the sample and the resonance condition in (3) are based on the assumption that the signal is due to an allowed transition that is sufficiently narrow that it is completely encompassed by the bandwidth of the resonator and that $S = \frac{1}{2}$. The conversion from χ'' to $\Delta\omega/\omega_0$ was based on a Lorentzian absorption line with T_2-determined linewidth. The detected voltage depends upon signal amplitude, but the number of spins is proportional to the area under the curve. For a Lorentzian and a Gaussian absorption line with the same linewidth at half height and the same peak height, the area under the Lorentzian curve is 1.476 times that for the Gaussian curve,[5] For a Lorentzian and a Gaussian first derivative line with the same peak-to-peak width and the same signal height, the area under the Lorentzian curve is 3.511 times that for the Gaussian curve.[6] Thus $N_{\min}$ depends upon the lineshape.

For a signal with hyperfine splitting, the resonance absorption is spread over multiple lines that may have different intensities. Typically only one of these lines will fall within the bandwidth of the resonator and the value of $N_{\min}$ is reduced by the ratio of the intensity of the observed line to the total intensity of the spectrum.

The case for $S > \frac{1}{2}$ requires replacement of the factor of 4 in (3) by $3/(S(S+1))$. Typically for $S > \frac{1}{2}$ the ESR signal intensity is spread over a wide range of resonant fields. Calculation of $N_{\min}$ then requires correction for the fraction of signal intensity that falls within the resonator bandwidth. In the extreme, if S is integer and the zero-field splitting is greater than the ESR quantum, there may be no observable transitions.

b. Power Saturation and Temperature Dependence

In (3) it is assumed that P is low enough so the spectrum is not saturated, i.e., that the signal response is in the regime that increases linearly with increase in $P^{1/2}$.

The ESR absorption intensity is proportional to the population difference of the two spin levels, which is given by the Boltzmann equation.[2] In the high temperature limit,[2] where $g\beta B \ll k_B T$, signal intensity is proportional to $1/T$. This limit is assumed in (2) and (3). In the low temperature limit,[2] where $g\beta B \gg k_B T$, signal intensity is independent of T. Intermediate temperature dependence is observed at intermediate temperatures (see the figure on p. 411 of ref. 2). The temperatures at which these limits apply depend upon the magnetic field. At X-band, the high temperature limit pertains above about 10 K, but higher temperatures are required to achieve this regime at the magnetic fields required to achieve resonance for $g = 2$ at higher microwave frequencies.

Based only on Boltzmann populations in the high-temperature limit, we would expect to improve sensitivity by operating at a lower sample temperature. However, it is also necessary to consider the temperature dependence of linewidths and relaxation times. A change in sample temperature that results in a decrease in linewidth, increases signal amplitude. Values of T_1 increase with decreasing temperature with a functional dependence that is strongly sample dependent.[7] T_2 is frequently independent of temperature at low temperatures, but decreases due to dynamic processes that average inequivalent electron spin environments and decreases with increasing temperature as T_1 approaches T_2. The increase in relaxation times with decreasing temperature usually requires a decrease in microwave power to avoid saturation, which may outweigh the increase in the Boltzmann population difference. Thus for some samples there is an optimum temperature at which to obtain maximum signal intensity, which is determined by the combined temperature dependence of linewidths, relaxation times, and Boltzmann populations.

2. Resonator

The calculation in (3) assumes a resonator with critical coupling and B_1 perpendicular to the external magnetic field. If the coupling is not critical, then a correction must be made for the coupling of incident power into the resonator and coupling of the signal out of the resonator to the detection system.[5,8,9] The value of Q can be measured, but the value of η is more difficult to determine. It requires evaluation of the magnetic field distribution in the resonator and its integration over the volume of the sample and of the resonator. For a loop gap resonator (LGR) this includes the gap and shield regions.

To describe the relative B_1 at the sample for different resonators at fixed incident power P, Hyde proposed a resonator efficiency parameter, $\Lambda = B_1/\sqrt{P}$.[10] If the sample does not saturate, then increased B_1 results in increased signal intensity. Increased B_1 for the same source power improves sensitivity if we are operating at maximum source power or if the source noise dominates, even if the increase in Λ is due to an increase in Q_L.

3. Detection System

It is often assumed that the noise in the spectrum is determined by thermal noise and by the bandwidth of the detection system. Before P_n in (3) is replaced by the expression for thermal noise, this condition must be verified. For example, in commercial X-band spectrometers with a rectangular resonator, the detection system noise (which may not be thermal noise) dominates below about 1 mW, but source noise dominates above about 1 mW.[11] The power at which this changeover occurs depends on spectrometer design.

Noise in a spectrum can be reduced by filtering the signal. An analog RC filter circuit is characterized by a time constant. Increasing the time constant decreases the noise in the output. However, if the time constant is too long relative to the rate of scan through a signal, the signal is distorted. As a general rule the time constant should be less than one-tenth of the time required to scan between the maximum and minimum points in a first derivative lineshape. For an analog system this time constant is used as the bandwidth of the detection system in (3).

4. Magnetic Field Modulation

Equation (3) does not take account of the improvement in signal-to-noise, and therefore in $N_{\min}$, that is achieved in many spectrometers by magnetic field modulation. To obtain relative estimates of sensitivity for comparison of two spectrometers, the impact of modulation amplitude can be taken into account by noting that signal amplitude increases proportional to modulation amplitude, provided the modulation amplitude is less than the linewidth. To include magnetic field modulation in the calculation of absolute numbers for $N_{\min}$ is more difficult. In addition to potential improvement in signal-to-noise due to phase-sensitive detection, we must also consider the $1/f$ noise in the detection system and eddy currents and/or microphonics in the resonator.

5. Additional Factors to Consider in Pulsed Experiments

In real spectrometers there is an instrumental "dead-time" following the pulse during which data cannot be recorded. Decay (often exponential) of signal amplitudes during the dead-time results in loss of signal intensity relative to the maximum at time zero. Mims concluded that the sensitivity of a spin echo experiment was approximately proportional to $Q^{1/2}$ if the length of the pulse was matched to the Q of the resonator.[12] Sensitivity also depends upon the duty cycle of the receiver and the digitizer sampling rate.[13] Wang *et al.* considered the S/N of an FID following a 90° pulse for two limiting cases.[14] If Q is proportional to the square root of frequency, then the S/N varies as $B_0^{7/2}$, and if Q is determined by the losses of the sample, S/N is proportional to B_0 assuming that the resistivity of the sample is frequency independent.[14]

For narrow lines, FT-ESR can give higher S/N than CW ESR in the same data recording time.[15] However, field-swept echo-detected spectra typically have poorer S/N than CW spectra because of duty-cycle differences. The relative time spent observing signal and baseline is the same for field-swept spectra, whether CW or pulsed. In the CW measurement there is 100% observation time. However, in the ESE experiment the echo is observed for only a small portion of the total experiment time, even if the experiment is not T_1 (and hence repetition time) limited. The signal can be digitized only for the duration of the echo (ca. 50 ns) but the measurement of each echo requires a time somewhat greater than 2τ, where τ is the time between pulses. In addition, the spectrometer receiver band width usually will be 20 to 50 MHz for a pulsed spectrometer, and 0.1 to 1 Hz for a CW spectrometer. The noise is proportional to the square root of the receiver bandwidth (unless the noise is dominated by some other part of the experiment).[16] There are other more mundane (and temporary) limits such as the speed of commercial digitizers.[15]

C. PREDICTED FREQUENCY DEPENDENCE OF ESR SIGNAL, NOISE, AND SIGNAL-TO-NOISE RATIO

To assess the frequency dependence of sensitivity and signal-to-noise requires evaluation of the frequency dependence of each of the terms in (2) and (3). We consider first the impact of sample and resonator on the signal and then consider sources of noise and their frequency dependence.

1. Sample

We continue to assume that we are dealing with the allowed transitions of an $S = \frac{1}{2}$ spin so that the sample contributes a factor of ω_0 to the frequency dependence of the signal intensity (2) or ω_0^{-1} to $N_{\min}$ (3). In (2) and (3) and in the following discussion it is assumed that the signal is encompassed entirely within the bandwidth of the resonator. As the frequency decreases, the resonator bandwidth (for the same Q) decreases, so for the same sample this assumption may be valid at X-band, but not valid at lower frequency. Anisotropy in g-values may make the assumption invalid at higher frequencies.

If the ESR signal extends over a wide magnetic field range, as for many transition metal complexes, the spins/Gauss factor is decreased and $N_{\min}$ is larger than for a sharp spectrum. Spectral dispersion increases with ω_0, which decreases $N_{\min}$. The breadth of the spectrum also results in less than optimal sensitivity in the usual magnetic field modulation schemes, since the maximum practical magnetic field modulation is much smaller than the maximum observable ESR linewidth.

Long electron spin relaxation times, including spectral diffusion times, may limit the B_1 that can be used in CW experiments and the repetition rate in pulsed experiments.

Interaction of the dielectric and conductivity properties of the sample with the microwaves also decreases S/N. Aqueous and conducting samples are commonly encountered, and they have frequency-dependent skin depths.[17,18] The dielectric properties can produce phase changes that can null signals from certain locations in samples.[18] The much larger penetration depth at low frequencies can result in many more spins in a large sample being visible at low frequency than at high frequency. This effect can reverse the general trend toward greater sensitivity at higher frequency. Although not yet tested in detail, heterogeneous samples appear to have greater penetration depth than homogeneous samples of the same type of material. The same physical interactions that result in the losses mentioned here also result in reduction of the resonator Q, and the detected ESR signal is proportional to resonator Q. The reduction in Q due to a given lossy sample decreases with frequency, even for the same filling factor. This effect also tends to favor lower-frequency ESR for lossy samples. All of these effects together could even reverse the expected trends of S/N with frequency, so that it is conceivable that the ultimate molar (spins per volume) sensitivity for biological samples and other lossy samples could increase as the frequency decreases. This remains to be demonstrated, but there is good basis for exploration in this direction.

The frequency dependence is different for forbidden transitions. For example, for the half-field transition, the transition probability increases with decreasing frequency.[19] Similarly, the depth of echo envelope modulation for $I = \frac{1}{2}$ increases with decreasing frequency.[20,21] In samples with large zero-field splitting, transitions might only be detectable using high-frequency ESR.[22]

2. Resonator

The frequency dependence of η or Q is a function of resonator and sample geometry. Up to this point the discussion has been valid for any resonator. To facilitate the following discussion we use, as an example, a right-cylindrical 1-loop–1-gap loop gap resonator (LGR) as described by Hyde and Froncisz.[23] Unlike a cavity where it was necessary to scale the volume of the resonator as $1/\omega_0^3$, an LGR is an example of a lumped element resonator where the volume of the resonator is less dependent on ω_0. The diameter of the loop is d and the length of the resonator is z.

a. Filling Factor

The filling factor, η, must be determined for each sample and resonator. This is not always a simple task, since it requires a detailed analysis of the electromagnetic fields in the resonator. However, we can often make an educated guess because the fields can usually be determined approximately. Since we are not able to accurately predict the change in filling factor for all cases of LGRs, we will assume as a first approximation that the filling factor will scale as if the microwave magnetic field were uniform and confined to the loop. For resonators with a constant loop size (e.g., 4 mm) and sample size, this approximation will overestimate η at higher frequency. This is because the gap spacing must increase with frequency for a constant loop size, and as the gap increases, the magnetic field will no longer be completely confined to the loop. In other words, for a constant size loop and sample, the filling factor will decrease as some function of ω_0 in this type of LGR. This effect is not included in the following equations.

b. Quality Factor Q

The frequency dependence of Q is given by (4) (see Section E.4 for derivation):

$$Q_L = \tfrac{1}{8}\sqrt{\omega_0}\sqrt{\mu_0}d\sqrt{2}\sqrt{\sigma}, \tag{4}$$

where σ is the conductivity of the material of the resonator and μ_0 is the permeability in a vacuum. Since σ is constant, the frequency dependence of Q is determined by the factor $\omega_0^{1/2}d$.

c. *Signal Intensity at Constant Incident Power*

Substitution of (4) into (2) gives

$$|\Delta E| = K_e \eta \, d\omega_0^{3/2}\sqrt{P} \qquad \text{where} \quad K_e = \frac{\chi_0\sqrt{2\mu_0\sigma Z_0}}{8\Delta\omega}. \tag{5}$$

The terms that do not depend on frequency are combined in K_e. Equation (5) indicates that the frequency dependence of the ESR signal is a function of η and d as well as $\omega_0^{3/2}$. We assume here, and throughout, that the detector response is linear in applied voltage, which is the normal case for a spectrometer with a biased crystal detector.

d. *Microwave Field B_1*

In samples that are saturable we need to consider the variation in B_1 with sample and resonator geometry. The frequency dependence is given by (see Section E.5 for derivation)

$$B_1 = K_b \frac{1}{\sqrt{zd}} \frac{\sqrt{P}}{\omega_0^{1/4}}, \tag{6}$$

where $K_b = k_1 \sqrt[4]{2\mu_0^3\sigma/\pi^2}$ and k_1 is the constant of proportionality between B_1, the magnitude of the circularly polarized component of the microwave magnetic field that is perpendicular to the external magnetic field and in phase with the Larmor precession, and B_{1t} which is the magnitude of the total microwave magnetic field in the resonator (see Section E.5).

Because B_1 was defined in terms of power dissipated in the resonator, this B_1 is the rms value, which is $1/\sqrt{2}$ times the peak value. To include this B_1 in the lineshape formula or to calculate the saturation factor, the term involving relaxation times would be given as $\frac{1}{2}\gamma^2 B_1^2 T_1 T_2$ where γ is the magnetogyric ratio and T_1, T_2 are electron spin relaxation times.

3. Frequency Dependence of Signal Intensity

We consider three cases: first the size of the sample is constant, second the size of the sample and resonator are scaled by $1/\omega_0$, and third the size of the sample is constant and the resonator is scaled with $1/\omega_0$. In cavity resonators, the size of the resonator necessarily is scaled as $1/\omega_0$. Results are summarized in Table 1.

a. *Case 1. Constant Sample Size, Constant Loop Size*

This would be the case, for example, for an LGR of a given loop size. In this case, the variation with frequency is that given in (4)–(6) with filling factor and resonator dimensions held constant. At constant P, B_1 would decrease as $\omega_0^{-1/4}$. To maintain a constant B_1 as ω_0 is increased linearly the input power must increase as $\omega_0^{1/2}$.

TABLE 1. Frequency dependence of parameters.[a]

	Case 1 Constant sample size Constant LGR size		Case 2 Sample size $\propto 1/\omega_0$ LGR size $\propto 1/\omega_0$		Case 3 Constant sample size LGR size $\propto 1/\omega_0$	
L	1	(28)	ω_0^{-1}	(7), (8), (28)	ω_0^{-1}	(7), (8), (28)
R	$\omega_0^{1/2}$	(29)	$\omega_0^{1/2}$	(29)	$\omega_0^{1/2}$	(29)
Q	$\omega_0^{1/2}$	(4)	$\omega_0^{-1/2}$	(4), (8)	$\omega_0^{-1/2}$	(4), (8)
η	1		1		ω_0^{3}	(11)
ESR signal, ΔE, at constant P	$\omega_0^{3/2}$	(5)	$\omega_0^{1/2}$	(9)	$\omega_0^{7/2}$	(12)
$B_1/\sqrt{P}$	$\omega_0^{-1/4}$	(6)	$\omega_0^{3/4}$	(10)	$\omega_0^{3/4}$	(10)
P to maintain ΔE, at constant B_1	$\omega_0^{1/2}$	(6)	$\omega_0^{-3/2}$	(10)	$\omega_0^{-3/2}$	(10)
ESR signal, ΔE, at constant B_1	$\omega_0^{7/4}$	(5), (6)	$\omega_0^{-1/4}$	(9), (10)	$\omega_0^{5/4}$	(10), (12)

[a] The equation numbers that are the bases for the various statements are given in parentheses. The table lists overall frequency dependence and does not include proportionality constants.

b. Case 2. Sample and Loop Sizes Scaled as $1/\omega_0$

This has been called the "unlimited" sample case. For this case the length z and diameter d of the resonator vary with frequency as given in (7) and (8),

$$z = \frac{k_2}{\omega_0}, \tag{7}$$

$$d = \frac{k_3}{\omega_0}. \tag{8}$$

Substituting (8) into (5) gives,

$$|\Delta E| = K_e \eta k_3 \sqrt{\omega_0 P}. \tag{9}$$

Substituting (7) and (8) into (6) gives,

$$B_1 = \frac{K_b}{\sqrt{k_2 k_3}} \omega_0^{3/4} \sqrt{P}. \tag{10}$$

Equation (10) indicates that in order to maintain a constant B_1 as ω_0 is increased, the input power must decrease as $\omega_0^{-3/2}$.

c. Case 3. Constant Sample Size, Loop Scaled as $1/\omega_0$

This case has been denoted as a "limited" sample. For this case, since the sample volume is constant and the resonator volume varies as ω_0^{-3}, the frequency dependence of the filling factor is

$$\eta = k_4 \omega_0^3. \tag{11}$$

Substituting (8) and (11) into (5) gives

$$|\Delta E| = K_e k_3 k_4 \omega_0^{7/2} \sqrt{P}. \tag{12}$$

The expression for B_1 in this case is the same as for Case 2, and again, in order to maintain a constant B_1, as ω_0 is increased linearly the input power must decrease as $\omega_0^{-3/2}$.

The results given in Table 1 were derived for the case of an LGR, however they are also applicable to other resonators, including cavities, provided the assumptions for each case are matched. In the frequently cited discussion of $N_{\min}$ in ref. 2 an additional factor of $1/\omega$ was added to (3) based upon a set of assumptions concerning the frequency dependence of the relationship between incident microwave power in the waveguide outside a cavity and B_1 in the cavity. The relationships in Table 1 do not bear out this assumption. The dependence of signal at constant power[24,25] (limited sample, Case 2) on $\omega_0^{1/2}$, of signal intensity for constant sample (Case 3) at constant power[4,24,25,26] on $\omega_0^{7/2}$, and of constant sample[26] (Case 3) at constant B_1 on ω_0 agree with predictions in the references cited, among others. If the sample is very lossy, it is desirable to position it in the nodal plane of the resonator, which gives a different frequency dependence for rectangular and cylindrical resonators.[5]

The range of exponents in Table 1 indicates that we need to exercise great care in selecting the appropriate relationships to predict the frequency dependence of signal intensity for a particular situation. We also need to consider practical realities. Scaling a resonator design over a wide frequency range may not be possible because of machining tolerances, or because gaps become too small to prevent arcing during high-power pulses. In addition, resonator dimensions may become so much smaller at higher frequencies that it is not possible to maintain a constant sample size. Thus the practical need for different resonators at different frequencies may prevent us from taking advantage of theoretically predicted advantages. We also need to consider sources of noise as discussed below.

d. Comparison with NMR

Since the fundamental spin resonance is the same, we expect agreement between predicted frequency dependence of signals for NMR and ESR. In NMR the signal is typically represented as a current induced in a coil by precession of the spins.[16,27,28,29,30] The signal is proportional to ω_0^2, where one factor of ω_0 comes from χ'' (as in (1) and (2)) and the second factor of ω_0 comes from Lenz's law. To compare this frequency dependence with that described in Section 1, note that since $Q_L = \omega_0 L/2R$, (2) can be written as $|\Delta E| = \chi_0(\omega_0^2/\Delta\omega)\eta(L/2R)\sqrt{Z_0 P}$. The typical NMR case would be case 1, i.e., constant sample size and constant loop size, for which L is constant. At NMR frequencies (e.g., 0.4 GHz) and for typical NMR coils skin effects are likely to be small so R also would be constant for constant loop size. The result derived from (1) and (2) then gives an ω_0^2 dependence in agreement with the NMR approach.

4. Noise

Thermal noise is the ultimate limit on S/N. It is not evident that this limit of $F = 1$ has been reached in any ESR measurement. Hence, we have to seek in each experimental case the actual limiting noise source. Many textbooks include a discussion of the noise of amplifiers and related components.[31–37] There are many papers discussing noise and signal-to-noise (S/N) in magnetic resonance. Early papers by Strandberg[38] and Buckmaster[39–43] included special emphasis on the noise of the source. Source and detection system noise were compared by Meijer.[44] A recent paper on noise in ESR, by Hyde and coworkers,[45] focused on cryogenic microwave preamplifiers. This section examines only a subset of factors relevant to the task of estimating the advantages of improving S/N by making various changes in the design and construction of an ESR spectrometer. The following listing of potential limiting noise sources cautions that S/N scaling laws such as those in (2) or (3) may not be relevant to practical spectroscopic problems in some cases, and that we should consider signal and noise separately.

Estimates of thermal noise are usually based upon the high-temperature limit, $h\nu \ll k_B T$. However, this is not satisfied in high-frequency, low-temperature ESR. The two sides of this "inequality" are equal at 4 K, 35 GHz. At 95 GHz and higher frequencies, we should use more complete expressions for the noise rather than the limiting form.[36,46]

a. *Source Noise*

Based on a consideration of source AM and FM noise, Strandberg concluded that the usual reflection mode cavity system would be source noise limited above about 1 mW incident power.[38] Most commercial CW ESR spectrometers are source noise limited for high power incident on the resonator.[11,39–43] At low powers incident on the resonator the spectrometer might be detector-noise limited. The sample determines the operating domain. For a klystron, frequency modulation usually contributes more to noise than amplitude modulation.[47] Noise output from several X-band sources has been compared.[33,48]

b. *Resonator*

A resonator converts source phase noise into amplitude noise. In a reference arm bridge tuned to absorption, phase noise power from the source causes a voltage variation at the detector. Hyde[10,47] concluded that this effect increases as Q^2, which implies that for the same source noise there is less resulting noise at the detector with a low-Q LGR than with a higher Q resonator. We have not been able to verify this relationship. With higher Q we observe higher S/N absorption spectra even when source noise dominates.

It also was reported that source phase noise increases with Q^2 if the bridge is tuned to dispersion.[47] Dispersion ESR measurements with a high-Q, reflection resonator are almost always limited by demodulation of source phase noise.[49] The use of lower-Q higher η LGRs results in less demodulation of the phase noise and hence better S/N for dispersion.[49] The use of a bimodal[50] or crossed-loop resonator[51,52] isolates the detected ESR signal from the source noise and enhances the S/N for both absorption and dispersion.

Spectrometers that use magnetic field modulation (most CW spectrometers) commonly have some range of normal operating parameters in which the limiting noise is microphonic and is due to interaction of the modulated magnetic field with the conducting parts of the resonator (e.g., the metallic tip on the iris adjustment screw or the inductive or capacitive probe in an LGR) and other metallic structures.

c. Detection System

For a series of amplification steps, the noise adds as described in Section E.3. In such a series, if the gain of the first stage is relatively high, the noise factor for the series is determined predominantly by the first-stage amplifier. For this reason work to improve the ESR detection system focuses on the first stage. Spaeth and coworkers[53] cited the values shown in Table 2 for X-band ESR spectrometer noise figures.

In a JEOL X-band spectrometer the use of a microwave preamplifier improved the experimental S/N by about a factor of 3 at low power.[54] In a Varian E-102 bridge installing a 2.8 dB noise figure GaAsFET preamplifier and adjusting the AFC gain improved spectrometer sensitivity by a factor of 2.[55]

One way to improve the noise factor of an amplifier is to cool it. Hyde and coworkers considered the effect of a cooled microwave preamplifier.[45] With a 35 K cryogenic microwave preamplifier CW performance was limited by losses in the cables and circulator. The benefits could be realized only if the resonator was cooled. When the resonator was overcoupled for a pulsed experiment, noise from the load on port 1 of the circulator contributed to the system noise and reduced the benefit of the cryogenic amplifier.[45,48] To achieve ultimate S/N it would be necessary to either cool the load or switch the input of the circulator to a cooled load during data collection.[45]

The simplest way to detect ESR is with a diode, but diodes exhibit $1/f$ noise, necessitating use of higher-frequency (such as 100 kHz) modulation and lock-in techniques, or heterodyne detection.[48] The spectral density of $1/f$ noise "increases without limit as frequency decreases."[56] With a $1/f$ noise-limited dc amplifier "measurement accuracy" cannot be improved by increasing the length of the measurement time.[56]

TABLE 2. ESR spectrometer noise figures.[53]

Spectrometer type	First-stage noise figure (dB)	Overall detector noise figure (dB)
Superheterodyne	$F_{\text{mixer}} = 3.5$	$F = 4$
Homodyne	$F_{\text{diode}}(100\,\text{KHz}) = 3.5$ $F(10\,\text{KHz}) = 16.6,$ $F(1\,\text{KHz}) = 132$	$F = 4.5$
Homodyne with GaAsFET	$F_{\text{amplifier}} = 2.25$	$F = 2.25$

The Bruker spectrometer used an MA40074 detector diode with $F = 3.5\,\text{dB}$.

The easiest way to minimize the noise in a spectrometer is to narrow the detection bandwidth. For a single RC filter, the effective noise bandwidth is $(4RC)^{-1}$. The wider bandwidth for time-domain experiments than for CW experiments contributes to poorer S/N in the time domain.

d. Sample

Flowing lossy samples may have turbulence that results in the limiting noise. Small inanimate samples will rarely be the major source of noise in an ESR measurement, but large conducting samples and living animals may well be the limiting noise source. ESR of live animals is almost always limited by the motion of the animal.

D. PERFORMANCE OF REAL SYSTEMS

1. ~9.2 GHz, 3400 G, X-Band

Equation (3) provides an estimate of the minimum number of detectable spins, without taking into account the improvement in signal-to-noise that occurs due to magnetic field modulation. Assume a rectangular resonator with a sample volume of $0.14\,\text{cm}^3$, $Q_L = 3600$, an estimated filling factor, $\eta = 0.01$, linewidth $= 1\,\text{G}$ ($10^{-4}\,\text{T}$, 0.1 mT), an ideal noise factor $F = 1$, and a detection time constant of 1 s, which is a noise equivalent bandwidth of $\frac{1}{8}$ Hz when the detection system has a 12 dB/octave rolloff filter. For incident power of 1 mW, $N_{\min} = 2.8 \times 10^{10}$ spins. If the signal does not saturate so that we can use $P = 200\,\text{mW}$, then $N_{\min} = 0.2 \times 10^{10}$ spins. Bruker Instruments reports that with the ESP300 spectrometers the observed signal-to-noise of 360 for weak pitch correspond to $N_{\min} = 0.8 \times 10^{10}$ spins/1 G linewidth at 200 mW.[57] The standard S/N test uses peak-to-peak noise, ignoring occasional noise spikes, whereas this calculation assumed standard deviation noise. This difference makes precise comparison difficult, but state-of-the-art X-band spectrometers are within a factor of two-to-four of the thermal limit. Note that magnetic field modulation has no effect on thermal noise, but helps decrease some low-frequency noise contributions. Modulation can also cause microphonic noise.

Hyde and Froncisz[10] examined the relative sensitivities of rectangular cavities and LGRs for various classes of samples. For samples with "limited" quantity or for flow experiments the LGR offers significant advantages. For samples with unlimited quantities that are not lossy, the higher Q and larger volume of the cavity is a substantial advantage.[10] For lossy samples, such as aqueous samples, the sensitivities of LGR and cavities were found to be similar on a molar basis.[10] For pulsed ESR experiments that are limited by source power, the higher $B_1/\sqrt{P}$ of an LGR can result in higher S/N.

2. Frequencies Lower than X-Band

a. Below 1 GHz, Lossy Samples

Experimental ESR data in this frequency range indicated that sensitivity scaled approximately linearly with ω_0.[17,58] Power absorbed in lossy samples reduces Q and lowers the signal-to-noise in accordance with (3). This effect is sometimes called "sample noise," and can be an important factor in the ESR signal-to-noise. Andrew[28] considered frequency dependence in MRI and spectroscopy in living systems and concluded that the NMR signal voltage is proportional to ω_0^2. Noise power in the radiofrequency coil probe circuit is proportional to $\omega_0^{1/2}$, while losses in the patient's body are proportional to ω_0^2. The voltage signal-to-noise ratio may be written as $\omega_0^2/(a\omega_0^{1/2} + b\omega_0^2)^{1/2}$ where a and b are constants.[28] At low frequencies where losses are small,[28] the S/N varies as $\omega_0^{7/4}$. At higher frequencies, where the coil losses are less important, the S/N varies linearly in ω_0.[28] This relationship has been extended to take into consideration the size of the sample and the coil.[59]

b. 1–4 GHz, L- and S-Bands

Froncisz and Hyde compared an S-band (3.8 GHz) LGR with an X-band (8.9 GHz) 2-gap LGR.[23] The data at X-band were recorded at 12.5 mW and the S-band data at 2 mW. For the Varian weak pitch sample extrapolation to the maximum system powers (200 mW at X-band and 32 mW at S-band) gave S/N values of 1450 at X-band and 142 at S-band. Subsequently Hyde and Froncisz reported that "the Varian pitch sample at 3 GHz yields about the same signal as in a TE_{102} cavity at 9.5 GHz, because of the improved filling factor, and the signal is about 3 times less at 1 GHz."[10] For an LGR that accommodates a 4 mm tube "there is a tradeoff between low Q and high filling factor that maintains the S/N within a factor of about 3 of an X-band spectrometer."[60]

In a 2.7 GHz spin-echo spectrometer the S/N for coal or irradiated fused quartz was about a factor of 30 to 50 poorer than at X-band.[61] The filling factor was about the same and the over-coupled Q at S-band and X-band was 20 and 100, respectively. Comparison with predicted frequency dependence is complicated because the signal at each frequency depends upon the degree of over-coupling of the resonator,[62] which was not stated.

3. Frequencies Higher than X-Band

a. 13.7 GHz (P-Band)

S/N for ESE at 9.5 and 13.7 GHz were compared.[63] At X-band a 0.031 cm^3 sample and LGR were used. At 13.7 GHz a 0.054 cm^3 sample and a Mims-type transmission cavity were used. Although the receiver noise was higher at P-band than at X-band the overall sensitivity was slightly higher at P-band which was attributed, at least in part, to the larger volume of sample.

b. 35 GHz (Q-Band)

Hyde and coworkers[64] showed that installing a mixer and low-noise microwave preamplifier in a Varian Q-band bridge reduced the overall receiver noise figure by 24.6 dB. This should caution against comparing signal-to-noise performance as a function of frequency without detailed measurement of the noise performance of each spectrometer.

In a pulsed spectrometer the S/N at Q-band was predicted to be a factor of 6 greater than at X-band for broad signals.[65] The observed improvement was greater than 10. The better than expected performance of the Q-band system was attributed to the extreme over-coupling of the X-band resonator.

c. 95 GHz (W-Band)

On a CW spectrometer with klystron source and Fabry–Perot resonator the sensitivity determined for nitroxide in fluid solution was 1×10^9 spins/G.[66] This is within a factor of 5 of the theoretical limit calculated by Lebedev for a Fabry–Perot resonator.[67] Use of a TE_{011} resonator and improvements in the spectrometer resulted in a sensitivity of about 4×10^7 spins/G (4×10^8 spins/mT) in both CW and pulsed mode.[68] On a pulsed spectrometer with a Fabry–Perot resonator N_{min}[69] was 5×10^{11}. The use of a cylindrical TE_{011} resonator instead of a Fabry–Perot resonator resulted in about a factor of 10 improvement in sensitivity for a point sample,[70] which is consistent with the improvement predicted by Lebedev.[67]

d. 140 GHz

On a 140 GHz spectrometer with an extended interaction klystron and Fabry–Perot resonator the absolute sensitivity was 10^9 spins/G for a sample of fluoranthenyl hexafluorophosphate.[71] The deviation from expected sensitivity of 10^7 spins/G was attributed to source noise. The S/N was expected to be lower for pulsed measurements than for CW. The observation of a factor of 2 better S/N for pulsed ESR than for CW ESR was attributed to the lower sensitivity of the pulse experiments to source noise, phase noise of the reference oscillator, microphonics, and drift.

e. 170 GHz

On a CW spectrometer with quasioptical reflection bridge and Fabry–Perot resonator[8,72] the observed N_{min} was about 10^{11}. Ways to improve N_{min} for this spectrometer to $\sim 10^8$ have been discussed.[8] This would be about the limit predicted by Lebedev for a Fabry–Perot resonator.[67]

f. 250 GHz

On a CW spectrometer with far-infrared laser source and Fabry–Perot resonator[73] the minimum number of detectable spins was estimated as 10^{12}/G.

E. DEFINITIONS AND DERIVATIONS

1. Definitions of Q, η, and χ''

The quality factor of the resonator, Q, can be expressed in a variety of ways, of which the following are particularly useful in the present context:

$Q_L = \omega$(energy stored/power dissipated);
$Q_L = \nu/\Delta\nu$, where ν is the microwave frequency and $\Delta\nu$ is the half-power bandwidth;
$Q_L = 2\pi\nu t_{\text{ringdown}}$, where t_{ringdown} is the cavity power ringdown time constant in a pulsed experiment;
$Q_L = Q_U/2$, where Q_L is the loaded Q, i.e., the Q of the resonator critically coupled to a transmission line and Q_U is the unloaded Q, i.e., for an isolated resonator; and
$\eta =$ $\int_{V_s} B_1^2\,dV / \int_{V_R} B_{1t}^2\,dv$, where V_S, V_R are the volumes of the sample and resonator, respectively.

B_1 is the magnitude of the component of the microwave magnetic field that is perpendicular to the external magnetic field and in phase with the Larmor precession, and B_{1t} is the magnitude of the total microwave magnetic field in the resonator volume, including that in the sample.

At resonance $\chi'' = (\omega_0/\Delta\omega)(Ng^2\beta^2/4V_S k_B T_S)$ for $S = \frac{1}{2}$, where N is number of spins in volume V_S and T_S is the temperature of the sample.[2]

2. Derivation of the Fundamental Detection Equation

To derive (1) we start with the loaded Q of a resonator without ESR absorption:

$$Q_L = \frac{\omega_0 \int_{V_R} B_{1t}^2\,dV}{\mu_0 P_T}, \tag{13}$$

where P_T is the power dissipated due to all losses including the source impedance, B_{1t} is the magnitude of the total microwave magnetic field, and V_R is the volume of the resonator. The loaded Q of the resonator, including power absorbed in the sample at resonance, denoted as Q_{Le}, can be expressed as (ref. 4, Eqs. 2 and 3):

$$Q_{Le} = \frac{(\omega_0/\mu_0)\int_{V_R} B_{1t}^2\,dV}{P_T + (\omega_0/\mu_0)\chi''\int_{V_S} B_1^2\,dV}, \tag{14}$$

where χ'' is the imaginary part of the volume RF magnetic susceptibility and V_S is the volume of the sample. Solving (13) for P_T, substituting into (14), and using the definition of η gives

$$Q_{Le} = \frac{Q_L}{(1 + \chi''\eta Q_L)}. \tag{15}$$

Since ESR absorption changes Q_L only slightly, we can assume that $\chi''\eta Q_L \ll 1$ and Q_{Le} can be approximated as

$$Q_{Le} = Q_L(1 - \chi''\eta Q_L). \tag{16}$$

Let $\Delta Q_L = Q_L - Q_{Le}$, the incremental change in Q_L due to ESR absorption, and we obtain the basic expression (see p. 382 of ref. 2):

$$\left|\frac{\Delta Q_L}{Q_L}\right| = \chi''\eta Q_L. \tag{17}$$

The ESR signal in a reflection resonator is produced by the change in the voltage reflected by the resonator as the magnetic field is swept through ESR. Without ESR, the voltage reflection coefficient is

$$\Gamma_V = \frac{R - n^2 Z_0}{R + n^2 Z_0}. \tag{18}$$

In (18), R is the resistance of the resonator, Z_0 is the characteristic impedance of the line, and n is a factor of the coupling network (e.g., the iris assembly in a cavity resonator) that transforms the line impedance into the effective impedance seen by the resonator. For critical coupling (the case were $\Gamma_V = 0$), $n^2 = R/Z_0$. If $n^2 > R/Z_0$ the resonator is over-coupled, and if $n^2 < R/Z_0$ the resonator is under-coupled. Experimentally, when critical coupling is judged in an ESR spectrometer by constancy of detector current upon varying microwave power incident on the resonator, this coupling, in practice, actually includes the resonator, microwave circulator, and associated transmission line, because of the leakage through the circulator and losses in the line and connectors.

When ESR occurs, power is absorbed which is equivalent to a change ΔR in the resonator resistance, R. The voltage reflection coefficient with ESR, Γ_{Ve} is

$$\Gamma_{Ve} = \frac{(R + \Delta R) - n^2 Z_0}{(R + \Delta R) + n^2 Z_0}. \tag{19}$$

Since Γ_V is the ratio of reflected voltage to the voltage, E, incident on the resonator, the ESR signal voltage ratio, $|\Delta E/E|$ is the difference between (18) and (19)

$$\left|\frac{\Delta E}{E}\right| = |\Delta\Gamma_{Ve}| = |\Gamma_V - \Gamma_{Ve}| = \frac{\Delta R}{(R + n^2 Z_0)}, \tag{20}$$

assuming $\Delta R \ll R$ in the denominator of (19).

In a reflection spectrometer, the resonator/circulator should be critically coupled to the line when there is no ESR, otherwise the large reflected power from the resonator would saturate the detector and mask the ESR signal. It can be shown that for all types of detection, critical coupling also provides the maximum ESR signal. At critical coupling, $n^2 Z_0 = R$. Substituting this into (20) gives

$$\left|\frac{\Delta E}{E}\right| = \frac{\Delta R}{2R}. \tag{21}$$

Alternatively we can express the ESR signal in terms of the change in Q of the resonator. At critical coupling, the loaded Q is

$$Q_L = \frac{\omega_0 L}{2R}, \tag{22}$$

ΔQ_L is the change in Q_L (17) due to ΔR that results from ESR. Then

$$\Delta Q_L = \frac{\omega_0 L}{2R + \Delta R} - \frac{\omega_0 L}{2R} = -\tfrac{1}{2}\omega_0 L \frac{\Delta R}{(R(2R + \Delta R))}. \tag{23}$$

Again, using $\Delta R \ll R$ in the denominator and combining with (22)

$$\left|\frac{\Delta Q_L}{Q_L}\right| = \frac{1}{2}\frac{\Delta R}{R}. \tag{24}$$

Finally, combining (17), (21), and (24) gives

$$\left|\frac{\Delta E}{E}\right| = \left|\frac{\Delta Q_L}{Q_L}\right| = \frac{1}{2}\frac{\Delta R}{R} = \chi''\eta Q_L. \tag{25}$$

This expression summarizes the fact that ESR is a change in effective resistance in the resonator, which causes a change in resonator Q that is detected as a change in reflected voltage. The E in (25) is physically significant, but difficult to define, especially in a waveguide. A more practical expression for the voltage term is

$$E = \sqrt{Z_0 P}, \tag{26}$$

where P is the microwave power delivered to the resonator and Z_0 is the characteristic impedance of the line. Substitution of (26) into (25) gives the fundamental detection expression, (1) of Section B:

$$|\Delta E| = \chi''\eta Q_L \sqrt{Z_0 P}. \tag{1}$$

3. Noise Factor for a Series of Amplification Steps

The noise factor for a two-port network such as an amplifier is the ratio of the signal-to-noise of the input to the signal-to-noise of the output, when the input termination impedance is specified and is at standard temperature, and it is commonly designated as F.[74,75] The noise figure, NF, is $10\log_{10}(F)$, although this too is denoted as F in some books and papers. Thus a "perfect" amplifier that adds no noise would have $F = 1$ and $NF = 0\,\text{dB}$.

One approach to the addition of noise in a cascade is as follows. Define,[76] Δf is the bandwidth, ΔN_0 is the available noise power at the output of the network, and ΔN_{ideal} is the output noise of an ideal noiseless network. $\Delta N_0 = \Delta N_{\text{ideal}} + \Delta N$, where ΔN is the noise added by the amplifier. Then, using the definitions in the previous paragraph,

$$F = \frac{\Delta N_0}{\Delta N_{\text{ideal}}} \quad \text{is the noise factor.}$$

When there are multiple stages, the noise caused by sources in each stage is called ΔN_1, ΔN_2, etc., and the power gains of each stage are G_1, G_2, etc. Then the overall noise factor (called noise figure in ref. 76) is related to the noise factor of each stage by

$$F = F_1 + \frac{F_2 - 1}{G_1} + \frac{F_3 - 1}{G_1 G_2} + \cdots. \tag{27}$$

An example of the calculation of F for an ESR bridge is given on p. 538 of ref. 2, first Ed.

4. Frequency Dependence of Q for LGR, Derivative of Equation (4)

To derive (4) we start with Q_L for an LGR, given by (22). The inductance L, is related to the length z and the cross-sectional area $A = \pi d^2/4$ of the loop, by

$$L = \tfrac{1}{4}\mu_0\pi\frac{d^2}{z}. \tag{28}$$

The resistance, including the skin effect[77] (skin depth $\delta = [\omega\mu_0\sigma/2]^{-1/2}$), of the resonator loop is

$$R = \frac{\pi d}{z}\sqrt{\omega_0\frac{\mu_0}{2\sigma}}. \tag{29}$$

The conductivity, σ, of the resonator is assumed to be constant with frequency. More accurate expressions are available for L and R that include effects of the finite-loop length and the contributions to R from the resonator gap.[9] However, the additional corrections will not affect the frequency dependence of L and R. Equations (28) and (29) are sufficient to analyze sensitivity as a function of frequency. Substituting (28) and (29) into (22) gives (4)

$$Q_L = \tfrac{1}{8}\sqrt{\omega_0}\sqrt{\mu_0}d\sqrt{2}\sqrt{\sigma}. \tag{4}$$

5. Frequency Dependence of B_1, Derivation of Equation (6)

B_1 is proportional to B_{1t}, which varies with frequency and resonator size

$$B_{1t} = \frac{\phi}{A}, \tag{30}$$

where ϕ is the total magnetic flux in the loop and $A = \pi d^2/4$. From the definition of inductance

$$\phi = LI, \tag{31}$$

where I is the current in the walls of the resonator that generates the magnetic flux. Combining (30) and (31)

$$B_{1t} = L\frac{I}{A}, \tag{32}$$

I is given by

$$I = \sqrt{\frac{P}{R}}, \tag{33}$$

where P is the microwave power dissipated in the resistance R of the resonator, i.e., the power from the spectrometer bridge. Substituting (28), (29), and (33) into (32) gives

$$B_{1t} = \frac{\mu_0^{3/4}}{\sqrt{\pi}\sqrt{d}\sqrt{z}} \sqrt{P} \, \frac{2^{1/4}\sigma^{1/4}}{\omega_0^{1/4}} \tag{34}$$

or, since B_1 is proportional to B_{1t},

$$B_1 = K_b \frac{1}{\sqrt{zd}} \frac{\sqrt{P}}{\omega_0^{1/4}}, \tag{6}$$

where $K_b = k_1 \sqrt[4]{2\mu_0^3\sigma/\pi^2}$ and k_1 is the constant of proportionality between B_1 and B_{1t}. Because of the definition in (33), both B_1 and B_{1t} are rms values.

ACKNOWLEDGMENTS

GRE and SSE have benefited from numerous discussions with Richard W. Quine (University of Denver) and with Dr Howard J. Halpern (University of Chicago) concerning sensitivities at low frequencies and from various presentations and discussions at scientific meetings by Professors Jack H. Freed (Cornell University), R. Linn Belford (University of Illinois), and others concerning sensitivity at high frequencies.

REFERENCES

[1] S. S. Eaton and G. R. Eaton, Bull. Magn. Reson. **16**, 149 (1995).
[2] C. P. Poole, Jr., *Electron Spin Resonance: A Comprehensive Treatise on Experimental Techniques* (Wiley-Interscience, New York, 1967), Ch. 14; 2nd ed., 1983, Ch. 11.
[3] D. J. E. Ingram, *Biological and Biochemical Applications of Electron Spin Resonance* (Plenum Press, New York, 1969), pp. 44–64.
[4] G. Feher, Sensitivity Considerations in Microwave Paramagnetic Resonance Absorption Techniques. Bell System Tech. J. **36**, 449 (1957).
[5] T. M. Wilmshurst, *Electron Spin Resonance Spectrometers* (Plenum Press, New York, 1968).
[6] C. P. Poole, Jr., *Electron Spin Resonance: A Comprehensive Treatise on Experimental Techniques* (Wiley-Interscience, New York, 1983), 2nd ed., Ch. 12.
[7] I. Bertini, G. Martini, and C. Luchinat, *in Handbook of Electron Spin Resonance*, edited by C. P. Poole, Jr. and H. A. Farach (American Institute of Physics, New York, 1994), Ch. 3.
[8] K. A. Earle, D. E. Budil, and J. H. Freed, Adv. Magn. Optical Reson. **19**, 253 (1996).
[9] G. A. Rinard, R. W. Quine, S. S. Eaton, G. R. Eaton, and W. Froncisz, J. Magn. Reson. **A108**, 71 (1994).
[10] J. S. Hyde and W. Froncisz, Loop Gap Resonators, in *Advanced EPR: Applications in Biology and Biochemistry*, edited by A. J. Hoff (Elsevier, Amsterdam, 1989), p. 277.
[11] Varian Spectrometer Manual 87-125-052, p. 4-3.
[12] W. B. Mims, in *Electron Paramagnetic Resonance*, edited by S. Geschwind (Plenum Press, New York, 1972), Ch. 4.
[13] J. Gorchester and J. H. Freed, J. Chem. Phys. **88**, 4678 (1988).
[14] Z. Wang, D.-J. Wang, E. A. Noyszewski, A. R. Bogdan, J. C. Haselgrove, R. Reddy, R. A. Zimmerman, and J. S. Leigh, Magn. Reson. Med. **27**, 362 (1992).
[15] T. F. Prisner, M. Rohrer, and K. Möbius, Appl. Magn. Reson. **7**, 167 (1994).
[16] A. Makovski, Magn. Reson. Med. **36**, 494 (1996).
[17] H. J. Halpern and M. K. Bowman, in *EPR Imaging and in vivo EPR*, edited by G. R. Eaton, S. S. Eaton, and K. Ohno (CRC Press, Boca Raton, FL, 1991), Ch. 6.

[18] M. Sueki, G. A. Rinard, S. S. Eaton, and G. R. Eaton, J. Magn. Reson. **A118**, 173 (1996).
[19] G. R. Eaton and S. S. Eaton, Biol. Magn. Reson. **8**, 339 (1989).
[20] D. J. Singel, in *Advanced EPR: Applications in Biology and Biochemistry*, edited by A. J. Hoff (Elsevier, Amsterdam, 1989), p. 119.
[21] J. H. Hankiewicz, C. Stenland, and L. Kevan, Rev. Sci. Instrum. **64**, 2850 (1993).
[22] S. S. Eaton and G. R. Eaton, Chapter XI in this volume on high frequency EPR.
[23] W. Froncisz and J. S. Hyde, J. Magn. Reson. **47**, 515 (1982).
[24] A. Abragam and B. Bleaney, *Electron Paramagnetic Resonance of Transition Ions* (Oxford University Press, Oxford, 1970).
[25] G. K. Fraenkel, in *Technique of Organic Chemistry*, Vol. 1, Part IV, *Physical Methods of Organic Chemistry*, 3rd ed., edited by A. Weissberger (Interscience, New York, 1960), Ch. XLII.
[26] Varian Spectrometer Manual 87-125-052, p. 3-8.
[27] A. Abragam, *The Principles of Nuclear Magnetism* (Oxford University Press, Oxford, 1961).
[28] E. R. Andrew, *Magnetic Resonance and Related Phenomena* (Elsevier, Amsterdam, 1989). 24th Ampere Congress, Poznan, 1988, p. 45.
[29] D. L. Holt, in *Encyclopedia of NMR*, edited by D. M. Grant and R. K. Harris, **7**, 4256 (1996).
[30] D. L. Holt and R. E. Richards, J. Magn. Reson. **24**, 71 (1976).
[31] T. H. Wilmshurst, *Signal Recovery from Noise in Electronic Instrumentation* (Adam Hilger IOP, Bristol, England, 1990).
[32] E. R. Davies, *Electronics Noise, and Signal Recovery* (Academic Press, New York, 1993).
[33] W. P. Robins, *Phase Noise in Signal Sources (Theory and Applications)*. IEE Telecommunications Series 9 (Peter Peregrinus, on behalf of the Institute of Electrical Engineers, London, 1982).
[34] C. D. Motchenbacher and J. A. Connelly, *Low-Noise Electronic System Design* (Wiley, New York, 1993).
[35] H. R. Raemer, *Statistical Communication Theory and Applications* (Prentice-Hall, Englewood Cliffs, NJ, 1969).
[36] R. E. Ziemer and W. H. Tranter, *Principles of Communications* (Houghton-Mifflin, Boston, 1985), 2nd ed. App. A.
[37] W. P. Jolly, *Low Noise Electronics* (American Elsevier, New York, 1967).
[38] M. W. P. Strandberg, Rev. Sci. Instrum. **43**, 307 (1972).
[39] H. A. Buckmaster and J. C. Dering, Canad. J. Phys. **43**, 1088 (1965).
[40] H. A. Buckmaster and J. C. Dering, J. Sci. Instrum. **43**, 554 (1966).
[41] H. A. Buckmaster and J. C. Dering, IEEE Trans. Instrum. Meas. **16**, 13 (1967).
[42] H. A. Buckmaster and J. C. Dering, IEEE Trans. Instrum. Meas. **18**, 237 (1969).
[43] H. A. Buckmaster and J. D. Skirrow, Canad. J. Phys. **50**, 1092 (1972).
[44] G. C. M. Meijer, J. Appl. Sci. Engng. **A1**, 129 (1975).
[45] S. Pfenninger, W. Froncisz, and J. S. Hyde, J. Magn. Reson. **A113**, 32 (1995).
[46] F. Pobell, *Matter and Methods at Low Temperature* (Springer-Verlag, New York, 1992).
[47] R. S. Alger, *Electron Paramagnetic Resonance: Techniques and Applications* (Interscience, New York, 1968), p. 79.
[48] H. E. Atlink, T. Gregorkiewicz, and C. A. J. Ammerlaan, Rev. Sci. Instrum. **63**, 5742 (1992).
[49] J. S. Hyde and A. Kusumi, Rev. Sci. Instrum. **53**, 1934 (1982).
[50] D. T. Teaney, M. P. Kline, and A. M. Portis, Rev. Sci. Instrum. **32**, 721 (1961).
[51] G. A. Rinard, R. W. Quine, B. T. Ghim, S. S. Eaton, and G. R. Eaton, J. Magn. Reson. **A122**, 50 (1996).
[52] G. A. Rinard, R. W. Quine, B. T. Ghim, S. S. Eaton, and G. R. Eaton, J. Magn. Reson. **A122**, 58 (1996).
[53] Ch. Hoentzsch, J. R. Niklas, and J. M. Spaeth, Rev. Sci. Instrum. **49**, 1100 (1978).
[54] G. Grampp, Rev. Sci. Instrum. **56**, 2050 (1985).
[55] S. L. Dexheimer and M. P. Klein, Rev. Sci. Instrum. **59**, 764 (1988).
[56] C. D. Motchenbacher and J. A. Connelly, *Low-Noise Electronic System Design* (Wiley, New York, 1993), pp. 27, 37.
[57] Bruker brochure, "ESP 300 Series Electron Spin Resonance", dated 1/21/90, p. 14.
[58] L. G. Stoodley, J. Elect. Control **14**, 531 (1963).
[59] R. D. Black, T. A. Early, P. B. Roemer, O. M. Mueller, A. Mogro-Campero, L. G. Turner, and G. A. Johnson, Science **259**, 793 (1993).

[60] R. Basosi, W. E. Antholine, and J. S. Hyde, Biol. Magn. Reson. **13**, 103 (1993).
[61] M. Romanelli, V. Kurshev, and L. Kevan, Appl. Magn. Reson. **7**, 427 (1994).
[62] G. A. Rinard, R. W. Quine, S. S. Eaton, G. R. Eaton, and W. Froncisz, J. Magn. Reson. **A108**, 71 (1994).
[63] B. E. Sturgeon and R. D. Britt, Rev. Sci. Instrum. **63**, 2187 (1992).
[64] J. S. Hyde, M. E. Newton, R. A. Strangeway, T. G. Camenisch, and W. Froncisz, Rev. Sci. Instrum. **62**, 2969 (1991).
[65] C. E. Davoust, P. E. Doan, and B. M. Hoffman, J. Magn. Reson. **A119**, 38 (1996).
[66] E. Haindl, K. Möbius, and H. Oloff, Z. Naturforsch. **40a**, 169 (1985).
[67] Ya. S. Lebedev, in *Modern Pulsed and Continuous-Wave Electron Spin Resonance*, edited by L. Kevan and M. K. Bowman (Wiley, New York, 1990), p. 365.
[68] K. Möbius, Appl. Magn. Reson. **9**, 389 (1995).
[69] R. T. Weber, J. A. M. Disselhorst, L. J. Prevo, J. Schmidt, and W. Th. Wenkebach, J. Magn. Reson. **81**, 129 (1989).
[70] J A. J. M. Disselhorst, H. van der Meer, O. G. Poluektov, and J. Schmidt, J. Magn. Reson. **A115**, 183 (1995).
[71] T. F. Prisner, S. Un, and R. G. Griffin, Israel J. Chem. **32**, 357 (1992).
[72] K. A. Earle, D. S. Tipikin, and J. H. Freed, Rev. Sci. Instrum. **67**, 2502 (1996).
[73] W. B. Lynch, K. A. Earle, and J. H. Freed, Rev. Sci. Instrum. **59**, 1345 (1988).
[74] K. Chong, Microwaves and RF, 91 (Feb. 1994).
[75] C. D. Motchenbacher and J. A. Connelly, *Low-Noise Electronic System Design* (Wiley, New York, 1993).
[76] J. F. Reintjes and G. T. Coate, *Principles of Radar*, 3rd ed. (McGraw-Hill, New York, 1952).
[77] *Reference Data for Radio Engineers*, 5th ed. (Howard W. Sams, Indianapolis, 1968), pp. 6–4.

CHAPTER II

FIELD-SWEPT VERSUS FREQUENCY-SWEPT SPECTRA

Y. C. Zhong and J. R. Pilbrow

Department of Physics, Monash University, Clayton, Victoria 3168, Australia

This chapter summarizes the major issues relating to the distinction between field-swept and frequency-swept ESR. Expressions are given that should form the basis for computer simulation of ESR spectra.

CONTENTS

A. INTRODUCTION

Normally, ESR spectra are measured by slowly sweeping the magnetic field B whilst keeping the microwave frequency ν, or ω $(= 2\pi\nu)$, constant. The ESR signal amplitude is traditionally written as $Y(B)$ (e.g., see classic references[1–5]). In particular, the validity of the familiar transformation, $\omega = \gamma B$, at other than the center of a line remained unquestioned.[6–9] To obtain a theoretically consistent description of field-swept ESR spectra, it turns out that the general expression for the signal amplitude should be formulated as a two-variable function,[10–12] $Y(B, \omega)$, of both magnetic field and microwave angular frequency. Further, some quantum mechanical details, that may well be neglected by simply using a lineshape function, must also be included.[13,14]

B. LINESHAPES

1. ESR Signal Amplitudes

Consider a slow-passage ESR measurement in which a spin system can be *completely* described by a spin Hamiltonian $\mathcal{H}_0$, containing a Zeeman term and additional terms arising from hyperfine and other interactions. The eigenstates and eigenvalues of $\mathcal{H}_0$ are denoted by $|a\rangle, |b\rangle, \ldots$; and $E_a, E_b, \ldots$, respectively. To proceed further, the spin Hamiltonian $\mathcal{H}_0$ is now divided into two parts,

$$\mathcal{H}_0 = \mathcal{H}_1 + \mathcal{H}_2, \tag{1}$$

where $\mathcal{H}_1$ stands for the Hamiltonian actually used for calculating an ESR spectrum in which all spin-dependent coupling terms under consideration are treated as constants. In (1), $\mathcal{H}_2$ in effect gives a realistic account of *line-broadening* phenomena which can be either homogeneous or inhomogeneous or some mixture of both. To be specific, $\mathcal{H}_1$ is applied to determine each *resonance center* for positioning a chosen lineshape function. Here, the eigenstates and eigenvalues of $\mathcal{H}_1$ are denoted as $|\alpha\rangle, |\beta\rangle, \ldots; E_\alpha, E_\beta, \ldots$, respectively, in order to distinguish them from those of $\mathcal{H}_0$.

For slow-passage ESR measurements under the high-temperature approximation, i.e., $\hbar\omega \ll k_B T$, the general expression for the signal amplitude $Y(\omega)$, detected in the absorption mode, can be derived in the form[13,14]

$$Y(\omega) = \frac{2C\eta Q_0 N\hbar\omega}{Zk_B T} \sum_{a,b} |\langle a|\mu_x|b\rangle|^2 \delta(\omega - \omega_{ba}), \tag{2}$$

where the summation is over all possible pairs of states $|a\rangle, |b\rangle$ of Hamiltonian $\mathcal{H}_0$.

In any computer simulation of ESR spectra, nonetheless, it is unlikely that we could obtain $\mathcal{H}_0$ completely. Instead, it is customary to formulate $\mathcal{H}_1$ and introduce a lineshape function to describe such details as unresolved hyperfine splitting. Besides, (2) implies a *frequency-swept* EPR experiment in which B is kept constant. Thus further derivation from (2) should be made by using the following relation,[13,14]

$$\sum_{a,b} |\langle a|\mu_x|b\rangle|^2 \delta(\omega - \omega_{ba}) = \sum_{\alpha,\beta} |\langle \alpha|\mu_x|\beta\rangle|^2 F_\omega[\omega - \omega_0(B), \sigma_\omega], \tag{3}$$

where F represents the lineshape function which can be chosen, as appropriate, from functions such as Gaussian or Lorentzian. In general, F is a two-variable function of the experimentally controlled B and ω. Equation (3) provides an accurate definition of the lineshape function. Substituting (3) into (2), a general expression for the ESR signal amplitue $Y(B, \omega)$, irrespective of whether either B or ω is swept, can be written as

$$Y(B, \omega) = \frac{2C\eta Q_0 N\hbar\omega}{Zk_B T} \sum_{\alpha,\beta} |\langle \alpha|\mu_x|\beta\rangle|^2 F_\omega[\omega - \omega_0(B), \sigma_\omega], \tag{4}$$

in which the resonance center is given by

$$\omega_0(B) = \frac{E_\beta - E_\alpha}{\hbar}, \tag{5}$$

where both E_α and E_β can be expressed as functions of B due to the effect of the Zeeman interaction. For $S = \frac{1}{2}$ systems in the absence of hyperfine splitting, we obtain the familiar form, $\omega_0 = g\beta B/\hbar$ and the summation in (4) disappears. In (4), η and Q_0 are the cavity or resonator filling factor and quality factor, respectively.

2. Distinction Between Resonance Condition and Resonance Center

In essence, (4) requires that the distinction between the *resonance condition* and *resonance center* must be made, viz.,

$$\hbar\omega = E_b - E_a \qquad \text{(resonance condition)}, \tag{6}$$

$$\hbar\omega_0 = E_\beta - E\alpha \qquad \text{(resonance center)}. \tag{7}$$

Equation (6) describes all resonant transitions associated with *one* ESR line without invoking a lineshape function whilst (7) allows us to position a lineshape function that is used to account for an ESR line with a finite linewidth σ_ω. In computer simulations of ESR spectra, of course, it is always the *resonance center* that must be calculated. In general, the distinction between the two concepts strictly forbids use of the convenient, but incorrect, transformation, $\omega = \gamma B$, when seeking to relate field and frequency variables across the whole of an ESR line or spectrum.

C. APPLICATIONS

1. Adoption of $1/g$ Scaling in Lineshape Function and Linewidth

Notice that the general form of a lineshape function F in (3) and (4) is written as an explicit function of the microwave frequency ω, i.e., F_ω, in which B is an implicit variable. In the case of field-swept ESR experiments, therefore, it is necessary to express F as an explicit function of B, i.e., $F_B(B - B_0, \sigma_B)$ where the microwave frequency is fixed. If both F_ω and F_B are normalized to unity, then both the lineshape function F_B and the related linewidth σ_B must be scaled in comparison with F_ω and σ_ω, respectively and this requirement can be readily shown for $S = \frac{1}{2}$ when F is a δ-function.[11] In the absence of hyperfine splitting, this can simply be denoted as $1/g$ scaling because

$$F_\omega(\omega - \omega_0, \sigma_\omega) = \frac{\hbar}{g\beta} F_B(B - B_0, \sigma_B). \tag{8}$$

Also,

$$\sigma_\omega = \frac{g\beta}{\hbar}\sigma_B. \tag{9}$$

In fact (8) represents the self-consistent explanation for the origin of the Aasa–Vanngard $1/g$ factor.[6] On the other hand, (9) indicates that g-anisotropy should be taken explicitly into account when relating values of σ_B and σ_ω. An example for which σ_ω is very nearly isotropic but σ_B is very anisotropic is found in an early article by one of the authors.[15]

Field-swept lineshapes may sometimes be asymmetric, caused in the first instance by "g-strain" where the linewidth, $\sigma_\omega \propto \omega$, which for $S = \frac{1}{2}$ means that it varies linearly with magnetic field, B. On the other hand, (8) does not provide an explanation for asymmetric field-swept ESR lines arising from Cr^{3+} in ruby single crystals,[16] where the spectra depend on both nonlinearity in the eigenvalue spacings and variation in the transition matrix element as a function of magnetic field. Further, we note that asymmetric lines have first derivatives that do not cross the base-line at the true line center. This can lead to discrepancies in spin Hamiltonian parameters determined from field-swept and zero field, frequency-swept spectra.[12] When $S > \frac{1}{2}$, in any event, $\partial B/\partial\omega_0$ may vary across an ESR line and so the transformation from field to frequency variables is unnecessarily complicated[6–9] and in view of the generality of (4) that approach is no longer recommended.

To date, it has become a prerequisite in many computer simulations of disordered ESR spectra to adopt the $1/g$ scaling (see, e.g., references[6–9,17–19]). It is also evident that both (8) and (9) should also be applied when seeking to describe single-crystal ESR spectra quantitatively.[20] Generally speaking, of course, such scaling should involve anisotropic hyperfine interactions as well, and some appropriate modification of (8) and (9) should be made accordingly whenever the hyperfine splitting can be resolved experimentally.

2. Utilization of the Boltzmann Factor

From (4), it now becomes certain that the Boltzmann population factor in the general expression for the ESR absorption signal $Y(B, \omega)$ is determined by the microwave frequency, since $N[1 - \exp(\hbar\omega/k_B T)]$ reduces to $N\hbar\omega/Zk_B T$. In fact, the frequency-dependent Boltzmann factor also provides a reasonable asymptotic form for the spin-temperature theory by Provotorov[21] in the absence of saturation. Incidentally, an inappropriate field-dependent Boltzmann factor would be obtained should the second component of the Hamiltonian, $\mathscr{H}_2$, be ignored.[10,22]

In practice, this conclusion should be taken into account when using the results of many early publications on magnetic resonance theory because the Boltzmann factor was inconsistently, and somewhat indiscriminately, written either as $\hbar\omega/k_B T$ or $\hbar\omega_0/k_B T$. For example, Pryce and Stevens[23] derived a frequency-dependent Boltzmann factor but dropped it out from the transition rate when calculating moments as prescribed by Kambo and Usui[24] who, themselves, retained the frequency-dependent Boltzmann factor in their expressions of moments. Obviously, the nth-moment in the latter is in fact the $(n+1)$th-moment in the former as also shown by McMillan and Opechowski,[26] but this statement should not be directly applied in field-swept ESR as the frequency is fixed. To avoid using a Boltzmann factor in moment calculations,

Richards[27] redefined the lineshape function. Other examples can be found in papers by Portis,[5] Castner,[28] and Smith *et al.*[29]

3. Relation Between Field Modulation and Frequency Modulation

Using the two-variable formulation for ESR spectra, the authors[14] also provided some new insights into the classical debate on the equivalence, or inequivalence, of field modulation and frequency modulation previously discussed within the framework of using a single-variable function.[30–35] When using phase-sensitive-detection where the first derivative is approximated by the "in-phase" first harmonic, the relevant equation, involving the partial derivatives related to the ESR spectra detected by the two modulation schemes, can be expressed as

$$\frac{\partial F_\omega}{\partial B} = -\left(\frac{\partial \omega_0}{\partial B}\right)\frac{\partial F_\omega}{\partial \omega}. \tag{10}$$

Clearly, (10) indicates that the ESR spectrum detected by using field-modulation must be scaled by the factor $\partial\omega_0/\partial B$ in order to be identical to that detected by using frequency-modulation. Besides, (10) for a single ESR line can be readily extended to powder cases.[13,14] Interestingly enough, for symmetric lineshape functions normalized to unity, the factor $\partial\omega_0/\partial B$ is the reciprocal of the Aasa–Vanngard $1/g$ factor.[11] Still, this relation does not alter the two basic expressions for scaling as shown by (8) and (9).

4. Explanation of Echo-Detected Two-Dimensional FT ESR

Recently, a new two-dimensional echo-detected Fourier transformed ESR spectroscopy[36] has been developed based on detection of the entire profiles of electron spin echoes as a function of the magnetic field. It is of special interest to note that although the original motivation of this pulsed ESR study was to eliminate the instrumental line-broadening which inevitably exists in conventional echo-detected ESR spectroscopy, the outcome of this study also provided the first experimental observation of the two-dimensional ESR spectra implied in (4). Naturally, care must be taken as the pulsed ESR detection stands apart from CW ESR detection through different dependence on relaxation times. In any case, the new pulsed ESR spectra would remain unexplained unless the two-variable formulation for ESR signal amplitude $Y(B,\omega)$ had been recognized.

D. CONCLUDING REMARKS

It has been demonstrated in this chapter that a consistent theoretical description of ESR spectra can be made by using a two-variable function and taking into account all of the transitions associated with a single ESR line. It is strongly recommended that (4) should always be used as the basis for calculations of field-swept ESR spectra. In particular, the familiar relation $\omega = \gamma B$ is valid only at the center of a resonance line. It is important to realize that the two-variable formulation for ESR spectra is not an alternative, but rather, it is a necessity to achieve theoretical consistency. The formulas given here are of practical importance for computer simulation of field-swept ESR spectra. In any case, the two variables, B and ω, cannot be interchanged in magnetic resonance theory. Besides, the possibility of asymmetric ESR lines occurring naturally in field sweep makes it impossible, *a priori*, to guess the correct lineshape function $F_B(B - B_0, \sigma_B)$. Finally, it also must be pointed out that the basic formulae are valid for general situations irrespective of whether an ESR line is homogeneously broadened or inhomogeneously broadened, and no restrictions are placed on the precise mathematical form of the lineshape function $F_\omega(\omega - \omega_0, \sigma_\omega)$.

REFERENCES

[1] R. Kubo and K. Tomita, J. Phys. Soc. Japan. **9**, 888 (1952).
[2] A. Abragam, *The Principles of Nuclear Magnetism* (Oxford University Press, New York, 1961).
[3] C. P. Slichter, *Principles of Magnetic Resonance* (Harper & Row, New York, 1963).
[4] F. Bloch, Phys. Rev. **70**, 460 (1946).
[5] A. M. Portis, Phys. Rev. **91**, 1070 (1953).
[6] R. Aasa and T. Vanngard, J. Magn. Reson. **19**, 308 (1975).
[7] G. Van Veen, J. Magn. Reson. **30**, 91 (1978).
[8] C.-H. Chang, D. Foster, and A. H. Kahn, J. Res. Nat. Bar. Standards **83**, 1333 (1978).
[9] R. C. Stevenson, J. Magn, Reson. **57**, 24 (1984).
[10] R. E. Coffman, J. Phys. Chem. **79**, 1129 (1975).
[11] J. R. Pilbrow, J. Magn. Reson. **58**, 186 (1984).
[12] J. R. Pilbrow, *Transition Ion Electron Paramagnetic Resonance* (Oxford University Press, New York, 1990).
[13] Y. C. Zhong, Ph.D. Thesis, Monash University, 1990.
[14] Y. C. Zhong and J. R. Pilbrow, J. Magn. Reson. **93**, 447 (1991).
[15] J. R. Pilbrow, J. Magn. Reson. **31**, 479 (1978).
[16] J. R. Pilbrow, G. R. Sinclair, D. R. Hutton, and G. J. Troup, J. Magn. Reson. **52**, 386 (1983).
[17] S. Cannistraro and G. Giugliarelli, Molecular Phys. **58**, 173 (1986).
[18] C. X. Wang, G. Giugliarelli, S. Cannistraro, and C. Fini, Il Nuovo Cimento, **8**, 76 (1986).
[19] A. R. Bizzarri and S. Cannistraro, Molecular Phys. **85**, 913 (1995).
[20] D. M. Wang and E. de Boer, J. Chem. Phys. **92**, 4698 (1990).
[21] B. N. Provotorov, Soviet Phys. JETP **31**, 1012 (1970).
[22] E. J. Soulie and P. C. Lesieur, J. Chem. Soc. Faraday Trans. 1, **85**, 4053 (1989).
[23] M. H. L. Pryce and K. W. H. Stevens, Proc. Phys. Soc. A **63**, 36 (1955).
[24] K. Kambo and T. Usui, Progr. Theoret. Phys. **8**, 302 (1952).
[25] F. T. Hedgcock and T. W. Raudorf, Canad. J. Phys. **50**, 579 (1972).
[26] M. McMillan and W. Opechowski, Canad. J. Phys. **38**, 1168 (1960).
[27] P. M. Richards, Phys. Rev. **142**, 186 (1996).
[28] T. G. Castner, Jr., Phys. Rev. **115**, 1506 (1959).
[29] M. R. Smith, H. A. Buckmaster, and R. Chatterjee, J. Magn. Reson. **25**, 485 (1977).
[30] N. Bloembergen, E. M. Purcell, and R. V. Pound, Phys. Rev. **73**, 679 (1948).
[31] K. Halbach, Phys. Rev. **119**, 1230 (1960).
[32] H. Primas, Helv. Phys. Acta **31**, 17 (1958).
[33] Y. W. Kim, J. Magn. Reson. **46**, 193 (1982).
[34] A. Dulcic and B. Rakvin, J. Magn. Reson. **52**, 323 (1983).
[35] T. Islam and I. Miyagawa, J. Magn. Reson. **51**, 383 (1982).
[36] C. J. Noble, Y. C. Zhong, J. R. Pilbrow, and D. R. Hutton, J. Magn. Reson. A **105**, 323 (1993).

CHAPTER III
RESONATORS

Charles P. Poole, Jr. and Horacio A. Farach

Department of Physics and Astronomy, University of South Carolina, Columbia, South Carolina 29208, USA

The characteristics of various resonators are discussed, including coupling structures, characteristic impedance, mode configurations, filling factor η, and quality factor Q. Resonator types include rectangular and cylindrical cavities, loop gap resonators, bimodal cavities, and several others. Many formulas of practical use are given.

CONTENTS

A. INTRODUCTION

Most ESR spectrometers employ a resonant cavity to concentrate the microwave energy at the sample and bring about an increase in sensitivity. This chapter will provide details for some commonly used resonators. In addition to standard cavities some investigators have used loop-gap cavities and other resonator types, and several of these will also be discussed. Much of the material in this chapter has appeared elsewhere in more detail.[1,2]

B. WAVEGUIDES AND CAVITIES

In a typical spectrometer a microwave bridge channels the microwaves from the generator to the cavity, and then from the cavity to the detector. The electromagnetic wave modes in waveguides are similar to those in cavity resonators, except for their phases. When a generator such as a klystron transmits power along a waveguide to a load under impedance-matched conditions there is no reflected wave, and all the power goes in one direction. The electric field vector E and magnetic field vector H are in phase to maximize Poynting's vector $E \times H$ which provides the power flow per unit area. This is shown in Fig. 1 which presents sketches of several rectangular waveguide modes.[3]

In cavity resonators the electric and magnetic field vectors are $\pi/2$ out of phase so Poynting's vector $E \times H$ averages to zero where $H = B/\mu_0$. This may be seen by comparing the TE_{102} cavity mode configuration sketched in Fig. 2 with the TE_{10} waveguide mode on the upper left of Fig. 1.

When a transmission line such as a waveguide is not matched there is a transmitted wave of electric field amplitude E_t moving forward and a reflected wave of amplitude E_r moving backward, their ratio being the reflection coefficient Γ

$$\Gamma = \frac{E_r}{E_t}, \tag{1}$$

where $0 \leq |\Gamma| \leq 1$ with $\Gamma = 0$ for perfect match, i.e., no reflected wave in the waveguide. For a matched cavity resonator $E_r = E_t$ and $\Gamma = 1$ inside the resonator.

A transmission line that is not matched $(0 < \Gamma < 1)$ has positions where the transmitted and reflected wave amplitudes add and subtract, to give

$$E_{\max} = |E_t| + |E_r|, \tag{2a}$$

$$E_{\min} = |E_t| - |E_r|. \tag{2b}$$

A common measure of the mismatch is the voltage standing wave ratio (VSWR)

$$\text{VSWR} = \frac{E_{\max}}{E_{\min}} = \frac{1+\Gamma}{1-\Gamma}, \tag{3}$$

which may be rearranged to

$$\Gamma = \frac{\text{VSWR} - 1}{\text{VSWR} + 1}. \tag{4}$$

For perfect match $\Gamma = 0$ and $\text{VSWR} = 1$ outside the cavity.

C. COUPLING STRUCTURES

In order to transfer efficiently the incident microwave power from the generator to the cavity and the ESR signal power from the cavity to the detector in the microwave bridge it is necessary to match the impedance of the system. Early vintage spectrometers employed a magic tee and later spectrometers made use of a circulator to channel the power from generator to cavity to detector. The bridge ordinarily contains an isolator to impedance match the generator to the waveguide, and the detector mount has matching devices incorporated in it.

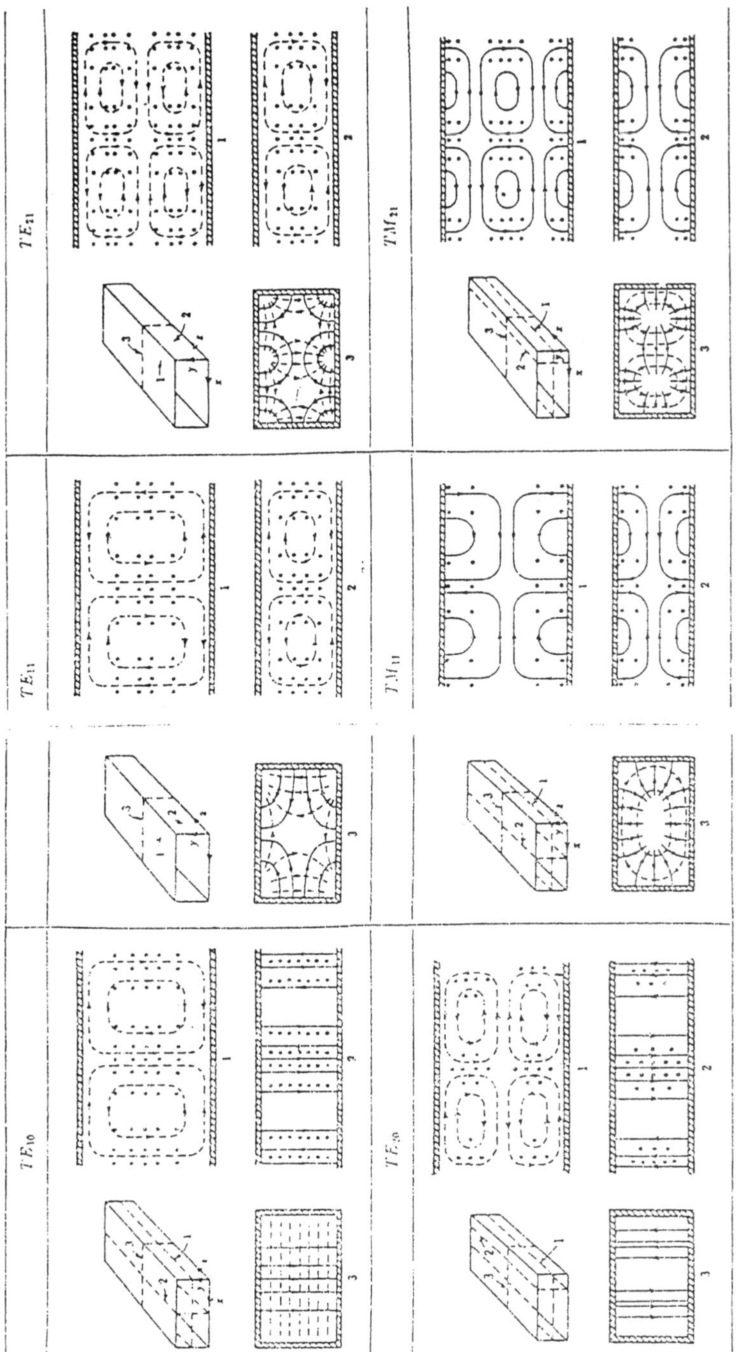

FIG. 1. Summary of wave types for rectangular guides. The symbol (•) denotes a vector directed up out of the paper and (○) a vector into the paper. Dotted lines refer to **H** and solid lines to **E** (ref. 3).

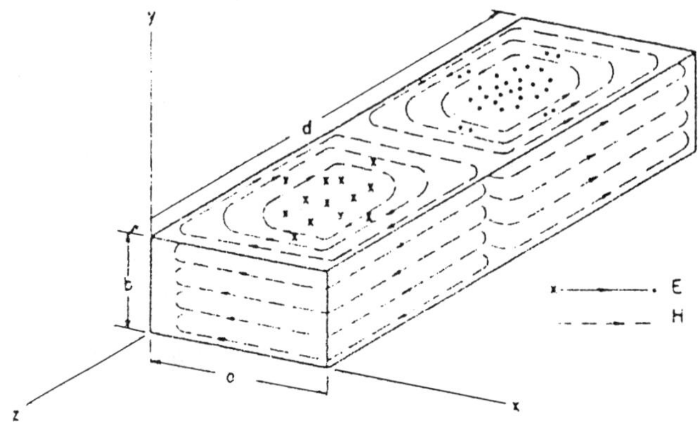

FIG. 2. Electromagnetic field configurations in a TE_{102} mode rectangular resonant cavity of dimensions a, b, and d (ref. 1, p. 129).

The cavity needs an iris or its equivalent to permit the entrance and exit of microwave power, and to provide a way to regulate the amount of coupling. An iris adjustment, a slide screw tuner, or another device can deliberately establish a small mismatch so power will be reflected to the detector.[1] Hyde[4] describes several microwave bridges with a variety of matching/tuning arrangements.

D. VELOCITY AND IMPEDANCE

A plane transverse electromagnetic (TEM) wave with respective electric and magnetic field amplitudes $E_m = E_x$ and $B_m = B_y$, respectively, propagating in free space, travels at the speed c,

$$c = \frac{1}{(\mu_0 \varepsilon_0)^{1/2}} = 2.9979 \times 10^8 \text{ m/s}, \tag{5}$$

and the ratio of E_x to B_y/μ_0 provides the characteristic impedance Z_0 of free space

$$Z_0 = \mu_0 \frac{E_x}{B_y} = (\mu_0/\varepsilon_0)^{1/2} = 120\pi \text{ ohms}. \tag{6}$$

The energy density of the electric and magnetic fields is equal to

$$\tfrac{1}{2}\varepsilon_0 E_x^2 = \frac{1}{2\mu_0} B_y^2. \tag{7}$$

A transverse electromagnetic (TEM) wave can propagate along a coaxial cable, but waveguides cannot support TEM waves. A waveguide can support transverse electric (TE) waves which have a magnetic field component B_z along the z-direction of propagation and $E_z = 0$, and transverse magnetic (TM) waves with a component E_z

along the propagation direction and $B_z = 0$. For the transverse electric case we have the energy balance equation

$$\tfrac{1}{2}\varepsilon_0\langle E_x^2\rangle = \frac{1}{2\mu_0}\langle B_y^2 + B_z^2\rangle. \tag{8}$$

Since only E_x and B_y contribute to Poynting's vector for the energy flow, we define the characteristic impedance

$$Z_{\mathrm{TE}} = \frac{\mu_0 E_x}{B_y} = \sqrt{\frac{\mu_0}{\varepsilon_0}}\frac{\lambda_g}{\lambda_0}, \tag{9}$$

where λ_0 is the free-space wavelength

$$\lambda_0 \nu = c \tag{10}$$

and the guide wavelength $\lambda_g > \lambda_0$, which is the distance between successive maxima along the waveguide z-direction, is obtained from the expression

$$\frac{1}{\lambda_0^2} = \frac{1}{\lambda_g^2} + \frac{1}{\lambda_c^2}, \tag{11}$$

where λ_c is the cutoff wavelength, and $\nu_c = c/\lambda_c$ is the lowest frequency of the particular mode that can propagate in the waveguide. Since λ_g is greater than λ_0 it follows that Z_{TE} is greater than Z_0 of free space. For TM modes the analogous expressions are

$$\tfrac{1}{2}\varepsilon_0\langle E_x^2 + E_z^2\rangle = \frac{1}{2\mu_0}\langle B_y^2\rangle, \tag{12}$$

$$Z_{\mathrm{TM}} = \mu_0\frac{E_x}{B_y} = \sqrt{\frac{\mu_0}{\varepsilon_0}}\frac{\lambda_0}{\lambda_g}, \tag{13}$$

with $Z_{\mathrm{TM}} < Z_0 < Z_{\mathrm{TE}}$, and (11) applying here also.

E. RECTANGULAR RESONANT CAVITY

A rectangular resonator with dimensions a, b, d along the x-, y-, z-axes, respectively, is depicted in Fig. 2. The propagation constant k

$$k = (k_x^2 + k_y^2 + k_z^2)^{1/2} = (k_c^2 + k_z^2)^{1/2} \tag{14}$$

has the components

$$k_x = \frac{m\pi}{a}, \qquad k_y = \frac{n\pi}{b}, \qquad k_z = \frac{p\pi}{d}. \tag{15}$$

The resonant frequency $2\pi f_0 = \omega_0$ is given by

$$k = \omega_0(\mu\varepsilon)^{1/2} = \frac{2\pi}{\lambda_0} = \pi\left(\frac{m^2}{a^2} + \frac{n^2}{b^2} + \frac{p^2}{d^2}\right)^{1/2}, \tag{16}$$

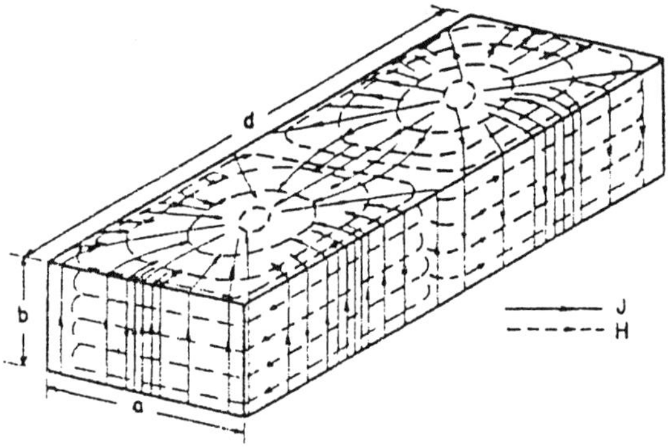

FIG. 3. Current distribution J in a TE_{102} mode rectangular resonant cavity with dimensions a, b, and d (ref. 1, p. 133).

where λ_0 is the free-space wavelength. The guide wavelength λ_g and cutoff wavelength λ_c are related to λ_0 through (11), where for this cavity

$$\frac{1}{\lambda_0} = \frac{k}{2\pi}, \tag{17}$$

$$\frac{1}{\lambda_g} = \frac{k_z}{2\pi}, \tag{18}$$

$$\frac{1}{\lambda_c} = \frac{k_c}{2\pi} = \frac{(k_x^2 + k_y^2)^{1/2}}{2\pi}. \tag{19}$$

The general mode configurations were written down elsewhere.[1]

Figure 2 gives the configurations of the electric and magnetic fields in the TE_{102} mode, and Fig. 3 shows the current flow in the cavity walls. The electric field begins and ends on induced charges on the broad face of the cavity, and the magnetic field H_t tangential to the cavity walls induces electric currents in the walls perpendicular to the H_t direction, as shown in Fig. 3. The circuit for the electric current flow is completed by the displacement current density $J_{\text{disp}} = D = \varepsilon_0\, dE/dt$ that flows through the centers of the magnetic field loops, and induces a charge density σ on the wide surface (xz-plane) near the centers of these loops. These charges reverse polarity every half-cycle, and the ohmic losses result from electrons rushing back and forth along the cavity walls every half-cycle to build up the charge.

Figure 4 shows the customary arrangement for inserting a sample into a rectangular TE_{102} cavity which is oriented in the magnet so the applied dc field B_0 is perpendicular to the plane of the page. In this configuration the rf magnetic field B_1, at the sample is perpendicular to the applied field B_0 and the sample in the middle of the tube is at a point of maximum B_1 and minimum (close to zero) E field.

The TE_{102} mode (sometimes called TE_{012}) is the dominant mode, and is also the

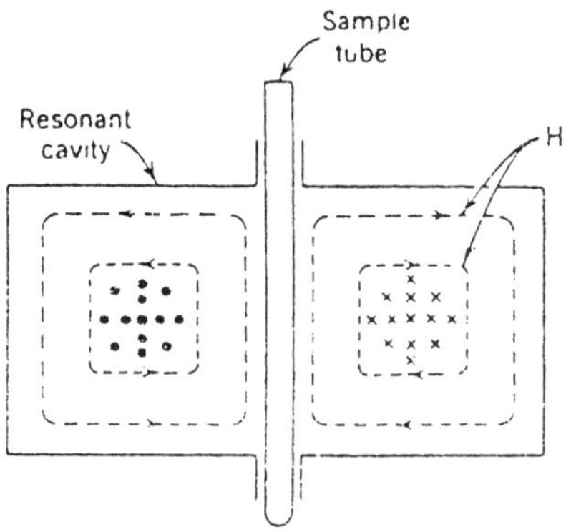

FIG. 4. ESR sample tube in rectangular TE_{102} resonant cavity (ref. 1, p. 134).

most important one used in ESR spectrometers. If we set $m = 1$, $n = 0$, and $p = 2$ in (15) and (16) we obtain for the only nonvanishing field components

$$B_x = \frac{B_0}{[1 + (d/2a)^2]^{1/2}} \sin\frac{\pi x}{a} \cos\frac{2\pi z}{d}, \tag{20}$$

$$B_z = \frac{-B_0}{[1 + (2a/d)^2]^{1/2}} \cos\frac{\pi x}{a} \sin\frac{2\pi z}{d}, \tag{21}$$

$$E_y = i\left(\frac{\mu}{\varepsilon}\right)^{1/2} B_0 \sin\frac{\pi x}{a} \sin\frac{2\pi z}{d}. \tag{22}$$

The coefficients of these wave configurations [(20)–(22)] are related through the characteristic impedance Z_{TE}. For this mode the resonant frequency f is independent of the (narrow) b dimension of the cavity, although, of course, the unloaded Q is a function of all three cavity dimensions. The lack of dependence of f on b means that the resonant cavity may be made as narrow as we wish to accommodate a small magnet gap. This is not possible with cylindrical cavities, so in general a cylindrical cavity requires a wider gap. The b dimension may be increased to accommodate variable temperature apparatus within the cavity.

Table 1 provides expressions for the quality factor of the TE_{102} cavity of Fig. 2 which is one guide wavelength long and has two mode cycles along the z-direction, and of the analogous TE_{101} cavity with one mode cycle and a length along z of $\lambda_g/2$. The table gives expressions for the quantity $Q_u\delta/\lambda_0$ where the skin depth δ,

$$\delta = \left(\frac{2}{\omega\mu_0\sigma}\right)^{1/2}, \tag{23}$$

for several metals, has the frequency dependence shown in Fig. 5. The unloaded Q arises from the losses produced by the rf current flowing in the walls in the manner

TABLE 1. Expressions for the unloaded quality factor Q_u of four commonly used resonators, where δ is the skin depth and λ_0 is the free-space wavelength.[1,2]

Rectangular TE_{101}	$Q_u \frac{\delta}{\lambda_0} = \frac{\frac{1}{2}b(a^2+d^2)^{3/2}}{d^3(a+2b)+a^3(d+2b)}$
Rectangular TE_{102}	$Q_u \frac{\delta}{\lambda_0} = \frac{4b(a^2+(\frac{1}{2}d)^2)^{3/2}}{d^3(a+2b)+4a^3(d+2b)}$
Cylindrical TE_{011} $(k_c a)'_{01} = 3.832$	$Q_u \frac{\delta}{\lambda_0} = \frac{\frac{1}{2\pi}\left[(k_c a)'^2_{01} + \left(\frac{\pi a}{d}\right)^2\right]^{3/2}}{(k_c a)'^2_{01} + 4\pi^2\left(\frac{a}{d}\right)^3}$
Cylindrical TE_{111} $(k_c a)'_{11} = 1.841$	$Q_u \frac{\delta}{\lambda_0} = \frac{\frac{1}{2\pi}\left[1-\left(\frac{1}{(k_c a)'_{11}}\right)^2\right]\left[(k_c a)'^2_{11} + \left(\frac{\pi a}{d}\right)^2\right]^{3/2}}{(k_c a)'^2_{11} + 4\pi^2\left(\frac{a}{d}\right)^3 + \left(1-\frac{2a}{d}\right)\left(\frac{\pi a}{(k_c a)'_{11} d}\right)^2}$

shown in Fig. 3, and penetrating into the metal by a depth given by δ. The losses occur because of the skin effect resistance R_s

$$R_s = \frac{1}{\sigma\delta}. \tag{24}$$

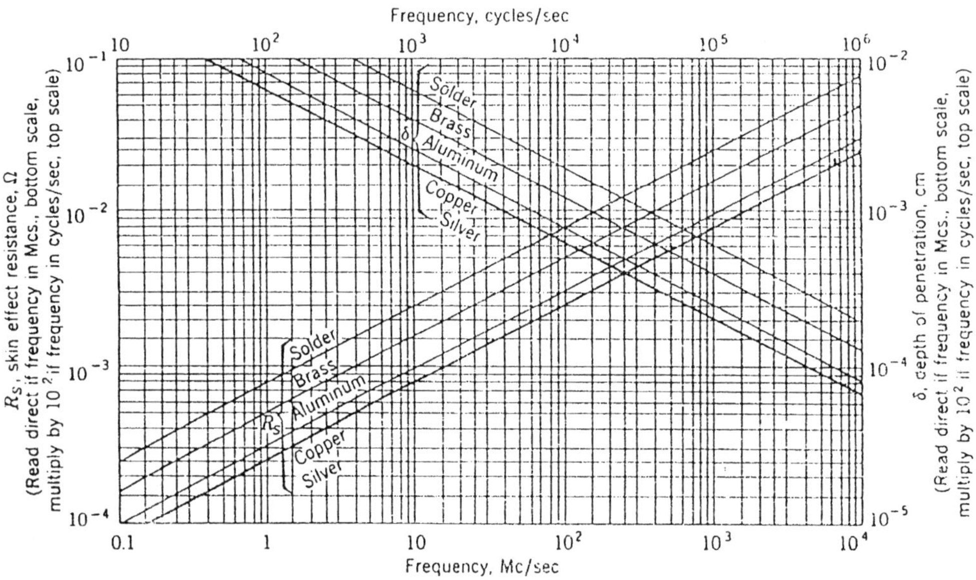

FIG. 5. Skin effect resistance R_s and skin depth δ for several plane conductors as a function of frequency (ref. 3, p. 252).

A knowledge of mode configurations and current flow in resonant cavities is particularly useful for designing special purpose resonators without appreciably decreasing their Q. For example, it is possible to cut slots in the bottom of the rectangular TE_{10p} cavity parallel to the directions of current flow shown in Fig. 3 without appreciably degrading the Q, and these slots may be used for UV irradiation.

F. CYLINDRICAL RESONANT CAVITY

In analogy to rectangular resonant cavities, cylindrical cavities can support both TE_{mnp} and TM_{mnp} modes, where the subscripts m, n, and p refer to the number of half-cycle variations in the angular (ϕ), radial (r), and longitudinal (z) directions, respectively. These modes are derived from the TE_{mn} and TM_{mn} cylindrical waveguide modes discussed elsewhere.[1]

Let the resonant cavity radius be a, and let the length be d. In analogy to the rectangular cavity case, we have

$$k = \omega_0(\mu\varepsilon)^{1/2} = (k_c^2 + k_z^2)^{1/2} = \left[\frac{(k_c a)_{mn}^2}{a^2} + \left(\frac{p\pi}{d}\right)^2\right]^{1/2}, \tag{25}$$

where $(k_c a)_{mn}$ is a Bessel function root, since cylindrical waveguide modes have besseloid radial variations and $k_z = p\pi/d$. For transverse magnetic waves TM_{mnp}, the quantity $(k_c a)_{mn}$ is the nth root of the mth-order Bessel function $J_m(k_c r)$

$$J_m(k_c a) = 0. \tag{26}$$

For transverse electric modes the first derivative of the Bessel function $J_m'(k_c r)$ must vanish at the surface

$$\frac{d}{d(k_c r)} J_m(k_c r)|_{r=a} = J_m'(k_c a) = 0 \tag{27}$$

and the nth root of this equation for the mth-order Bessel function is denoted by $(k_c a)_{mn}'$. The most used Bessel roots $(k_c a)_{mn}'$ and $(k_c a)_{mn}$ are listed in columns 2 and 4, respectively, of Table 2.

The dependence of the frequency $f = \omega_0/2\pi$ on the radius a and the length d of a cylindrical cavity in a vacuum (or air) is easily presented graphically by rearranging (25) to the form

$$(2af)^2 = \left(\frac{c(k_c a)_{mn}}{\pi}\right)^2 + \left(\frac{cp}{2}\right)^2\left(\frac{2a}{d}\right)^2, \tag{28}$$

where $c = 1/(\mu_0\varepsilon_0)^{1/2}$ is the velocity of light in vacuo. For convenience we plot $(2af)^2$ against $(2a/d)^2$ in Fig. 6. The slopes of the lines on this figure are proportional to p^2, where $p = 0, 1, 2$ for the modes shown. The intercepts at $(a/d) = 0$ are determined by

TABLE 2. The nth roots of the mth-order Bessel functions $J'_m(k_c a)$ and $J_m(k_c a)$, and the corresponding TE_{mnp} and TM_{mnp} cylindrical cavity modes (ref. 2, p. 299).

TE_{mnp} Mode	$J'_m(k_c a)$	TM_{mnp} Mode	$J_m(k_c a)$
11 p	1.841	01 p	2.405
21 p	3.054	11 p	3.832
01 p	3.832	21 p	5.136
31 p	4.201	02 p	5.520
41 p	5.318	31 p	6.380
12 p	5.332	12 p	7.016
51 p	6.415	41 p	7.588
22 p	6.706	22 p	8.417
02 p	7.016	03 p	8.654
61 p	7.501	51 p	8.772
32 p	8.016	32 p	9.761
13 p	8.536	61 p	9.936
71 p	8.578	13 p	10.174
42 p	9.283		
81 p	9.648		
23 p	9.970		
03 p	10.174		

the mn Bessel function roots. For designing a resonant cavity it is best to select values of a and d so that there are no extraneous modes that have resonant frequencies near the design point. For TE modes, $(k_c a)'_{mn}$ replaces $(k_c a)_{mn}$ in (25) and (28).

The quality factor Q for a cavity resonator defined by

$$Q = \frac{2\pi(\text{energy stored})}{\text{energy dissipated per cycle}} \tag{29}$$

$$= \frac{\omega}{2\mu P_0}\int B_1^2\, dV_c, \tag{30}$$

depends on the mode and on the dimensions. The Q for an empty cavity with an infinitely small coupling hole, designated by Q_u, is called the unloaded Q. Table 1 gives specific expressions of $Q_u\delta/\lambda_0$ for the TE_{111} and TE_{011} cylindrical cavity modes. A TE_{0np} mode has a maximum value for $Q_u\delta/\lambda_0$ when the cavity length d equals the diameter $2a$.

As Fig. 6 indicates, the dominant mode (smallest cavity size for a particular wavelength) in a cylindrical cavity, namely the TE_{111} mode, has the smallest $(2af)$ values, not counting the anomalous TE_{mno} modes whose frequencies are independent of the ratio a/d. The TE_{111} configuration corresponds to half of the TE_{112} mode which is sketched in Fig. 7(b). It is analogous to the TE_{101} rectangular cavity mode,[5] and the two may be derived from one another by deforming the cylinder into a rectangular shape and vice versa. Since this mode is dominant, it may be used in situations where

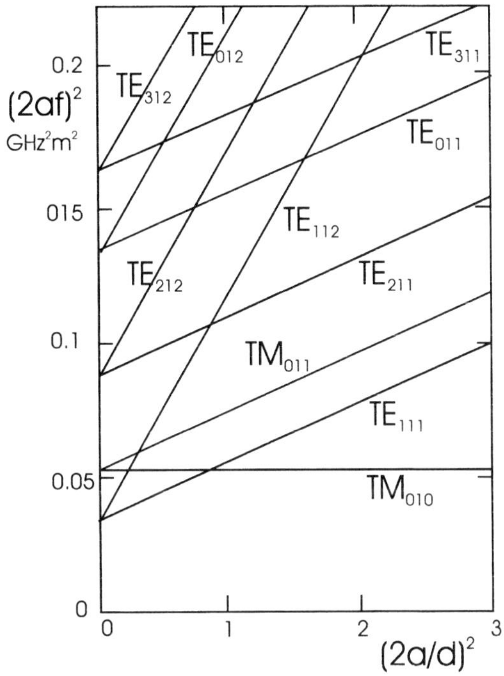

FIG. 6. Mode chart for right circular cylinder of radius a, length d, and resonant frequency f (ref. 2, p. 298).

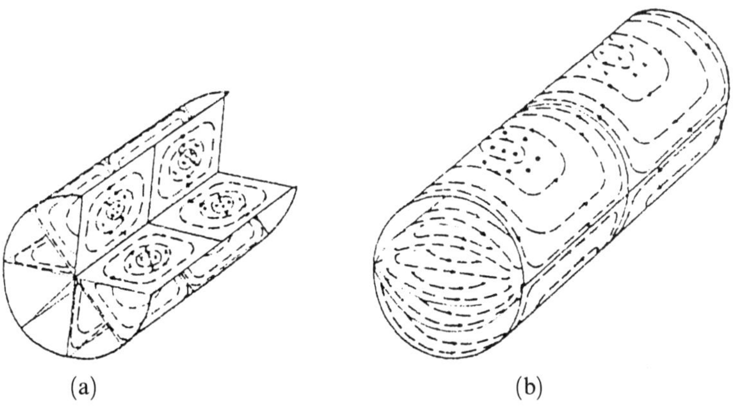

FIG. 7. Diagrammatic sketch of the TE_{012} (left) and TE_{112} (right) cylindrical resonant cavity modes (adapted from ref. 5, pp. 608 and 610).

the magnet pole gap is too small to support larger cylindrical cavities. Unfortunately, the TE_{111} mode has the lowest Q factor of any transverse electric mode.

Most frequency meters and many sample cavities employ the cylindrical TE_{011} mode with half the length of the TE_{012} mode sketched in Fig. 7(a) because it has a fairly high Q factor, and because there is no rf current flow between the cylinder walls and the end plate. In fact, there is no electric current flow in either the radial (r) or longitudinal (z) directions, but only in the angular (ϕ) direction. This property enables us to use a piston-like tuning plunger on the end plate, so that the frequency may be varied by screwing the end plate in and out. The current flow property renders it unnecessary to achieve good electric contact between the end plate and the cylinder. If we deliberately leave a finite gap between the piston edges and the walls, then other modes will be suppressed because they require current flow across this gap. In particular, this will suppress the TM_{111} mode (not shown in Fig. 6) that is degenerate with the TE_{011} mode. The cylindrical TE_{011} mode is particularly useful for a sample cavity since H_z is very strong along the cavity axis. It is possible to have a very large hole in the end without appreciably decreasing the Q so this cavity may be used with a larger sample than the rectangular TE_{102} cavity. There is ample space for positioning cryogenic apparatus along the cavity axis.

G. SPECIAL PURPOSE CAVITIES

The rectangular and cylindrical cavity modes that we have described are commonly used in general purpose resonators with different sample arrangements. A number of resonators have been designed with mode configurations suitable for special applications, and we will describe some of them.

The rectangular TE_{102} resonator of Fig. 4 has the sample tube inserted in the plane between the two B_1-field loops. If the same cavity is made twice as long corresponding to the TE_{104} mode, then there are four B_1-field loops and two sample positions analogous to the one shown in Fig. 4, one for an unknown sample and one for a reference sample. If each position is provided with separate modulation coils, then individual scans can be made to record the two specimens, or they can be recorded simultaneously, depending on how the modulation coils are activated. By using different modulation frequencies for each coil we can obtain simultaneous, independent recordings.

A typical cavity such as the one illustrated in Fig. 4 has the microwave magnetic field B_1 perpendicular to the large applied field B_0. Cavities have also been designed[5] to operate with B_1 parallel to B_0, and to have the capability of easily switching between the two methods of detection.

A number of bimodal cavity designs have appeared in the literature, in which two microwave modes are excited simultaneously. An example of such a cavity has two rectangular TE_{103} modes degenerate in frequency excited perpendicular to each other with two-thirds of their mode cycles superimposed.[7] Reactive and resistive paddles are employed to tune the modes to the same frequency. They are needed because the insertion of the sample tends to detune the modes. For the usual transverse magnet arrangement one mode will have a perpendicular B_1 and the other a parallel B_1 field.

A slotted tube resonator[8,9] and been found useful in pulsed-ESR. A stripline resonator has been employed in biological ESR,[10] and has functioned as a transmission cavity in spin echo ESR.[11] Helical resonators[12] could be applied in ESR, as could re-entrant cavities.[1]

A loop-gap resonator[13,14] consists of distinguishable inductances and capacitances about 0.1λ in size, and it may be considered as a structure intermediate in character between a lumped circuit and a distributed circuit. It has a high filling factor η which considerably increases its sensitivity relative to that of conventional cavities when the sample size is limited, even though its Q is a factor of 10 lower than that of traditional resonators. Froncisz and Hyde[13] reported a six- to seven-times gain in sensitivity for a loop-gap resonator at X-band. These resonators have the attributes that make them ideally suited for pulsed ESR, namely a low Q for fast-pulsed power dissipation and a high filling factor to provide a large B_1 field at the sample position.[14] Loop-gap resonators have appeared in several forms,[4] and have been applied to pulsed ESR,[15] ENDOR,[16] spin echo,[17] and saturation transfer.[18]

H. QUALITY AND FILLING FACTORS

The ESR signal at resonance is proportional to the amount of power absorbed by the sample, and this in turn is proportional to the average value of the microwave magnetic field $\langle B_1^2\rangle_s^{1/2}$ at the sample, where in general,

$$\langle B_1^2\rangle_s = \frac{\int B_1^2\,dV}{\int dV} = \frac{1}{V}\int B_1^2\,dV. \tag{31}$$

More precisely, B_1 is the component of the rf magnetic field perpendicular to the static applied field. If the sample has an rf susceptibility χ'' and a volume V_s, then the amount of power absorbed is proportional to $\chi'' V_s\langle B_1^2\rangle_s$. The resonance absorption will decrease the Q of the resonant cavity, and this may be expressed mathematically by adding the reciprocal of the sample loss quality factor Q_χ to the corresponding terms for the unloaded (Q_u), dielectric loss (Q_ε) and radiation loss through coupling holes (Q_r) as follows:

$$\frac{1}{Q} = \frac{1}{Q_u} + \frac{1}{Q_\varepsilon} + \frac{1}{Q_r} \qquad \text{off resonance,} \tag{32}$$

$$\frac{1}{Q} = \frac{1}{Q_u} + \frac{1}{Q_\varepsilon} + \frac{1}{Q_r} + \frac{1}{Q_\chi} \qquad \text{at resonance,} \tag{33}$$

where Q_χ is the ratio of the power stored in the overall cavity volume V_c to the power dissipated in the sample

$$Q_\chi = \frac{\frac{1}{2}\mu_0 \int_{\text{cavity}} B_1^2\,dV}{\frac{1}{2}\mu_0 \int_{\text{sample}} \chi'' B_1^2\,dV} \tag{34}$$

$$= \frac{V_c\langle B_1^2\rangle_c}{\chi'' V_s\langle B_1^2\rangle_s}. \tag{35}$$

Equation (35) may be expressed in terms of the filling factor η,

$$Q_\chi = \frac{1}{(\chi''\eta)}, \tag{36}$$

where η, defined by

$$\eta = \frac{\int_{\text{sample}} B_1^2\, dV}{\int_{\text{cavity}} B_1^2\, dV} = \frac{V_s\langle B_1^2\rangle_s}{V_c\langle B_1^2\rangle_c}, \tag{37}$$

is a measure of the extent to which the microwave energy density $B_1^2/2\mu_0$ is present at the sample. In the absence of saturation effects the sensitivity of an ESR spectrometer is directly proportional to η, as explained in Section B of Chapter I.

It is usually easy to obtain the volume ratio V_s/V_c, but the ratio $\langle B_1^2\rangle_s/\langle B_1^2\rangle_c$ requires the integration of B_1^2 over both the sample and the cavity. Representative calculations of this ratio have been published elsewhere.[1] In particular it was found that for a rectangular TE_{102} mode cavity of length d, height a, and width b with a small sample at its center the filling factor η is given by

$$\eta = \frac{4}{1+(d/2a)^2}\frac{V_s}{V_c}, \tag{38}$$

independent of the cavity's narrowest dimension b. For a typical cavity $d \sim 2a \sim 4b$ to give

$$\eta \approx 2V_s/V_c \tag{39}$$

corresponding to an average squared magnetic field at the sample equal to twice that in the overall cavity. For a flat cell or a cylindrical ESR sample tube, oriented as shown in Fig. 4, the filling factor is

$$\eta = \frac{V_s}{V_c}. \tag{40}$$

For a cylindrical TE_{011} resonant cavity of length d and radius a with a small sample located in the center, the filling factor is

$$\eta = \frac{12.33}{1+(0.82a/d)^2}\frac{V_s}{V_c}. \tag{41}$$

For a filled cylindrical sample tube of radius $r \ll a$ located along the axis of such a cavity, the filling factor η is

$$\eta = \frac{6.16}{1+(0.82a/d)^2}\frac{V_s}{V_c}. \tag{42}$$

The coefficients in (41) and (42) arise from Bessel function roots.

It will be instructive to give more specific details for these two cavity resonators. For the rectangular cavity let $d = 2a = 4.5$ cm and $b = 1$ cm so that V_c is close to 10 cm^3. The cylindrical cavity has $a = 2.05$ cm, $a/d \sim \frac{1}{2}$, and $V_c \sim 54$ cm^3. For comparison purposes a second cylindrical cavity of the same frequency (9.7 GHz) but with an a/d ratio of 3/2 will also be considered. Its radius may be determined from Fig. 6, and the ratio of the Q_L values for the two cylindrical cavities may be deduced from Fig. 5-8 of ref. 1, 2nd ed. The data on these three cavities are listed in Table 3. This table provides a comparison of three samples: a small piece of DPPH of volume $V_s = 10^{-3}$ cm^3; a filled cylindrical tube 6 cm long with radius $r = 0.2$ cm; and the entire cavity itself, e.g., filled with gas. The ratios of these sample volumes to each cavity volume are listed in Table 3, and the filling factors are included for each case. The loaded quality factors Q_L for these three cases were calculated previously.

There are several interesting things to be learned from Table 3. In the first place, the cylindrical cavities have much higher Q's than the rectangular cavity, and thus store more microwave energy. However, this energy is spread over a larger volume so that the average microwave magnetic field strength is about the same in all three cases. The cylindrical cavity concentrates B_1 to a greater extent in the center (particularly for a small a/d ratio), but for cylindrical sample tubes extending the length of the cavity, some of the total B_1 is ineffective because the radial component is, in general, not perpendicular to the large applied magnetic field, whereas all of the microwave magnetic field components in the rectangular TE_{102} cavity are perpendicular to the applied

TABLE 3. Comparison of a cylindrical TE_{011} mode resonator and a rectangular TE_{102} mode resonator at X-band with a power of about $P_w = 100$ mW incident upon the cavity (ref. 1, p. 169).

		Cylindrical cavities		
Quantity	Sample No.[a]	$a/d = 1/2$	$a/d = 3/2$	Rectangular cavity ($d = 2a$)
Dimensions, cm		$a = 2.05$	$a = 3.0$	
		$d = 4.1$	$d = 2.0$	$1 \times 2.25 \times 4.5$
V_c, cm^3	—	54	57	10
Q_L	—	12.000	8000	4000
$(\langle B_1^2\rangle_c)^{1/2}$ mT	—	0.048	0.037	0.063
V_s/V_c	1	1.85×10^{-5}	1.7×10^{-5}	10^{-4}
η	1	1.6×10^{-4}	8.2×10^{-5}	2×10^{-4}
$(\langle B_1^2\rangle_c)^{1/2}$ mT	1	0.14	0.080	0.090
V_s/V_c	2	9.5×10^{-3}	4.3×10^{-3}	2.8×10^{-2}
η	2	4.2×10^{-2}	1×10^{-2}	2.8×10^{-2}
$(\langle B_1^2\rangle_c)^{1/2}$ mT	2	0.10	0.060	0.064
V_s/V_c	3	1.00	1.00	1.00
η	3	0.85	0.69	1.00
$(\langle B_1^2\rangle_c)^{1/2}$ mT	3	0.045	0.031	0.064

[a] Sample No. 1: small piece of DPPH ($V_s = 10^{-3}$ cm^3). Sample No. 2: cylindrical tube of radius $r = 0.2$ cm and effective length $= d$ ($V_s = 0.126d$ cm^3). Sample No. 3 fills the entire cavity ($V_s = V_c$).

field. Note that the sensitivity for the cylindrical cavity is improved by decreasing its radius to length ratio a/d. The Q is a maximum when $d = 2a$, so the optimum sensitivity will be obtained with $d > 2a$.

These considerations neglect the manner in which the real and imaginary parts of the sample's dielectric constant perturb the value of the rf magnetic field (i.e., $\langle B_1^2 \rangle_s$) at the sample. To take account of the extent to which a sample of high dielectric constant "draws in" the rf field B_1 we may define a dielectric enhancement factor[6] $K_d \geq 1$,

$$B_{1\varepsilon} = K_d B_1, \tag{43}$$

where B_1 is the field in the absence of the sample, and $B_{1\varepsilon}$ is the field in the sample. Dielectric enhancement factors for several commonly used organic solvents, lie in the range from 1.05 for ethanol and 2.35 for dimethylglycol[6] and they are dependent on the sample cell configuration.

Hyde[20] noted from (30) that it is useful to define the resonator efficiency parameter $\Lambda(x, y, z)$ mentioned in Section B.2 of Chapter I

$$\Lambda(x, y, z) = \frac{B_1(x, y, z)}{P_0^{1/2}}, \tag{44}$$

which characterizes the effectiveness of the cavity mode in concentrating the incoming power at a particular position x, y, z in the cavity. This parameter is an index of the efficiency of the cavity in optimizing the sensitivity.

REFERENCES

[1] C. P. Poole, Jr., *Electron Spin Resonance* (Wiley-Interscience, New York, 1st ed., 1967; 2nd ed., 1983; reprinted Dover, New York, 1996), cf. especially Resonant Cavity chapter.

[2] C. G. Montgomery, editor, *Technique of Microwave Measurements* (McGraw-Hill, New York, 1947; reprinted Boston Publications, Boston, 1964).

[3] S. Ramo, J. R. Whinnery, and T. van Duzer, *Fields and Waves in Communications Electronics* (Wiley, New York, 1965).

[4] J. M. Hyde, *Handbook of Microwave Technology* (Academic Press, New York, 1995), Vol. 2, p. 365.

[5] J. F. Reintjes and G. T. Coate, *Principles of Radar* (McGraw-Hill, New York, 1952).

[6] K. D. Kramer and W. Müller-Warmuth, Z. Angew. Phys. **16**, 281 (1963).

[7] J. M. Hyde, J. C. W. Chien, and J. H. Freed, J. Chem. Phys. **48**, 4211 (1968).

[8] H. J. Schneider and P. Dullenkopf, Rev. Sci. Instrum. **48**, 68 (1977).

[9] M. Mehring and F. Freysoldt, J. Phys. E: Sci. Instrum. **13**, 894 (1980).

[10] B. Johansson, S. Haroldson, L. Petterson, and O. Beckman, Rev. Sci. Instrum. **45**, 1444 (1974).

[11] J. L. Davis and W. B. Mims, Rev. Sci. Instrum. **49**, 1095 (1978).

[12] R. H. Webb, Rev. Sci. Instrum. **33**, 732 (1962).

[13] W. Froncisz and J. S. Hyde J. Magn. Reson. **47**, 515 (1982).

[14] R. H. Robinson, H. Thomann, A. H. Beth, P. Fajer, and L. R. Dalton, in *EPR and Advanced EPR Studies of Biological Systems*, edited by L. R. Dalton (CRC Press, Boca Raton, FL, 1985), Ch. 3.

[15] S. Pfenninger, J. Forrer, A. Schweiger, and Th. Weiland, Rev. Sci. Instrum. **59**, 752 (1988).

[16] J. R. Anderson, R. A. Venters, M. K. Bowman, A. E. True, and B. M. Hoffman, J. Magn. Reson. **65**, 165 (1985).

[17] R. D. Britt and M. P. Klein, J. Magn. Reson. **74**, 535 (1987).

[18] D. D. Thomas, C. H. Wendt, W. Froncisz, and J. S. Hyde, Biophys. J. **43**, 131 (1983).

CHAPTER IV
LINESHAPES

Charles P. Poole, Jr. and Horacio A. Farach

Department of Physics and Astronomy, University of South Carolina, Columbia, South Carolina 29208, USA

Lorentzian and Gaussian lineshapes of individual lines are described, and then the discussion is extended to overlapping hyperfine structure, convolutions, anistropies, and powder patterns. More specialized cases like small spin clusters, semirandom distributions, conduction electrons, exchange, and saturation are also treated.

CONTENTS

A. INTRODUCTION

The shape and width of a resonant line sometimes reveal important information on relaxation times and line broadening mechanisms. The power dependence of the amplitude and width can provide estimates of the spin–lattice and spin–spin relaxation times, T_1 and T_2, respectively. Unsymmetrical features of the line shape usually reflect anisotropies in the spin Hamiltonian parameters. Hyperfine coupling constants can often be extracted from barely resolved spectra. Spin label lineshapes provide the extent of tumbling motion. Instrumental conditions are also able to affect the line shape and width. We saw in Section B of Chapter I how magnetic field modulation increases the amplitude of a resonant line, and we will see below the influence of high power levels.

B. LORENTZIAN AND GAUSSIAN SHAPES

The two most common lineshapes are Lorentzian and Gaussian[1] with the following analytical forms for the absorption $Y_{(B)}$, absorption first derivative $Y'_{(B)}$, absorption second derivative $Y''_{(B)}$, dispersion $d_{(B)}$, and dispersion first derivative $d'_{(B)}$. The Lorentzian expressions are

$$Y_{(B)} = \frac{y_m}{1 + [(B - B_0)/\frac{1}{2}\Delta B_{1/2}]^2}, \tag{1a}$$

$$Y'_{(B)} = \frac{16 y'_m \dfrac{B - B_0}{\frac{1}{2}\Delta B_{\mathrm{pp}}}}{\left[3 + \left(\dfrac{B - B_0}{\frac{1}{2}\Delta B_{\mathrm{pp}}}\right)^2\right]^2}, \tag{1b}$$

$$Y''_{(B)} = (27 y''_m) \frac{1 - \dfrac{(B - B_0)^2}{(\frac{1}{2}\Delta B_{\mathrm{pp}})^2}}{\left[3 + \left(\dfrac{B - B_0}{\frac{1}{2}\Delta B_{\mathrm{pp}}}\right)^2\right]^3}, \tag{1c}$$

$$d_{(B)} = \frac{2\dfrac{(B - B_0)}{(\frac{1}{2}\Delta B_{1/2})}}{1 + \left(\dfrac{B - B_0}{\frac{1}{2}\Delta B_{1/2}}\right)^2}, \tag{2a}$$

$$d'_{(B)} = \frac{9 - 3\left(\dfrac{B - B_0}{\frac{1}{2}\Delta B_{\mathrm{pp}}}\right)^2}{\left[3 + \left(\dfrac{B - B_0}{\frac{1}{2}\Delta B_{\mathrm{pp}}}\right)^2\right]^2}. \tag{2b}$$

Some corresponding Gaussian lineshape expressions are

$$Y_{(B)} = y_m \exp\left[-\left(\frac{B - B_0}{\frac{1}{2}\Delta B_{1/2}}\right)^2 \ln 2\right], \tag{3a}$$

$$Y'_{(B)} = y'_m e^{1/2}\left(\frac{B - B_0}{\frac{1}{2}\Delta B_{\mathrm{pp}}}\right) \exp\left\{-\frac{1}{2}\left(\frac{B - B_0}{\frac{1}{2}\Delta B_{\mathrm{pp}}}\right)^2\right\}, \tag{3b}$$

$$Y''_{(B)} = y''_m \left[\left(\frac{B - B_0}{\frac{1}{2}\Delta B_{\mathrm{pp}}}\right)^2 - 1\right] \exp\left[-\frac{1}{2}\left(\frac{B - B_0}{\frac{1}{2}\Delta B_{\mathrm{pp}}}\right)^2\right]. \tag{3c}$$

In these expressions, y_m is the amplitude of the absorption line, y'_m is the peak-to-peak amplitude of the first derivative, $\Delta B_{1/2}$ is the full linewidth between half amplitude points, and ΔB_{pp} is the full peak-to-peak linewidth.

Figure 1 and Tables 1 and 2 give data and comparisons between these two lineshapes. Table 1 lists values of the zeroth (i.e., the area A), second, and fourth moments ($n = 0, 2, 4$, respectively) defined by

$$A = \int_0^\infty Y\,dB = -\int_0^\infty (B - B_0)Y'\,dB, \tag{4}$$

$$\langle B^n \rangle = \frac{1}{A}\int_0^\infty (B - B_0)^n Y\,dB = -\frac{1}{A(n+1)}\int (B - B_0)^{n+1} Y'\,dB, \tag{5}$$

where the "center field" B_0 is chosen so the first moment vanishes

$$\int_0^\infty (B - B_0)Y\,dB = \int_0^\infty (B - B_0)^2 Y'\,dB = 0. \tag{6}$$

Parameters that characterize a symmetric second derivative resonant line are defined in Fig. 2 and appear in Table 1. The above lineshape expressions were normalized to unit amplitude. The line widths of a Lorentzian lineshape are related to the spin–spin relaxation time T_2 as follows:

$$\Delta\omega_{1/2} = \gamma\Delta B_{1/2} = \sqrt{3}\Delta\omega_{pp} = \sqrt{3}\gamma\Delta B_{pp} = \frac{2}{T_2}, \tag{7}$$

C. MOMENTS AND WIDTHS OF RESONANT LINES

The Van Vleck[2] theory of moments assumes that the line broadening arises from the Hamiltonian

$$H = g\mu_B B \sum_j S_{zj} + \sum_{jk} A_{jk}\mathbf{S}_j \cdot \mathbf{S}_k + \frac{\mu_0 g^2 \mu_B^2}{4\pi} \sum_{k>j}\left[\frac{\mathbf{S}_j \cdot \mathbf{S}_k}{r_{jk}^3} - 3\frac{(\mathbf{r}_{jk}\cdot\mathbf{S}_j)(\mathbf{r}_{jk}\cdot\mathbf{S}_k)}{r_{jk}^5}\right], \tag{8}$$

where r_{jk} is the distance between the jth and kth spins, A_{jk} is proportional to the exchange integral J_{jk}, and the term in square brackets is the dipolar interaction.

For a powder sample with cubic symmetry the second moment for like spins is given by

$$\langle B^2 \rangle = \left(\tfrac{3}{5}\right)\left(\frac{\mu_0 g\mu_B}{4\pi}\right)^2 S(S+1)\left[\sum_j \left(\frac{1}{r_{jk}}\right)^6 + 4\pi\rho\int_{a_0}^\infty \frac{r^2\,dr}{r^6}\right], \tag{9}$$

where the nearby spins are taken into account in the summation and the integration begins at the distance a_0 of the next nearest spin, where ρ is the number of spins per unit volume. If the broadening is from unlike spins such as electron centers broadened

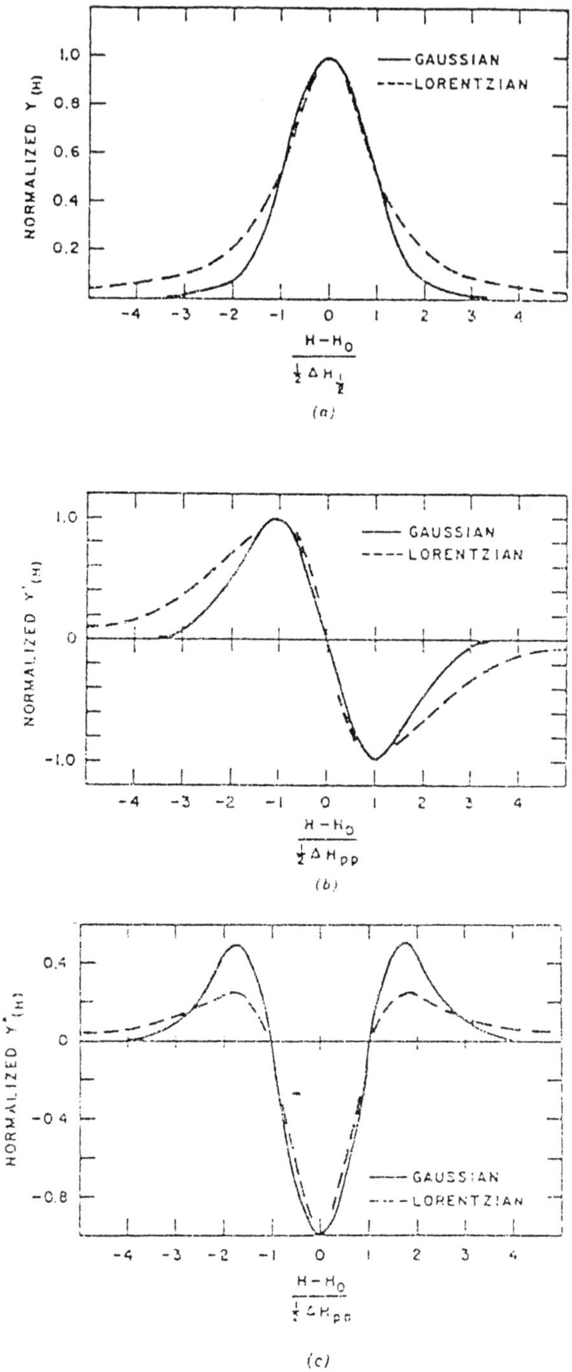

FIG. 1. Comparison of Gaussian and Lorentzian lineshapes showing: (a) absorption lines plotted with the same half amplitude linewidth; (b) first derivative absorption lines plotted with the same peak-to-peak linewidth; and (c) second derivative absorption lines plotted with the same peak-to-peak linewidth (from ref. 1, p. 480).

TABLE 1. Comparison of Gaussian and Lorentzian lineshapes (ref. 1, p. 476). The factors B_i'' are defined in Fig. 2 as H_i''.

Parameter	Gaussian shape	Lorentzian shape
$\Delta B_{1/2}/\Delta B_{pp}$	$(2\ln 2)^{1/2} = 1.1776$	$3^{1/2} = 1.7321$
$y_m(y_m'\Delta B_{pp})$	$e^{1/2}/2 = 0.8244$	$4/3 = 1.3333$
$y_m(y_m''\Delta B_{pp}^2)$	$\frac{1}{4} = 0.2500$	$3/8 = 0.3750$
$A/(y_m\Delta B_{1/2})$	$(\frac{1}{2})(\pi/\ln 2)^{1/2} = 1.0643$	$\pi/2 = 1.5708$
$A/(y_m'\Delta B_{pp}^2)$	$(\frac{1}{2})(\pi e/2)^{1/2} = 1.0332$	$2\pi/3^{1/2} = 3.6276$
$\langle B^2\rangle/(\Delta B_{1/2})^2$	$1/(8\ln 2) = 0.1803$	∞
$\langle B^4\rangle/(\Delta B_{1/2})^4$	$3/[64(\ln 2)^2] = 0.0974$	∞
$\langle B^2\rangle/(\Delta B_{pp})^2$	$\frac{1}{4} = 0.2500$	∞
$\langle B^4\rangle/(\Delta B_{pp})^4$	$\frac{3}{16} = 0.1875$	∞
y_1''/y_2''	$\frac{1}{2}e^{3/2} = 2.2409$	$64^{1/3} = 4.0$
$B_1''/\Delta B_{pp}$	0.626	0.567
$B_2''/\Delta B_{pp}$	$3^{1/2} = 1.7321$	$3^{1/2} = 1.7321$
$B_3''/\Delta B_{pp}$	2.52	$81^{1/4} = 3.0$
For solid DPPH, $A/[y_m'(\Delta B_{pp})^2 = 2.2$		
$e = 2.7183$	$\ln 2 = 0.6931$	
$\pi = 3.1416$		
$2^{1/2} = 1.4142$	$\pi^{1/2} = 1.7724$	
$3^{1/2} = 1.7321$	$(\ln 2)^{1/2} = 0.8326$	
$e^{1/2} = 1.6487$		

TABLE 2. Amplitudes $Y'(B)$ of Gaussian and Lorentzian absorption first derivative lineshapes at multiples of $\frac{1}{2}\Delta B_{pp}$ from the center (ref. 1, p. 483).

$\left(\frac{B-B_0}{\frac{1}{2}\Delta B_{pp}}\right)$	Gaussian	Lorentzian[a]
0.0	0.0000	0.0000
0.5	0.7275	0.7574
1.0	0.0000	1.0000
1.5	0.8029	0.8701
2.0	0.4461	0.6531
3.0	0.0549	0.3333
4.0	0.0022	0.1773
5.0	3×10^{-5}	0.1020
6.0	1.5×10^{-7}	0.0631
7.0		0.0414
8.0		0.0285
9.0		0.0204
10.0		0.0151

[a] When $|B - B_0| > 10\Delta B_{pp}$ we may set $Y'(B) \sim 16Y_m'[(B-Bo)/\frac{1}{2}\Delta B_{pp}]^{-3}$ for the Lorentzian lineshape.

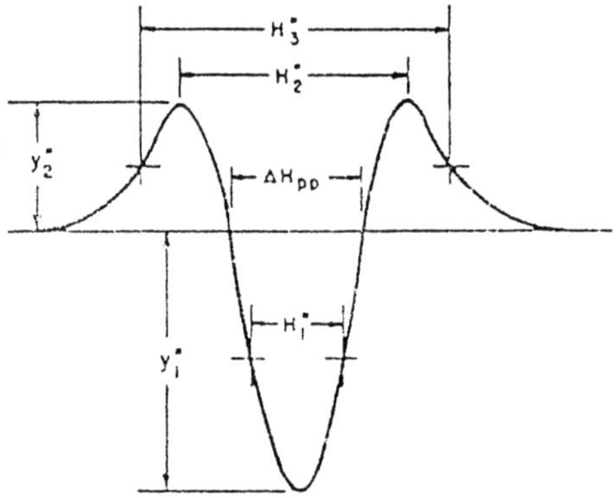

FIG. 2. Second derivative resonant absorption line showing the definitions of various parameters (from ref. 1, p. 479). The symbol H is replaced by B in Table 1.

by surrounding protons, then the factor $\frac{3}{5}$ must be replaced by $\frac{4}{15}$, and $g\mu_B$ is replaced by its proton counterpart.

For a symmetrical line odd moments vanish and for the Gaussian and Lorentzian cases we have for the even moments

$$\langle B^n \rangle = (n-1)!!(\tfrac{1}{2}\Delta B_{\mathrm{pp}})^n, \quad \text{Gaussian}, \tag{10a}$$

$$\langle B^n \rangle = \infty, \quad \text{Lorentzian}, \tag{10b}$$

where the double factorial is defined by $(n-1)!! = (n-1)(n-3)\ldots(5)(3)(1)$.

We see from Table 1 that for a Gaussian lineshape

$$\Delta B_{\mathrm{pp}} = 2\langle B^2 \rangle^{1/2} = 0.8492\Delta B_{1/2}. \tag{11}$$

This gives, respectively, from (9), the following expressions for the peak-to-peak linewidth in the units millitesla (10 G = 1 mT) for like electron spins, unlike electron spins and nuclei causing the broadening of a Gaussian shaped line

$$\Delta B_{\mathrm{pp}} = 1.437 \times 10^3 g[S(S+1)]^{1/2} \left[\sum_j \left(\frac{1}{r_{jk}} \right)^6 + 4\pi\rho \int_{a_0}^{\infty} \frac{dr}{r^4} \right]^{1/2} \quad \text{like electron spins}, \tag{12a}$$

$$\Delta B_{\mathrm{pp}} = 0.958 \times 10^3 g[S(S+1)]^{1/2} \left[\sum_j \left(\frac{1}{r_{jk}} \right)^6 + 4\pi\rho \int_{a_0}^{\infty} \frac{dr}{r^2} \right]^{1/2} \quad \text{unlike electron spins}, \tag{12b}$$

$$\Delta B_{\mathrm{pp}} = 0.5217 \left(\frac{\mu}{\mu_N} \right) [I(I+1)]^{1/2} \left[\sum_j \left(\frac{1}{r_{jk}} \right)^6 + 4\pi\rho \int_{a_0}^{\infty} \frac{dr}{r^4} \right]^{1/2} \quad \text{nuclear spins}. \tag{12c}$$

Values for the ratio μ/μ_N of the nuclear moment μ to the nuclear magneton μ_N are listed in the NMR table on page 21 of Volume 1. The length coordinates in the square brackets of (12) are in angstrom (Å) units.

D. OVERLAPPING RESONANCES

When hyperfine structure is present the spectrum is a multiplet, and if the line width is comparable to the line separation the adjacent component lines will overlap. The spectrum can be simulated by placing a lineshape function of suitable amplitude and width at each spectral line position and adding them with the aid of a computer.

A nuclear spin I produces $2I + 1$ hyperfine components and n nuclear spins I produce $(2I + 1)^n$ lines of equal intensity if they are all inequivalent, i.e., if they all have different and incommensurate coupling constants. If there are n nuclear spins of individual value I_j with totally inequivalent coupling constants so no lines overlap, then the total number of lines N_{total} in the hyperfine pattern, all of equal intensity, is given by the product

$$N_{\text{total}} = \prod_{j=1}^{n} (2I_j + 1). \tag{13}$$

As an example, consider the coupling together of three protons and two nitrogen nuclei with the values

$$I_p = \tfrac{1}{2}, \qquad N_p = 3, \qquad I_{\text{N}} = 1, \qquad N_{\text{N}} = 2. \tag{14}$$

If all those spins are inequivalent to each other (13) gives for the total intensity

$$N_{\text{total}} = [2(\tfrac{1}{2}) + 1]^3 (2 + 1)^2 = 72. \tag{15}$$

Now consider the particular case in which the three protons are equivalent, and this produces the four line hyperfine multiplet with the intensity ratio 1 : 3 : 3 : 1. Two equivalent nitrogens produce a five line spectrum 1 : 2 : 3 : 2 : 1. The coupling constant case $A_p \gg A_{\text{N}}$ produces a quartet of pentets and the case $A_p \ll A_{\text{N}}$ produces a pentet of quartets as shown in Fig. 3. When $A_p = A_{\text{N}}$ an eight line spectrum is obtained with the intensity ratios 1 : 5 : 12 : 18 : 18 : 12 : 5 : 1 as shown.

Figure 4 displays the Lorentzian and Gaussian broadened hyperfine patterns for $N_p = 0$ to $N_p = 4$ equivalent protons ($I = \frac{1}{2}$) for various ratios of component width to component separation. We see from the figure that for the same separation to width ratio a Gaussian multiplet is better resolved than a Lorentzian one. For these equally coupled protons the relative intensities follow the binomial distribution

$$\begin{array}{ccccccccc l}
 & & & & 1 & & & & & N_p = 0, \\
 & & & 1 & & 1 & & & & N_p = 1, \\
 & & 1 & & 2 & & 1 & & & N_p = 2, \\
 & 1 & & 3 & & 3 & & 1 & & N_p = 3, \\
1 & & 4 & & 6 & & 4 & & 1 & N_p = 4.
\end{array} \tag{16}$$

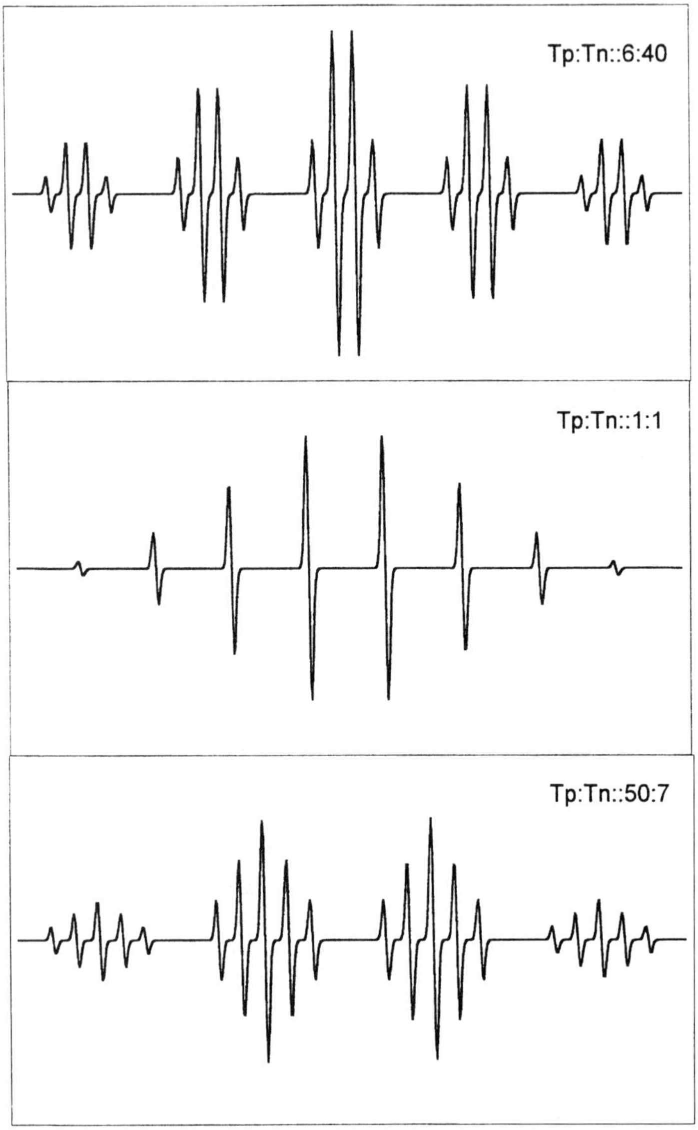

(a)

FIG. 3. Hyperfine patterns for two equally coupled $I = 1$ nuclei and three equally coupled $I = \frac{1}{2}$ nuclei with the respective coupling constants A_N for nitrogen and A_p for protons. Spectra are shown for the respective ratios $A_p/A_N = 3/20$ (top), 1 (middle), and 50/7 (bottom) for (a) Gaussian and (b) Lorentzian lineshapes. It is clear that the resolution here and in Fig. 4 is greater for the Gaussian case (a).

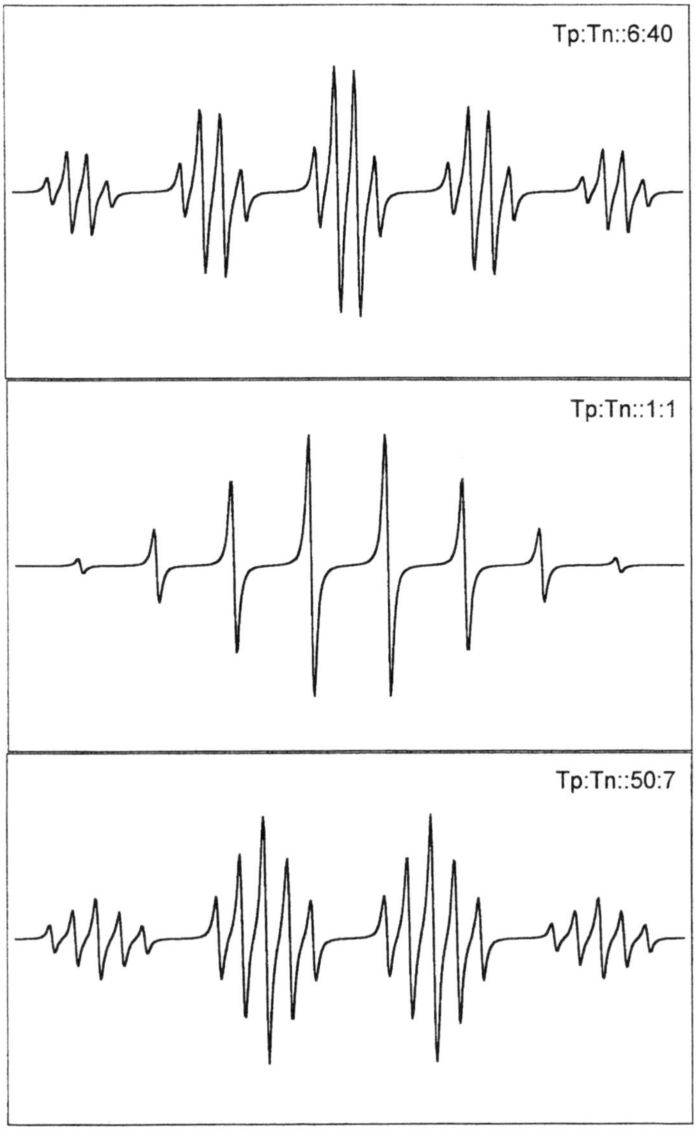

(b)

FIG. 3 (*continued*)

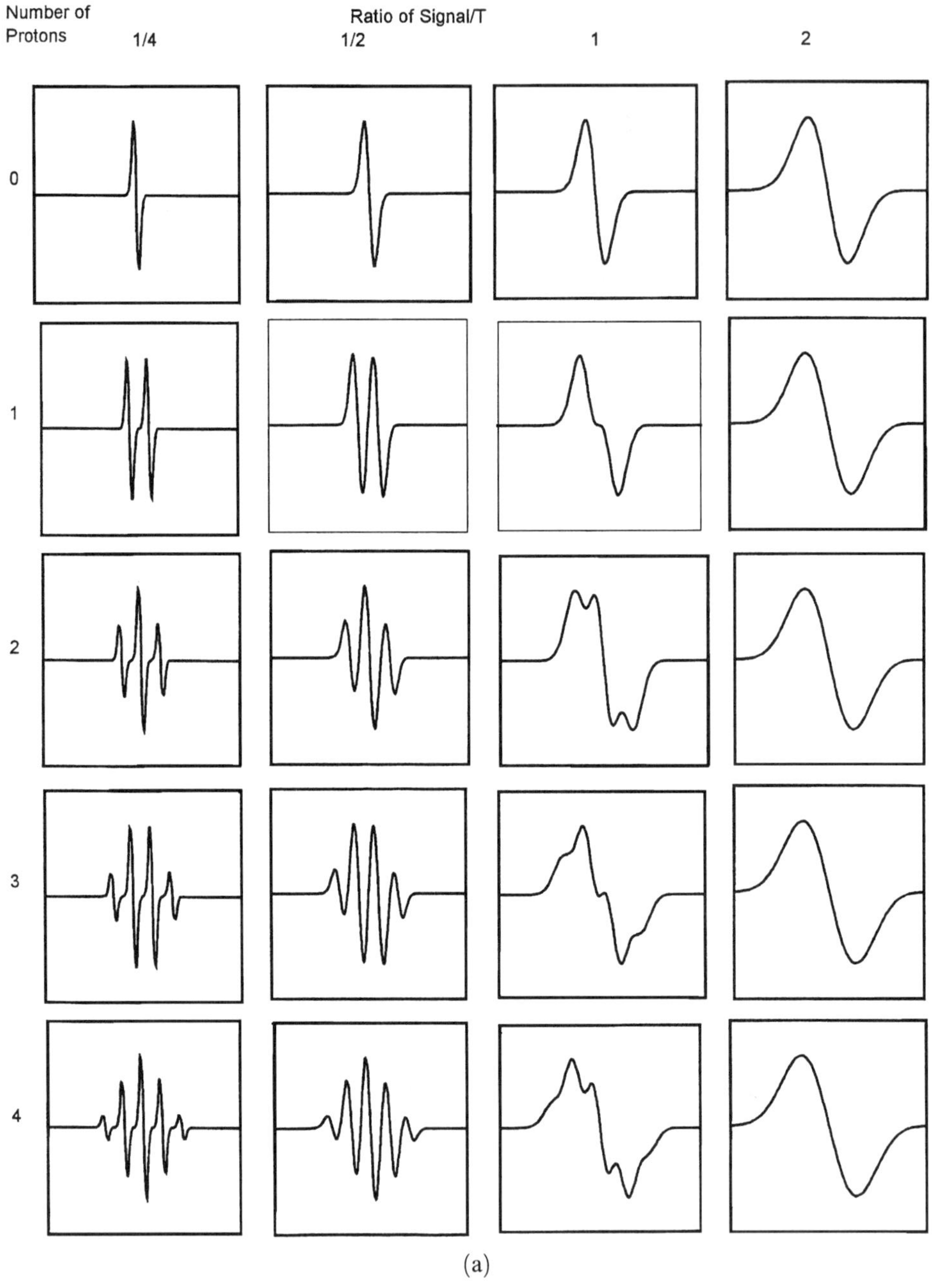

FIG. 4. Hyperfine patterns calculated for (a) Gaussian and (b) Lorentzian lineshapes using four ratios of the component width to the component separation (1/4, 1/2, 1, 2) for the respective number of protons and intensity ratios 0(1), 1(1 : 1), 2(1 : 2 : 1), 3(1 : 3 : 3 : 1), and 4(1 : 4 : 6 : 4 : 1).

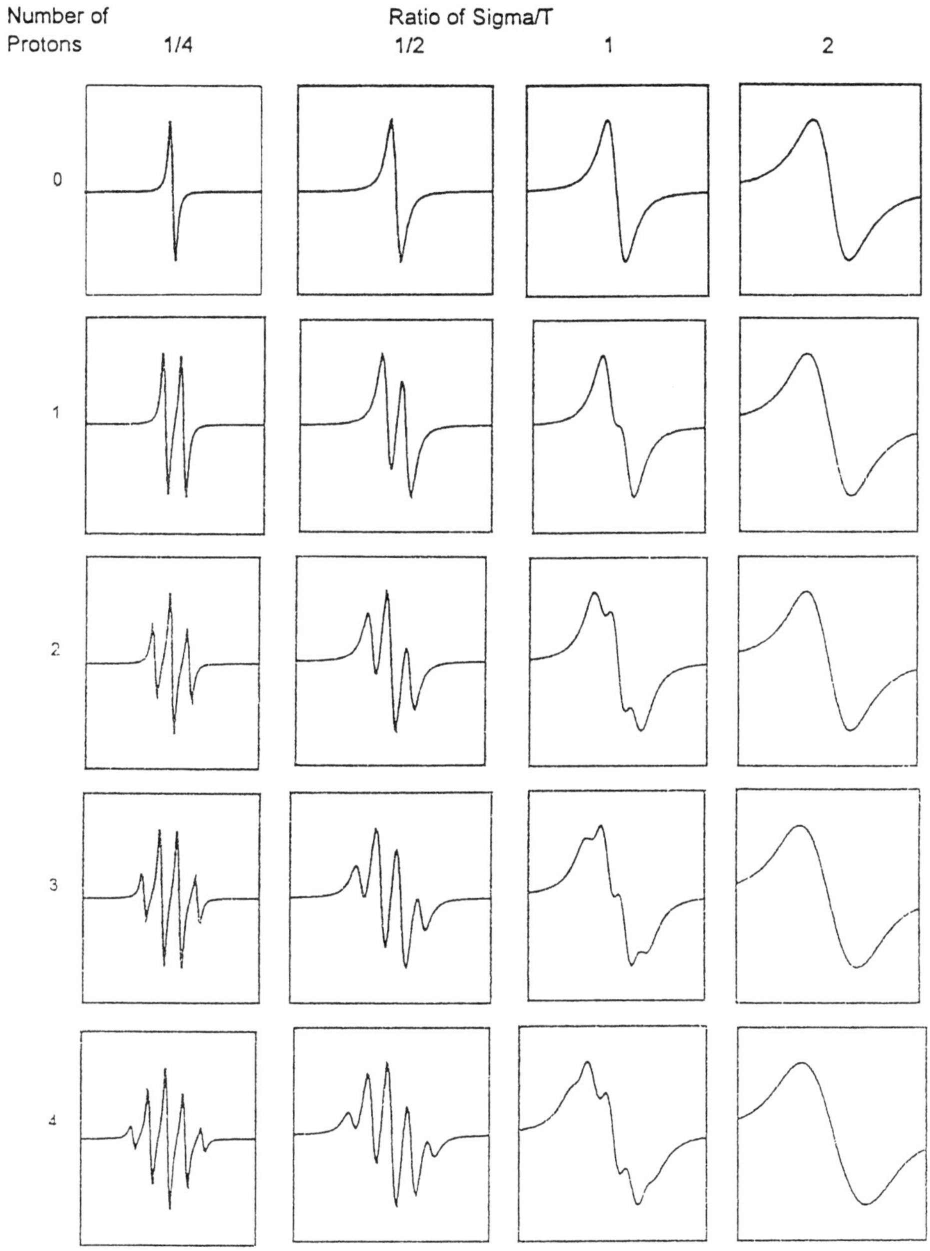

(b)

FIG. 4 (*continued*)

The total intensity N_{total} can be obtained from the expression $(2I+1)$ via (13) with $I=\frac{1}{2}$, or by summing the intensity ratios (16). For $N_p=3$ we have

$$N_{\text{total}} = (2I+1)^3 = 2^3 = 8, \qquad N_{\text{total}} = 1+3+3+1 = 8. \tag{17}$$

E. CONVOLUTION SHAPE

When the spectral line is broadened independently by both Gaussian and Lorentzian effects, the absorption amplitude is given by the following convolution:[3,4]

$$Y_{(v,b)} = \frac{b}{\pi}\int_{-\infty}^{\infty} \frac{e^{-x^2}\,dx}{b^2+(u-x)^2}, \tag{18}$$

where b is a measure of the ratio of the Lorentzian linewidth ${}^{\text{L}}\Delta B_{1/2}$ to the Gaussian width ${}^{\text{G}}\Delta B_{1/2}$,

$$b = (\ln 2)^{1/2}\frac{{}^{\text{L}}\Delta B_{1/2}}{{}^{\text{G}}\Delta B_{1/2}} \tag{19}$$

and u is the distance from the line center B_0 in units of ${}^{\text{G}}\Delta B_{1/2}/\ln 2$, as follows:

$$u = 2(\ln 2)^{1/2}\frac{(B-B_0)}{{}^{\text{G}}\Delta B_{1/2}}. \tag{20}$$

This Voigt profile function (18) has been tabulated for various values of b and u.[3,5] The convolution lineshape is applicable to the inhomogeneous broadening case of a Gaussian-shaped envelope formed from narrow Lorentzian spin packets.

F. ANISOTROPIC *g*-FACTORS

If the g-factor is totally anisotropic with the principal values g_1, g_2, and g_3 along the x-, y-, z-axis then it has the following directional dependence in spherical coordinates:

$$g(\theta,\phi) = (g_1^2\sin^2\theta\sin^2\phi + g_2^2\sin^2\theta\cos^2\phi + g_3^2\cos^2\theta)^{1/2}, \tag{21}$$

which reduces to

$$g(\theta) = (g_\perp^2\sin^2\theta + g_{11}^2\cos^2\theta)^{1/2} \tag{22}$$

for the case of axial symmetry where

$$g_\perp = g_1 = g_2, \tag{23a}$$

$$g_{11} = g_3. \tag{23b}$$

For complete anisotropy $g_1 > g_2 > g_3$ we adopt the convention $B_1 < B_2 < B_3$ for the magnetic field values along the three principal directions, where

$$B_1 g_1 = B_2 g_2 = B_3 g_3. \tag{24}$$

The powder pattern lineshape function $Y_{(B)}$ of an infinitely narrow (delta function) component line averaged over all angles θ, ϕ for the completely anisotropic g-factor of (21) may be expressed in terms of complete elliptic integrals of the first kind K as follows:

$$Y_{(B)} = \frac{2B_1 B_2 B_3}{\pi B^2 b_B} K(a_B/b_B), \qquad B_1 \leq B \leq B_2, \tag{25a}$$

$$Y_{(B)} = \frac{2B_1 B_2 B_3}{\pi B^2 a_B} K(b_B/a_B), \qquad B_2 \leq B \leq B_3, \tag{25b}$$

where B is the applied magnetic field and

$$a_B = [(B_3^2 - B_2^2)(B^2 - B_1^2)]^{1/2}, \tag{26a}$$

$$b_B = [(B_2^2 - B_1^2)(B_3^2 - B^2)]^{1/2}. \tag{26b}$$

We should note that $a_B = b_B$ for the special case $B = B_2$ and recall the limiting behavior

$$\lim K_{(x)} = \infty \qquad \text{for} \quad x \to 1, \tag{27}$$

to see that the powder pattern has an infinity when the applied field B is scanned through the value B_2. For a finite but narrow component width $\Delta B_{1/2} \ll B_3 - B_1$ the amplitude at the point $B = B_2$ is very large but finite, as shown in Fig. 5(a). The corresponding first derivative powder pattern is sketched in Fig. 5(b).

For axial symmetry either $B_1 = B_2$ or $B_2 = B_3$ and we have for the case $B_{11} < B_\perp$

$$Y_{(B)} = \frac{B_{11}(B^2 + B_{11}^2)}{2B^2} \left(\frac{g_{11}^2 - g_\perp^2}{g_{11}^2 B_{11}^2 - g_\perp^2 B^2} \right)^{1/2}, \tag{28}$$

which has the limiting value $Y(B) = 1$ for $B = B_{11}$ and becomes infinite for $B = B_\perp$ since, from (24) we have $g_{11} B_{11} = g_\perp B_\perp$. This powder pattern and its derivative are sketched in Fig. 5.

G. HYPERFINE POWDER PATTERNS

The spectrum for the case of an axially symmetric g-factor and hyperfine structure with the same axis of symmetry, corresponding to the Hamiltonian

$$H = g_{11}\mu_B S_z B_z + g_\perp \mu_B (S_x B_x + S_y B_y) + A_{11} S_z I_z + A_\perp (S_x I_x + S_y I_y) \tag{29}$$

consists of $2I + 1$ powder patterns. Each M_I assumes integral or half integral values in the range $-I \leq M_I \leq I$ and the $I = 1$ spin label case is illustrated in Fig. 6. Each M_I

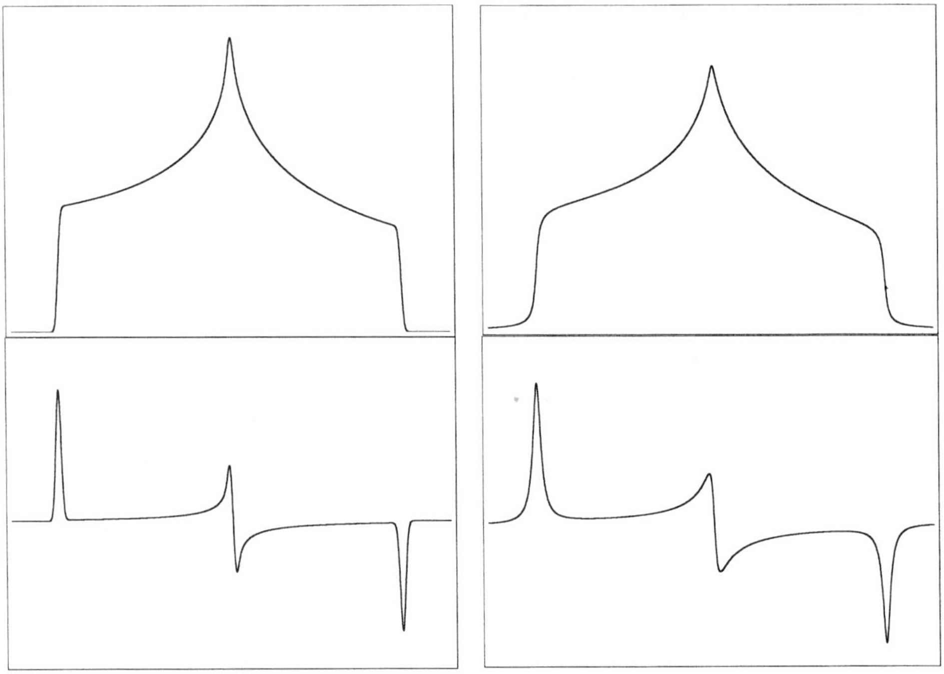

(a) (b)

FIG. 5. Absorption (top) and first derivative (bottom), (a) Gaussian and (b) Lorentzian powder pattern lineshapes for a completely anisotropic g-factor ($g_{xx} > g_{yy} > g_{zz}$) for a relatively narrow linewidth that is 10% of the total spread of the spectrum.

component line stretches from an infinity at $(B_0 g_0 - A_{\perp} M_I)/g_{\perp}$ to the point $(B_0 g_0 - A_{11} M_I)/g_{11}$ on the magnetic field scale as shown in Fig. 12-27 of ref. 1. The $M_I = 0$ center line for an integer spin is a narrow singlet, as shown in Fig. 6. Individual hyperfine component line powder patterns can overlap as is often the case with Cu^{2+} for which $I = \frac{3}{2}$. When hyperfine structure is present, a delta-function powder pattern can have extra singularities in addition to the ones discussed above.

H. SMALL SPIN CLUSTERS

In a small cluster of spins the main line broadening mechanism is the dipole–dipole interaction H_{dd} arising from the magnetic field of each spin interacting with the other spins. For two spins S_1 and S_2 separated by the distance r we have the Hamiltonian

$$H_{dd} = \frac{\mu_0}{4\pi}\left(\frac{g^2\mu_B^2}{r^3}\right)\left[\mathbf{S}_1 \cdot \mathbf{S}_2 - \frac{3(\mathbf{r}\cdot\mathbf{S}_1)(\mathbf{r}\cdot\mathbf{S}_2)}{r^2}\right], \tag{30}$$

where the factor $\mu_0/4\pi$ is present when MKS (or SI) units are used, and is omitted for cgs units. The typical applied field B_{app} of 0.32 tesla is so much stronger than the local fields responsible for the dipolar broadening that for like spins $S_1 = S_2$ this

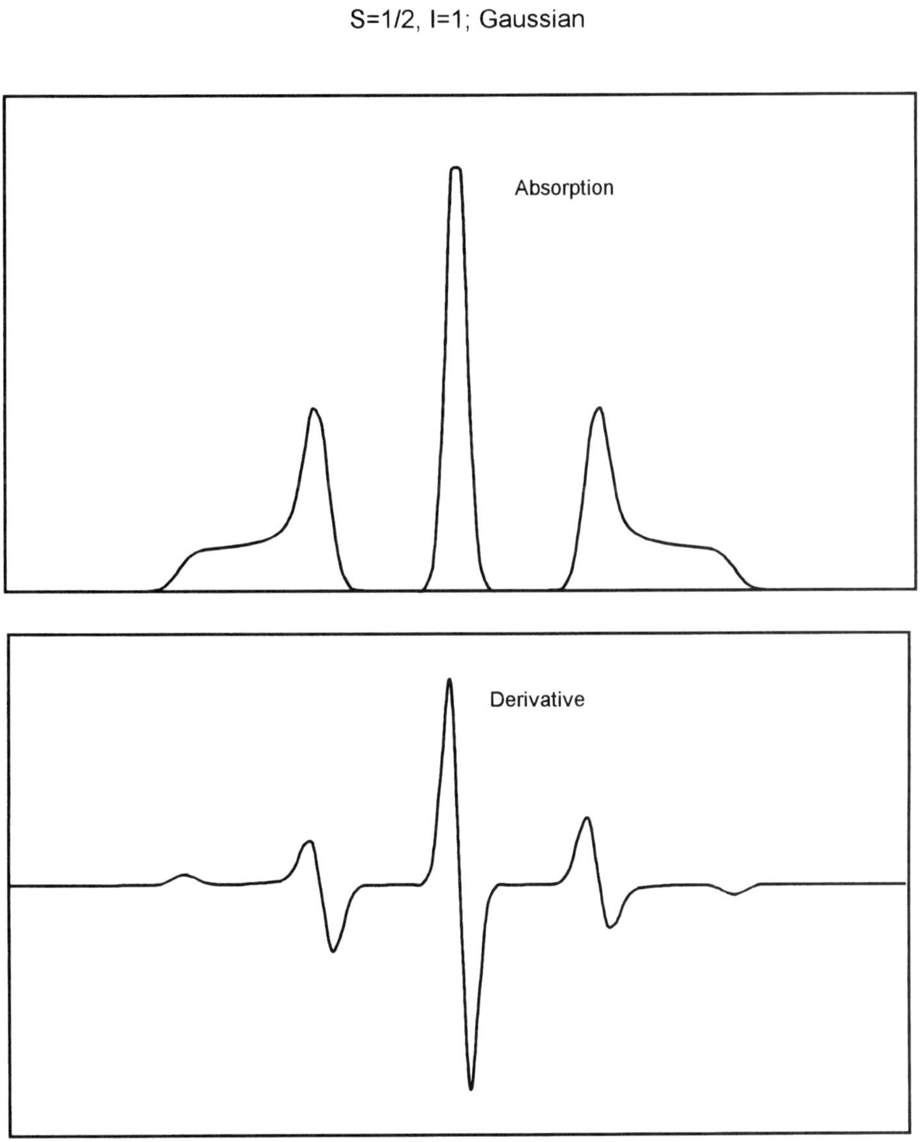

FIG. 6. Absorption (top) and derivative (bottom), Gaussian powder pattern of the $I = 1$ nitrogen hyperfine triplet.

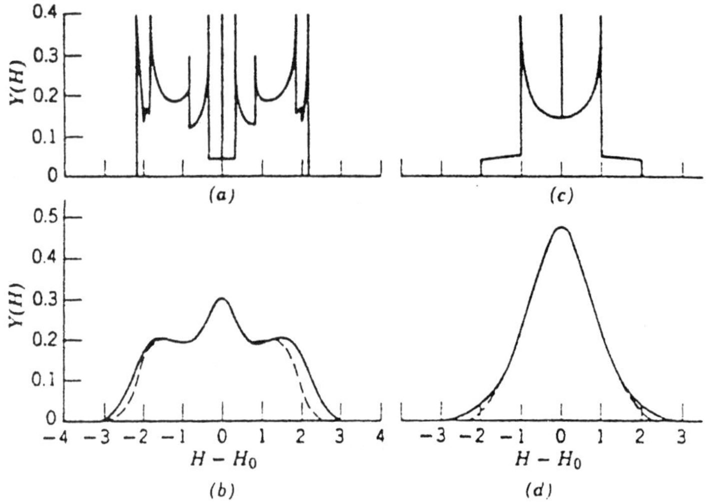

FIG. 7. Powder pattern lineshapes of spin $\frac{1}{2}$ nuclei. The patterns on the right are for two interacting nuclei, and those on the left are for three interacting nuclei at the apices of an equilateral triangle. The upper patterns are delta-function shapes, and those at the bottom are for a finite component width (E. R. Andrew and R. Bersohn, J. Chem. Phys. **18**, 159 (1950)).

Hamiltonian reduces to

$$H_{\mathrm{dd}} = S(S+1)\left(\frac{\mu_0}{4\pi}\right)\left(\frac{g^2\mu_B^2}{r^3}\right)(1-3\cos^2\theta), \tag{31}$$

where θ is the angle of B_{app} relative to the radius vector $\mathbf{r}$ between the two spins.

Figure 7 shows the powder pattern shape functions (a) for three spins at the vertices of an equilateral triangle and (b) for two spins when there are no other sources of broadening. The spectra (c) and (d) at the bottom of the pattern are obtained when the component lines have a finite width.

I. AMORPHOUS AND SEMIRANDOM DISTRIBUTIONS

Until now we have been discussing lineshapes found with randomly oriented microcrystallites of spin systems with definite Hamiltonian parameters. These situations occur with an ionic or molecular crystal in which the paramagnetic species has a definite site symmetry. There is another type of material, namely a glassy or amorphous substance, in which the paramagnetic species is randomly oriented and in addition exhibits a range of Hamiltonian parameters. The powder spectrum of such a material is then an average over both angles and Hamiltonian parameters. Several techniques are available for the calculation of powder spectra of amorphous solids, and these have been reviewed elsewhere.[6]

Another case involves microcrystallites whose axes have a preferred alignment relative to a certain direction in the sample. A distribution function $\rho(\delta)$ describes this,

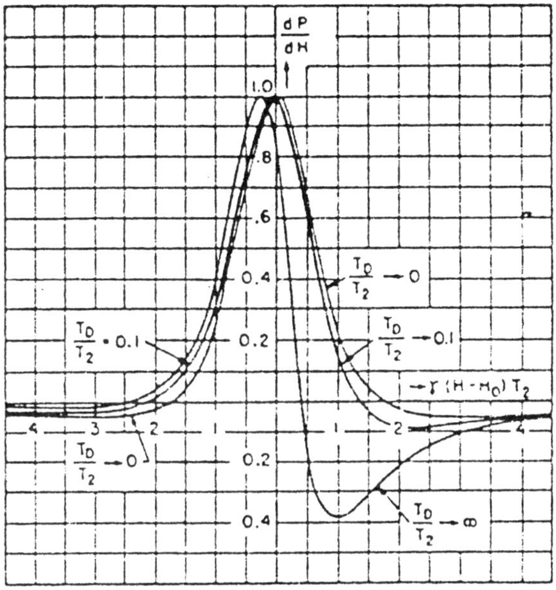

FIG. 8. Conduction ESR lineshapes obtained with thick metal films for different ratios of the diffusion time T_D to the spin–spin relaxation time T_2 (from ref. 8; see ref. 1, p. 495).

where δ is the angle between the microscrystalline axis and the preferred direction. For this situation, ESR measurements can be carried out with the magnetic field set at various angles ψ relative to this preferred direction, and each spectrum $Y(B, \psi)$ will be an average over microcrystallite distributions, and hence powder-like in appearance.[7]

J. CONDUCTION ELECTRONS

The ESR of conduction electrons in metals provides lineshapes called Dysonian[8] and some typical examples[9] are shown in Fig. 8. These shapes depend on the electron mean free path Λ, the time T_D that it takes an electron to diffuse through the skin depth δ, the time T_T needed to traverse the sample, and the usual relaxation times T_1 and T_2 where, for metals $T_1 = T_2$. The normal $(\Lambda \ll \delta)$ and anomalous $(\Lambda \gg \delta)$ skin depth cases have been studied. Experimentally recorded lineshapes can be analyzed to provide estimates of the various characteristic times.[8–10]

K. EXCHANGE

For spin $\frac{1}{2}$ the Anderson[11] exchange theory gives the eigenvalues λ:

$$\lambda = i\Delta\omega + P \tag{32}$$

$$= i\omega_e + \omega_0 \pm \sqrt{T^2 - \omega_e^2}, \tag{33}$$

where the real part P is the line position and the imaginary part $\Delta\omega$ is the width, ω_0 is the Larmor frequency, ω_e is the exchange frequency, and T is the hyperfine interaction frequency. For $T \gg \omega_e$ we have

$$\Delta\omega = \omega_e, \tag{34}$$

$$P = \omega_0 \pm T(1 - \omega_e^2/2T^2), \tag{35}$$

$$= \omega_0 \pm T_{\text{eff}}, \tag{36}$$

and for $T \ll \omega_e$ we have

$$\Delta\omega - T^2/2\omega_e, \tag{37}$$

$$P = \omega_0. \tag{38}$$

Figure 9 shows how the effective hyperfine splitting depends on the exchange, and Fig. 10 presents the spectra in the range from zero to strong exchange.[11,12]

L. SATURATION

If the microwave power level is taken into account, a Lorentzian line Y and its first derivative Y' have the form

$$Y = \frac{B_1 y_m^0}{1 + (B - B_0)^2\gamma^2 T_2^2 + B_1^2\gamma^2 T_1 T_2}, \tag{39}$$

$$Y' = \frac{16(B - B_0)\gamma T_2 B_1 y_m^{0'}}{3^{3/2}[1 + (B - B_0)^2\gamma^2 T_2^2 + B_1^2\gamma^2 T_1 T_2]^2}, \tag{40}$$

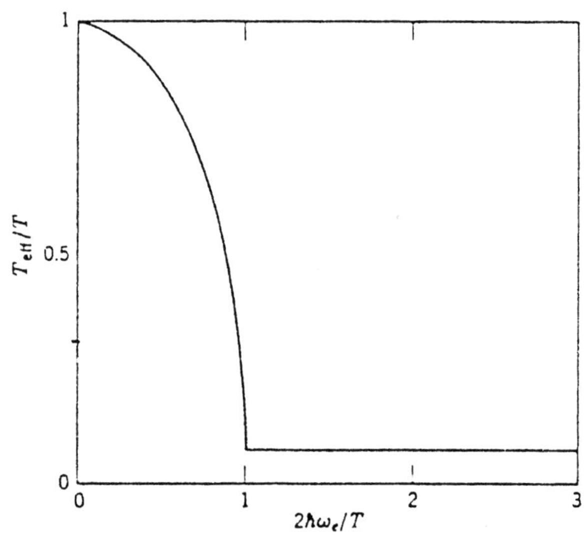

FIG. 9. Dependence of the effective hyperfine coupling T_{eff} on the exchange frequency ω_e. The abscissa and ordinate are normalized relative to the hyperfine coupling constant $T/2$ (from ref. 12, p. 177).

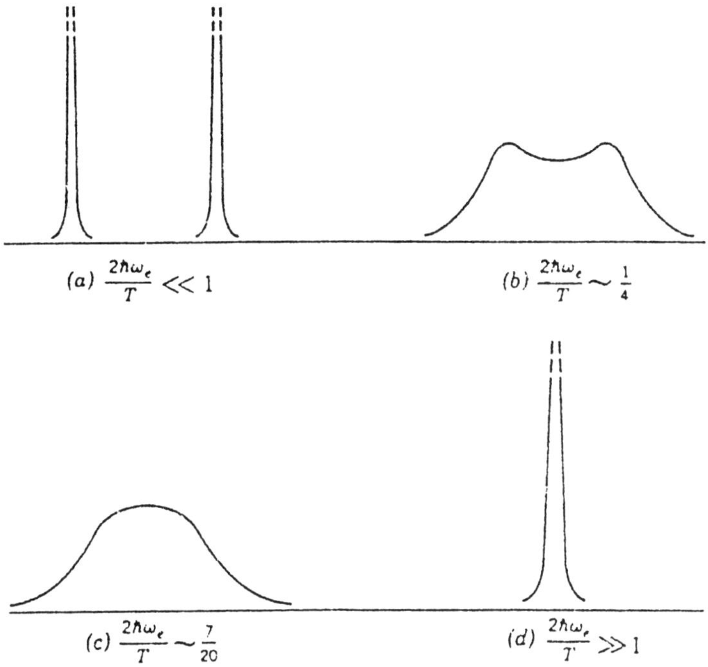

FIG. 10. Spin $\frac{1}{2}$ hyperfine doublet, (a) in the absence of exchange, (b) with weak exchange, (c) with moderate exchange, and (d) with strong exchange (from ref. 12, p. 178).

where the quantities $Y_m^0 B_1$ and $Y_m^{0'} B_1$ are the respective maximum amplitudes before saturation $(B_1^2 \gamma^2 T_1 T_2) \ll 1$ which occurs at $B = B_0$ and γ is the gyromagnetic ratio given by

$$\gamma = \frac{g\mu_B}{\hbar} = 0.87934 \times 10^8 \text{ g} \quad \text{Hz/mT}, \tag{41}$$

and the spin–spin relaxation time T_2 is related to the half amplitude and peak-to-peak linewidths (in mT) below saturation via (7):

$$T_2 = \frac{2}{\gamma \Delta B_{1/2}^0} = \frac{2}{3^{1/2} \gamma \Delta B_{\text{pp}}^0} = \frac{1.3131 \times 10^{-8}}{g \Delta B_{\text{pp}}^0}. \tag{42}$$

The shape functions may be expressed in terms of the saturation factor S,

$$S = \frac{1}{1 + \gamma^2 B_1^2 T_1 T_2}, \tag{43}$$

and we have

$$Y'_{(B)} = \frac{16(B - B_0)\gamma T_2 S^2 B_1 y_m^{0'}}{3^{3/2}[1 + S(B - B_0)^2 \gamma^2 T_2^2]^2}. \tag{44}$$

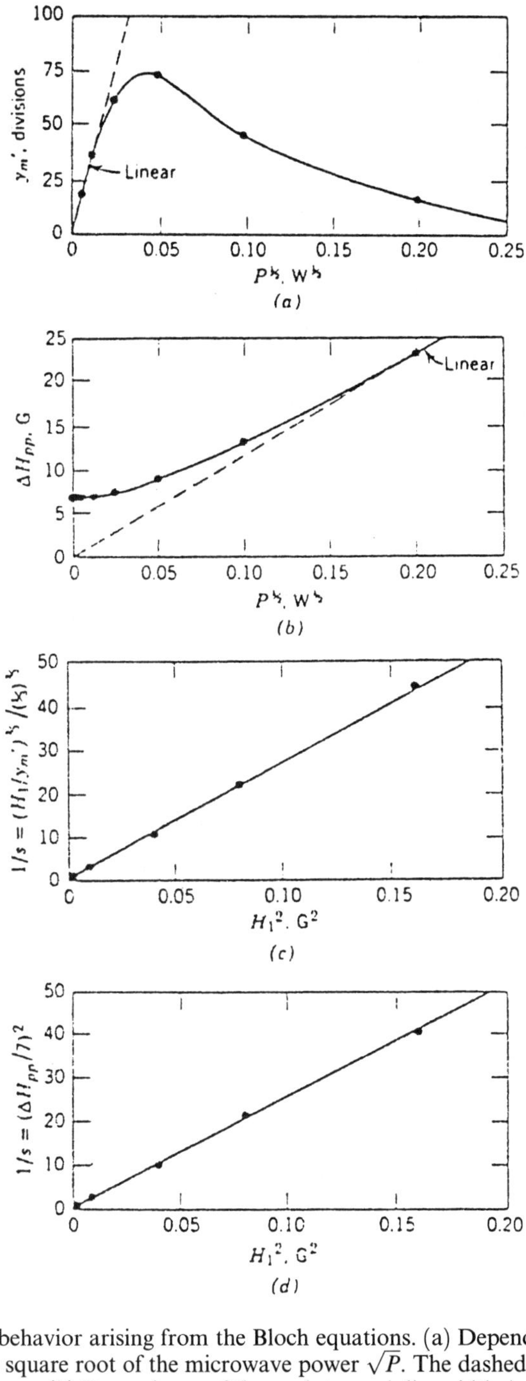

FIG. 11. Saturation behavior arising from the Bloch equations. (a) Dependence of peak-to-peak amplitude y'_m on the square root of the microwave power $\sqrt{P}$. The dashed line is the asymptotic behavior for low powers. (b) Dependence of the peak-to-peak linewidth $\Delta\omega_{pp}$ on the square root of the microwave power $\sqrt{P}$. The dashed line is the asymptotic behavior at high powers. (c) Linear dependence of the amplitude factor $(y'_m/B_1)^{-2/3}$ on the square of the microwave field strength B_1^2. (d) Linear dependence of the square of the peak-to-peak linewidth $(\Delta B_{pp})^2$ on the square of the microwave field strength B_1^2 (from ref. 1, p. 593).

The linewidth under saturating conditions is

$$\Delta B_{\text{pp}} = (\Delta B^{0}_{\text{pp}}) S^{-1/2}. \tag{45}$$

The peak amplitude Y'_m of $Y'_{(B)}$ occurs for $B - B_0 = \pm\frac{1}{2}\Delta B_{\text{pp}}$, and has the value

$$Y'_m = (B_1 y^{0'}_m) S^{3/2}. \tag{46}$$

Figures 11(a) and (b), respectively, show plots of Y'_m from (46) and ΔB_{pp} from (45) versus the square root of the microwave power P, where

$$P = (\text{const}) B_1^2 \tag{47}$$

and Figs. 11(c) and 11(d) show the linear plots of $1/S$ from (46) and (45) respectively, versus B_1^2.

The peak in the curve Y'_m versus B_1 shown in Fig. 11(a) comes at $S = \frac{2}{3}$ and provides us with the expression for the spin lattice relaxation time

$$T_1 = \frac{0.49 \times 10^{-8} \Delta B^{0}_{\text{pp}}}{(gB_1)^2}, \tag{48}$$

where the fields are in mT, and B_1 is the rf field that produces the respective maxima.

REFERENCES

[1] C. P. Poole, Jr., *Electron Spin Resonance* (Wiley-Interscience, New York, 1983), 2nd ed.
[2] J. H. Van Vleck, Phys. Rev. **74**, 1168 (1948).
[3] D. W. Posener, Aust. J. Phys. **12**, 184 (1959).
[4] H. A. Farach and H. Teitelbaum, Can. J. Phys. **45**, 2913 (1967).
[5] S. Ganapathy and V. Parthasarathi, Indian J. Pure Appl. Phys. **15**, 63 (1977).
[6] C. P. Poole and H. A. Farach, Bull. Magn. Reson. **1**, 162 (1980).
[7] H. Panepucci and H. A. Farach, Med. Phys. **4**, 46 (1977).
[8] F. J. Dyson, Phys Rev. **98**, 349 (1955).
[9] G. Feher and A. Kip, Phys. Rev. **98**, 337 (1955).
[10] Ref. 1, p. 494.
[11] P. W. Anderson, J. Phys. Soc. Japan **9**, 316 (1954).
[12] C. P. Poole, Jr. and H. A. Farach, *Theory of Magnetic Resonance* (Wiley, New York, 1987), 2nd ed.

CHAPTER V

ELECTRON SPIN ECHO ENVELOPE MODULATION

John McCracken

Department of Chemistry, Michigan State University, East Lansing, Michigan 48824, USA

The electron spin echo envelope modulation technique of pulsed ESR spectroscopy is described and compared to the complementary technique of electron nuclear double resonance. The hyperfine sublevel correlation spectroscopy technique is then explained. The chapter concludes with some remarks on the outlook for the future.

CONTENTS

A. INTRODUCTION

The Electron Spin Echo Envelope Modulation (ESEEM) technique of pulsed EPR spectroscopy has emerged as a powerful method for the measurement of electron–nuclear hyperfine couplings that are often masked by inhomogeneous broadening of the EPR lineshape. It provides an alternative, and somewhat complementary, method to the Electron Nuclear Double Resonance (ENDOR) technique covered in the first volume of this series. The development of ESEEM methodology and its applications has proceeded steadily over the past 15 years, and several excellent review articles and specialty books that cover this topic have appeared.[1–6] In addition, the quantum mechanical details of the ESEEM experiment have also been covered in two recently published EPR textbooks.[7,8] The goal of this chapter is to provide an ESEEM primer for scientists who may be interested in applying the method to chemical problems. The presentation will begin with a simple semiclassical description of two-pulse ESEEM for an $S = \frac{1}{2}$, $I = \frac{1}{2}$ spin system and then focus on parallels between the semiclassical and quantum mechanical descriptions of the experiment. It is hoped that the reader will gain an understanding of the strengths and shortcomings of the technique and of the goals of current method development research being carried out in this field.

B. QUALITATIVE DESCRIPTION OF THE ESEEM EXPERIMENT

In an ESEEM experiment, electron spin echoes are generated by the application of a sequence of resonant microwave pulses to the sample. The electron spin echo decay envelope is then measured by plotting the integrated intensity of these echoes as a function of the time between two of the pulses in the sequence. The result shows an overall decay of the electron spin magnetization that is often modulated by electron–nuclear hyperfine interactions. Figure 1 shows the result of a two-pulse, 90°–τ–180°, ESEEM experiment done on the Ni(III)(CN)$_4$(H$_2$O)$_2^-$ complex in an aqueous glass at a magnetic field position close to the $g_\perp$ feature of the CW–EPR spectrum.[9] The data show an overall decay characterized by a phase memory time, T_M or T_2^* of approximately 1.5 μs, and modulations that arise from hyperfine coupling of the ^{14}N and ^{1}H nuclei of the ligands and surrounding solvent molecules to the Ni(III) ion.

The origin of these modulations and the information that they carry regarding hyperfine interactions can be most easily understood by combining the classical picture of spin echo formation with the quantum mechanical treatment of an unpaired electron spin coupled to a spin $\frac{1}{2}$ nucleus. Figure 2 shows the classical picture of electron spin echo formation for a two-pulse, 90°–τ–180°, spin echo experiment. The coordinate system is chosen to rotate about the z-axis at an angular frequency, ω_0, corresponding to the microwave pulse frequency. The laboratory magnetic field, $\mathbf{B_0}$, is directed along $\mathbf{z}$ as is the bulk magnetization of the sample, $\mathbf{M}$, when the spin system is at thermal equilibrium prior to the application of the first microwave pulse (Fig. 2(a)). The geometry of the sample probe is such that the magnetic field associated with the micro-

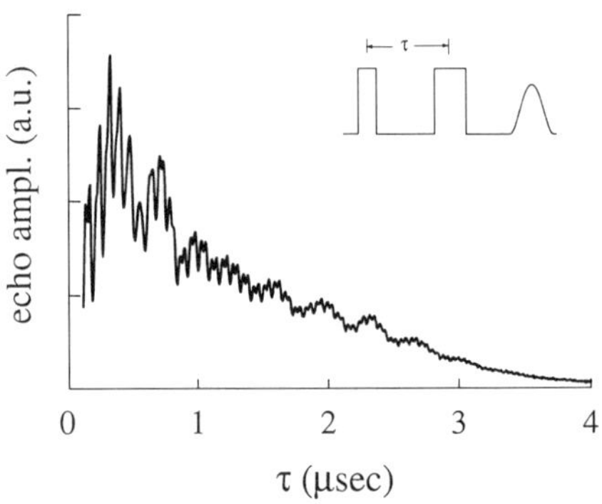

FIG. 1. Two-pulse ESEEM data collected for the Ni(III)(CN)$_4$(H$_2$O)$_2^-$ complex in a 50% v/v water/ethylene glycol glass. Conditions for this measurement were: microwave frequency, 8.935 GHz; microwave pulse power, 49 dBm; magnetic field strength, 3005 G; pulse duration (90°), 15 ns FWHM; and pulse repetition rate, 7 Hz. Each point represents the average of 10 events.

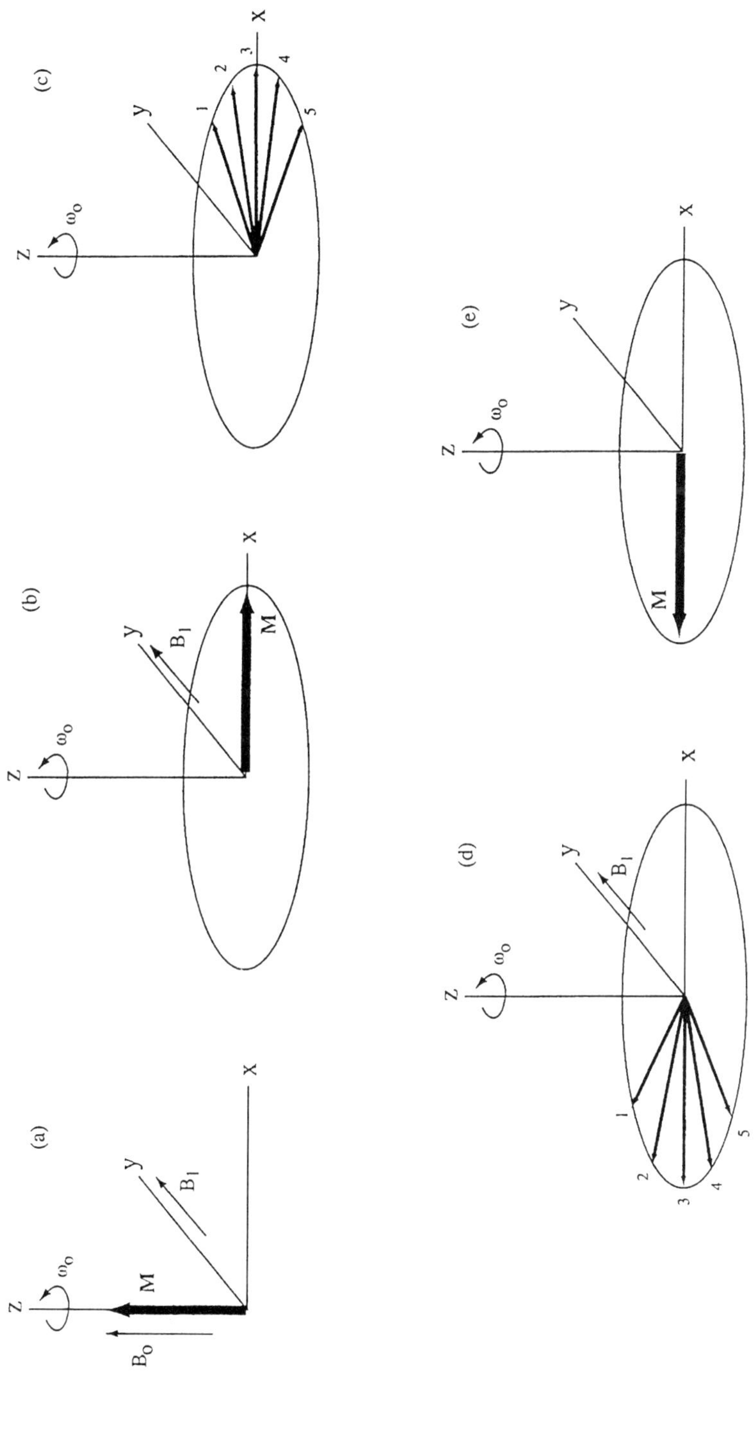

FIG. 2. Vector model for spin echo formation as a consequence of a $90°-\tau-180°$ sequence of resonant pulses.

wave pulses, $\mathbf{B_1}$, is directed perpendicular to the laboratory field and will be stationary along **y** in the rotating frame. When the first pulse is switched on, the $\mathbf{B_1}$ field torques the magnetization about **y** until the pulse ceases at the completion of a 90° rotation (Fig. 2(b)). Because the magnetization vector consists of a number of spin packets, or families of spins that see different local magnetic environments, it will fan out in the transverse plane during the free precession period of duration τ (Fig. 2(c)). The spin packets labeled 1 and 2, in Fig. 2(c), see a larger effective magnetic field and precess faster than ω_0. These spins develop a positive phase angle, with respect to the x-axis during free precession, while those that see smaller local fields, packets 4 and 5, develop a phase of the opposite sense. The second pulse torques the packet magnetization vectors through an angle of 180° about **y** (Fig. 2(d)) and leads to the refocusing of the magnetization to form the spin echo at time 2τ after the start of the pulse sequence (Fig. 2(e)).

The spin Hamiltonian for an $S = \frac{1}{2}$, $I = \frac{1}{2}$ coupled spin system consists of electronic Zeeman, nuclear Zeeman, and electron–nuclear hyperfine interaction terms. For the case of an isotropic electron g-matrix and an axial hyperfine interaction, this Hamiltonian can be conveniently written in the laboratory reference frame, using the schematic representation of Fig. 3, and is given by

$$\mathscr{H}/\hbar = \omega_S \hat{S}_Z + A_{zz}\hat{S}_Z\hat{I}_Z + A_{xz}\hat{S}_Z\hat{I}_X - \omega_I\hat{I}_Z, \tag{1}$$

where $A_{zz} = A = A_{\parallel}\cos^2\theta + A_{\perp}\sin^2\theta$, $A_{xz} = B = (A_{\parallel} - A_{\perp})\cos\theta\sin\theta$, $\omega_S = g\beta B_0/\hbar$, and $\omega_I = g_n\beta_n B_0/\hbar$. $A_{\parallel}$ and $A_{\perp}$ are the principal values of an axially symmetric hyperfine tensor that can be expressed in terms of a Fermi contact coupling, A_{iso}, and a dipole–dipole coupling term, $T = gg_n\beta\beta_n/r^3$, with $A_{\parallel} = A_{\text{iso}} + 2T$ and $A_{\perp} = A_{\text{iso}} - T$. The angle θ describes the orientation of the principal axis of the hyperfine tensor with respect to the laboratory field, B_0. The remaining terms in the above equations are: g,

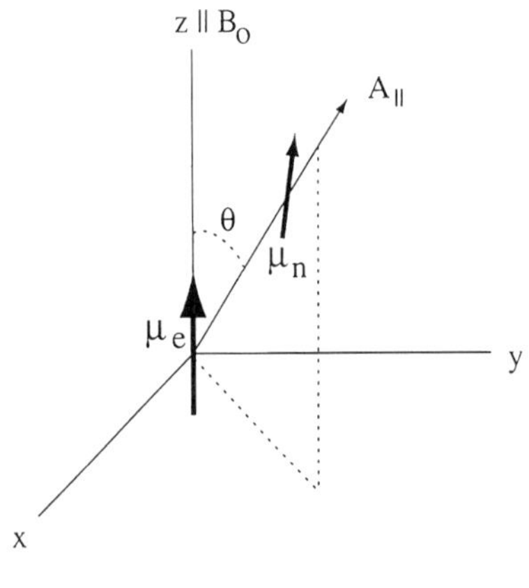

FIG. 3. Schematic representation of the electron–nuclear hyperfine interaction in the laboratory axis system.

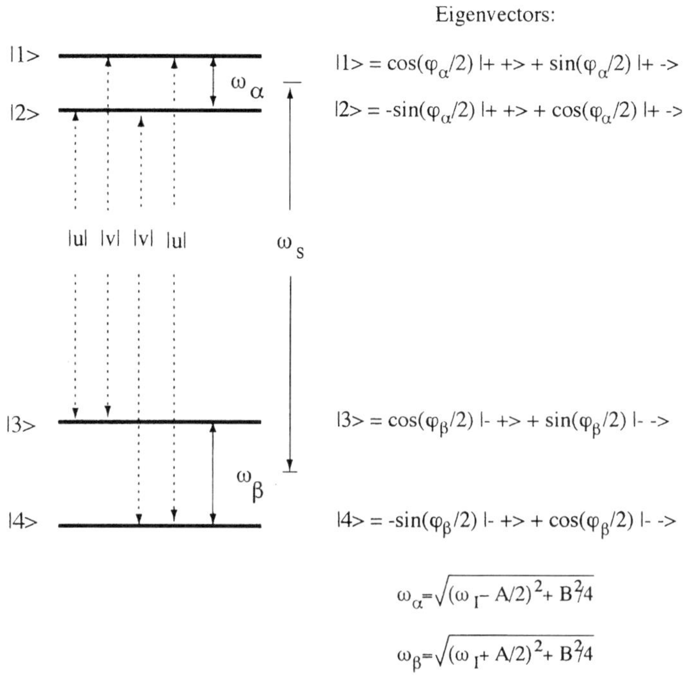

FIG. 4. Energy level diagram for an $S = \frac{1}{2}, I = \frac{1}{2}$ spin system described by a spin Hamiltonian that includes electronic Zeeman, nuclear Zeeman, and electron–nuclear hyperfine interactions.

the electron g-factor; g_n, the nuclear g-value; β, the Bohr magneton; and β_n, the nuclear magneton. The Hamiltonian matrix is easily constructed in a general basis set consisting of electron and nuclear spin product states, $|m_S, m_I\rangle$. Diagonalization to yield the eigenvalues and eigenvectors of (1) can be carried out independently for the two electron spin manifolds since the only term in the operator that gives rise to off-diagonal matrix elements is that involving $\hat{I}_x$. The results are summarized in the energy level scheme of Fig. 4.

The normalized probability amplitudes for the EPR transitions marked $|u|$ and $|v|$ in Fig. 4 are given by

$$|u| = \langle 2|\hat{S}_x|3\rangle/(0.5g\beta B_1) = \sin\{(\varphi_\alpha - \varphi_\beta)/2\} \tag{2}$$

and

$$|v| = \langle 1|\hat{S}_x|3\rangle/(0.5g\beta B_1) = \cos\{(\varphi_\alpha - \varphi_\beta)/2\}. \tag{3}$$

The angles φ_α and φ_β define the axes of quantization for the α and β electron spin manifolds, respectively, and are given by $\sin\varphi_\alpha = B/2\omega_\alpha$ and $\sin\varphi_\beta = B/2\omega_\beta$, where ω_α and ω_β are defined in Fig. 4. In general, all four of the possible EPR transitions for the $S = \frac{1}{2}$, $I = \frac{1}{2}$ spin system of Fig. 4 are allowed. These transitions will be excited simultaneously in a pulsed EPR experiment, provided that the microwave pulses have sufficient bandwidth.

The origin of ESEEM can now be understood by putting the quantum mechanical results, summarized in Fig. 4, together with the classical picture of echo formation in Fig. 2. This semiclassical picture is shown in Fig. 5, where we take the microwave frequency, ω_0, to be equal to ω_S of Fig. 4, and focus on the response of a single packet of spins that undergoes a transition from $|3\rangle$ to $|2\rangle$ in response to the first, 90°, pulse. After the first free precession period, τ (Fig. 5(a)), this packet has fallen behind the frame precession and developed a phase, $(\omega_{23} - \omega_0)\tau$, with respect to the **x**-axis. In response to the π-pulse, our spin packet is torqued 180° about **y** and a portion of it is split into a packet that will now precess at ω_{24} as a result of undergoing the transition from $|2\rangle$ to $|4\rangle$. Because ω_{24} is larger than ω_0, this new packet will precess in the opposite direction from the portion that maintains the ω_{23} frequency (Fig. 5(b)) and interfere with echo formation at time, τ, after the second pulse (Fig. 5(c)). The extent of interference that results from this "branching" of the EPR transitions depends on the projection of the "branched" spin packet onto the x-axis (Fig. 5(c)), $\cos\{(\omega_{23} - \omega_{24})\tau\}$, weighted by the probability for the "branching." Thus, the interference is modulated as we vary the time between the 90° and 180° microwave pulses with a frequency of $|\omega_{23} - \omega_{24}| = \omega_\beta$.

The concepts that can be derived from the simplified semiclassical model pictured in Fig. 5 are readily apparent in the analytical expressions for the two-pulse ESEEM expected for an $S = \frac{1}{2}$, $I = \frac{1}{2}$ spin system. The form of the $E_{\text{mod}}(\tau)$ that is particularly convenient for this purpose is

$$E_{\text{mod}}(\tau) = |u|^4 + |v|^4 + |u|^2|v|^2[2\cos\omega_\alpha\tau + 2\cos\omega_\beta\tau - \cos(\omega_\alpha - \omega_\beta)\tau - \cos(\omega_\alpha + \omega_\beta)\tau]. \tag{4}$$

Equation (4) shows that modulations of the two-pulse echo amplitude occur at the fundamental hyperfine frequencies and at their sum and difference combination frequencies. The amplitude of the modulations is given by the product of the transition probabilities for the two different transitions associated with "branching," $|u|^2|v|^2$, while the nonmodulated portion of the echo envelope depends only on the product of the transition probabilities for the "nonbranching" spins, $|u|^4$ or $|v|^4$. Substituting the expressions for $|u|$ and $|v|$ given by (2) and (3) into (4) yields

$$E_{\text{mod}}(\tau) = 1 - (k/2)[1 - \cos\omega_\alpha\tau - \cos\omega_\beta\tau + 0.5\cdot\cos(\omega_\alpha - \omega_\beta)\tau + 0.5\cdot\cos(\omega_\alpha + \omega_\beta)\tau], \tag{5}$$

where

$$k = \sin^2(\varphi_\alpha - \varphi_\beta) = [\omega_I B/(\omega_\alpha\omega_\beta)]^2 \tag{6}$$

is often referred to as the modulation depth parameter.

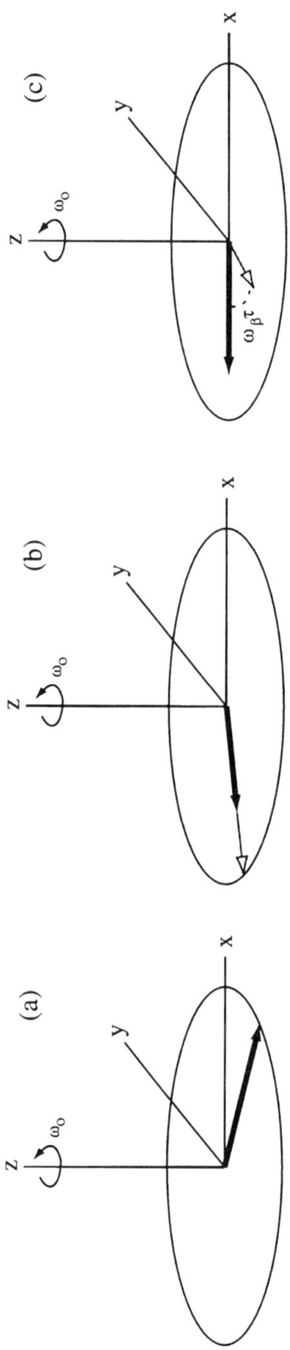

FIG. 5. Vector model showing the interference between electron spin packets that result from transition "branching" in a two-pulse ESEEM experiment.

From the above discussion, it is clear that observation of ESEEM requires that "branching" of the EPR transitions be affected by the microwave pulses. This places a quantum mechanical constraint on the ESEEM experiment, in that each energy level must be involved in at least two different microwave transitions, and an experimental constraint that requires the microwave pulse bandwidth to cover the spread in frequencies needed to fully excite the "branching." The experimentally observed ESEEM function is a product of the quantum mechanically derived modulation function and a decay function that describes the loss of magnetization due to spin relaxation. These decay functions are typically modeled with exponential forms: $\exp(-\tau/\tau_0)^n$ where $n = 1, 2$, or 0.5.[12] For a 90°–τ–180° or 2-pulse echo experiment, $\tau_0 = T_2^*$, a time that is typically on the order of 1 μs, as evidenced by the data shown for the Ni(III) model complex in Fig. 1. This rapid background decay serves to reduce the frequency resolution of the two-pulse measurement. A further complexity of the ESEEM experiment is introduced when multiple nuclei contribute to the modulation of a single paramagnetic center. The modulation function in this instance is given by the product of the individual modulation functions, due to each coupled nucleus giving an overall ESEEM function of the form

$$E(\tau) = V_{\text{decay}} \prod_{i=1}^{N} E_{\text{mod}}^{i}(\tau), \tag{7}$$

where N represents the number of coupled nuclei.[10,13] Equation (7) shows that the two-pulse ESEEM data, or the corresponding ESEEM spectrum obtained from Fourier transformation of the time domain data,[14] can become complex when only a few nuclei contribute. Not only will each nucleus give rise to fundamental, $\Delta m_I = \pm 1$, frequencies and combination frequencies, but the product of (7) will lead to frequencies that are combinations of the frequencies arising from different nuclei. This increased complexity, when combined with the reduced resolution that results from a rapid phase memory time, can severely hamper two-pulse ESEEM analysis.

Some of the above problems are alleviated by use of the three-pulse or 90°–τ–90°–T–90° sequence where the amplitude of the stimulated echo, occurring at τ after the third pulse, is normally monitored as a function of T (Fig. 6). The background decay for this sequence is governed by electron spin-lattice relaxation, electron cross relaxation, and nuclear spin–spin relaxation processes. All of these processes are much longer than the electron spin–spin relaxation time and lead to better frequency resolution than typically found for the two-pulse experiment. The following analytical expression (8) for the three-pulse modulation function of our model $S = \frac{1}{2}$, $I = \frac{1}{2}$ system

$$\begin{aligned} E_{\text{mod}}(\tau, T) = {} & |u|^4 + |v|^4 + |u|^2|v|^2[\cos(\omega_\alpha \tau) + \cos\omega_\beta \tau \\ & + 2\sin^2(\omega_\alpha \tau/2)\cos\{\omega_\beta(\tau + T)\} + 2\sin^2(\omega_\beta \tau/2)\cos\{\omega_\alpha(\tau + T)\}] \end{aligned} \tag{8}$$

has a simplified form, when compared to its two-pulse counterpart, in that only modulations at the fundamental hyperfine frequencies are observed. Equation (8) also shows that the τ-value chosen for a particular measurement will affect the amplitudes

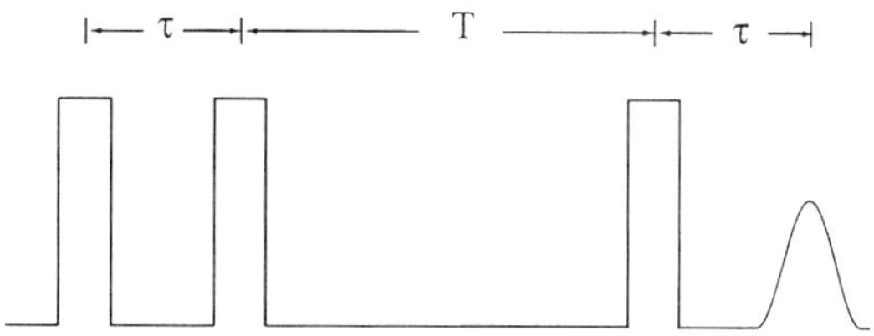

FIG. 6. Three-pulse, or stimulated echo pulse sequence.

of modulations and allow us to enhance or suppress the contribution from an electron spin manifold by adjusting τ over a range of values determined by the period of the complementary hyperfine frequency. This effect is known as the τ-suppression effect and can be useful in making spectral assignments. The product rule governing the form of the overall modulation function, when multiple nuclei are coupled to the paramagnetic center, is also simplified when compared to that of the two-pulse experiment in that products are taken only between frequencies belonging to the same electron spin manifold[2,15]

$$E(\tau, T) = (V_{\text{decay}}/2)\left[\prod_{i=1}^{N} E_{\alpha}^{i}(\tau, T) + \prod_{j=1}^{N} E_{\beta}^{j}(\tau, T)\right]. \tag{9}$$

C. COMPARISON OF ESEEM AND ENDOR

The above presentation indicates that ESEEM spectra, obtained by Fourier[14] or parametric[16] analysis of the time domain data, offer information on hyperfine interactions that is identical to that obtained in an ENDOR experiment. The constraints placed on ESEEM by the requirements of nuclear state mixing and microwave pulse bandwidth, make the method particularly useful for measuring weaker hyperfine couplings of nuclei with small magnetic moments. Because ENDOR experiments on such systems are typically difficult, the two methods are often considered to be complementary to each other. It should be noted that exceptions to this general trend are common and that the utility of both methods has been steadily extended over the past 20 years! In general, the ESEEM method has two chief advantages over ENDOR: (1) the data collection process is systematic; one optimizes the spectrometer for the detection of spin echoes and collects the decay envelope; and (2) the amplitudes of the frequency components observed in an ESEEM experiment can be used to quantitate the number of nuclei giving rise to an interaction once the hyperfine coupling parameters have been determined.[17–19] The primary disadvantage of ESEEM, when compared to powder ENDOR, is that the data are typically difficult to analyze and often require computer simulation to extract all of the information on magnetic couplings contained in the spectrum.

Part of the reason for the increased complexity of the analysis of ESEEM versus ENDOR data is illustrated in Fig. 7. The powder ENDOR spectrum expected for an

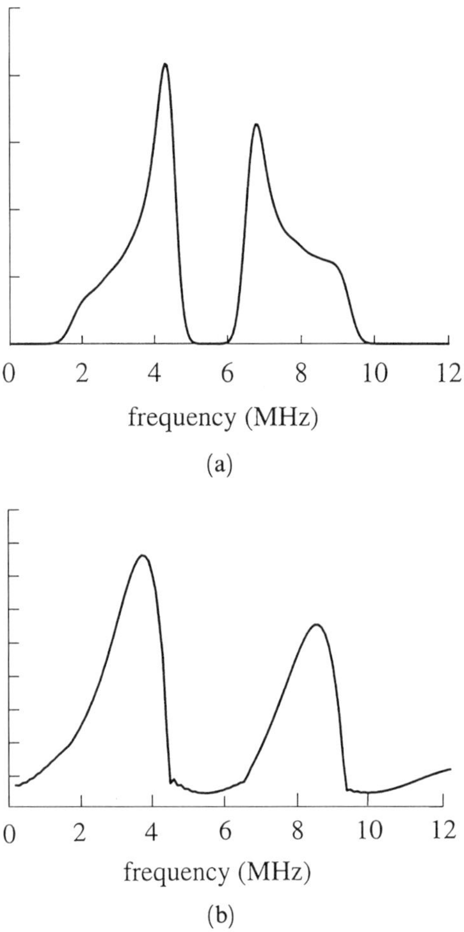

FIG. 7. Simulations of (a) a powder ENDOR spectrum and (b) a three-pulse ESEEM spectrum for an $S = \frac{1}{2}, I = \frac{1}{2}$ spin system described by the following spin Hamiltonian parameters: g, 2.004; g_n, 2.261 (^{31}P); A_{iso}, 4.0 MHz; r, 2.6 Å; B_0, 3200 G; and the number of theta increments of over 90° was 180. For the ESEEM simulation (b), a τ-value of 200 ns and overall dead-time of 240 ns were used.

$S = \frac{1}{2}$, $I = \frac{1}{2}$ system where the hyperfine coupling is of axial symmetry, described by $g_n = 2.261$ (^{31}P), $A_{\text{iso}} = 4.0$ MHz and $r = 2.6$ Å ($A_{\parallel} = 7.1$ MHz and $A_{\perp} = 2.4$ MHz), and the electron spin characterized by an isotropic g-matrix with a resonant magnetic field strength of 3200 G, is shown in trace a. The spectrum shows the characteristic powder lineshape where the tensor symmetry and principal values can be determined from the frequencies of the turning points and lineshape singularities.[20] The corresponding ESEEM spectrum, obtained using the same parameters, is shown in trace b. The lineshape features are absent from this simulation because the ESEEM spectrum is a product of the orientation dependence of the hyperfine coupling that gives rise to the powder lineshape of trace *a*, with the modulation depth parameter, *k* (6). The depth parameter goes to zero at the canonical orientations (the turning points and lineshape singularities) because one of the EPR transition probability amplitudes, $|u|$ or $|v|$ in Fig. 4, goes to zero.[21] Typically, computer simulations of the ESEEM data shown in trace *b* would be needed to determine the principal values of the hyperfine coupling tensor because this information is contained in the frequencies, amplitudes, and damping characteristics of the modulations.

An example that illustrates these points and leads into current methodology developments aimed at alleviating some of the shortcomings of ESEEM is that of measuring the hyperfine couplings of the 3,5 ring protons of a neutral tyrosine radical in a $NaOH/H_2O$ glass (inset, Fig. 8). This species is a typical π-radical with the majority of

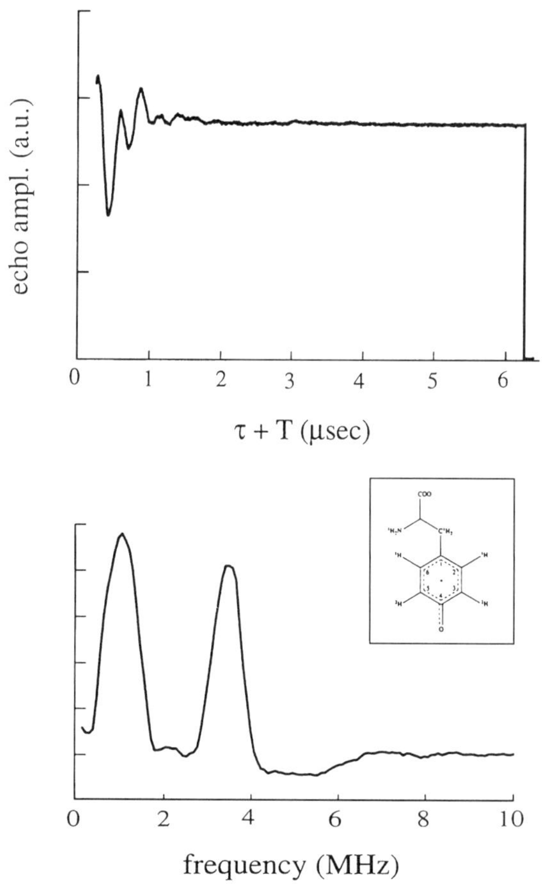

FIG. 8. Three-pulse ESEEM data and Fourier transform for 3,5 ^{2}H-tyrosine neutral radical in an NaOH glass. Measurement conditions were: microwave frequency, 9.132 GHz; pulse power, 40 W; pulse duration, 20 ns FWHM; magnetic field strength, 3265 G; pulse repetition rate, 10 Hz; sample temperature, 4.2 K; and 10 events were averaged per point.

the unpaired electron spin density positioned on the 1,3 and 5 ring carbons and the oxygen of the phenol side chain.[22] The hyperfine couplings for the 3,5 α-protons are expected to be rhombic and cover a large frequency range in the ENDOR spectrum.[24] As discussed in Volume 1 of this series, such couplings are typically difficult to resolve in powder ENDOR studies.[23] ESEEM measurements are facilitated by substitution of the 3,5 protons with deuterium that serves to optimize EPR transition probabilities for the branching required by the method, reduce the hyperfine spectrum to a frequency range that can easily be covered by the microwave pulse bandwidth, and narrow the linewidths of the hyperfine peaks so that their damping times are of sufficient length with respect to the instrument's response, or dead-time, to permit detection and faithful reproduction in the frequency domain. Figure 8 shows ESEEM time domain and corresponding Fourier transform spectra for the 3,5 deuterons at X-band obtained using a three-pulse spin echo sequence with $\tau = 214$ ns. The data show deep modulations (the modulation depths are nearly 50% of the overall amplitude) that damp quickly. The frequency spectrum consists of a pair of featureless peaks centered about the deuterium Larmor frequency of 2.13 MHz. Taken alone, these data can be analyzed, in terms of a spin Hamiltonian that models the hyperfine tensor, as having either axial or rhombic symmetry because the role of the hyperfine anisotropy in the simulations is to define the amplitudes of the peaks. Recent work has shown that a detailed study of the three-pulse ESEEM spectrum over a range of τ-values from 0.2 to 1.3 μs coupled with spectral simulations provides enough information to determine the hyperfine tensor symmetry[24] and measure its principal values. Key to this work were the extra constraints placed on computer simulations by the deuterium double quantum ($\Delta m_I = \pm 2$) transitions and the sculpting of the fundamental ($\Delta m_I = \pm 1$) deuterium lineshapes observed at longer τ-values.[25] This analysis was tedious in that it required computer simulation of several discrete three-pulse ESEEM data sets.

D. HYSCORE

The above discussion immediately leads to the proposal that three-pulse ESEEM data be systematically collected in both τ- and T-dimensions to aid in spectral assignment and to formulate a rational approach to data analysis. The information contained in many one-dimensional experiments could then be contained in a single, two-dimensional Fourier transform with much higher precision. Furthermore, the ω_α and ω_β frequencies that originate from the same hyperfine coupling will give rise to cross peaks, simplifying peak assignments and summarizing the lineshape information of the one-dimensional data set in cross-peak "contours." Such two-dimensional ESEEM experiments were first proposed and carried out by Merks and DeBeer in 1981.[26] However, the disparity in relaxation times that govern the decay of the electron spin echo amplitude in these two time dimensions severely limits the use of the three-pulse sequence for two-dimensional ESEEM. An alternative method for performing these experiments is the HYSCORE (hyperfine sublevel correlation spectroscopy) technique developed by Höfer *et al.*[27] The simplest pulse scheme for HYSCORE utilizes four pulses to generate stimulated echoes according to the sequence $90°$–τ–

$90°$–t_1–$180°$–t_2–$90°$. Modulation of the stimulated echo amplitude, occurring at τ after the third $90°$ microwave pulse, is then studied at a fixed value of τ while systematically varying both t_1 and t_2 to build up a two-dimensional data set. This is a variant of the standard three-pulse stimulated echo experiment except that the π-pulse mixes, or correlates, hyperfine frequencies that arise from the same coupling. Experimental HYSCORE data have only recently started to appear in the literature, but they have already shown much promise in resolving some of the problems with ESEEM measurements. Gemperle *et al.*[28] derived an expression for ESEEM obtained using the general four-pulse scheme of HYSCORE for an $S = \frac{1}{2}$, $I = \frac{1}{2}$ spin system. The amplitudes of the cross peaks for this case are given by

$$S(t_1, t_2) = \tfrac{1}{4}kC[c^2 \cos(\omega_\beta t_2 + \omega_\alpha t_1 + \phi_+) + c^2 \cos(\omega_\alpha t_2 + \omega_\beta t_1 + \phi_+)$$
$$+ s^2 \cos(\omega_\beta t_2 - \omega_\alpha t_1 - \phi_-) + s^2 \cos(\omega_\alpha t_2 - \omega_\beta t_1 - \phi_-)],$$

where

$$C = -2 \sin(\omega_\alpha \tau/2) \sin(\omega_\beta \tau/2),$$

$$\phi_\pm = (\omega_\alpha \pm \omega_\beta)\tau/2,$$

$$s^2 = |\omega_I^2 - (\omega_\alpha + \omega_\beta)^2/4|/(\omega_\alpha + \omega_\beta),$$

$$c^2 = |\omega_{I^2} - (\omega_\alpha - \omega_\beta)^2/4|/(\omega_\alpha \omega_\beta),$$

and $k = 4c^2s^2$. This expression suggests a simple way to separate the ESEEM contributions of strongly coupled nuclei from those that show weak electron–nuclear hyperfine coupling. If the data are processed by taking a cosine FT in the t_2 dimension followed by a complex FT in the t_1 dimension, the cross peaks will separate into different frequency quadrants according to the sign of the dominant phase of $S(t_1, t_2)$. For weak hyperfine couplings $\omega_\alpha \cong \omega_\beta \cong \omega_1, s^2$ approaches zero, while c^2 goes to one. The cross peaks will then be separated by the complex FFT in the t_1 domain along with contributions that show a positive phase or positive ω_1 frequency. For strong electron nuclear couplings, the situation is reversed, with cross peaks being found in the quadrant where a negative phase is operative along the ω_1 axes. For an axial hyperfine coupling without appreciable nuclear quadrupole interaction, if $A_\parallel > 2\omega_I$, cross peaks will be found at $(-\omega_\alpha, \omega_\beta)$ and $(-\omega_\beta, \omega_\alpha)$. When $A_\parallel < 2\omega_I$, the cross peaks are found at $(\omega_\alpha, \omega_\beta)$ and $(\omega_\beta, \omega_\alpha)$. This facet of HYSCORE was first demonstrated by Shane *et al.*[29] and has been utilized recently to isolate the ESEEM contributions of weakly coupled nitrogens from those of nitrogen atoms that showed stronger couplings characterized by "deep" or intense ESEEM in a bacterial photosynthetic reaction center[30] and in an iron–sulfur protein.[31]

HYSCORE also provides for systematic data analysis to recover the hyperfine tensor principal values and tensor orientation with respect to the other magnetic tensors associated with a paramagnetic center. This information is contained in contour plots of the cross peaks and can be easily visualized by considering a first-order hyperfine interaction under the strong- and weak-coupling cases outlined above. For the weak-coupling condition, $A_{\parallel} < 2\omega_I$, the first-order powder ENDOR spectrum for our $S = \frac{1}{2}$, $I = \frac{1}{2}$ system consists of a pair of peaks whose lineshape characteristics, singularities, and turning points are symmetric about the nuclear Larmor frequency. This type of spectrum is drawn along the f_1 and f_2 frequency axes of the schematic HYSCORE contour plot shown in Fig. 9(a). If we construct cross peaks for these idealized spectra by connecting the corresponding lineshape features of the ν_α peak along f_2 axis with the ν_β peak along the f_1 axes and vice versa, they form "ridges" (shaded areas of Fig. 9) perpendicular to the frequency diagonal. The extremes of these

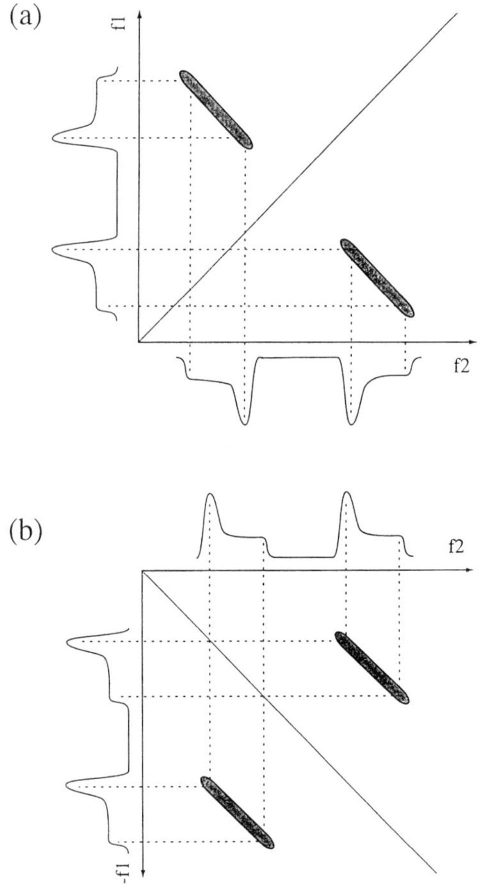

FIG. 9. HYSCORE contour plots of cross peaks that would arise from an $S = \frac{1}{2}$, $I = \frac{1}{2}$ spin system considered at the level of first-order perturbation theory. For (a) $2\nu_n > A_{\parallel}$ and for (b) $2\nu_n < A_{\parallel}$.

ridges provide a measure of the parallel and perpendicular components of the hyperfine coupling. The situation for the strong-coupling case, Fig. 9(b), differs in that the powder ENDOR spectra are no longer symmetrically positioned about a center frequency, with respect to their parallel turning points and perpendicular singularities. Still, the cross peaks can be constructed in the same manner as for the weak-coupling case and give rise to ridges that form lines parallel to the frequency diagonal.[29] When second-order effects are included in the hyperfine interactions, the ridges shown in Fig. 9 develop a curvature.[32] Recently, Dikanov and Bowman have developed analytical expressions for the cross-peak contours predicted for disordered $S = \frac{1}{2}$, $I = \frac{1}{2}$ spin systems and presented a simple graphical means for determination of the principal elements of the hyperfine tensor. Briefly, plotting of the cross peaks on an axis system where the frequencies, f_1 and f_2, are squared results in "straight-line" contours where tensor elements can be derived from the slope and intercept. These authors have also addressed the case of rhombic hyperfine couplings, where the arc-shaped contours found for the axial case give way to wedge-shaped cross peaks. The analysis of these more complex shapes can still be managed by plotting the contours in an f_1^2 versus f_2^2 axis system and analyzing the resulting triangular contours to extract the principal elements of the hyperfine tensor.[33]

E. OUTLOOK

The availability of a commercial pulsed EPR instrument since the late 1980s has fueled a rapid increase in the number of publications using ESEEM to characterize the structures of paramagnetic centers. These applications are beginning to play a significant role in driving the development of new experimental methods in the field. Recent examples include a one-dimensional pulse experiment to improve the resolution of broad hyperfine lines,[34] more sophisticated methods for enhancing the quality of HYSCORE spectra,[35] more efficient detection methods for recording ESEEM,[36–38] and graphical data analysis strategies like the analysis of HYSCORE cross-peak contours mentioned in the previous section.[33] It is hoped that this brief introduction has removed some of the "mystery" surrounding ESEEM and will serve to provide motivation for potential users to start learning about this important method of EPR spectroscopy.

REFERENCES

[1] S. A. Dikanov and Y. D. Tsvetkov, *Electron Spin Echo Envelope Modulation Spectroscopy* (CRC Press, Boca Raton, FL, 1992).

[2] A. Schweiger, Angew. Chem. Int. Ed. Engl. **30**, 265–292 (1991).

[3] L. Kevan, in *Modern Pulsed and Continuous Wave Electron Spin Resonance*, edited by L. Kevan and M. K. Bowman (Wiley-Interscience, New York, 1990), Ch. 231.

[4] (a) W. B. Mims and J. Peisach, in *Advanced EPR*, edited by A. Hoff (Elsevier, Amsterdam, 1989), Ch. 1; (b) S. A. Dikanov and A. V. Ashtashbin, *ibid.*, Ch. 2.

[5] C. D. Keijzers, E. J. Reijerse, and J. Schmidt, *Pulsed EPR: A New Field of Applications* (North-Holland, Amsterdam, 1989).

[6] A. Ponti and A. Schweiger, Appl. Magn. Reson. **7**, 363–403 (1994).
[7] N. M. Atherton, *Principles of Electron Spin Resonance* (Ellis-Harwood PTR, Prentice Hall, Englewood Cliffs, NJ, 1993), Ch. 11.
[8] J. R. Pilbrow, *Transition Ion Electron Paramagnetic Resonance* (Oxford Science, Oxford, 1990), Ch. 10.
[9] (a) T. L. Pappenhagen and D. W. Margerum, J. Amer. Chem. Soc. **107**, 4576–4577 (1985);
(b) J. McCracken and S. Friedenberg, J. Phys. Chem. **98**, 467–473 (1994).
[10] W. B. Mims, Phys. Rev. B **5**(7), 2409–2419 (1972).
[11] W. B. Mims, Phys. Rev. B **6**(9), 3543–3545 (1972).
[12] W. B. Mims, Rev. Sci. Instrun. **36**, 1472 (1965).
[13] L. G. Rowan, E. L. Hahn, and W. B. Mims, Phys. Rev. **137** (1A), A61–A71 (1965).
[14] W. B. Mims, J. Magn. Reson. **59**, 291–306 (1984).
[15] S. A. Dikanov, A. A. Shubin, and V. N. Parmon, J. Magn. Reson. **42**, 474–487 (1981).
[16] R. DeBeer, D. Van Ormondt, W. W. F. Punappel, and J. W. C. van der Veen, Israel. J. Chem. **28**, 249–261 (1988).
[17] L. Kevan, M. K. Bowman, P. A. Narayana, and R. K. Boeckman, J. Chem. Phys. **63**(1), 409–416 (1975).
[18] W. B. Mims, J. Peisach, and J. L. Davis, J. Chem. Phys. **66**, 5536–5550 (1977).
[19] J. McCracken, S. Pember, S. J. Benkovic, J. J. Villafranca, R. J. Miller, and J. Peisach, J. Amer. Chem. Soc. **110**, 1069–1074 (1988).
[20] S. M. Blinder, J. Chem. Phys. **33**(3), 748 (1960).
[21] A. DeGroot, R. Evelo, and A. J. Hoff, J. Magn. Reson. **66**, 331–343 (1986).
[22] B. A. Barry and G. T. Babcock, Proc. Nat. Acad. Sci. U.S.A. **84**, 7099–7103 (1987).
[23] L. Piekara-Sady and L. D. Kispert, in *Handbook of Electron Spin Resonance*, Vol. I, edited by C. P. Poole and H. A. Farach (AIP Press, Woodbury, NY, 1994), p. 351.
[24] W. Gordy, Theory and Applications of Electron Spin Resonance, Vol. XV, *Techniques of Chemistry*, edited by W. West (Wiley, New York, 1980), pp. 206–238.
[25] K. Warncke and J. McCracken, J. Chem. Phys. **101**(3), 1832 (1994).
[26] R. P. J. Merks and R. DeBeer, J. Magn. Reson. **37**, 305 (1980).
[27] P. Höfer, A. Grupp, H. Nebenführ, and M. Mehring, Chem. Phys. Lett. **132**(3), 279–282 (1986).
[28] C. Gemperle, G. Aebli, A. Schweiger, and R. R. Ernst, J. Magn. Reson. **88**, 241–256 (1990).
[29] J. J. Shane, P. Höfer, E. J. Reijerse, and E. DeBoer, J. Magn. Reson. **99**, 596–604 (1992).
[30] H. Käss, J. Rautter, B. Bönigk, P. Höfer, and W. Lubitz, J. Phys. Chem. **99** 436–448 (1995).
[31] J. K. Shergill, C. L. Joannou, J. R. Mason, and R. Cammack, Biochemistry **34**(51), 16533–16542 (1995).
[32] P. Höfer, J. Magn. Reson., Ser. A **111**, 77–86 (1994).
[33] S. A. Dikanov and M. K. Bowman, J. Magn. Reson., Ser. A **116**, 125–128 (1995).
[34] A. Ponti and A. Schweiger, J. Chem. Phys. **102**(13), 5207–5219 (1995).
[35] R. Song, Y. C. Zhong, C. J. Noble, J. R. Pilbrow, and D. R. Hutton, Chem. Phys. Lett. **237**, 86–90 (1995).
[36] M. K. Bowman, Israel J. Chem. **32**, 339 (1992).
[37] E. Hoffmann and A. Schweiger, Chem. Phys. Lett. **220**, 467–477 (1994).
[38] O. Burghaus and M. P. Klein, Chem. Phys. Lett. **243**, 480–485 (1995).

CHAPTER VI
TRANSITION SERIES IONS

Sushil K. Misra

Physics Department, Concordia University, Montreal, Quebec H3G 1M8, Canada

Forms of the spin Hamiltonians are given for transition ions with unfilled d and f shells. Characteristics of these ions are also given. The s-state ions d^5 and f^7 are covered at the end.

CONTENTS

A. INTRODUCTION

The transition-metal ions are those having unfilled $3d$ electron shells (iron group; first-transition series), unfilled $4d$ electron shells (palladium group), unfilled $5d$ electron shells (platinum group), unfilled $4f$ electron shells (rare-earth group), and unfilled $5f$ electron shells (actinide group).

Table 1[1] gives the electron configurations of the neutral ions in the various transition groups. The s- and p-electrons outside the d- and f-shells are the valence electrons of the neutral atoms, which are not present in the ions that are studied in solids. The S-state ions are those for which the unfilled shell is half-full, i.e., with d^5 and f^7 configurations. For an appropriate interpretation of EPR spectra it is important to know the spin Hamiltonians of the various transition ions;[1–9] these are described in the following sections. General details of the spin Hamiltonian for all point group site symmetries are given in the next chapter, while the procedures to evaluate spin-Hamiltonian parameters are described in Chapter IX.

TABLE 1.[1] Ground state configuration of the electron shells of the neutral atoms belonging to various transition groups.

Iron group (First transition group)			Palladium group (Second transition group)			Platinum group (Third transition group)		
Atomic No.	Atom	Configuration	Atomic No.	Atom	Configuration	Atomic No.	Atom	Configuration
21	Sc	$(Ar)3d4s^2$	39	Y	$(Kr)4d5s^2$	71	Lu	$(La)5d6s^2$
22	Ti	$3d^24s^2$	40	Zr	$4d^25s^2$	72	Hf	$5d^26s^2$
23	V	$3d^34s^2$	41	Nb	$4d^45s$	73	Ta	$5d^36s^2$
24	Cr	$3d^54s$	42	Mo	$4d^55s$	74	W	$5d^46s^2$
25	Mn	$3d^54s^2$	43	Tc	$4d^55s^2$	75	Re	$5d^56s^2$
26	Fe	$3d^64s^2$	44	Ru	$4d^75s$	76	Os	$5d^66s^2$
27	Co	$3d^74s^2$	45	Rh	$4d^85s^2$	77	Ir	$5d^76s^2$
28	Ni	$3d^84s^2$	46	Pd	$4d^{10}$	78	Pt	$5d^96s$
29	Cu	$3d^{10}4s$	47	Ag	$4d^{10}5s$	79	Au	$5d^{10}6s$
30	Zn	$3d^{10}4s^2$	48	Cd	$4d^{10}5s^2$	80	Hg	$5d^{10}6s^2$

Rare-earth group (Lanthanides)			Actinide group		
Atomic No.	Atom	Configuration	Atomic No.	Atom	Configuration
57	La	$(Xe)5d6s^2$	89	Ac	$(Rn)6d7s^2$
58	Ce	$4f^26s^2$	90	Th	$6d^27s^2$
59	Pr	$4f^36s^2$	91	Pa	$5f^26d7s^2$
60	Nd	$4f^46s^2$	92	U	$5f^35d7s^2$
61	Pm	$4f^56s^2$	93	Np	$5f^45d7s^2$
62	Sm	$4f^66s^2$	94	Pu	$5f^67s^2$
63	Eu	$4f^76s^2$	95	Am	$5f^77s^2$
64	Gd	$4f^75d6s^2$	96	Cm	$5f^76d7s^2$
65	Tb	$4f^85d6s^2$	97	Bk	$5f^86d7s^2$
66	Dy	$4f^{10}6s^2$	98	Cf	$5f^{10}7s^2$
67	Ho	$4f^{11}6s^2$	99	Es	$5f^{11}7s^2$
68	Er	$4f^{12}6s^2$	100	Fm	$5f^{12}7s^2$
69	Tm	$4f^{13}6s^2$			
70	Yb	$4f^{14}6s^2$			

B. FIRST-TRANSITION SERIES (3d^n, IRON GROUP) IONS

Table 2[1] gives a list of the valence of the ions, the ground state of the free ions, the spin–orbit coupling constant of the free ions, and the spectroscopic properties. The nuclear properties of the ions belonging to the first-transition series ions can be found in refs. 10 and 11. A list of the orbital degeneracy in the lowest state in an octahedral symmetry, the spin degeneracy, and the total degeneracy of the lowest level is also included. The schemes of lowest energy levels due to different crystal-field terms, spin–orbit coupling, and the Zeeman effect for the various ions d^1 to d^9 are depicted in Fig. 1.[4,9] The various ions are discussed below in order of increasing number of unpaired 3d electrons.

3d^1 (VO^{2+}, Ti^{3+}, V^{4+}, Cr^{5+}, Mn^{6+}). In an octahedral crystal field the ground state, $^2T_2$, is split by spin–orbit coupling into a lower quartet Γ_8 and a higher doublet Γ_7. The g-factor for the Γ_8 level is zero, since there is no first-order Zeeman effect. The energy levels of the three Kramers doublets, into which the ground state $^2T_2$ splits due to an axial field and spin–orbit coupling in a field of predominantly octahedral symmetry, superimposed on a field of tetragonal symmetry, are three doublets $E_\pm$ and E_0 with energies

$$E_\pm = -\tfrac{1}{6}d - 28\mu_B \pm \tfrac{1}{2}W \qquad \text{and} \qquad E_0 = \tfrac{1}{3}\delta - 28\mu_B - \tfrac{1}{2}\lambda. \tag{1}$$

In (1), σ represents the splitting of the orbital levels in the absence of spin–orbit coupling, μ_B is the Bohr magneton and λ is the spin–orbit coupling constant. In particular, W and δ are given by

$$W = [(\delta + \tfrac{1}{2}\lambda)^2 + 2\lambda^2]^{1/2}, \qquad \delta = -(60B + 9A), \tag{2}$$

where A and B are the coefficients of the operators $O_2^0(L_z)[= 3L_z^2 - L(L+1)]$ and $O_4^0(L_z)[= 35L_z^4 - 30L(L+1)L_z^2 + 25L_z^2 - 6L(L+1) + 3L^2(L+1)^2]$, respectively,[9] in the spin Hamiltonian describing the tetragonal field. For $\delta > 0$, the doublet E_- lies the lowest, while for $\delta < 0$, E_0 lies the lowest. When E_0 lies the lowest, $g_\parallel = g_\perp = 0$, whereas when E_- lies the lowest

$$g_\parallel = \left(\frac{\delta + \frac{1}{2}\lambda}{S}\right) - 1 \qquad \text{and} \qquad g_\perp = \left(\frac{\delta - \frac{3}{2}\lambda}{S}\right) + 1. \tag{3}$$

For a weak trigonal distortion of the octahedral field, the results are the same as those given by (3), provided that δ is appropriately defined. When the trigonal field is not much weaker than the cubic field, one no longer has the g-factors corresponding to the E_0 doublet equal to zero; in fact, g may be appreciably different from zero.

The separation between the doublets is $\sim 100\ \text{cm}^{-1}$ for slightly distorted octahedral complexes. Therefore, due to broadening of the EPR line by rapid spin-lattice relaxation EPR has only been observed at very low temperatures. On the other hand, in complexes of low molecular symmetry, as in solutions containing Ti^{3+} ions in the presence of complexing agents or in vanadyl and chromyl compounds, EPR may be observed even at room temperature due to transitions between components of the

TABLE 2.[1] Spectroscopic electronic properties of the iron group ($3d^n$) ions. (The ε and γ orbitals represent the energy levels corresponding to the threefold degenerate d-wave functions xy/r^2, yz/r^2, zx/r^2, and twofold degenerate wave functions $(x^2-y^2)/r^2$ and $(3z^2-r^2)/r^2$, respectively, in a cubic field.)

1. Configuration	$3d^1$	$3d^2$	$3d^3$	$3d^4$	$3d^5$	$3d^6$	$3d^7$	$3d^8$	$3d^9$
2. Valence state of the ion	VO^{2+} Ti^{3+} V^{4+} Cr^{5+} Mn^{6+}	Ti^{2+} V^{3+} Cr^{4+} Mn^{5+} Fe^{6+}	V^{2+} Cr^{3+} Mn^{4+}	Cr^{2+} Mn^{3+} Fe^{4+}	Cr^{+} Mn^{2+} Fe^{3+} Co^{4+}	Mn^{+} Fe^{2+} Co^{3+}	Mn^{0} Fe^{+} Co^{2+} Ni^{3+} Cu^{4+}	Fe^{0} Co^{+} Ni^{2+} Cu^{3+}	Ni^{+} Cu^{2+}
3. Ground state of free ion	${}^2D_{3/2}$	${}^3F_2$	${}^4F_{3/2}$	${}^5D_0$	${}^6S_{5/2}$	${}^5D_4$	${}^4F_{9/2}$	${}^3F_4$	${}^5D_{5/2}$
4. Free ion spin–orbit coupling constants (cm^{-1})	154	104	87 55	85 57		−100	−180	−335	−852
5. Orbital ground state in octahedral symmetry	Γ_5 Triplet	Γ_4 Triplet	Γ_2 Singlet	Γ_3 Doublet		Γ_5 Triplet	Γ_4 Triplet	Γ_2	Γ_3
6. Total electronic spin	$\frac{1}{2}$	1	$\frac{3}{2}$	2	$\frac{5}{2}$	2	$\frac{3}{2}$	1	$\frac{1}{2}$
7. Degeneracy of the ground state	4	3	4	1	either 4 or 2	3	2	3	2
8. Nature of spin-lattice relaxation rate	Fast	Fast	Slow	Slow	Slow	Fast	Fast	Slow	Slow
9. Number of electrons in the ε and γ orbitals (in accordance with Pauli's principle and Hund's rule)	ε	$(\frac{9}{5})\varepsilon(\frac{1}{5})\gamma$	$(\varepsilon)^3$	$(\varepsilon)^3\gamma$	$(\varepsilon)^3\gamma^2$	$(\varepsilon)^4\gamma^2$	$(\frac{24}{5})\varepsilon(\frac{11}{5})\gamma$	$(\varepsilon)^6\gamma^2$	$(\varepsilon)^6\gamma^3$

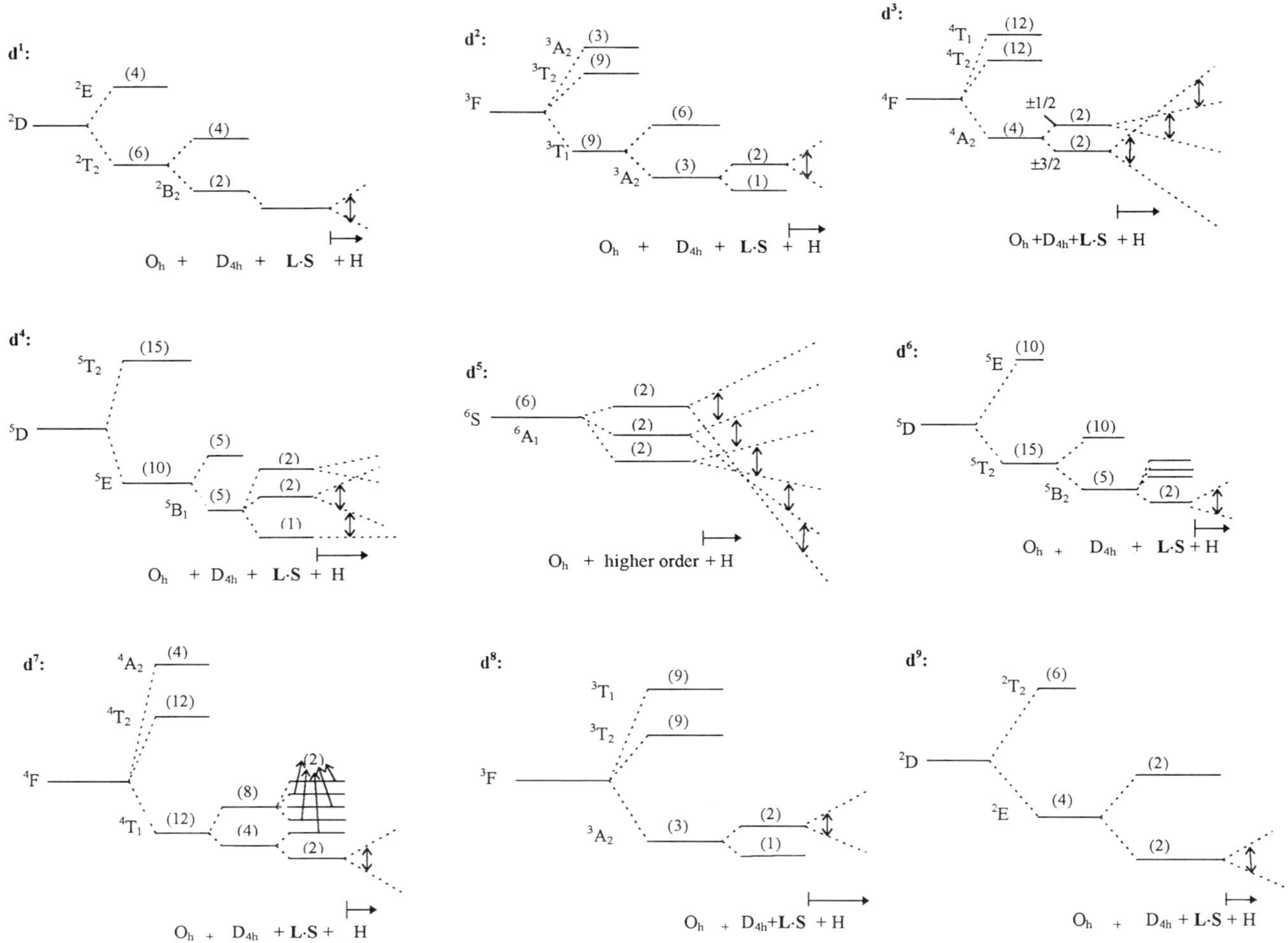

FIG. 1. The ground-state splittings for the various $3d^n$ ions (arbitrary scale) under the action of octahedral (O_h), tetragonal (D_{4h}), spin–orbit ($\mathbf{L} \cdot \mathbf{S}$), and an external magnetic field ($\mathbf{H}$), as described by König and König[4] and by Abragam and Bleaney.[9] Vertical double arrows indicate EPR transitions that can be observed.

lowest doublets, described by $S = \frac{1}{2}$ and the spin Hamiltonian

$$\mathcal{H} = g_{\parallel}\mu_B H_z S_z + g_{\perp}\mu_B(H_x S_x + H_y S_y) + A S_z I_z + B(S_x I_x + S_y I_y). \tag{4}$$

In some cases Ti^{3+} and VO^{2+} complexes show a pair–exchange interaction between Ti^{3+} and VO^{2+} ions, respectively, resulting in a lower singlet and a higher triplet state. For these cases $S = 1$, and the following spin Hamiltonian applies:

$$\mathcal{H} = g_{\parallel}\mu_B H_z S_z + g_{\perp}\mu_B(H_x S_x + H_y S_y) + b_2^0(S_z^2 - S(S+1)/3) + A S_z I_z + B(S_x I_x + S_y I_y). \tag{5}$$

An alternative approach to this exchange interaction effect, particularly applicable to VO^{2+} ions, is discussed below in context with the Cu^{2+} ion ($3d^9$).

For the case of the Mn^{6+} ion $S = \frac{1}{2}$ and $I = \frac{5}{2}$, the ground state is a doublet in a tetrahedral field. The applicable spin Hamiltonian is that described by (5).

EPR of the VO^{2+} ion has been reviewed by Kohin[6] and by Chand, Jain, and Upreti.[7] The ground-state configuration of both V^{4+} and the molecular ion VO^{2+}, and its complexes, is $3d^1$; however, a multiple bond exists between V^{4+} and O^{2-}. The ground state, in a tetragonally-distorted octahedral field, is described by the ${}^2B_{2g}(d_{xy})$ atomic orbital. The effect of the ligand electrons is taken into account by considering, in addition, to crystal-field theory, the molecular–orbital (MO) approach; since the crystal-field theory alone is not sufficient.

The molecular–orbital approach, in which the individual electron orbitals are constructed to be linear combinations of the atomic orbitals, is well described by Kohin.[6] The EPR spectra of the VO^{2+} ion $S = \frac{1}{2}$, $I = \frac{7}{2}$ are described by the following spin Hamiltonian

$$\mathcal{H} = \mu_B \mathbf{S} \cdot \tilde{\mathbf{g}} \cdot \mathbf{H} + \mathbf{S} \cdot \tilde{\mathbf{A}} \cdot \mathbf{I} + \mathbf{I} \cdot \tilde{\mathbf{Q}} \cdot \mathbf{I} - \mu_n \mathbf{H} \cdot \tilde{\mathbf{g}}_n \cdot \mathbf{I}. \tag{6}$$

Here the successive terms represent the electronic Zeeman, the hyperfine, the quadrupole, and the nuclear–Zeeman interactions, respectively. When the principal axes (x, y, z) of the $\tilde{\mathbf{g}}$, $\tilde{\mathbf{A}}$, $\tilde{\mathbf{Q}}$, and $\tilde{\mathbf{g}}_n$ matrices coincide, the spin Hamiltonian (4) simplifies to

$$\mathcal{H} = \sum_{i=x,y,z} (\mu_B g_i H_i S_i + A_i S_i I_i - \mu_n g_n H_i I_i) + Q'[I_z^2 - I(I+1)/3] + Q''(I_x^2 - I_y^2). \tag{7}$$

The solution of (7) to second-order perturbation can be found in Abragam and Bleaney.[9] The fitting of EPR line positions, using the least-squares technique and second-order perturbed eigenvalues, to evaluate the spin-Hamiltonian parameter is described by Misra,[12] for the general case where the principle axes of the $\tilde{\mathbf{g}}$ and $\tilde{\mathbf{A}}$ matrices are not coincident. More details are given in Chapter IX.

EPR provides information on the symmetry and orientation of the vanadyl complex in the host lattice. Usually, the vanadyl complex takes either the form $[VOX_4]^{n-}$, or $[VOX_5]^{n-}$, where the fivefold coordination about the metal ion is square pyramidal and the axial vanadyl oxygen atom remains significantly closer to the metal ion. The five-coordinated species can weakly add a sixth axial ligand *trans* to the vanadyl oxygen atom to form the $[VOX_5]^{n-}$ species; wherein, the effective symmetry appears to be C_{4v}, with both $\tilde{\mathbf{g}}$ and $\tilde{\mathbf{A}}$ matrices being axially symmetric. It is expected that the Z-axes of the $\tilde{\mathbf{g}}$ and $\tilde{\mathbf{A}}$ matrices coincide with the V=O bond of the vanadyl ion. When the symmetry is lower than C_{4v}, both $\tilde{\mathbf{g}}$ and $\tilde{\mathbf{A}}$ matrices become nonaxial.

The electronic energy and bonding parameters can be related to $\tilde{\mathbf{g}}$ and $\tilde{\mathbf{A}}$ matrices in transition-metal complexes. The following simplified expressions, appropriate to C_{4v} symmetry, neglecting small overlap terms, can be used for the estimation of bonding parameters[13,14]

$$\begin{aligned}
\Delta g_{\parallel} &= \frac{8\lambda\beta_1^{*2}\beta_2^{*2}}{\Delta E(b_2^* \rightarrow b_1^*)},\\
\Delta g_{\perp} &= \frac{2\lambda e_\pi^{*2}\beta_2^{*2}}{\Delta E(b_2^* \rightarrow e_\pi^*)},\\
\frac{-A_{\parallel}}{P} &= \beta_2^{*2}\left(K+\frac{4}{7}\right) + \Delta g_{\parallel} + \frac{3}{7}\Delta g_{\perp},\\
\frac{-A_{\perp}}{P} &= \beta_2^{*2}\left(K-\frac{2}{7}\right) + \frac{11}{14}\Delta g_{\perp}.
\end{aligned} \tag{8}$$

In (7) $\Delta g_{\parallel,\perp} = 2.0023 - g_{\parallel,\perp}$; β_1^{*2}, β_2^{*2}, and e_π^{*2} are the molecular–orbital (MO) bonding parameters, ΔE are the energy splittings of the corresponding orbitals, P $(= 2.0023\, g_n\mu_B\mu_n\langle r^{-3}\rangle)$ is the dipolar coupling parameter, λ is the spin–orbit coupling constant for the vanadyl ion, and K is the Fermi-contact interaction constant. Because of the uncertainties in the various quantities, EPR data do not provide an unambiguous determination of the electronic energy levels. The parameters β_1^{*2} and β_2^{*2} indicate the in-plane σ bonding and the in-plane π bonding, respectively; while e_π^{*2} indicates the out-of-plane π bonding. β_1^{*2}, depending upon the σ-donor strength of the ligand, decreases as the covalent bonding increases. The β_2^{*2}, being affected by the delocalization of the electron on to the ligands, decreases as the covalent bonding increases. Finally, the π-bond strength decreases as a result of the increased charge on the metal, and, in general, the stronger the in-plane donor atom the less covalent is the vanadyl V=O bond. The parameter K is sensitive to the 4s character of the orbitals. Additional information about bonding can be obtained by an analysis of the ligand hyperfine structure.

3d^2 (Ti^{2+}, V^{3+}, Cr^{4+}, Mn^{5+}, Fe^{6+}). This configuration corresponds to non-Kramers ions. The ^{3}F free-ion ground state is split by an octahedral field with an orbital triplet ($^3T_1$) lying lowest, which is further split by a trigonal field, resulting in an orbital-singlet ground state as shown in Fig. 1. The threefold spin degeneracy of this singlet is lifted by the combined action of the trigonal field and the spin–orbit coupling. EPR is observed at a field B_0 for the transition within the $M = \pm 1$ doublet with $g_{\perp} = 0$, and $g_{\parallel} = h\nu/2B_0\mu_B\cos\theta$, where θ is the angle between the crystal-field axis and the Zeeman field, $\vec{B}_0$.

The spin–orbit interaction splits the $^3T_1$ octahedral ground state into $\Gamma_1 + \Gamma_3 +$

$\Gamma_4 + \Gamma_5$, of which the lowest triplet Γ_4 splits into a lowest-lying singlet ($m = 0$) and a doublet ($m = \pm 1$) lying several cm^{-1} higher in energy. For the triplet $S = 1$, the applicable spin Hamiltonian is the same as that given by (5). The resonance involves the levels $m = \pm 1$, and is thus weak. The best-known examples are those of V^{3+} ions in the lattices of Al_2O_3 and CdS single crystals.

In the case of a tetrahedral field, the $^3A_2$ ground state is also split into a singlet and a doublet, resulting in $S = 1$, characterized by the spin Hamiltonian (5). EPR is easily observed even at room temperature, since the lowest level is an orbital singlet; for example, Cr^{4+} in tetralkyls, such as $Cr(CH_3)_4 \cdot [MnO_4]^{3-}$ is an inorganic ion of tetrahedral symmetry in this configuration. The hyperfine term $\mathbf{S} \cdot \tilde{\mathbf{A}} \cdot \mathbf{I}$ applies to the ions with nonzero nuclear magnetic moment, i.e., V^{3+}, Cr^{4+}, Mn^{5+}.

$3d^3$ (V^{2+}, Cr^{3+}, Mn^{4+}). The octahedral field produces an $^4A_2$ term lowest, which is split by axial fields into two Kramers doublets, with the total spin $S = \frac{3}{2}$, which are described by the spin Hamiltonian

$$\mathscr{H} = g_{\parallel}\mu_B H_z S_z + g_{\perp}\mu_B(H_x S_x + H_y S_y) + b_2^0[S_z^2 - S(S+1)/3] + b_2^2(S_x^2 - S_y^2) + A(S_z I_z + S_y I_y + S_z I_z). \tag{9}$$

The g-value is almost isotropic. The hyperfine term $\mathbf{S} \cdot \tilde{\mathbf{A}} \cdot \mathbf{I}$ applies to all the ions with nonzero nuclear spin, i.e., V^{2+}, Cr^{3+}, and Mn^{4+}.

$3d^4$ (Cr^{2+}, Mn^{3+}, Fe^{4+}). The ground state for these ions in the free state is ^5D. The strength of the ligand field determines the multiplicity of the ground state for ions in coordination compounds. In weak ligand field $S = 2$, whereas in strong ligand fields $S = 1$.

(i) **Weak ligand fields** ($S = 2$). A cubic (octahedral) field causes an orbital doublet ^5E to lie the lowest for this non-Kramers configuration as shown in Fig. 1. A tetragonal distortion splits this doublet further, only the fivefold spin degeneracy remains. EPR may be observed between the five singlets, which are produced by the splitting of these level, anywhere from 1–10 cm^{-1}, by the joint effect of a rhombic crystal field and spin–orbit coupling. Thus $S = 2$, with the spin Hamiltonian

$$\mathscr{H} = g\mu_B(H_x S_x + H_y S_y + H_z S_z) + b_2^0[S_z^2 - S(S+1)/3] + b_2^2(S_x^2 - S_y^2). \tag{10}$$

An example is the Cr^{2+} ion in $CrSO_4{\cdot}5H_2O$.

For a slightly distorted tetragonal field (D_{2d} symmetry), the total spin $S = 2$ is characterized by the spin Hamiltonian

$$\mathscr{H} = g\mu_B(H_x S_x + H_y S_y + H_z S_z) + b_2^0(S_z^2 - S(S+1)/3) + b_2^2(S_x^2 - S_y^2) + b_4^0[35S_z^4 - 30S(S+1)S_z^2 + 25S_z^2] + \tfrac{1}{6}a(S_1^4 + S_2^4 + S_3^4). \tag{11}$$

The cubic axes 1, 2, 3 and the principal axes x, y, z coincide for cubic crystals, as well as $b_2^2 = 0$ in (11), for example, CrS doped ZnS and CdS.

In a distorted tetrahedral field, the orbital triplet lies the lowest; the combined effect of the lower symmetry fields and the spin–orbital coupling results in a slightly split non-Kramers doublet below all other levels. The EPR for this doublet ($S = \frac{1}{2}$) is described by the following spin Hamiltonian

$$\mathcal{H} = g_{\parallel}\mu_B B_z S_z + \Delta S_x, \tag{12}$$

where Δ is the separation within the lowest doublet.

(ii) **Strong ligand fields** ($S = 1$). There is hardly any ion that is found to belong to this case, so it will not be discussed.

3*d*⁵ (Cr^+, Mn^{2+}, Fe^{3+}, Co^{4+}). These are S-state ions, described in detail in Section G.

3*d*⁶ (Mn^+, Fe^{2+}, Co^{3+}). The ground state of this free ion is $^5D_4$. In coordination compounds, the strength of the ligand fields determines the multiplicity of the ground state. $S = 2$ in weak ligand fields, $S = 0$ in strong ligand fields, and $S = 1$ in intermediate spin situations.

(i) **Weak ligand fields** ($S = 2$). An octahedral field produces the ground state $^5T_2$. Spin–orbit coupling along with an axial field are responsible for a singlet ($M = 0$) and a doublet ($M = \pm 1$) lying lowest. The doublet is further split by a rhombic field. Mostly, EPR signals are observed at low temperatures since lines are broadened by spin-lattice relaxation at high temperatures. The lowest doublet is described by the effective spin $S = \frac{1}{2}$ with the spin Hamiltonian given by (11); an example being FeF_2.

(ii) **Strong ligand fields** ($S = 0$). No EPR is observed since this case corresponds to diamagnetism as $S = 0$.

(iii) **Intermediate spin cases** ($S = 1$). EPR is not observed since the singlet ($M = 0$) lies lowest separated by a large zero-field splitting from the doublet ($M = \pm 1$).

3*d*⁷ (Mn^0, Fe^+, Co^{2+}, Ni^{3+}, Cu^{4+}). The ground state for free $3d^7$ ions is ^4F. The multiplicity in coordination compounds is determined by the strength of the ligand field: $S = \frac{3}{2}$ in weak fields and $S = \frac{1}{2}$ in strong fields.

(i) **Strong ligand fields** ($S = \frac{1}{2}$; low spin). The EPR spectra are analyzed using the spin Hamiltonian

$$\mathcal{H} = \mu_B(g_x H_x S_x + g_y H_y S_y + g_z H_z S_z) + A_x S_x I_x + A_y S_y I_y + A_z S_z I_z. \tag{13}$$

In general, $g_{\parallel} \sim g_e = 2.0023$ and $g_{\perp} > g_{\parallel}$ since $\lambda < 0$; for example, planar complexes of Co^{2+} and Ni^{3+}, like dithiolates. Quadrupole nuclear interactions are important in Co^{2+} ($I = \frac{7}{2}$), and the spin Hamiltonian for axial symmetry becomes

$$\begin{aligned}\mathcal{H} = {} & g_{\parallel}\mu_B H_z S_z + g_{\perp}\mu_B(H_x S_x + H_y S_y) + A S_z I_z + B(S_x I_x + S_y I_y) \\ & + Q'[I_z^2 - I(I+1)/3] - g_n\mu_n(H_x I_x + H_y I_y + H_z I_z).\end{aligned} \tag{14}$$

The unpaired electron is essentially in the $d_{3z^2-r^2}$ orbital, which is perpendicular to the plane of the complex. For the strong-field configuration ($t_2^6 e$) the ground state is an orbital doublet in an octahedral field split by the tetragonal field into two Kramers doublets, described by the g-factors

$$g_{\parallel} = 2 - \frac{8\lambda}{\Delta}, \qquad g_{\perp} = 2 - \frac{2\lambda}{\Delta} \quad (d_{x^2-y^2} \text{ configuration}), \tag{15}$$

and

$$g_{\parallel} = 2, \qquad g_{\perp} = 2 - \frac{6\lambda}{\Delta} \quad (d_{3z^2-r^2} \ \text{configuration}). \tag{16}$$

In the above equations, Δ is the cubic field splitting. The sign of the tetragonal field determines which one of the above doublets lies lowest.

In a trigonal field, or in a cubic field, the orbital doublet is not expected to split. However, the Jahn–Teller effect causes a spontaneous distortion to reduce the energy by splitting the orbital levels. At temperatures above a certain value (critical temperature), the dynamic Jahn–Teller effect occurs, interpreted to be due to the ionic complex oscillating between several distorted configurations of equal energy. This results in an isotropic g-value, which is the mean of the distorted values given by (15) and (16), i.e., $g = 2 - (4\lambda/\Delta)$. Below the critical temperature, the distortions are frozen-in, giving anisotropic spectra, consistent with the $d_{x^2-y^2}$ doublet being the lowest.

(ii) **Weak ligand fields** ($S = \frac{3}{2}$; high spin). In an octahedral field a $^4T_1$ state lies lowest, split by spin–orbit interaction. The lowest level is a Kramers doublet, with the effective spin $\frac{1}{2}$, associated with an isotropic EPR spectrum. In a field of lower symmetry, there are six Kramers doublets that contribute to the spectrum, becoming very anisotropic, characterized by $S = \frac{1}{2}$. The spectra are described by the spin Hamiltonian (13). The g-factor, to first order, is predicted to be 4.33 in an octahedral field since an effective spin is used. When distortions to the octahedral field, e.g., in MgO and CaO, are considered, much different g-values are expected, e.g., Co^{2+} in TiO_2 has $g_{xx} = 2.090$, $g_{yy} = 3.725$, and $g_{zz} = 5.860$.

In octahedral symmetry Fe^+ and Co^{2+} ($t_2^5e^2$) experience the weak crystal-field interaction. The lowest doublet of the $t_2^5e^2$ configuration in an axially distorted octahedral field is described by the following g-values ($S = \frac{1}{2}$), in terms of a parameter x ($0 \leq x \leq \infty$):

$$g_{\parallel} = 2 + 4(\alpha + 2)\left[\left(\frac{3}{x^2} - \frac{4}{(x+2)^2}\right)\Big/\left(1 + \frac{6}{x^2} + \frac{8}{(x+2)^2}\right)\right], \tag{17}$$

$$g_{\perp} = 4\left[\left(1 + \frac{2\alpha}{x+2} + \frac{4}{x(x+2)}\right)\Big/\left(1 + \frac{6}{x^2} + \frac{8}{(x+2)^2}\right)\right], \tag{18}$$

where the trigonal, or tetragonal, splitting parameter δ is given by

$$\delta = \alpha\lambda\left[\frac{x+3}{2} - \frac{3}{x} - \frac{4}{x+2}\right]. \tag{19}$$

Here α is expected to lie between 1.0 and 1.5. Co^{2+} also exhibits a hyperfine structure described by $\mathbf{S} \cdot \tilde{\mathbf{A}} \cdot \mathbf{I}$; the matrix $\tilde{\mathbf{A}}$ is expected to be quite anisotropic.

In a ligand field of tetrahedral symmetry, the $^4A_2$ ground state is split into two Kramers doublets in fields of lower symmetry, e.g., in Cs_3CoCl_5 and Cs_3CoBr_5. The applicable spin Hamiltonian is ($S = \frac{3}{2}$)

$$\mathcal{H} = g_{\parallel}\mu_B H_z S_z + g_{\perp}\mu_B(H_x S_x + H_y S_y) + b_2^0[S_z^2 - S(S+1)/3], \quad (20)$$

Ni^{3+} experience the strong field case due to its extra change ($t_2^6 e$).

3*d*[8] (Fe^0, Co^+, Ni^{2+}, Cu^{3+}). An octahedral field splits the ^{3}F ground state of the free $3d^8$ ion into two triply degenerate states and the orbitally nondegenerate $^3A_2$ singlet state which lies the lowest. The spin degeneracy (Γ_5) is not removed by the spin–orbit coupling. The lowest level is split by axial fields into a singlet and a doublet, and a rhombic field splits it into tree singlets. The EPR is described the same way as that for the $3d^3$ configuration using the spin Hamiltonian (10), except that here $S = 1$, rather than 2. The Co^+ and Ni^{2+} spectra sometimes exhibit "double quantum" transitions, for which $\Delta M = \pm 2$, M being the electronic magnetic quantum number. EPR of $3d^8$ ions can be observed at 77 K, or even at room temperature, since there are no excited states close to the ground state. The broad EPR lines of the Ni^{2+} ions may sometimes be difficult to detect.

3*d*[9] (Ni^+, Cu^{2+}). The details of the spin Hamiltonian, ground-state wave functions, and the Cu^{2+} Jahn–Teller (JT) effect have been well described in a review article on Cu^{2+} by Misra and Wang.[8] The ground state of the configuration is $t_2^6 e^3$ in an octahedral field, which may be regarded as a single hole in an e orbital. The theory is the same as that for a $3d^7$ ion in a strong octahedral field for which the g-factors have been given by (17) and (18). The JT distortion also takes place in the same way as that for a $3d^7$ ion in a trigonal, or cubic, symmetry. Thus, orbital degeneracy is lifted by the JT effect in the case of trigonal symmetry although the ^{2}E state is not split. The JT effect may distort the molecular complex at higher temperatures. At low temperature, on the other hand, the JT effect leads to a static distortion, resulting into a spectrum due to three magnetic complexes, each characterized by tetragonal symmetry. The applicable spin Hamiltonian is ($S = \frac{1}{2}$):

$$\begin{aligned}\mathcal{H} = {} & \mu_B(g_x H_x S_x + g_y H_y S_y + g_z H_z S_z) + b_2^0[S_z^2 - S(S+1)/3] \\ & + b_2^2(S_x^2 - S_y^2) + Q'[I_z^2 - I(I+1)/3] + Q''(I_x^2 - I_y^2). \end{aligned} \quad (21)$$

In a rhombic symmetry with sixfold coordination the ground-state wave function is a supposition of the $|x^2 - y^2\rangle$ and $|3z^2 - r^2\rangle$ states. In an octahedral sixfold coordination the ^{2}D orbital splits into an E_g (Γ_3) doublet and a T_{2g} (Γ_5) triplet, the ground state being the T_{2g} state, which can be further split by the JT effect. In a tetragonally distorted octahedral field, there are found admixtures of the excited states $|xy\rangle$, and $\{|yz\rangle, |zx\rangle\}$ in the ground-state wave function, which is either $|x^2 - y^2\rangle$, or $|3z^2 - r^2\rangle$. The principal g-values satisfy the relations $g_{\parallel} > g_{\perp} > g_e$ for an elongated tetragonal distortion and $g_{\perp} > g_{\parallel} \approx g_e$ for a compressed-tetragonal distortion, g_e being the free-electron value. In a square-planar symmetry, which is the limiting case of an elongated tetragonally distorted octahedron, the ground-state wave function is predominantly $|x^2 - y^2\rangle$. In a tetrahedral symmetry, with fourfold coordination, the orbital state of the Cu^{2+} ion splits into a E_g (Γ_3) doublet and a T_{2g} (Γ_5) triplet, with the ground state being T_{2g}. In a trigonally distorted tetrahedral field the T_{2g} orbital of Cu^{2+} splits further into a ^2E doublet and $^2B_2$ singlet, the ground state being the ^2E doublet, for which a JT splitting can take place.

In octahedral symmetry the $\tilde{\mathbf{g}}$ and $\tilde{\mathbf{A}}$ matrices have the following principal values

$$g_{\|} = g_e - \frac{8\lambda}{\Delta}, \qquad A_{\|} = p\left\{-K - \tfrac{4}{7} - \frac{6\lambda}{7\Delta_2} - \frac{8\lambda}{\Delta_1}k\right\},$$
$$g_{\perp} = g_e - \frac{2\lambda}{\Delta}, \qquad A_{\perp} = p\left\{-K + \tfrac{2}{7} - \frac{11\lambda}{7\Delta_2}\right\}, \tag{22}$$

while for octahedral symmetry with a compressed tetragonal distortion

$$g_{\|} = g_e, \qquad A_{\|} = p\left\{-K + \tfrac{4}{7} + \frac{6\lambda}{7\Delta_3}\right\},$$
$$g_{\perp} = g_e - \frac{6\lambda}{\Delta_3}, \qquad A_{\perp} = p\left\{-K - \tfrac{2}{7} - \frac{45\lambda}{7\Delta_3}\right\}. \tag{23}$$

In (22) and (23) Δ, Δ_1, Δ_2, and Δ_3 are the energy splittings of the Cu^{2+} ion in a tetragonally distorted octahedral crystal field, $p = 2g_n\mu_B\mu_n\langle r^{-3}\rangle$; where $\langle r^{-3}\rangle$ is the average of the inverse-cube radius of the Cu^{2+} ion, R is the orbital-reduction factor, and K is the Fermi-contact interaction constant.

The JT effect of the Cu^{2+} ion can be described in terms of a CuL_6 complex, where L is a surrounding ligand ion. Bates and Chandler[15,16] have reviewed the JT experimental data on the Cu^{2+} ion at tetrahedral sites.

EPR spectra of the Cu^{2+} ion sometimes exhibit an exchange interaction between isolated Cu^{2+} ion pairs, resulting in a lower singlet and a higher triplet state. The spin Hamiltonian is described as follows ($S = 1$):

$$\mathscr{H} = \mu_B(g_xH_xS_x + g_yH_yS_y + g_zH_zS_z) + A_xS_xI_x + A_yS_yI_y$$
$$+ A_zS_zI_z + Q'[I_z^2 - I(I+1)/3] + Q''(I_x^2 - I_y^2). \tag{24}$$

Alternatively, the interaction between the pair of ions 1 and 2 can be expressed by the spin Hamiltonian

$$\mathscr{H} = \mathscr{H}_1 + \mathscr{H}_2 + \mathscr{H}_{\text{int}}, \tag{25}$$

where $\mathscr{H}_1$ and $\mathscr{H}_2$ are the spin Hamiltonians of the isolated ions, and $\mathscr{H}_{\text{int}}$ represents the interaction between them. For the case of the two ions in identical sites of axial symmetry with their z-axes parallel to the internuclear vector $\mathbf{r}$, we have

$$\mathscr{H} = g_{\|}\mu_BH_z(S_{1z} + S_{2z}) + g_{\perp}\mu_B[H_x(S_{1x} + S_{2x}) + H_y(S_{1y} + S_{2y})]$$
$$+ A(S_{1z}I_{1z} + S_{2z}I_{2z}) + B(S_{1x}I_{1x} + S_{2x}I_{2x} + S_{1y}I_{1y} + S_{2y}I_{2y}) + \mathscr{H}_D, \tag{26}$$

where

$$\mathscr{H}_D = (\mu_B^2/r^3)[g_x^2 S_{1x}S_{2x} + g_y^2 S_{1y}S_{2y} - 2g_z^2 S_{1z}S_{2z}]. \tag{27}$$

In (27) only the magnetic dipole–dipole interaction is operative. Diagonalization of the corresponding Hamiltonian matrix yields, in addition to the conventional EPR parameters $g_\parallel$, $g_\perp$, A, B, a value for the inter-nuclear distance, r, between the metal ions.

C. SECOND- AND THIRD-TRANSITION SERIES ($4d^n$, PALLADIUM AND $5d^n$, PLATINUM GROUP) IONS

There are much less EPR data reported for the $4d$ and $5d$ groups, since they are much less abundant in nature, and because often low temperatures are required for observation of EPR. It is not so easy, as compared to the iron-group ions, to find suitable crystal lattices into which they can be introduced. These ions tend to form strong covalent bonds with neighboring ions, giving rise to a strong crystal field. Thus, for ions containing up to three unpaired electrons, the resonance results are similar to those for the corresponding $3d$ ions, but for the d^n ions with $n \geq 4$, there always exists the strong-field configurations, similar to those for the Fe^{3+} and Ni^{2+} ions. The majority of the these ions possess isotopes with nonzero nuclear spins. Table 3[2] contains the details of nuclear data for those ions, for which EPR have been observed. An earlier review of the $4d$, $5d$ elements has been published by Griffiths, Owen, and Ward.[17]

$4d^1$ (Nb^{4+}, Mo^{5+}), **$5d^1$** (Ta^{4+}, W^{5+}, Re^{6+}). The details of EPR for these configurations are the same as those for the $3d^1$ ion. An isolated Kramers doublet lies lowest in a distorted octahedral symmetry; the g-factor is expected to be in the range 1.0–2.0. Low temperatures are usually, but not always, required to observe EPR. In general, the complexes show axial symmetry in the various samples studied. The applicable spin Hamiltonian is, thus, described by (4).

$4d^2$ (Mo^{4+}), **$5d^2$** (Re^{5+}, Ru^{6+}). The features of EPR resonance for these ions are similar to those of $3d^2$ ions. For example, the $(ReO_4)^{3-}$ complex has a configuration similar to the $3d^2$ ion complex $(MnO_4)^{3-}$.

$4d^3$ (Mo^{3+}, Tc^{4+}), **$5d^3$** (Re^{4+}). In an octahedral field, an orbital-singlet lies the lowest, the threefold spin degeneracy being raised by the spin–orbit and lower symmetry terms. In the case of the Mo^{3+} ion and the majority of d^3 ions in an external field the spin Hamiltonian (9) applies. In order to analyze EPR data in a cubic (octahedral) symmetry site, e.g., for the Re^{4+} ion, extra terms are needed in the spin Hamiltonian ($S = \frac{3}{2}, I = \frac{3}{2}$):

$$\begin{aligned}\mathscr{H} = {} & g\mu_B \mathbf{H}\cdot\mathbf{S} + \mu_B u[H_x S_x^3 + H_y S_y^3 + H_z S_z^3 - \tfrac{1}{5}(\mathbf{H}\cdot\mathbf{S})(3S^2 - 1)] \\ & + A\mathbf{S}\cdot\mathbf{I} + U[I_x S_x^3 + I_y S_y^3 + I_z S_z^3 - \tfrac{1}{5}(\mathbf{S}\cdot\mathbf{I})(3S^2 - 1)].\end{aligned} \tag{28}$$

$4d^5$ (Mo^{+}, Ru^{3+}), **$5d^5$** (W^{+}, Re^{2+}, Os^{3+}, Ir^{4+}). These are S-state ions; the details are given in Section F.

TABLE 3.[2] Nuclear properties of the palladium ($4d^n$) and platinum ($5d^n$) groups. (I is nuclear spin.)

Configuration	Ion	Isotope	% Abundance	I
$4d^1$	Nb^{4+}	Nb^{93}	100	$\frac{9}{2}$
	Mo^{5+}	Mo^{95}	15.7	$\frac{5}{2}$
		Mo^{97}	9.6	$\frac{5}{2}$
$4d^3$	Mo^{3+}	Mo^{95}	15.7	$\frac{5}{2}$
		Mo^{97}	9.6	$\frac{5}{2}$
	Tc^{4+}	Tc^{99}	~100	$\frac{9}{2}$
$4d^5$	Ru^{3+}	Ru^{99}	12.8	$\frac{5}{2}$
		Ru^{101}	17.0	$\frac{5}{2}$
$4d^7$	Rh^{2+}	Rh^{103}	100	$\frac{1}{2}$
$4d^9$	Rh^0	Rh^{103}	100	$\frac{1}{2}$
	Ag^{2+}	Ag^{107}	51.3	$\frac{1}{2}$
		Ag^{109}	48.6	$\frac{1}{2}$
$5d^1$	Ta^{4+}	Ta^{181}	100	$\frac{9}{2}$
	W^{5+}	W^{183}	14.4	$\frac{1}{2}$
$5d^3$	Re^{4+}	Re^{185}	37.1	$\frac{5}{2}$
		Re^{187}	62.9	$\frac{5}{2}$
$5d^5$	Ir^{4+}	Ir^{191}	38.5	$\frac{3}{2}$
		Ir^{193}	61.5	$\frac{3}{2}$
$5d^7$	Os^+	Os^{187}	1.6	$\frac{1}{2}$
		Os^{189}	16.1	$\frac{3}{2}$
	Pt^{3+}	Pt^{195}	33.8	$\frac{1}{2}$
$5d^9$	Tl^{2+}	Tl^{203}	29.5	$\frac{1}{2}$
		Tl^{205}	70.5	$\frac{1}{2}$
	Pb^{3+}	Pb^{207}	22.6	$\frac{1}{2}$

***4d*^{7} (Rh^{2+}, Pd^{3+}), *5d*^{7} (Pt^{3+}).** In a trigonal symmetry a JT splitting is experienced by this configuration, similar to that for a $3d^7$ ion. Because of the occurrence of the dynamical JT effect, at higher temperatures an isotropic spectrum ($g_\parallel = g_\perp$) is observed, while at low temperatures, the JT distortions are frozen-in resulting in an anisotropic spectrum ($g_\parallel \neq g_\perp$). In monoclinic symmetry, the orbital doublet for this configuration is split, thus no JT effect is observed. Most of the compounds in which EPR has been observed are dithiolates and related complexes. The applicable spin Hamiltonian is described by (13) with $S = \frac{1}{2}$.

***4d*^{9} (Rh^0, Ag^{2+}), *5d*^{9} (Tl^{2+}, Pb^{3+}).** Dynamic and static JT effects are expected for this configuration at high and low temperatures, respectively. The theory to interpret EPR spectra is the same as that described for $3d^9$ ions.

D. RARE-EARTH ($4f^n$) IONS

Introduction

For these ions the $4f$ shell contains the magnetic electrons. The configuration consists of the terms characterized by well-defined L and S values, which are coupled by the spin–orbit interaction to a well-defined value of the total angular momentum J. In some rare-earth ions a partial breakdown of Russel–Saunders coupling causes admixture of levels having the same angular momentum J. The values of the spin–orbit coupling constants for the rare-earth group ions, as listed in Table 4(a),[1] are much larger than those for the iron-group ions. However, the crystal fields seen by the rare-earth ions, which are smaller than the spin–orbit couplings, are, in turn, very much smaller than those in the iron-group ions. This is due to:

(i) The average values, $\langle r^n \rangle$, for the f^n, rare earth, wave functions, are smaller than those for the d^n, iron-group, wave functions, since the $4f^n$ functions do not extend very far unlike the $3d^n$ functions;
(ii) The rare-earth $4f$ electrons are screened by the outer electron which, in turn, attenuates the effect of the crystal-field potential;
(iii) the crystals of most rare-earth ions possess low symmetries, so that the axial and cubic contributions to the electrostatic crystal-field potentials are partially, or even fully, suppressed.

For a $4f^n$ ($n = 1$ to 13) rare-earth ion in a crystal, the ground state is that determined by Hund's rule with the total angular momentum J, which is a good quantum number since the separation between adjacent J levels (100–1000 cm^{-1}) due to the spin–orbit interaction generally exceeds the overall splitting of the ground-state manifold due to the crystal field by at least an order of magnitude. The crystalline electric field (CEF) Hamiltonian within the ground-state (J) multiplet of a rare-earth ion, in terms of Stevens' operators, is[9]

$$\mathscr{H}_{\text{CEF}} = \sum_{l,m} B_l^m O_l^m, \tag{29}$$

where $l = 2, 4, 6$ ($|m| \leq l$).

The CEF parameters, B_l^m, can be, alternatively, expressed as

$$B_l^m = A_l^m \langle r^l \rangle \alpha_l, \tag{30}$$

where the α_l are Stevens' operator-equivalent factors[9] and $\langle r^l \rangle$ are the radial integrals which have been calculated for the most rare-earth ions using the Hartree–Fock procedure.[9] A_l^m are often assumed to be the same for all the rare-earth ions in an isostructural series of compounds.[9] The required α_l and $\langle r^l \rangle$ values for the various trivalent $4f^n$ rare-earth ions[9] are listed in Table 4(b).[18]

In general, for a rare-earth ion in a crystal, an electron paramagnetic resonance transition may be observed at microwave frequencies at low temperatures between those two lowest levels of the ground-state J multiplet, which are significantly populated and possess a nonzero magnetic moment corresponding to an effective spin

TABLE 4(a).[1] Spectroscopic and nuclear properties of rare-earth ions.

Trivalent ion	Ground state of free ion	Spin–orbit coupling in cm^{-1}	$\langle r^{-3} \rangle$ in $(A)^{-3}$	Isotope	Nuclear spin	Nuclear magnetic moment	Larmor frequency (MHz) at 10^4 Gauss	Quadrupole moment (Q) $\times 10^{-24}$ cm^{-2}
^{58}Ce	$^2F_{5/2}$	640.0	32.5					
^{59}Pr	$^3H_4$	800.0	37.0	141	$\frac{5}{2}$	3.92	11.95	-5.4×10^{-2}
^{60}Nd	$^4I_{9/2}$	900.0	42.0	143	$\frac{7}{2}$	−1.03	2.24	<1.2
				145	$\frac{7}{2}$	0.64	1.4	<1.2
^{61}Pm	$^5I_4$	(1070.0)	47.0					
^{62}Sm	$^6H_{5/2}$	1200.0	51.0	147	$\frac{7}{2}$	−0.83	1.8	0.72
				149	$\frac{7}{2}$	−0.68	1.5	0.72
^{63}Eu	$^7F_0$	1410.0	57.0	151	$\frac{5}{2}$	3.4	10.0	~1.2
				153	$\frac{5}{2}$	1.5	4.6	~2.5
^{64}Gd	$^8S_{7/2}$	1540.0	62.0	155	$\frac{3}{2}$	−0.24	1.2	1.1
	$^2F_{5/2}$			157	$\frac{3}{2}$	−0.32	1.6	1.0
^{65}Tb	$^7F_6$	1770.0	68.0	159	$\frac{3}{2}$	1.52	7.72	
^{66}Dy	$^6H_{15/2}$	1860.0	74.0	161	$\frac{5}{2}$	0.38	1.2	
				163	$\frac{5}{2}$	0.53	1.6	
^{67}Ho	$^5I_8$	2000.0	80.0	165	$\frac{7}{2}$	3.29	7.17	2.0
^{68}Er	$^4I_{15/2}$	2350.0	86.0	167	$\frac{7}{2}$	0.48	1.04	~10.0
^{69}Tm	$^3H_6$	2660.0	92.0	169	$\frac{1}{2}$	−0.20	3.05	
^{70}Yb	$^2F_{7/2}$	2940.0	98.0	171	$\frac{1}{2}$	0.43	6.6	
				173	$\frac{5}{2}$	−0.6	1.8	3.9

TABLE 4(b).[18] List of the Hund's rule ground state $^{2S+1}L_J$ (J, L, S denote the total, orbital, and spin-angular momentum, respectively), g_J (as defined by (31)), α_l and $\langle r^l \rangle$ values for the various $4f^n$ rare-earth ions. The last column gives the references from where the various $\langle r^l \rangle$ values have been obtained. More details are given in refs. 9 and 18.

Ion	$4f^n$	$^{2S+1}L_J$	g_J	α_2	α_4	α_6	$\langle r^2 \rangle$	$\langle r^4 \rangle$	$\langle r^6 \rangle$
Ce^{3+}	$4f^1$	$^2F_{5/2}$	$\frac{6}{7}$	$\frac{-2}{5\cdot 7}$	$\frac{2}{3^2\cdot 5\cdot 7}$	0	1.200	3.455	21.226
Pr^{3+}	$4f^2$	$^3H_4$	$\frac{4}{5}$	$\frac{-2^2\cdot 13}{3^2\cdot 5^2\cdot 11}$	$\frac{-2^2}{3^2\cdot 5\cdot 11^2}$	$\frac{2^4\cdot 17}{3^4\cdot 5\cdot 7\cdot 11^2\cdot 13}$	1.086	2.822	15.726
Nd^{3+}	$4f^3$	$^4I_{9/2}$	$\frac{8}{11}$	$\frac{-7}{3^2\cdot 11^2}$	$\frac{-2^3\cdot 17}{3^3\cdot 11^3\cdot 13}$	$\frac{-5\cdot 17\cdot 19}{3^3\cdot 7\cdot 11^3\cdot 13^2}$	1.001	2.401	12.39
Pm^{3+}	$4f^4$	$^5I_4$	$\frac{3}{5}$	$\frac{2\cdot 7}{3\cdot 5\cdot 11^2}$	$\frac{2^3\cdot 7\cdot 17}{3^3\cdot 5\cdot 11^3\cdot 13}$	$\frac{2^3\cdot 17\cdot 19}{3^3\cdot 7\cdot 11^3\cdot 13^2}$	0.937[b]	2.124[b]	10.480
Sm^{3+}	$4f^5$	$^6H_{5/2}$	$\frac{2}{7}$	$\frac{13}{3^2\cdot 5\cdot 7}$	$\frac{2\cdot 13}{3^3\cdot 5\cdot 7\cdot 11}$	0	0.883	1.897	8.775
Eu^{3+}	$4f^6$	$^7F_0$		0	0	0	0.816[b]	1.695[b]	7.429[b]
Gd^{3+}	$4f^7$	$^8S_{7/2}$		0	0	0	0.785	1.515	6.281
Tb^{3+}	$4f^8$	$^7F_6$	$\frac{3}{2}$	$\frac{-1}{3^2\cdot 11}$	$\frac{2}{3^3\cdot 5\cdot 11^2}$	$\frac{-1}{3^4\cdot 7\cdot 11^2\cdot 13}$	0.739[b]	1.150[b]	5.238[b]
Dy^{3+}	$4f^9$	$^6H_{15/2}$	$\frac{4}{3}$	$\frac{-2}{3^2\cdot 5\cdot 7}$	$\frac{-2^3}{3^3\cdot 5\cdot 7\cdot 11\cdot 13}$	$\frac{2^2}{3^3\cdot 7\cdot 11^2\cdot 13^2}$	0.726	1.322	5.102
Ho^{3+}	$4f^{10}$	$^5I_8$	$\frac{5}{4}$	$\frac{-1}{2\cdot 3^2\cdot 5^2}$	$\frac{-1}{2\cdot 3\cdot 5\cdot 7\cdot 11\cdot 13}$	$\frac{-5}{3^3\cdot 7\cdot 11^2\cdot 13^2}$	0.692	1.221	4.489
Er^{3+}	$4f^{11}$	$^4I_{15/2}$	$\frac{6}{5}$	$\frac{2^2}{3^2\cdot 5^2\cdot 7}$	$\frac{2}{3^2\cdot 5\cdot 7\cdot 11\cdot 13}$	$\frac{2^3}{3^3\cdot 7\cdot 11^2\cdot 13^2}$	0.666	1.126	3.978
Tm^{3+}	$4f^{12}$	$^3H_6$	$\frac{7}{6}$	$\frac{1}{3^2\cdot 11}$	$\frac{2^3}{3^4\cdot 5\cdot 11^2}$	$\frac{-5}{3^4\cdot 7\cdot 11^2\cdot 13}$	0.646	1.119	3.646
Yb^{3+}	$4f^{13}$	$^2F_{7/2}$	$\frac{8}{7}$	$\frac{2}{3^2\cdot 7}$	$\frac{-2}{3\cdot 5\cdot 7\cdot 11}$	$\frac{2^2}{3^3\cdot 7\cdot 11\cdot 13}$	0.613	0.960	3.104

$S = \frac{1}{2}$. The effective g-values corresponding to these levels ($|+\rangle$ and $|-\rangle$) can be calculated from the matrix elements of the components of the angular momentum operators $\mathbf{J}_x$, $\mathbf{J}_y$, $\mathbf{J}_z$ within these two levels,

$$g_\alpha \sigma_\alpha = g_J \begin{pmatrix} \langle + \|J_\alpha\| + \rangle & \langle + \|J_\alpha\| - \rangle \\ \langle - \|J_\alpha\| + \rangle & \langle - \|J_\alpha\| - \rangle \end{pmatrix}; \qquad \alpha = x, y, z, \tag{31}$$

where

$$g_J = \langle J\|\Lambda\|J\rangle = 1 + \frac{J(J+1) + S(S+1) - L(L+1)}{2J(J+1)} \tag{32}$$

is the Landé factor, and $|\pm\rangle$ denote the orthonormal eigenvectors corresponding to the two lowest levels considered for the calculation of the magnetic moment, and σ_α are Pauli spin matrices. Finally, the effective g-values are given by[9]

$$g_x = g_J \varepsilon_x, \qquad g_x = g_J \varepsilon_x, \qquad g_x = g_J \varepsilon_x. \tag{33}$$

In (33), ε_x, ε_y, ε_z denote, respectively, the differences of the eigenvalues of $\mathbf{J}_x$, $\mathbf{J}_y$, and $\mathbf{J}_z$ for these two levels. The values of g_J as calculated using (32) for different rare-earth ions are also listed in Table 4(b).[15] As for the $|\pm\rangle$ levels, we choose the set of the two levels which lies the lowest and for which all ε_α ($\alpha = x, y, z$) are not zero. For Kramers ions, these belong to the lowest-lying doublet. On the other hand, for non-Kramers ions, we have to carefully choose these two levels. When they belong to different eigenvalues the $|+\rangle$ and $|-\rangle$ eigenvectors should be weighted by appropriate

Boltzmann factors, which are, respectively, $(1/N)\exp(-E_+/2kT)$ and $(1/N)\exp(-E_-/2kT)$, where $N = \sqrt{[\exp(-E_+/kT) + \exp(-E_-/kT)]/2}$. The g-values are then temperature-dependent.

The crystal-field splittings for the rare-earth ions are rather small ($<k_BT$) as indicated by the almost nearly equal values of the magnetic susceptibitlity of the rare-earth ions in crystal and those of the free ions at room temperature.

Odd number of 4*f* electrons. The rare-earth ions with an odd number of electrons, except for the S-state ions Eu^{2+} and Gd^{3+}, have a single Kramers doublet lying lowest in an axial, or lower, symmetry field. Their EPR spectra are described by the spin Hamiltonian

$$\mathscr{H} = \mu_B g \mathbf{H} \cdot \mathbf{S} + A\mathbf{I} \cdot \mathbf{S}. \tag{34}$$

Even number of 4*f* electrons. The Jahn–Teller effect is expected to remove the degeneracy of the splitting of each J level into singlets and doublets in C_{3h} or C_{3v} symmetry. However, this effect is rather small since the $4f$ electrons are screened, being inner electrons. It is the characteristic of accidental (non-Kramers) doublets that a first-order Zeeman effect is observed if the magnetic field is applied parallel to the axis of symmetry. On the other hand, only a second-order effect is expected if it is applied normal to the symmetry axis. The non-Kramers rare-earth ions, with an even number of electrons, may show EPR in an axial symmetry, which is due to the unsplit doublet. In that case a Δ term is included in the spin Hamiltonian to take into account small departures from axial symmetry due to crystalline imperfections:

$$\mathscr{H} = \mu_B g \mathbf{H} \cdot \mathbf{S} + A\mathbf{S} \cdot \mathbf{I} + \Delta_x S_x + \Delta_y S_y + Q'[I_z^2 - I(I+1)/3] - \mu_n g_n \mathbf{H} \cdot \mathrm{I}, \tag{35}$$

with $S = \frac{1}{2}$. Usually a parameter Δ is used: $\Delta \equiv (\Delta_x^2 + \Delta_y^2)^{1/2}$, and the condition of resonance is

$$hv = \sqrt{(g_z \mu_B H_z + Am)^2 + (\Delta_x^2 + \Delta_y^2)}. \tag{36}$$

There is expected a distribution in the values of Δ_x, Δ_y, which in turn determines the line shape. This is due to random variations of the crystal-field parameters of lower symmetry. These transitions are sometimes quite intense.

Excepting the S-state ions, the rare-earth ions require a temperature of 20 K, or lower, for observation of their EPR spectra. The nuclear data of the rare-earth ions are given in Table 4(a).

The details of EPR spectra for the various $4f^n$ ions are described as follows:

$4f^1$ ($^2F_{5/2}$, Ce^{3+}). For a tetragonal symmetry, the lowest doublet is described as $(a|\pm\frac{5}{2}\rangle + b|\mp\frac{3}{2}\rangle)$ and the g-values are mostly close to $g_\parallel \approx 3.0$ and $g_\perp \approx 1.5$. Sometimes the states $|\pm\frac{1}{2}\rangle$ are also admixed into the lowest doublet by the crystal field.

In threefold symmetry, the EPR is either observed from the doublet $|\pm\frac{5}{2}\rangle$ for which $g_\parallel \approx 4.0$ and $g_\perp \approx 0$, or form the doublet $|\pm\frac{1}{2}\rangle$, for which $g_\parallel \approx 0.95$ and $g_\perp \approx 2.185$. For a trigonal symmetry $g_\parallel \approx 2.4$ and $g_\perp \approx 0.1$.

In cubic symmetry, the ground state is fourfold degenerate (Γ_8), of which the levels are unequally spaced in a magnetic field, for $\mathbf{H}$ parallel to a cubic axis the energies are

$\pm\frac{1}{2}g_L\mu_B H$ and $\pm\frac{11}{6}g_L\mu_B H$, where g_L is the Lande g-factor ($=\frac{6}{7}$ for Ce^{3+}) and H is the intensity of the magnetic field.

In an axial field, there is a splitting of the $J=\frac{5}{2}$ state into three Kramers doublets. The corresponding g-factors are: $g_\parallel=\frac{6}{7}$, $g_\perp=\frac{18}{7}$; $g_\parallel=\frac{18}{7}$, $g_\perp=0$; and $g_\parallel=\frac{30}{7}$, $g_\perp=0$ for the doublets $J_z=\pm\frac{1}{2}$, $\pm\frac{3}{2}$, and $\pm\frac{5}{2}$, respectively.

In the case of ethyl sulfates, there are two levels, the lower one characterized by $g_\parallel=0.955$, $g_\perp=2.185$ corresponding to the doublet $J_z=\pm\frac{1}{2}$, while the upper one is characterized by $g_\parallel=3.72$, $g_\perp=0.2$ corresponding to the doublet $J_z=\pm\frac{5}{2}$. For a better description, the effect of the crystal-field term A_6^6 should be taken into account. For double nitrates, the crystal-field potential terms A_4^3 and A_6^3 produce considerable admixtures of $\Delta J_z=\pm 3$. Thus, $J_z=\pm\frac{5}{2}$ level becomes $N[\cos\theta|\frac{5}{2},\frac{7}{2}\rangle+\sin\theta|\frac{5}{2},-\frac{5}{2}\rangle]$ $+a|\frac{7}{2},\frac{7}{2}\rangle+b|\frac{7}{2},-\frac{5}{2}\rangle+c|\frac{7}{2},\frac{1}{2}\rangle$. For certain trichlorides, the lowest doublet is characterized by $J_z=\pm\frac{5}{2}$, for which $g_\parallel=\frac{30}{7}$ and $g_\perp=0$. The best fits are obtained for the ground state doublet $a|\pm 2,\pm\rangle-b|3,\mp\rangle+c|\mp 3,\mp\rangle$, with $a=0.373$, $b=0.925$, $c=0.071$.[1]

$4f^2$ ($^3H_4$, Pr^{3+}). The experimental data suggest that the lowest levels are described by $\cos\theta|4,\pm 2\rangle-\sin\theta|4,\mp 4\rangle$ for C_{3h} site symmetry (ethyl sulfates) with the g-values given by $\frac{4}{5}(4\cos^2\theta-8\sin^2\theta)$; and by $a|4,\pm 2\rangle+b|4,\mp 4\rangle+c|4,\mp 1\rangle$ for C_{3v} site symmetry (for double nitrates and trichlorides). These expressions do not take into account the admixture from the $J=5$ manifold.

$4f^3$ ($^4I_{9/2}$, Nd^{3+}). In a tetragonal symmetry the ground-state doublet of this Kramers ion is described by $(a|\pm\frac{9}{2}\rangle+b|\pm\frac{1}{2}\rangle+c|\mp\frac{7}{2}\rangle)$ ($J=\frac{9}{2}$ manifold lies the lowest), while in a threefold symmetry the ground state is $(a|\pm\frac{7}{2}\rangle+b|\mp\frac{5}{2}\rangle)$. When the symmetry is cubic, the lowest level is fourfold degenerate (Γ_8), the same as that for a $4f^1$ ion.

$4f^4$ ($^5I_4$, Pm^{3+}). In a threefold symmetry the ground-state doublet is described by $(a|J=\frac{5}{2},J_z=\pm\frac{1}{2}\rangle+b|J=\frac{7}{2},J_z=\pm\frac{1}{2}\rangle)$, the g-values being nearly isotropic. The third-order crystal-field terms mix the state $|\pm\frac{1}{2}\rangle$ with $|\pm\frac{5}{2}\rangle$, rendering the $\tilde{g}$ tensor anisotropic.

$4f^5$ ($^6H_{5/2}$, Sm^{3+}). The experimental results are described by the doublet $\cos\theta|\frac{5}{2},\pm\frac{1}{2}\rangle\pm\sin\theta|\frac{7}{2},\pm\frac{1}{2}\rangle$ for ethyl sulfates, characterized by predominantly axial symmetry, with $\theta\sim 4°$, since the ratio $Bg_\parallel/Ag_\perp$ is ~ 4 revealing considerable admixture of the various J levels. On the other hand, in the case of double nitrates, characterized by a predominant threefold symmetry, the doublet is described by $N[\cos\theta|\frac{5}{2},\pm\frac{1}{2}\rangle\pm\sin\theta|\frac{5}{2},\pm\frac{1}{2}\rangle]+a|\frac{7}{2},\pm\frac{1}{2}\rangle+b|\frac{7}{2},\mp\frac{5}{2}\rangle+c|\frac{7}{2},\pm\frac{7}{2}\rangle$, where N is the normalization constant.

$4f^7$ ($^8S_{7/2}$; Eu^{2+}, Gd^{3+}). These S-state ions are described in detail in Section G.

$4f^8$ ($^7F_6$, Tb^{3+}). In an axial symmetry a non-Kramers doublet lies lowest, described predominantly by the states $|J_z=\pm 6\rangle$ with an admixture of the $|0\rangle$ state: $a|\pm 6\rangle+b|0\rangle+c|\mp 6\rangle$ for which the g-values are $g_\parallel=18$, $g_\perp\approx 0$. The doublet is split somewhat due to the A_6^6 crystal-field term, which mixes the $J_z=\pm 6$ and $J_z=0$ states.

$4f^9$ ($^6H_{15/2}$, Dy^{3+}). For this Kramers ion, in a cubic symmetry, a Γ_8 quartet lies lowest, responsible for EPR. In addition, the first excited doublet also gives an isotropic EPR line. In an octahedral coordination (cubic symmetry) a doublet level lies lowest, for which the g-value is expected to be 6.60. However, this value is modified due to covalency effects, which are, however, small for rare-earth ions due to the screening of the $4f$ electrons by the outer electrons. In the double nitrate $g_{\|} = 4.281$, $g_{\perp} = 8.923$, described by the doublet $a|\pm\frac{13}{2}\rangle + b|\pm\frac{7}{2}\rangle) + c|\pm\frac{1}{2}\rangle + d|\mp\frac{5}{2}\rangle + e|\mp\frac{11}{2}\rangle$. The admixture from the $J = \frac{13}{2}$ manifold is expected to be small since $Ag_{\perp}/Bg_{\|} \sim 1$.

$4f^{10}$ ($^5I_8$, Ho^{3+}). For this ion in a tetragonal (fourfold), or a threefold symmetry, $g_{\|}$ lies in the range 13.5–16.0 and $g_{\perp} = 0$. The hyperfine splitting is very large, of the order of an X-band microwave quantum ($\sim$9.5 GHz).

The complexity of the Ho^{3+} EPR spectrum originates from two factors:

(i) The ground state is the doublet $(a|\pm 7\rangle + b|\pm 1\rangle) + c|\mp 5\rangle)$ situated close to a singlet $(a|+6\rangle + b|-6\rangle) + c|0\rangle)$. Thus, transitions between a doublet level to the singlet level can be observed, in addition to the transitions between the doublet levels. This results, to some extent, in a spectrum explained by the spin Hamiltonian characteristic of axial symmetry that splits the singlet away from the doublet for an effective spin $S = 1$:

$$\begin{aligned}\mathscr{H} &= \mu_B[g_{\|}H_zS_z + g_{\perp}(H_xS_x + H_yS_y)] + b_2^0[S_z^2 - S(S+1)/3] \\ &\quad + AS_zI_z + B(S_xI_x + S_yI_y) + Q'[I_z^2 - I(I+1)/3] \\ &\quad - R_{\|}I_zH_z - R_{\perp}(I_xH_x + I_yH_y) - g_N\mu_N(I_xH_x + I_yH_y + I_zH_z). \qquad (37)\end{aligned}$$

(ii) Due to large hyperfine structure, additional complexity arises from the appearance of hyperfine transitions $\Delta m = \pm 1$, where m is the nuclear magnetic quantum number.

The salient features of Ho^{3+} EPR spectra are as follows:

(a) When the microwave magnetic field, $\mathbf{H}_{rf}$, is parallel to the z-axis, $\Delta M = 2$ (forbidden) transitions, where M is the electronic magnetic quantum number, are observed, associated with eight hyperfine lines each. The line positions are described by[1]

$$h\nu = 2g_{\|}\mu_B H + (2A - B^2/D)m - (B^2 h\nu/2D^2)[I(I+1) - m^2]. \qquad (38)$$

(b) When $\mathbf{H}\|$ z-axis and $\mathbf{H}_{\text{rf}}\perp$ z-axis, seven hyperfine lines, each placed midway between pairs described in (a) above, are observed. This corresponds to forbidden hyperfine transitions $\Delta m = \pm 1$. The line positions are described by[1]

$$h\nu = 2g_{\|}\mu_B H + 2(A - Q')(m + \tfrac{1}{2}) - [B^2 h\nu/2D^2][I(I+1) + \tfrac{1}{4} - (m + \tfrac{1}{2})^2]. \quad (39)$$

(c) When $\mathbf{H}\perp$ and $\mathbf{H}_{\text{rf}}\|$ to z-axis, six unequally spaced lines are observed, corresponding to transition between $|\pm\rangle$ levels. These are further split as a second-order effect when $\mathbf{H}\perp$ z-axis. The hyperfine structure is rather large, of the same order as the energy difference between levels.

(d) When $\mathbf{H}\|$ z-axis, a number of lines are observed at high frequencies and high magnetic fields. The hyperfine line separation becomes lesser at higher magnetic fields. There exists a second group of lines whose separation decreases toward the low field. These transitions correspond to those between $|1\rangle$ and $|0\rangle$ levels. Of the two sets, one appears before these levels cross. The transitions are described by[1]

$$hv = \pm b_2^0 - g_{\parallel}\mu_B H - Am[1 + B^2/2(hv)^2 \pm mH^2/(\pm 4D - 2hv)$$
$$+ B^2[I(I+1) - 3m^2][1/hv + (\pm 4D - 2hv)^{-1}]. \quad (40)$$

The + and − signs are used depending on the sign of b_2^0 being negative or positive, respectively.

4f^{11} ($^4I_{15/2}$, Er^{3+}, Ho^{2+}). In a tetragonal site symmetry the ground-state doublet is described by $|J_z = \pm\frac{1}{2}\rangle$, while in a trigonal (threefold) symmetry the ground-state doublet is of the form $(a|\pm\frac{7}{2}\rangle + b|\mp\frac{5}{2}\rangle)$. In a cubic symmetry, either a Γ_7 doublet lies lowest, described by $(a|\pm\frac{13}{2}\rangle + b|\pm\frac{5}{2}\rangle + c|\mp\frac{3}{2}\rangle + d|\mp\frac{11}{2}\rangle)$, corresponding to an isotropic g-value of 6.772, or a Γ_6 doublet lies lowest described by $(a|\pm\frac{15}{2}\rangle + b|\pm\frac{7}{2}\rangle + c|\pm\frac{1}{2}\rangle + d|\pm\frac{9}{2}\rangle)$, with the g-value ≈ 5.9. In an octahedral coordination (cubic field) a Γ_8 quartet lies lowest.

4f^{13} ($^2F_{7/2}$; Yb^{3+}, Tm^{2+}). This configuration corresponds to a single hole in the 4f shell. In an eightfold cubic coordination a Γ_7 doublet lies the lowest, described by $(\sqrt{3/2}|\pm\frac{5}{2}\rangle - \frac{1}{2}|\pm\frac{3}{2}\rangle)$, for which the expected g-value is 3.429. In a tetragonal field, a Γ_7 doublet is expected to lie lowest. In an octahedral symmetry a Γ_6 doublet lies lowest, with the theoretical g-value of 2.667. Covalency effects may modify this value. In a tetragonal site symmetry, the ground-state doublet is probably $(a|\pm\frac{5}{2}\rangle + b|\pm\frac{3}{2}\rangle)$.

E. ACTINIDE (5f^n) IONS

For these ions the unpaired electrons are in the 5f shell, behaving like a second rare-earth group. The electronic properties of these ions are influenced by actinide contraction and by relativistic effects. The contraction causes the energy separation of the 5f, 6d, and 7s electrons to be small, thus a strong competition of these electrons takes place in determining the electronic properties of the actinides. Because of the relativistic effects the Coulomb interelectronic repulsion descreases and the spin–orbit coupling is increased. This leads to a severe breakdown of the Russell–Saunders coupling. The crystal-field effects are stronger for the actinide ions as compared with those of the lanthanide ions, because of larger ionic radii of the actinide ions than those of the lanthanide ions. EPR results indicate that the 5f wave function, unlike the 4f wave function, extends on the surrounding diamagnetic neighbors.

The EPR data on actinide ions are meager. This is because it is difficult to grow single crystals with elements that are strongly radioactive, since the intense radioactivity causes destruction of the crystal in the vicinity of the paramagnetic ion, which gives rise to EPR of free radicals, obscuring the EPR spectrum of the paramagnetic ion.

Table 5 gives the valencies of the actinide group.

A review article including the EPR results on some uranium compounds has been published by Ursu and Lupei.[3]

5f^1. This configuration corresponds to the ions Np^{3+} and U^{5+}.

TABLE 5. Valences of the actinide group.[1]

No. of *f* electrons	Ac	Th	Pa	U	Np	Pu	Am	Cm	Bk	Cf
0	Ac^{3+}	Th^{4+}	Pa^{5+}	U^{6+}						
				$\underline{UO_2}^{2+}$						
1		Th^{3+}?	Pa^{4+}	U^{5+}	Np^{6+}					
				UO_2^{+}	$\underline{NpO_2}^{2+}$					
2		Th^{2+}	Pa^{3+}	$\underline{U}^{4+}$	Np^{5+}	Pu^{6+}				
					NpO_2^{+}	$\underline{PuO_2}^{2+}$				
3				$\underline{U}^{3+}$	$\underline{Np}^{4+}$	Pu^{5+}	Am^{6+}			
						PuO_2^{+}	AmO_2^{2+}			
4				U^{2+}	Np^{3+}	Pu^{4+}	Am^{5+}			
							AmO_2^{+}			
5						Pu^{3+}	Am^{4+}			
6							Am^{3+}			
7							Am^{2+}	$\underline{Cm}^{3+}$	Bk^{4+}	Cf^{5+}
8									Bk^{3+}	Cf^{4+}
9										Cf^{3+}

Ions with question marks denote doubtful valencies.
Ions which are underlined are those for which paramagnetic resonance has been reported.

Neptunyl ion. This ion has one magnetic electron. The states $m = M = \pm 3, \pm 2, \pm 1, 0$, and $M = m = \frac{1}{2}$ are possible (m refers to the individual electrons). The lowest energy, in the presence of an axial field, is for $M = \pm\frac{3}{2}$, followed next by $M = \pm 2$ which lies higher by about 10^{-4} cm^{-1}. The spin–orbit coupling splits the fourfold degeneracy into two Kramers doublets with $M_j = M + m = \pm\frac{7}{2}$ and $\pm\frac{5}{2}$. The resulting lowest state is a doublet, for which we can employ a spin Hamiltonian of the form

$$\begin{aligned}\mathscr{H} &= \mu_B[g_\parallel H_z S_z + g_\perp(H_x S_x + H_y S_y)] + b_2^0[S_z^2 - S(S+1)/3] \\ &\quad + AS_z I_z + B(S_x I_x + S_y I_y) + Q'[I_z^2 - I(I+1)/3] \\ &\quad - R_\parallel I_z H_z - R_\perp(I_x H_x + I_y H_y) - g_N \mu_N (I_x H_x + I_y H_y + I_z H_z),\end{aligned}$$

with $S' = \frac{1}{2}$. The g-values of the lowest-lying doublet are given by the matrix elements

$$g_\parallel = 2\langle a|kL_z + 2S_z|a\rangle, \qquad g_\perp = 2\langle a|kL_z + 2S_x|b\rangle, \tag{41}$$

where $|a\rangle$ $(= |3, -\frac{1}{2}\rangle)$ and $|b\rangle$ $(= |-3, \frac{1}{2}\rangle)$ are the basic states describing the doublet. Thus, $g_\parallel = 6k - 2$. In the presence of a threefold symmetry the g-factors given by (E. 1) are modified

$$\begin{aligned} g_\parallel &= \frac{k(6 - 6p^2 + 4q^2) - (2 + 2p^2 - 2q^2)}{1 + p^2 + q^2}, \\ g_\perp &= \frac{4p - 2\sqrt{6}kpq}{1 + p^2 + q^2}. \end{aligned} \tag{42}$$

In a first approximation, assuming $k = 1$ (no covalency), it is found that $p \approx 0.1$ and $q \geq 0.1$.

U^{5+} ions. The first excited state of this ion ($6d^1$) lies at about 10,000 cm^{-1} above the ground state ($5f^1$). The spin–orbit coupling splits the ground configuration into two levels, $^2F_{5/2}$ and $^2F_{7/2}$, the former being the lowest. The interaction of the $5f$ electron with the crystal field depends strongly on the coordination of the ions in the crystal. Two major cases have been found for cubic symmetry:

(a) For an eightfold, or tetrahetral, coordination, the crystal-field interaction is weaker than the spin–orbit coupling. In a cubic crystal field, the following splitting of levels takes place:

$$\begin{aligned} ^2F_{7/2} &\rightarrow \Gamma_6(\text{doublet}) + \Gamma_7(\text{doublet}) + \Gamma_8(\text{quartet}), \\ ^2F_{5/2} &\rightarrow \Gamma_6(\text{doublet}) + \Gamma_8(\text{quartet}), \end{aligned} \tag{43}$$

the crystal-field components belonging to each electronic level are closely bunched together, which enhances relaxation. Thus, EPR can be detected only at very low temperatures. The ground crystal-field level can be Γ_6, or Γ_8, depending on the strength of interaction, resembling the situation for Ce^{3+}.

(b) In an octahedral coordination, the crystal-field interaction is stronger than the spin–orbit coupling, it affects only the ^2F term of the $5f^1$ configuration

$$^2F(5f^1) \rightarrow \Gamma_2(\text{singlet}) + \Gamma_4(\text{triplet}) + \Gamma_5(\text{triplet}). \tag{44}$$

Additional splitting is caused by the spin–orbit coupling; in addition, lower symmetry causes further splitting. The ground level of the crystal field is usually Γ_7, originating from the Γ_2 singlet, containing admixtures from all the Γ_7 states. Since the ground level is sufficiently far apart from the excited levels, the relaxation processes are weak and EPR can only be observed at liquid nitrogen, or slightly higher, temperatures.

UO_2^+ constitutes another class of U^{5+} complexes; here the uranium ion is strongly bonded to two oxygens. The electronic structure of this complex is similar to that of the neptunyl complex NpO_2^+. Here, EPR can, in principle, be observed only at low temperatures.

$5f^2$. This configuration corresponds to plutonyl $(PuO_2)^{2+}$ and U^{4+} ions.

Plutonyl ions. The lowest states of $2f$-electron configuration, consisting of two electrons with parallel spins ($M_s = \pm 1$) and in the orbits of $m_L = \pm 3$, $m_L = \pm 2$, respectively, are given by $M_L = \pm 5$, $M_s = \pm 1$. Here M_L and M_s are the magnetic quantum numbers corresponding to the total orbital and spin-angular momenta, while m_L, m_s correspond to the individual orbital and spin-angular momenta. Spin–orbit coupling splits these two states, and the ground state is $^3H_4$. The principal g-factors are

$$
\begin{aligned}
g_{\parallel} &= 2(kM_L + 2M_s) = 10k - 4, \\
g_{\perp} &= 0.
\end{aligned}
\tag{45}
$$

The off-diagonal matrix elements of the nonaxial crystal-field terms and the orbital reduction factor k cause a reduction in the value of the g-factor.

The non-Kramers ground-state doublet does not exhibit the usual microwave absorption, because $g_{\perp} = 0$. However, if the microwave and Zeeman fields are parallel to each other, and applied along the symmetry axis, we could observe EPR.

Tetravalent uranium (U^{4+}) ion. For this ion, the ground level is $^3H_4$. The spin–orbit coupling is stronger than the crystal field, regardless of the symmetry and coordination. In a cubic crystal field the ground level splits as follows:

$$
^2H_4 \to \Gamma_1(\text{singlet}) + \Gamma_3(\text{doublet}) + \Gamma_4(\text{triplet}) + \Gamma_5(\text{triplet}). \tag{46}
$$

In an octahedral symmetry the ground state is always a Γ_1 singlet. The Γ_4 level lies more than 1000 cm^{-1} above this; thus, the Γ_1 level gets populated only at low temperatures. No EPR is expected for this case.

In cases of eightfold and tetrahedral (cubic) symmetry, the ground level can be a Γ_1 singlet, or a Γ_5 triplet, depending on the ratio of the fourth- to sixth-order terms in the crystal-field potential; in lower symmetries the ground level can be either a singlet, or a non-Kramers doublet. The crystal-field levels are closely bunched, the relaxation processes are strong, and EPR can be observed only at very low temperatures.

$5f^3$ ($^4I_{9/2}$; U^{3+}, Np^{4+}).

U^{3+} ions. The ground level for this ion is $^4I_{9/2}$. The spin–orbit coupling is stronger than the crystal-field interaction, regardless of the symmetry and coordination. A cubic field causes the splitting

$$
^4I_{9/2} \to \Gamma_6(\text{doublet}) + 2\Gamma_8(\text{quartet}). \tag{47}
$$

A distortion of an axial, or a lower, symmetry splits the quartets in doublets. EPR can always be observed, except for those situations where the ground state is a nonmagnetic doublet with $g_\perp = 0$, e.g., a weak trigonal distortion.

The hyperfine-structure of U^{235} and U^{233} shows the $\Delta m = \pm 1$ and $\Delta m = \pm 2$ transitions. Hyperfine structure of U^{3+} has been used mainly to determine the nuclear moments of odd uranium isotopes.

The EPR spectra of U^{3+} and Np^{3+} are very similar. (Np^{3+} has the configuration $4f^3$, with the same ground state, $^4I_{9/2}$.) Under a C_{3h} symmetry, the ground state is

$$\cos\theta|J_z = \pm\tfrac{7}{2}\rangle + \sin\theta|J_z = \mp\tfrac{5}{2}\rangle. \tag{48}$$

When there is axial symmetry around one of the cubic axes, the ground-state doublet is represented by

$$a|J_z = \pm\tfrac{9}{2}\rangle + b|J_z = \pm\tfrac{1}{2}\rangle + c|J_z = \mp\tfrac{7}{2}\rangle, \tag{49}$$

where c is expected to be small.

$5f^4$ Divalent Uranium Ion (U^{2+}). The ground-state configuration of divalent uranium has not been uniquely determined. The three possible configurations that have been proposed are $5f^4$, $5f^3 6d$, or $5f^3 7s$; these configurations have either the $^5I_4$, or $^5L_6$, as the lowest level. The ground level of $^5I_4$ is a singlet and that of $^5L_6$ is either a Γ_1 singlet, or a Γ_3 nonmagnetic doublet, in an eightfold cubic crystal field. EPR of these configurations cannot, thus, be recorded.

$5f^5$ ($^6H_{5/2}$; Pu^{3+}). In a field of threefold symmetry, EPR is observed from a single Kramers doublet $|J_z = \pm\tfrac{1}{2}\rangle$, similar to the spectrum of Sm^{3+} ($4f^5$ configuration). A doublet hyperfine structure from Pu^{239} can also be observed.

$5f^7$ ($^8S_{7/2}$). Cm^{3+} corresponds to an S-state configuration which is discussed in the next section.

F. S-STATE ($3d^5, 4d^5, 5d^5, 4f^7$) IONS

Introduction. The S-state transition ions have the configurations $3d^5$ (Ti^{-1}, V^0, Cr^+, Mn^{2+}, Fe^{3+}, Co^{4+}), $4d^5$, $5d^5$, $4f^7$ (Gd^{3+}, Eu^{2+}), and $5f^7$ (Cm^{3+}). For d^5 ions the ground state is $^6S_{5/2}$, while for f^7 ions it is $^8S_{7/2}$. When these ions are placed in a medium, or weak, crystal field, the ground state will remain an S state. Neither the crystal field, nor the spin–orbit coupling, can remove the sixfold or eightfold degeneracies of the d^5 and f^7 ions, respectively. Even the cubic field cannot remove these degeneracies according to group theory. Higher-order perturbations, including simultaneously the crystal-field and spin–orbit coupling, are necessary to split the S-state. Being of higher order, these perturbation produce a small splitting of the ground state. The spectra of S-state ions can be observed with ease even at room temperature; for this reason considerable EPR data have been reported on these ions. Nevertheless, the mechanisms responsible for the splitting of S-state ions have not yet been completely understood.

TABLE 6.[19] Nonvanishing zero-field operator coefficients B_k^q in the spin Hamiltonian for the various symmetries. Also included is the relationship between the principal values of the hyperfine tensor, describing the interaction $S \cdot \tilde{A} \cdot I = A_x S_x I_x + A_y S_y I_y + A_z S_z I_z$. For monoclinic symmetry C_2 represents the orientation of the twofold axis, while x, y, z represent the magnetic axes.

Cubic:	B_4^0, $B_4^4 = 5B_4^0$, $B_6^4 = -21B_6^0$; $A_z = A_x = A_y$.
Axial:	B_2^0, B_4^0, B_6^0; A_z, $A_x = A_y$.
Hexagonal:	B_2^0, B_4^0, B_6^0, B_6^6; A_z, $A_x = A_y$.
Tetragonal:	B_2^0, B_4^0, B_4^4, B_6^0, B_6^4; A_z, $A_x = A_y$.
Trigonal:	B_2^0, B_4^0, B_4^3, B_6^0, B_6^3, B_6^6; A_z, $A_x = A_y$.
Orthohombic:	B_k^m; $k = 2, 4, 6$; $0 \leq m\,(\text{even}) \leq k$; A_z, A_y, A_x.
Monoclinic	Orthorhombic + B_1 where: $B_1(C_2 \| x) = B_k^{-(2n-1)}$, $B_1(C_2 \| y) = B_k^{(2n-1)}$, $B_1(C_2 \| z) = B_k^{-2n}$ and $k = 2, 4, 6$; $1 \leq n \leq k/2$; A_z, A_y, A_x
Triclinic	B_k^q; $k = 2, 4, 6$; $-k \leq q \leq k$; A_z, A_y, A_x.

Spin Hamiltonian. The spin Hamiltonian for the S-state ions, $S = \frac{5}{2}(d^5)$ and $\frac{7}{2}(f^7)$, can, for an arbitrary symmetry, be expressed as

$$\mathscr{H} = \mu_B \mathbf{S} \cdot \tilde{\mathbf{g}} \cdot \mathbf{H} + \sum_m \mathrm{B}_2^m \mathrm{O}_2^m + \sum_m \mathrm{B}_4^m \mathrm{O}_4^m + \sum_m \mathrm{B}_6^m \mathrm{O}_6^m, \tag{50}$$

where the last term, involving sixth-order spin operators, is to be included only for f^7 ions. In (50) O_n^m ($|m| \leq n$) are the spin operators as given by Abragam and Bleaney.[9] Not all B_n^m are nonzero for different symmetries. Table 6[19] gives a list of the nonzero B_n^m's for the various symmetries.

Sophisticated numerical techniques are now available for the determination of $\tilde{g}$ and B_ℓ^m from a knowledge of EPR line positions. They have been well discussed by Misra.[20] Here the EPR line positions, for several orientations of the Zeeman field, are simultaneously fitted in a least-squares manner, utilizing numerical diagonalization of the spin-Hamiltonian matrix on a digital computer. For more details see Chapter IX.

Theoretical Considerations. The experimental results on the S-state ions reveal the following:

(i) There are large variations of the ground-state splittings for any ion in different host lattices.
(ii) The magnitude of the axial-splitting parameter (B_2^0) generally varies considerably with temperature, increasing with temperature.
(iii) The g-factors deviate only slightly from the free-electron value ($g_e = 2.0023$). This deviation is always negative for the rare-earth ions (Gd^{3+}, Eu^{2+}), but can be positive for the iron-group ions (Mn^{2+}, Fe^{3+}).

(iv) A large anomalous hyperfine structure is found for Mn^{2+}, and smaller structure for Eu^{2+} and Gd^{3+}, the hyperfine-structure constant showing a regular variation with covalency.

The nature of the splitting of the S-state ions, which can explain the above observations, is not yet completely understood. A few possible mechanisms have been pointed out, which involve higher-order perturbations due to, e.g., spin–orbit, crystal field interactions, either within the d^5, or f^7, configuration or involving excited states of higher-lying configurations.

An overlap, or superposition, model has been applied for the *ab initio* calculation of the spin-Hamiltonian parameters (SHP), originally proposed by Newman.[21,22] It has been extensively applied to the estimation of SHP of Gd^{3+} ion.[5] The point-charge and induced-dipole models have also been successfully used for estimating SHP of the Gd^{3+} ion in some host lattices.[5] However, a clear-cut microscopic model is not yet available for the estimation of SHP.

As for the anomalous hyperfine structure, exhibited by the S-state ions, extensive hyperfine structures have been observed for all S-state ions, although a half-filled shell, should not show any hyperfine structure. The origin of the hyperfine structure has been proposed to lie in the configurational interaction of the form $3s3d^54s$, with $3s^23d^5$, for iron-group ions, and similarly that of $4s4f^55s$, with $4s^24f^7$, for the rare-earth ions. Only a small amount of admixture is necessary to explain the splitting, since unpaired s electrons contribute considerably to the hyperfine-structure splitting due to the Fermi–contact interaction. The contribution of the unpaired s-electron states can be expressed as $K\boldsymbol{S}\cdot\boldsymbol{I}$, where K is positive for iron-group ions. The sign of K has not been determined for the rare-earth group. The hyperfine-structure splitting varies from crystal to crystal, being larger the more ionic the crystal. The calculation of the hyperfine splitting from first principles is not easy.

$3d^5$ (Cr^+, Fe^{3+}, Mn^{2+}, Co^{4+}). For this half-filled 3d shell, the free ion ground state is $^6S_{5/2}$. There are two main cases to be considered: weak field ($S = \frac{5}{2}$) and strong field ($S = \frac{1}{2}$). In addition, there is an intermediate spin case $S = \frac{3}{2}$. They are discussed as follows.

(i) **Weak-field case.** The sixfold spin degeneracy ($S = \frac{5}{2}$) is lifted by the spin–orbit coupling and higher-order perturbations, resulting in three Kramers doublets separated by energies of ~ 0.1–$1.0\ \mathrm{cm}^{-1}$ in zero magnetic field. In general, there are $2S = 5$ allowed fine-structure ($\Delta M = \pm 1$) lines, and often forbidden lines $\Delta M = \pm 2, \pm 3$ are also observed. For Mn^{2+}, which shows a characteristic hyperfine spectrum of six components, there are thirty allowed hyperfine lines ($\Delta M = \pm 1$, $\Delta m = 0$) observed by EPR. The g-values for both Fe^{3+} and Mn^{2+} are also close to 2.0. Hyperfine splittings for Mn^{55} are usually in the range 50–$90 \times 10^{-4}\ \mathrm{cm}^{-1}$, while those for Fe^{57} (2% abundant) $\sim 10^{-3}\ \mathrm{cm}^{-1}$. The cubic-field splitting for the Cr^+ ion is $\sim 3 \times 10^{-4}\ \mathrm{cm}^{-1}$. Hyperfine structure from Cr^{53} (9.55% abundant) can also be observed ($A \sim 14 \times 10^{-4}\ \mathrm{cm}^{-1}$). As the ground state is only very loosely coupled to the crystal field the relaxation times are long. Thus, EPR is readily observed at room temperature.

For trigonal distortions, as in alums, the spin Hamiltonian has the form ($S = \frac{5}{2}$):

$$\mathscr{H} = g\mu_B(H_xS_x + H_yS_y + H_zS_z) + (b_4^4/15)(S_\xi^4 + S_\eta^4 + S_\zeta^4 - \tfrac{707}{16})$$
$$+ b_2^0(S_z^2 - \tfrac{35}{12}) + b_4^0[S_z^4 - (\tfrac{95}{14})S_z^2 + \tfrac{81}{16}] + A(S_xI_x + S_yI_y + S_zI_z). \quad (51)$$

Here the z-axis is considered to be along the [111] direction with respect to the axes (ξ, η, ζ), which is a set of mutually perpendicular axes. The last term in (51) applies to the Mn^{2+} ion for which the nuclear spin $I = \frac{5}{2}$. For the cases of the Fe^{3+} ion, the hyperfine interaction due to the Fe^{57} nucleus ($I = \frac{1}{2}$) is not observed except in compounds enriched with Fe^{57}. Further, the fourth-order terms in electron-spin operators in (51) are usually small. Thus, usually the spin Hamiltonian (50), without the fourth- and sixth-order terms, applies.

For orthorhombic symmetry, as in Tutton salts, the spin Hamiltonian (50) applies, but without the sixth-order terms $b_6^m O_6^m$. If b_2^0, b_2^2 are finite but small (0.001 to 0.1 cm^{-1}), five EPR lines are observed. When b_2^0, $b_2^2 \gg g\mu_B H$ as in the large majority of well-accounted-for EPR spectra of high spin d^5 compounds, the eigenvalues of the spin Hamiltonian in zero magnetic field corresponds to three Kramers doublets. There are two simplifying cases. If $b_2^0 \neq 0$, $b_2^2 = 0$, the lowest doublet can be characterized by the effective g-values $g_\parallel = 2$, $g_\perp = 6$. When $b_2^0 = 0$, $b_2^2 \neq 0$, the middle doublet exhibits an isotropic value $g = 4.29$.

For a general solution, the characteristic secular determinant of the matrix of the spin Hamiltonian

$$\mathscr{H} = \mu_B(g_xH_xS_x + g_yH_yS_y + g_zH_zS_z) + b_2^0(S_z^2 - \tfrac{35}{12}) + b_2^2(S_x^2 - S_y^2), \quad (52)$$

within the manifold of $S = \frac{5}{2}$, may be solved numerically. The g-tensor may be taken equal to 2.0, and we can calculate the eigenvalues by using b_2^0 and $\lambda = b_2^2/b_2^0$. (There is axial symmetry for $\lambda = 0$, while for $\lambda = \frac{1}{3}$ we have the maximum rhombic symmetry. Values of $\lambda > \frac{1}{3}$ represent a convergence toward axial symmetry.) Dowsing and Gibson[4,23] prepared a graphical display of computed EPR transitions as functions of b_2^0 and λ (in the range $0 \leq \lambda \leq \frac{1}{3}$) exhibiting the magnetic fields at X-band frequencies.

(ii) **Strong-field case.** When the crystal field is strong enough, e.g., Mn^{2+} in $K_4Fe(CN)_6$, Fe^{3+} in $K_3Co(CN)_6$, Co^{4+} in Al_2O_3, the coupling between the d electrons is broken down, resulting in a low-spin ground state. All five electrons occupy t orbitals; the system may be considered as a single hole in an orbital-triplet level. There is an isolated Kramers doublet which lies lowest in a field of low symmetry, and corresponds to an effective spin $S' = \frac{1}{2}$ with a strongly anistropic g-factor ($\sim$0.7–2.5). This situation is prevalent in the ions Ti^-, Cr^+, Mn^{2+}, and Fe^{3+} in the low-spin state. In a strong ligand field characterized by octahedral symmetry, the ground state is $^2T_2$. The following spin Hamiltonian applies ($S = \frac{1}{2}$):

$$\mathscr{H} = \mu_B(g_xH_xS_x + g_yH_yS_y + g_zH_zS_z) + A_xS_xI_x + A_yS_yI_y + A_zS_zI_z. \quad (53)$$

The rapid decrease of spin-lattice relaxation times of these ions with increasing temperature results in line broadening at temperatures >20 K. Thus, low temperatures are required to observe EPR.

(iii) **Intermediate spin case** ($S = \frac{3}{2}$). The spin Hamiltonian applicable in this case ($S = \frac{3}{2}$) is

$$\mathscr{H} = g\mu_B \mathbf{H} \cdot \mathbf{S} + b_2^0(S_z^2 - \tfrac{5}{4}) + b_2^2(S_x^2 - S_y^2). \tag{54}$$

The typical example is that of bis(diisopropyldithiocarbamato) Fe^{3+} chloride[4] for which the value $g = 2.00$ was assumed.

4d^5 (Mo^+, Ru^{3+}), ***5d***^{5} (W^+, Re^{2+}, Os^{3+}, Ir^{4+}). In a strong octahedral field this configuration may be considered as a single hole in a t_2 orbital. The theory is then similar to that for a d^1 ion, but with a negative spin–orbit coupling. The g-values are

$$\begin{aligned} g_{\parallel} &= -1 + \frac{3(\delta - \lambda/2)}{W}, \\ g_{\perp} &= 1 + \frac{(\delta + 3\lambda/2)}{W}, \end{aligned} \tag{55}$$

where $W = [(\delta - \lambda/2)^2 + 2\lambda^2]^{1/2}$. For an accurate interpretation, covalent effects should be taken into account. Ru^{3+} shows hyperfine structure from Ru^{99} and Ru^{101} with $A \sim 50 \times 10^{-4}$ cm^{-1}, whereas Ir^{4+} shows a four-line structure from Ir^{191} and Ir^{193} with $A \sim 25 \times 10^{-4}$ cm^{-1}. Temperature of 4 K, or less, are required for EPR to be observed.

4f^7 (Eu^{2+}, Gd^{3+}, Tb^{4+}). The ground state of this half-filled shell is $^8S_{7/2}$, which is retained in crystals, because the crystal field is weak for rare-earth ions. There are four Kramers doublets which are split in zero magnetic field (even in a cubic field) by energies of order 0.1 cm^{-1}. The effective spin $S = \frac{7}{2}$; thus there are observed seven allowed fine-structure transitions ($\Delta M = \pm 1$). The g-values are typically ~ 1.99. Both Gd^{3+} and Eu^{2+} show hyperfine structure, but the splitting parameters for Gd^{155} and Gd^{157} are usually rather small ($\sim 5 \times 10^{-4}$ cm^{-1}), and the structure is not always resolved. Typical values for Eu^{2+} are $A^{151} \sim 30 \times 10^{-4}$ cm^{-1}, $A^{153} \sim 15 \times 10^{-4}$ cm^{-1}. A clear-cut linear dependence of the values of the spin-Hamiltonian parameters of Gd^{3+} on the rare-earth host-ion radius has been found in a series of iso-structural hosts.[5]

The results of EPR studies on Gd^{3+} have been summarized in a review article by Misra and Upreti,[5] covering the literature from 1977 to 1983.

The spin-lattice relaxation times are usually long for $4f^7$ ions, similar to that for $3d^5$ ions, and EPR may be observed at room temperature. Tb^{4+} is an exception, temperatures less than 20 K being necessary for the observation of EPR.

5f^7 (Cm^{3+}). Earlier EPR measurements suggest that the splitting of the four spin doublets in zero magnetic field is small, similar to that for Gd^{3+} and Eu^{2+} ($4f^7$ configuration). However, recent measurements are in disagreement, indicating large differences in $g_{\parallel}$ and $g_{\perp}$; this has been interpreted as arising from the $\pm\frac{1}{2}$ doublet, which lies lower than the next doublet by an energy much greater than the measuring quantum. If the host crystal fluoresces strongly, e.g., $LaCl_3$, the crystal symmetry is destroyed within a few hours after crystal growth by the intense α radiation of Cm.

REFERENCES

[1] W. Low, *Paramagnetic Resonance in Solids* (Academic Press, New York, 1960).
[2] J. W. Orton, *Electron Paramagnetic Resonance* (Iliffe Books, London, 1968).
[3] I. Ursu and V. Lupei, Magn. Reson. Rev. **10**, 253 (1986).
[4] E. König and G. König, in *Electron Paramagnetic Resonance of Coordination and Organometallic Transition Compounds*, Landolt Börnstein, New Series, edited by K.-H. Hellewege and A. M. Hellewege (Springer-Verlag, Berlin, 1984), Vol. 12.
[5] S. K. Misra and G. C. Upreti, Magn. Reson. Rev. **10**, 333 (1986).
[6] R. P. Kohin, Magn. Reson. Rev. **5**, 75 (1979).
[7] P. Chand, V. K. Jain, and G. C. Upreti, Magn. Reson. Rev. **14**, 49 (1988).
[8] S. K. Misra and C. Wang, Magn. Reson. Rev. **14**, 157 (1990).
[9] A. Abragam and B. Bleaney, *Electron Paramagnetic Resonance of Transition Ions* (Clarendon Press, Oxford, 1970).
[10] I. P. Biryukov, M. G. Voronkov, and I. A. Safin, *Tables of Nuclear Quadrupole Resonance Frequencies* (IPST Press, Jerusalem, 1969).
[11] E. R. Andrew, *Nuclear Magnetic Resonance* (Cambridge University Press, Cambridge, 1969).
[12] S. K. Misra, Physica **B151**, 433 (1988).
[13] J. M. Assour, J. Goldmacher, and S. E. Harrison, J. Chem. Phys. **43**, 159 (1965).
[14] L. J. Boucher, E. C. Tynan, and T. F. Yen, in *Electron Spin Resonance of Metal Complexes*, edited by T. F. Yen (Plenum Press, New York, 1969).
[15] C. A. Bates and D. E. Chandler, J. Phys. **C4**, 2713 (1971).
[16] C. A. Bates and D. E. Chandler, J. Phys. **C6**, 1975 (1973).
[17] J. H. E. Griffiths, J. Owen, and I. M. Ward, Proc. Roy. Soc. (London) **A219**, 526 (1953).
[18] S. K. Misra, Y. Chang, and J. Felsteiner, J. Phys. Chem. Solids **58**, 1 (1997).
[19] S. Subramanian and S. K. Misra, J. Phys.: Conden. Matter **1**, 9483 (1989).
[20] S. K. Misra, Magn. Reson. Rev. **10**, 285 (1986).
[21] D. J. Newman, Adv. Phys. **24**, 197 (1971);
[22] D. J. Newman and W. Urban, Adv. Phys. **24**, 793 (1975).
[23] R. D. Dowsing and J. F. Gibson, J. Chem. Phys. **50**, 294 (1969).

CHAPTER VII
TRANSITION ION HAMILTONIANS

Sushil K. Misra

Physics Department, Concordia University, Montreal, Quebec H3G 1M8, Canada

Forms of the spin Hamiltonian for transition series ions in various site symmetries are given and their coefficients are tabulated for a range of electronic configurations.

CONTENTS

A. INTRODUCTION

The purpose of this chapter is to review the terms of the spin Hamiltonians appropriate to all the 32 point groups, and to provide the forms of the required Steven's operators, as well as their matrix elements applicable to all spin values $S \leq \frac{7}{2}$, covering almost all cases encountered in practice. The form of spin Hamiltonian for a transition ion depends on the point symmetry at its crystallographic site in the host crystal, rather than on the space group of the crystal. A number of publications have dealt with the forms of the spin Hamiltonian as well as with the various spin operators that are present in the spin Hamiltonian.[1-14] The most commonly used spin operators are known as Steven's operator equivalents,[15-22] denoted as Y_n^m ($|m| \leq n$). They are listed here in Appendix I, Section A. The matrix elements of these operators form the Hamiltonian matrix, which are also listed in Appendix I.[15,16] A systematic listing is included of the spin Hamiltonians of all 32 crystallographic site symmetries. The descent in symmetry interrelationships of the point groups are given in Appendix II. The tabulation of the transition ion site symmetries for some common host crystals is included in Appendix III. This information is important for phase transition studies. Review and evaluations of spin-Hamiltonian parameters, provided by Misra,[23-28] can be used to evaluate these parameters for the various point groups in crystals using the least-squares technique, fitting simultaneously a large number of EPR line positions recorded for several orientations of the external magnetic field we can either employ the eigenvalues and eigenvectors of the spin-Hamiltonian matrix as evaluated on a computer using numerical techniques, or use eigenvalues calculated to second order in perturbation. More details are given in Chapter IX dealing with numerical evaluation of spin-Hamiltonian parameters.

The different approaches of various authors have led to some confusion in the notation. Results are, therefore, given here in a common notation along with comparisons with other widely used ones. In addition, site symmetries at crystallographic sites are not always readily available since they are not included, for example, in Wyckoff's works,[29] but they can be found in the International Tables.[30,31]

B. SITE SYMMETRIES

The terms that can exist in the spin Hamiltonian for a transition ion metal or other paramagnetic species at a crystallographic site are determined by the point group symmetry of that site, the higher the symmetry the smaller the number of terms. Table 1 lists the point group symmetries associated with each of the eleven spin-Hamiltonian types. When the symmetry at a particular site is lowered, as by the passage through a structural phase transition, additional terms are added to the spin Hamiltonian. Of particular interest is what is called the descent of symmetry, namely the lowering of symmetry by infinitesimal movement of atoms away from the positions that they occupy in the more symmetrical arrangement. Appendix II describes descent of symmetry for distortion of:

(a) a cubic Hamiltonian along a fourfold (C_4 or S_4) axis;
(b) a cubic Hamiltonian along a threefold (C_3) axis;
(c) hexagonal Hamiltonian along a sixfold (C_6) axis; and
(d) a hexagonal Hamiltonian along a transverse twofold (C_2) axis.

Tetrahedral and octahedral coordinations are quite common because closely packed spheres such as oxygen ions have one octahedral and two tetrahedral interstitial sites per oxygen sphere, and many commonly studied crystals have their overall structure largely determined by the exact or approximate close-packing arrangement of large anions. Relatively small transition metal cations ordinarily occupy the interstitital sites between the larger anions. The site symmetries for transition ions in a number of commonly studied host crystals are included in Appendix III (Crystal Hosts).

C. SPIN HAMILTONIANS

The spin Hamiltonian may be written as a summation over products of terms of the form $B_n^m \hat{O}_n^m$ and $C_n^m \hat{\Omega}_n^m$ in zero magnetic field, where for a total spin $S \leq \frac{7}{2}$ the factor n is an even integer between 0 and 6 ($n \leq 2S$), and m lies in the range $0 \leq |m| \leq n$ for each n. The $\hat{O}_n^m$ and $\hat{\Omega}_n^m$ are equivalent spin operators which are listed in Appendix I, Section B, along with their matrix elements for various spin values ($\leq \frac{7}{2}$), and the coefficients B_n^m and C_n^m depend on the crystal field at the transition metal ion site. In

TABLE 1. Relationships of 10 derived point groups and 11 inversion point groups with their corresponding 11 proper point group for the seven crystal systems. Some derived groups are obtained by the direct product operation denoted by $\times$, and some are obtained by adding the symmetry element given in parentheses to the indicated subgroup. Schönflies symbols are used for the point groups with international symbols given in square brackets. Column 2 gives the spin Hamiltonian associated with each proper point group. (Some alternate notations are: $C_s = C_{1h}$, $C_i = S_2$, $C_{3i} = S_6$, $D_2 = V$, $D_{2d} = D_{2v} = V_d$, $D_{2h} = V_h$, $D_{3d} = D_{3v}$, cubic = isometric.) From ref. 33.

Crystal system	Spin Hamiltonian	Proper point group	Order (n)	Derived point group (order n)	Inversion point group (order $2n$)
Triclinic	$\mathscr{H}_C^{\text{tric}}$	C_1 [1]	1		$C_1 \times i = C_i$ [$\bar{1}$]
Monoclinic	$\mathscr{H}_C^{\text{mono}}$	C_2 [2]	2	$C_1 \times \sigma_h = C_s$ [m]	$C_2 \times i = C_{2h}$ [2/m]
Orthorhombic	$\mathscr{H}_D^{\text{orth}}$	D_2 [222]	4	$C_2 \times \sigma_v = C_{2v}$ [mm2]	$D_2 \times i = D_{2h}$ [mmm]
Trigonal	$\mathscr{H}_C^{\text{trig}}$	C_3 [3]	3		$C_3 \times i = C_{3i}$ [$\bar{3}$]
	$\mathscr{H}_D^{\text{trig}}$	D_3 [32]	6	$C_3(+\sigma_v) = C_{3v}$ [3 m]	$D_3 \times i = D_{3d}$ [$\bar{3}$m]
Tetragonal	$\mathscr{H}_C^{\text{tetr}}$	C_4 [4]	4	$C_2(+\sigma_h) = S_4$ [$\bar{4}$]	$C_4 \times i = C_{4h}$ [4/m]
	$\mathscr{H}_D^{\text{tetr}}$	D_4 [422]	8	$C_4(+\sigma_v) = C_{4v}$ [4 mm]	
				$D_2(+\sigma_d) = D_{2d}$ [$\bar{4}$2 m]	$D_4 \times i = D_{4h}$ [4/mmm]
Hexagonal	$\mathscr{H}_C^{\text{hexa}}$	C_6 [6]	6	$C_3(+\sigma_h) = C_{3h}$ [$\bar{6}$]	$C_6 \times i = C_{6h}$ [6/m]
	$\mathscr{H}_D^{\text{hexa}}$	D_6 [622]	12	$C_6(+\sigma_v) = C_{6v}$ [6 mm]	
				$D_3 \times \sigma_h = D_{3h}$ [$\bar{6}$2 m]	$D_6 \times i = D_{6h}$ [6/mmm]
Cubic	$\mathscr{H}_C^{\text{cube}}$	T [23]	12		$T \times i = T_h$ [m$\bar{3}$]
	$\mathscr{H}_D^{\text{cube}}$	O [432]	24	$T(+\sigma_d) = T_d$ [$\bar{4}$3 m]	$O \times i = O_h$ [m$\bar{3}$m]

TABLE 2. Nonzero coefficients of zero-field terms in the spin Hamiltonian (Note: Some common notations for spin-Hamiltonian parameters are as follows:

$$D = b_2^0 = 3B_2^0, \qquad E = b_2^2 = 3B_2^2, \qquad b_4^m = 60B_4^m, \qquad b_6^m = 1260B_6^m.$$

For cubic symmetry $a = \frac{2}{5}b_4^4$, where a describes the fourth-order term in the spin Hamiltonian:

$$(a/6)[S_x^4 + S_y^4 + S_z^4 - \tfrac{1}{5}S(S+1)(3S^2 + 3S - 1)],$$

while for axial-symmetry: $F = 2b_4^0$, where F describes the fourth-order term in the spin Hamiltonian:

$$\frac{F}{180}[35S_z^4 - 30S(S+1)S_z^2 + 25S_z^2 + 6S(+1) + 3S^2(S+1)^2].)$$

Spin Hamiltonian	Nonzero coefficients (B_l^m) of real operators O_l^m	Nonzero coefficients (C_l^m) of imaginary operators Ω_l^m
$C^{(1)} : \mathscr{H}_C^{\text{tric}}$	$B_2^0, B_2^1, B_2^2, B_4^0, B_4^1, B_4^2, B_4^3, B_4^4,$ $B_6^0, B_6^1, B_6^2, B_6^3, B_6^4, B_6^5, B_6^6$	$C_2^1, C_2^2, C_4^1, C_4^2, C_4^3, C_4^4$ $C_6^1, C_6^2, C_6^3, C_6^4, C_6^5, C_6^6$
$C^{(2)} : \mathscr{H}_C^{\text{mono}}; C_2 \| Z^*$	$B_2^0, B_2^2, B_4^0, B_4^2, B_4^4, B_6^0, B_6^2, B_6^4, B_6^6$	$C_2^2, C_4^2, C_4^4, C_6^2, C_6^4, C_6^6$
$C^{(2)} : \mathscr{H}_C^{\text{mono}}; C_2 \| Y^*$	$B_2^0, B_2^2, B_4^0, B_4^2, B_4^4, B_6^0, B_6^2, B_6^4, B_6^6,$ $B_2^1, B_4^1, B_4^3, B_6^1, B_6^3, B_6^5$	
$C^{(2)} : \mathscr{H}_C^{\text{mono}}; C_2 \| X^*$	$B_2^0, B_2^2, B_4^0, B_4^2, B_4^4, B_6^0, B_6^2, B_6^4, B_6^6$	$C_2^1, C_4^1, C_4^3, C_6^1, C_6^3, C_6^5$

$C^{(3)} : \mathscr{H}_D^{\text{ortho}}$	$B_2^0, B_2^2, B_4^0, B_4^2, B_4^4, B_6^0, B_6^2, B_6^4, B_6^6$	
$C^{(4)} : \mathscr{H}_C^{\text{tetr}}$	$B_2^0, B_4^0, B_4^4, B_6^0, B_6^4$	C_4^4, C_6^4
$C^{(5)} : \mathscr{H}_D^{\text{tetr}}$	$B_2^0, B_4^0, B_4^4, B_6^0, B_6^4$	
$C^{(6)} : \mathscr{H}_C^{\text{trig}}$	$B_2^0, B_4^0, B_4^1, B_4^3, B_6^0, B_6^1, B_6^3, B_6^6$	C_4^3, C_6^3, C_6^6
$C^{(7)} : \mathscr{H}_D^{\text{trig}}$	$B_2^0, B_4^0, B_4^1, B_4^3, B_6^0, B_6^1, B_6^3, B_6^6$	
$C^{(8)} : \mathscr{H}_C^{\text{hexa}}$	$B_2^0, B_4^0, B_6^0, B_6^6$	C_6^6
$C^{(9)} : \mathscr{H}_D^{\text{hexa}}$	$B_2^0, B_4^0, B_6^0, B_6^6$	
$C^{(10)} : \mathscr{H}_C^{\text{cube}}$	Four-fold symmetry axis: B_4^0, $B_4^4(= 5B_4^0)$, B_6^0, B_6^2, B_6^4 $(= -21B_6^0)$, B_6^6 $(= -B_6^2)$	
$C^{(11)} : \mathscr{H}_D^{\text{cube}}$	Four-fold symmetry axis: B_4^0, B_4^4 $(= 5B_4^0)$, B_6^0, B_6^4 $(= -21B_6^0)$ Three-fold symmetry axis**: B_4^0, B_4^3 $(= \pm 20\sqrt{2}B_4^0)$, B_6^0, B_6^3 $(= \mp(35\sqrt{2}B_6^0/4)$	

* C_2 axis could be parallel to any one of the magnetic X, Y, Z axes for monoclinic symmetry.

** The upper and lower signs refer to the "Watanabe" and "Hutchings" systems as discussed in detail by Rudowicz,[14] who also lists other transformations to symmetry axes.

TABLE 3. Nonzero coefficients of field-dependent terms in the spin Hamiltonian as listed by Al'tshuler and Kozyrev[16] and by Grekhov *et al.*[12] (Note: $\mu_B g_{xx} = B^1_{1x}$, $\mu_B g_{yy} = C^1_{1y}$, $\mu_B g_{zz} = B^0_{1z}$.)

Spin Hamiltonian	X-Component	Y-Component	Z-Component
$C^{(1)} : \mathscr{H}^{\text{tric}}_C$ ($g_{xx} \neq g_{yy} \neq g_{zz}$; $A_{xx} \neq A_{yy} \neq A_{yy}$)	$B^1_{1x}, B^1_{3x}, B^2_{3x}, B^3_{3x}, B^1_{5x}, B^2_{5x}, B^3_{5x}, B^4_{5x}, B^5_{5x}$	$C^1_{1y}, C^1_{3y}, C^2_{3y}, C^3_{3y}, C^1_{5y}, C^2_{5y}, C^3_{5y}, C^4_{5y}, C^5_{5y}$	$B^0_{1z}, B^0_{3z}, B^1_{3z}, B^2_{3z}, B^3_{3z}, B^0_{5z}, B^1_{5z}, B^2_{5z}, B^3_{5z}, B^4_{5z}, B^5_{5z}$
$C^{(2)} : \mathscr{H}^{\text{mono}}_C$ $C^{(3)} : \mathscr{H}^{\text{ortho}}_D$ ($g_{xx} \neq g_{yy} \neq g_{zz}$; $A_{xx} \neq A_{yy} \neq A_{yy}$)	$B^1_{1x}, B^1_{3x}, B^3_{3x}, B^1_{5x}, B^3_{5x}, B^5_{5x}$	$C^1_{1y}\ (= B^1_{1x}), C^1_{3y}\ (= B^1_{3x}), C^3_{3y}\ (= -B^3_{3x}), C^1_{5y}\ (= B^1_{5x}), C^3_{5y}\ (= -B^3_{5x}), C^5_{5y} (= B^5_{5x})$	$B^0_{1z}, B^0_{3z}, B^2_{3z}, B^0_{5z}, B^2_{5z}, B^4_{5z}$
$C^{(4)} : \mathscr{H}^{\text{tetr}}_C$ $C^{(5)} : \mathscr{H}^{\text{tetr}}_D$ ($g_{xx} = g_{yy} \neq g_{zz}$; $A_{xx} = A_{yy} \neq A_{zz}$)	$B^1_{1x}, B^1_{3x}, B^3_{3x}, B^1_{5x}, B^3_{5x}, B^5_{5x}$	$C^1_{1y}(= B^1_{1x}), C^1_{3y}(= B^1_{3x}), C^3_{3y}\ (= -B^3_{3x}), C^1_{5y}(= B^1_{5x}), C^3_{5y}\ (= -B^3_{5x}), C^5_{5y}\ (= B^5_{5x})$	$B^0_{1z}, B^0_{3z}, B^0_{5z}, B^4_{5z}$
$C^{(6)} : \mathscr{H}^{\text{trig}}_C$ $C^{(7)} : \mathscr{H}^{\text{trig}}_D$ ($g_{xx} = g_{yy} \neq g_{zz}$; $A_{xx} = A_{yy} \neq A_{zz}$)	$B^1_{1x}, B^1_{3x}, B^2_{3x}, B^1_{5x}, B^2_{5x}, B^4_{5x}, B^5_{5x}$	$C^1_{1y}\ (= B^1_{1x}), C^1_{3y}\ (= B^1_{3x}), C^2_{3y}\ (= -B^2_{3x}), C^1_{5y}\ (= B^1_{5x}), C^2_{5y}\ (= -B^2_{5x}), C^4_{5y}\ (= B^4_{5x}), C^5_{5y}\ (= -B^5_{5x})$	$B^0_{1z}, B^0_{3z}, B^3_{3z}, B^0_{5z}, B^3_{5z}$
$C^{(8)} : \mathscr{H}^{\text{hexa}}_C$ $C^{(9)} : \mathscr{H}^{\text{hexa}}_D$ ($g_{xx} = g_{yy} \neq g_{zz}$; $A_{xx} = A_{yy} \neq A_{zz}$)	$B^1_{1x}, B^1_{3x}, B^1_{5x}, B^5_{5x}$	$C^1_{1y}\ (= B^1_{1x}), C^1_{3y}\ (= B^1_{3x}), C^1_{5y}\ (= B^1_{5x}), C^5_{5y}\ (= -B^5_{5x})$	$B^0_{1z}, B^0_{3z}, B^0_{5z}$
$C^{(10)} : \mathscr{H}^{\text{cube}}_C$ $C^{(11)} : \mathscr{H}^{\text{cube}}_D$ ($g_{xx} = g_{yy} = g_{zz}$; $A_{xx} = A_{yy} = A_{zz}$)	$B^1_{1x}, B^1_{3x}, B^3_{3x}, B^1_{5x}, B^3_{5x}, B^5_{5x}$ With the following relations: $B^1_{1x} = C^1_{1y} = B^0_{1z}$; $B^3_{5x} = B^1_{5x} - \frac{1}{4}B^2_{5z}$; $B^3_{3x} = -\frac{5}{3}B^1_{3x} - \frac{2}{3}B^2_{3z}$; $B^5_{5x} = \frac{7}{5}B^1_{5x} - \frac{3}{5}B^3_{5x} - \frac{2}{5}B^2_{5z}$.	$C^1_{1y}, C^1_{3y}, C^3_{3y}, C^1_{5y}, C^3_{5y}, C^5_{5y}$ With the following relations: $C^1_{3y} = B^1_{3x} + \frac{1}{2}B^2_{3z}$; $C^3_{3y} = \frac{5}{3}B^1_{3x} + \frac{1}{6}B^2_{3z}$ $C^3_{5y} = -B^3_{5x} - \frac{1}{8}B^3_{5x}$; $C^5_{5y} = \frac{7}{5}B^1_{5x} - \frac{3}{5}B^3_{5x} - \frac{1}{40}B^2_{5z}$.	$B^0_{1z}, B^0_{3z}, B^2_{3z}, B^0_{5z}, B^2_{5z}, B^4_{5x}$ With the following relations: $B^0_{3z} = -\frac{4}{3}B^1_{3x} - \frac{1}{3}B^2_{3z}$; $B^0_{5z} = -(\sqrt{7}/5)[B^1_{5x} - \frac{2}{3}B^3_{5x} - \frac{1}{6}B^2_{5z}]$; $B^4_{5z} = \frac{7}{2}B^1_{5x} + 3B^3_{5x} - \frac{1}{4}B^2_{5z}$.

the presence of an externally applied Zeeman field we also have field-dependent terms of the type $H_\alpha B^m_{n\alpha}\hat{\Omega}^m_n$ and $H_y C^m_{ny}\hat{\Omega}^m_n$ ($\alpha = x, z$).

Finally,

$$\mathscr{H} = \sum_{n=1,3,5} \left(\sum_{m=0}^{n} \sum_{\alpha=x,z} H_\alpha B^m_{n\alpha} \hat{O}^m_n + \sum_{m=1}^{n} H_y C^m_{ny} \hat{\Omega}^m_n \right) + \sum_{n=0,2,4,6} \sum_{m=0}^{n} B^m_n \hat{O}^m_n + \sum_{n=2,4,6} \sum_{m=1}^{n} C^m_n \hat{\Omega}^m_n .$$

In the above spin Hamiltonian, the first two terms, containing $B^m_{n\alpha}$ and C^m_{ny}, are the Zeeman field (H)-dependent terms, while the last two terms are the zero-field terms.

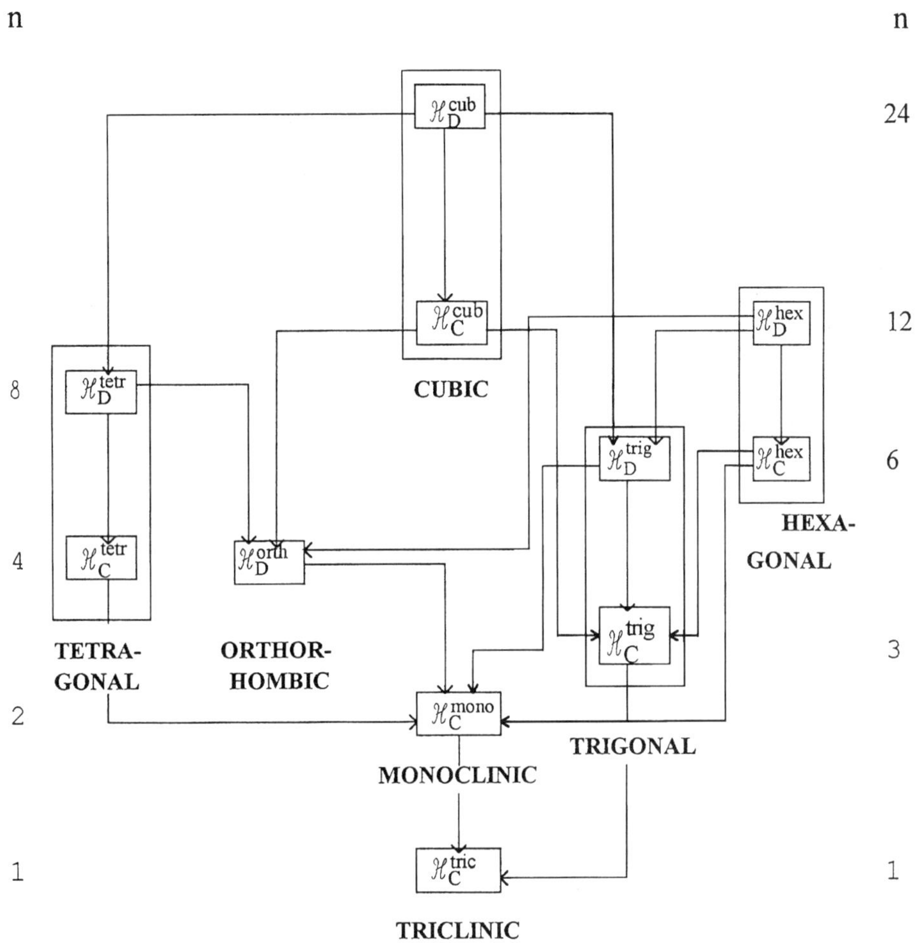

FIG. 1. Descent of symmetry relations for the 11 spin Hamiltonians grouped by crystal system. The numbers n of the scales on the left and right are the orders of the proper group associated with each Hamiltonian. From ref. 33.

There are 11 forms of the spin Hamiltonian, one for each of the 11 proper point groups. The relationships of the 10 derived point groups and the 11 inversion point groups with their corresponding 11 proper point groups for the seven crystal systems are listed in Table 1.[32] Table 2 lists the nonzero coefficients B_n^m and C_n^m in these 11 spin Hamiltonians, while Table 3 lists the nonzero coefficients B_{nx}^m, B_{nz}^m, and C_{ny}^m. The first row of Table 2 is for spin Hamiltonian $\mathscr{H}_C^{\text{tric}}$ which has all possible terms. The next six rows are for spin Hamiltonians with dominant twofold and fourfold axes, and the following four rows are for spin Hamiltonians with dominant threefold or sixfold axes. Results are given for the two cubic spin Hamiltonians for the quantization direction

z-axis chosen in the fourfold direction C_4, and also for z selected along a threefold direction C_3. The first column of Table 3 indicates how the g-factor and hyperfine tensors are isotropic ($g_{xx} = g_{yy} = g_{zz}$; likewise for A_{ij}) for the cubic point groups, how they are axially symmetric ($g_{xx} = g_{yy} \neq g_{zz}$; likewise for A_{ij}) for hexagonal, tetragonal, and trigonal site symmetries, and they are completely anisotropic ($g_{xx} \neq g_{yy} \neq g_{zz}$; likewise for A_{ij}) for the orthorhombic, monoclinic, and triclinic cases. Descent of symmetry expressions for the distortions of the cubic Hamiltonians $\mathscr{H}_D^{\text{cube}}$ and $\mathscr{H}_C^{\text{cube}}$ along fourfold (C_4) and threefold (C_3) axes, together with distortions of the hexagonal Hamiltonians $\mathscr{H}_D^{\text{hex}}$ and $\mathscr{H}_C^{\text{hex}}$ along threefold (C_3) and twofold (C_2) axes are given in Appendix II. These descent-of-symmetry expressions can be deduced with the help of Fig. 1, which describes how we arrive at lower symmetries starting from a higher symmetry.

Many of the details described in this chapter, except for a listing of the matrix elements, have been provided by Misra, Poole, Jr., and Farach.[33]

REFERENCES

[1] G. S. Painter and C. P. Poole, Jr., J. Phys. Chem. Solids **29**, 1754 (1968).
[2] G. F. Koster and H. Statz, Phys. Rev. **113**, 445 (1959).
[3] W. Hauser, *Paramagnetic Resonance* (Academic Press, New York, 1963), p. 297.
[4] A. M. Leushin, Paramagnitnyi rezonans (Kazan University, 1964), p. 42.
[5] W. J. C. Grant and M. W. P. Strandberg, J. Phys. Chem. Sol. **25**, 635 (1964).
[6] W. H. Huang, F. C. Lin, and J. K. Zhu, Proc. Phys. Soc. **84**, 661 (1964).
[7] T. Ray, Proc. Roy. Soc. **A227**, 76 (1964).
[8] A. Bieri and K. F. Kneubühl, Phys. Kond. Mater. **4**, 230 (1965).
[9] I. I. Geru, Ukrain Fiz. Zh. **10**, 726 (1965).
[10] A. B. Roitsin, Fiz. Tverd. Tela **4**, 2948 (1962); **5**, 151 (1963).
[11] N. I. Deryugina and A. B. Roitsin, Ukrain Fiz. Zh. **11**, 594 (1966).
[12] A. M. Grekhov and A. B. Roitsin, Phys. Stat. Sol. (b) **74**, 323 (1976).
[13] M. E. Rose, *Elementary Theory of Angular Momentum* (Wiley, New York, 1957).
[14] C. Rudowicz, Magn. Reson. Rev. **13**, 1 (1987).
[15] A. Abragam and B. Bleaney, *Electron Paramagnetic Resonance of Transition Ions* (Dover, New York, 1986).
[16] S. A. Al'tshuler and B. M. Kozyrev, *Electron Paramagnetic Resonance in Compounds of Transition Elements* (Wiley, New York, 1974).
[17] K. W. H. Stevens, Proc. Phys. Soc. **A65**, 209 (1952).
[18] H. A. Buckmaster, Canad. J. Phys. **40**, 1670 (1962).
[19] D. A. Jones, J. M. Baker, and F. D. Pope, Proc. Phys. Soc. **74**, 249 (1959).
[20] B. R. Judd, Proc. Roy. Soc. **A227**, 552 (1955).
[21] J. M. Baker, B. Bleaney, and W. Hayes, Proc. Roy. Soc. **A247**, 141 (1958).
[22] R. J. Elliott and K. W. H. Stevens, Proc. Roy. Soc. **A219**, 387 (1953).
[23] S. K. Misra, Magn. Reson. Rev. **10**, 333 (1986).
[24] S. K. Misra, J. Magn. Reson. **23**, 403 (1976).
[25] S. K. Misra and S. Subramanian, J. Phys. **C15**, 7199 (1982).
[26] S. K. Misra, Arab. J. Sci. and Engng. **13**, 255 (1988).
[27] S. K. Misra, Physica **121B**, 193 (1983).
[28] S. K. Misra, Physica **124B**, 53 (1984).
[29] R. W. G. Wyckoff, *Crystal Structures* (Wiley, New York, 2nd ed., Vol. 1, 1964; Vol. 2, 1965; Vol. 3, 1965).

[30] Internationale Tabellen zur Bestimmung von Kristalstrukturen, Vol. 1 (Grbrüder Borntrager, Berlin, 1935). Photo-lithoprint reproduction by Edwards, Ann Arbor, Michigan, 1944.
[31] *International Tables for X-Ray Crystallography*, Vol. 1, edited by N. F. M. Henry and Dame K. Lonsdale (Kynoch Press, Birmingham, England, 1959–1965); L. Jansen and M. Boon, *Theory of Finite Groups; Applications to Physics* (North-Holland, Amsterdam, 1967).
[32] S. I. Altmann and P. Herzig, *Point Group Theory Tables* (Clarendon Press, Oxford, 1994).
[33] S. K. Misra, C. P. Poole, Jr., and H. A. Farach, Appl. Magn. Reson. **11**, 29 (1996).

APPENDIX I. SPIN OPERATORS AND THEIR MATRIX ELEMENTS

(A) SPIN OPERATORS

The operators $\hat{Y}_n^m$ ($|m| \le n$) depending on the components $S_\pm (\equiv S_x \pm iS_y)$ and S_z of the spin angular momentum S are listed below:

$$\hat{Y}_0^0 = 1,$$

$$\hat{Y}_1^0 = S_z,$$

$$\hat{Y}_1^{\pm 1} = S_\pm,$$

$$\hat{Y}_2^0 = 3S_z^2 - S(S+1),$$

$$\hat{Y}_2^{\pm 1} = \tfrac{1}{2}(S_z S_\pm + S_\pm S_z),$$

$$\hat{Y}_2^{\pm 2} = S_\pm^2,$$

$$\hat{Y}_3^0 = 5S_z^3 - \{3S(S+1) - 1\}S_z,$$

$$\hat{Y}_3^{\pm 1} = \tfrac{1}{2}[\{5S_z^2 - S(S+1) - \tfrac{1}{2}\}S_\pm + S_\pm\{5S_z^2 - S(S+1) - \tfrac{1}{2}\}],$$

$$\hat{Y}_3^{\pm 2} = \tfrac{1}{2}(S_z S_\pm^2 + S_\pm^2 S_z),$$

$$\hat{Y}_3^{\pm 3} = S_\pm^3,$$

$$\hat{Y}_4^0 = 35S_z^4 - \{30S(S+1) - 25\}S_z^2 + 3S^2(S+1)^2 - 6S(S+1),$$

$$\hat{Y}_4^{\pm 1} = \tfrac{1}{2}[\{7S_z^3 - (3S(S+1)+1)S_z\}S_\pm + S_\pm\{7S_z^3 - (3S(S+1)+1)S_z\}],$$

$$\hat{Y}_4^{\pm 2} = \tfrac{1}{2}[\{7S_z^2 - S(S+1) - 5\}S_\pm^2 + S_\pm^2\{7S_z^2 - S(S+1) - 5\}],$$

$$\hat{Y}_4^{\pm 3} = \tfrac{1}{2}(S_z S_\pm^3 + S_\pm^3 S_z),$$

$$\hat{Y}_4^{\pm 4} = S_\pm^4,$$

$$\hat{Y}_5^0 = 63S_z^5 - \{70S(S+1) - 105\}S_z^3 + \{15S^2(S+1)^2 - 50S(S+1) + 12\}S_z,$$

$$\hat{Y}_5^{\pm 1} = \tfrac{1}{2}[\{21S_z^4 - 14S(S+1)S_z^2 + S^2(S+1)^2 - S(S-1) + \tfrac{3}{2}\}S_\pm$$
$$+ S_\pm\{21S_z^4 - 14S(S+1)S_z^2 + S^2(S+1)^2 - S(S-1) + \tfrac{3}{2}\}],$$

$$\hat{Y}_5^{\pm 2} = \tfrac{1}{2}[\{3S_z^3 - \{S(S+1)+5\}S_z\}S_\pm^2 + S_\pm^2\{3S_z^3 - \{S(S+1)+5\}S_z\}],$$

$$\hat{Y}_5^{\pm 3} = \tfrac{1}{2}[\{9S_z^2 - S(S+1) - \tfrac{33}{2}\}S_\pm^3 + S_\pm^3\{9S_z^2 - S(S+1) - \tfrac{33}{2}\}],$$

$$\hat{Y}_5^{\pm 4} = \tfrac{1}{2}(S_z S_\pm^4 + S_\pm^4 S_z),$$

$$\hat{Y}_5^{\pm 5} = S_\pm^5,$$

$$\hat{Y}_6^0 = 231S_z^6 - \{315S(S+1) - 735\}S_z^4 + \{105S^2(S+1)^2 - 525S(S+1) + 294\}S_z^2 - 5S^3(S+1)^3 + 40S^2(S+1)^2 - 60S(S+1),$$

$$\hat{Y}_6^{\pm 1} = [\{35S_z^5 - \{30S(S+1) - 15\}S_z^3 + \{5S^2(S+1)^2 - 10S(S+1) + 120\}S_z\}S_\pm + S_\pm\{35S_z^5 - \{30S(S+1) - 15\}S_z^3 + \{5S^2(S+1)^2 - 10S(S+1) + 120\}S_z\}],$$

$$\hat{Y}_6^{\pm 2} = \tfrac{1}{2}[\{33S_z^4 - \{18S(S+1) + 123\}S_z^2 + S^2(S+1)^2 + 10S(S+1) + 102\}S_\pm^2 + S_\pm^2\{33S_z^4 - \{18S(S+1) + 123\}S_z^2 + S^2(S+1)^2 + 10S(S+1) + 102\}],$$

$$\hat{Y}_6^{\pm 3} = \tfrac{1}{2}[\{11S_z^3 - \{3S(S+1) + 59\}S_z\}S_\pm^3 + S_\pm^3\{11S_z^3 - \{3S(S+1) + 59\}S_z\}],$$

$$\hat{Y}_6^{\pm 4} = \tfrac{1}{2}[\{11S_z^2 - S(S+1) - 38\}S_\pm^4 + S_\pm^4\{11S_z^2 - S(S+1) - 38\}],$$

$$\hat{Y}_6^{\pm 5} = \tfrac{1}{2}(S_z S_\pm^5 + S_\pm^5 S_z),$$

$$\hat{Y}_6^{\pm 6} = S_\pm^6.$$

(B) MATRIX ELEMENTS

The matrix elements of the operators $\hat{O}_n^m$ $(m \geq 0)$, defined as $\hat{O}_n^m = \frac{1}{2}(\hat{Y}_n^{+m} + \hat{Y}_n^{-m})$, appearing in the spin Hamiltonian, are listed below.

In order to obtain the unlisted nonzero matrix elements of $\hat{O}_n^m$, the following relations can be used

$$\langle M|\hat{O}_n^m|M-m\rangle = (-1)^{m+n}\langle -M|\hat{O}_n^m|-M+m\rangle = \langle M-m|\hat{O}_n^m|M\rangle.$$

In order to obtain the matrix elements of $\hat{\Omega}_n^m$, defined as $\hat{\Omega}_n^m = (1/2i)(\hat{Y}_n^{+m} - \hat{Y}_n^{-m})$, the matrix elements of $\hat{O}_n^m$ can be exploited as follows

$$\langle M|\hat{\Omega}_n^m|M-m\rangle = (-1)^{m+n+1}\langle -M|\hat{\Omega}_n^m|-M+m\rangle = -\langle M-m|\hat{\Omega}_n^m|M\rangle.$$

In general,

$$\langle M|\hat{\Omega}_n^m|M \pm m\rangle = \pm i\langle M|\hat{O}_n^m|M \pm m\rangle.$$

In all the tables below, the numbers in column F are factors multiplying the matrix elements appearing in the corresponding rows.

O_2^0

S	F	⟨0‖0⟩	⟨1‖1⟩	⟨2‖2⟩	⟨3‖3⟩	⟨4‖4⟩	⟨5‖5⟩	⟨6‖6⟩	⟨7‖7⟩	⟨8‖8⟩
1	1	−2	1							
2	3	−2	−1	2						
3	3	−4	−3	0	5					
4	1	−20	−17	−8	7	28				
5	3	−10	−9	−6	−1	6	15			
6	3	−14	−13	−10	−5	2	11	22		
7	1	−56	−53	−44	−29	−8	19	52	91	
8	3	−24	−23	−20	−15	−8	1	12	25	40

S	F	$\langle \frac{1}{2} \| \frac{1}{2} \rangle$	$\langle \frac{3}{2} \| \frac{3}{2} \rangle$	$\langle \frac{5}{2} \| \frac{5}{2} \rangle$	$\langle \frac{7}{2} \| \frac{7}{2} \rangle$	$\langle \frac{9}{2} \| \frac{9}{2} \rangle$	$\langle \frac{11}{2} \| \frac{11}{2} \rangle$	$\langle \frac{13}{2} \| \frac{13}{2} \rangle$	$\langle \frac{15}{2} \| \frac{15}{2} \rangle$
$\frac{3}{2}$	3	−1	1						
$\frac{5}{2}$	2	−4	−1	5					
$\frac{7}{2}$	3	−5	−3	1	7				
$\frac{9}{2}$	6	−4	−3	−1	2	6			
$\frac{11}{2}$	1	−35	−29	−17	1	25	55		
$\frac{13}{2}$	6	−8	−7	−5	−2	2	7	13	
$\frac{15}{2}$	3	−21	−19	−15	−9	−1	9	21	35

O_2^1

S	F	$\langle 1\|0\rangle$	$\langle 2\|1\rangle$	$\langle 3\|2\rangle$	$\langle 4\|3\rangle$	$\langle 5\|4\rangle$	$\langle 6\|5\rangle$	$\langle 7\|6\rangle$	$\langle 8\|7\rangle$
1	$\frac{1}{4}$	$\sqrt{(2)}$							
2	$\frac{1}{4}$	$\sqrt{(6)}$	6						
3	$\frac{1}{4}$	$2\sqrt{(3)}$	$3\sqrt{(10)}$	$5\sqrt{(6)}$					
4	$\frac{1}{4}$	$2\sqrt{(5)}$	$9\sqrt{(2)}$	$5\sqrt{(14)}$	$14\sqrt{(2)}$				
5	$\frac{1}{4}$	$\sqrt{(30)}$	$6\sqrt{(7)}$	$10\sqrt{(6)}$	$21\sqrt{(2)}$	$9\sqrt{(10)}$			
6	$\frac{1}{4}$	$\sqrt{(42)}$	$6\sqrt{(10)}$	30	$7\sqrt{(30)}$	$9\sqrt{(22)}$	$22\sqrt{(3)}$		
7	$\frac{1}{4}$	$2\sqrt{(14)}$	$9\sqrt{(6)}$	$25\sqrt{(2)}$	$14\sqrt{(11)}$	54	$11\sqrt{(26)}$	$13\sqrt{(14)}$	
8	$\frac{1}{4}$	$6\sqrt{(2)}$	$3\sqrt{(70)}$	$5\sqrt{(66)}$	$14\sqrt{(15)}$	$18\sqrt{(13)}$	$11\sqrt{(42)}$	$13\sqrt{(30)}$	60

S	F	$\langle \frac{1}{2}\| -\frac{1}{2}\rangle$	$\langle \frac{3}{2}\|\frac{1}{2}\rangle$	$\langle \frac{5}{2}\|\frac{3}{2}\rangle$	$\langle \frac{7}{2}\|\frac{5}{2}\rangle$	$\langle \frac{9}{2}\|\frac{7}{2}\rangle$	$\langle \frac{11}{2}\|\frac{9}{2}\rangle$	$\langle \frac{13}{2}\|\frac{11}{2}\rangle$	$\langle \frac{15}{2}\|\frac{13}{2}\rangle$
$\frac{3}{2}$	$\frac{1}{2}$	0	$\sqrt{(3)}$						
$\frac{5}{2}$	1	0	$\sqrt{(2)}$	$\sqrt{(5)}$					
$\frac{7}{2}$	$\frac{1}{2}$	0	$\sqrt{(15)}$	$4\sqrt{(3)}$	$3\sqrt{(7)}$				
$\frac{9}{2}$	1	0	$\sqrt{(6)}$	$\sqrt{(21)}$	6	6			
$\frac{11}{2}$	$\frac{1}{2}$	0	$\sqrt{(35)}$	$8\sqrt{(2)}$	$9\sqrt{(3)}$	$8\sqrt{(5)}$	$5\sqrt{(11)}$		
$\frac{13}{2}$	1	0	$2\sqrt{(3)}$	$3\sqrt{(5)}$	$3\sqrt{(10)}$	$2\sqrt{(33)}$	$5\sqrt{(6)}$	$3\sqrt{(13)}$	
$\frac{15}{2}$	$\frac{1}{2}$	0	$3\sqrt{(7)}$	$4\sqrt{(15)}$	$3\sqrt{(55)}$	$16\sqrt{(3)}$	$5\sqrt{(39)}$	$12\sqrt{(7)}$	$7\sqrt{(15)}$

O_2^2

S	F	$\langle 1\|-1\rangle$	$\langle 2\|0\rangle$	$\langle 3\|1\rangle$	$\langle 4\|2\rangle$	$\langle 5\|3\rangle$	$\langle 6\|4\rangle$	$\langle 7\|5\rangle$	$\langle 8\|6\rangle$
1	1	1							
2	1	3	$\sqrt{(6)}$						
3	1	6	$\sqrt{(30)}$	$\sqrt{(15)}$					
4	1	10	$3\sqrt{(10)}$	$3\sqrt{(7)}$	$2\sqrt{(7)}$				
5	1	15	$\sqrt{(210)}$	$2\sqrt{(42)}$	$6\sqrt{(3)}$	$3\sqrt{(5)}$			
6	1	21	$2\sqrt{(105)}$	$6\sqrt{(10)}$	$3\sqrt{(30)}$	$\sqrt{(165)}$	$\sqrt{(66)}$		
7	1	28	$6\sqrt{(21)}$	$15\sqrt{(3)}$	$5\sqrt{(22)}$	$6\sqrt{(11)}$	$3\sqrt{(26)}$	$\sqrt{(91)}$	
8	1	36	$6\sqrt{(35)}$	$\sqrt{(1155)}$	$3\sqrt{(110)}$	$2\sqrt{(195)}$	$\sqrt{(546)}$	$3\sqrt{(35)}$	$2\sqrt{(30)}$

S	F	$\langle \frac{3}{2}\|-\frac{1}{2}\rangle$	$\langle \frac{5}{2}\|\frac{1}{2}\rangle$	$\langle \frac{7}{2}\|\frac{3}{2}\rangle$	$\langle \frac{9}{2}\|\frac{5}{2}\rangle$	$\langle \frac{11}{2}\|\frac{7}{2}\rangle$	$\langle \frac{13}{2}\|\frac{9}{2}\rangle$	$\langle \frac{15}{2}\|\frac{11}{2}\rangle$
$\frac{3}{2}$	1	$\sqrt{(3)}$						
$\frac{5}{2}$	1	$3\sqrt{(2)}$	$\sqrt{(10)}$					
$\frac{7}{2}$	1	$2\sqrt{(15)}$	$3\sqrt{(5)}$	$\sqrt{(21)}$				
$\frac{9}{2}$	1	$5\sqrt{(6)}$	$3\sqrt{(14)}$	$2\sqrt{(21)}$	6			
$\frac{11}{2}$	1	$3\sqrt{(35)}$	$2\sqrt{(70)}$	$6\sqrt{(6)}$	$3\sqrt{(15)}$	$\sqrt{(55)}$		
$\frac{13}{2}$	1	$14\sqrt{(3)}$	$6\sqrt{(15)}$	$15\sqrt{(2)}$	$\sqrt{(330)}$	$3\sqrt{(22)}$	$\sqrt{(78)}$	
$\frac{15}{2}$	1	$12\sqrt{(7)}$	$3\sqrt{(105)}$	$5\sqrt{(33)}$	$2\sqrt{(165)}$	$6\sqrt{(13)}$	$\sqrt{(273)}$	$\sqrt{(105)}$

O_3^0

S	F	$\langle 0\|0\rangle$	$\langle 1\|1\rangle$	$\langle 2\|2\rangle$	$\langle 3\|3\rangle$	$\langle 4\|4\rangle$	$\langle 5\|5\rangle$	$\langle 6\|6\rangle$	$\langle 7\|7\rangle$	$\langle 8\|8\rangle$
2	6	0	−2	1						
3	30	0	−1	1	1					
4	6	0	−9	−13	−7	14				
5	6	0	−14	−23	−22	−6	30			
6	30	0	−4	−7	−8	−2	0	−11		
7	6	0	−27	−49	−61	−46	−35	13	91	
8	30	0	−7	−13	−17	−18	−15	−7	7	28

S	F	$\langle \frac{1}{2}\|\frac{1}{2}\rangle$	$\langle \frac{3}{2}\|\frac{3}{2}\rangle$	$\langle \frac{5}{2}\|\frac{5}{2}\rangle$	$\langle \frac{7}{2}\|\frac{7}{2}\rangle$	$\langle \frac{9}{2}\|\frac{9}{2}\rangle$	$\langle \frac{11}{2}\|\frac{11}{2}\rangle$	$\langle \frac{13}{2}\|\frac{13}{2}\rangle$	$\langle \frac{15}{2}\|\frac{15}{2}\rangle$
$\frac{3}{2}$	$\frac{1}{2}$	−9	3						
$\frac{5}{2}$	3	4	−7	9					
$\frac{7}{2}$	$\frac{15}{2}$	−3	−7	−15	7				
$\frac{9}{2}$	3	−12	−31	−35	−14	42			
$\frac{11}{2}$	$\frac{15}{2}$	−7	−133	−25	−21	−3	33		
$\frac{13}{2}$	3	−24	−67	−95	−98	−66	11	143	
$\frac{15}{2}$	$\frac{3}{2}$	−63	−179	−265	−301	−267	−143	91	455

O_3^1

S	F	$\langle 1\|0\rangle$	$\langle 2\|1\rangle$	$\langle 3\|2\rangle$	$\langle 4\|3\rangle$	$\langle 5\|4\rangle$	$\langle 6\|5\rangle$	$\langle 7\|6\rangle$	$\langle 8\|7\rangle$
2	2	$\sqrt{(6)}$	3						
3	10	$-\sqrt{(3)}$	0	$\sqrt{(6)}$					
4	6	$-3\sqrt{(5)}$	$-2\sqrt{(2)}$	$\sqrt{(14)}$	$7\sqrt{(2)}$				
5	2	$-7\sqrt{(30)}$	$-9\sqrt{(7)}$	$\sqrt{(6)}$	$24\sqrt{(2)}$	$18\sqrt{(10)}$			
6	10	$-2\sqrt{(42)}$	$-3\sqrt{(10)}$	-3	$\sqrt{(30)}$	$3\sqrt{(22)}$	$11\sqrt{(3)}$		
7	6	$-9\sqrt{(14)}$	$-11\sqrt{(6)}$	$-10\sqrt{(2)}$	$\sqrt{(11)}$	23	$8\sqrt{(26)}$	$13\sqrt{(14)}$	
8	10	$-21\sqrt{(2)}$	$-3\sqrt{(70)}$	$-2\sqrt{(66)}$	$\sqrt{(15)}$	$3\sqrt{(13)}$	$4\sqrt{(42)}$	$7\sqrt{(30)}$	42

S	F	$\langle \frac{1}{2}\|-\frac{1}{2}\rangle$	$\langle \frac{3}{2}\|\frac{1}{2}\rangle$	$\langle \frac{5}{2}\|\frac{3}{2}\rangle$	$\langle \frac{7}{2}\|\frac{5}{2}\rangle$	$\langle \frac{9}{2}\|\frac{7}{2}\rangle$	$\langle \frac{11}{2}\|\frac{9}{2}\rangle$	$\langle \frac{13}{2}\|\frac{11}{2}\rangle$	$\langle \frac{15}{2}\|\frac{13}{2}\rangle$
$\frac{3}{2}$	1	-3	$-\sqrt{(3)}$						
$\frac{5}{2}$	3	-4	$\sqrt{(2)}$	$-2\sqrt{(5)}$					
$\frac{7}{2}$	5	-6	$-\sqrt{(15)}$	$\sqrt{(3)}$	$3\sqrt{(7)}$				
$\frac{9}{2}$	1	-60	$-19\sqrt{(6)}$	$-2\sqrt{(21)}$	42	84			
$\frac{11}{2}$	15	-7	$-\sqrt{(35)}$	$-2\sqrt{(2)}$	$\sqrt{(3)}$	$3\sqrt{(5)}$	$3\sqrt{(11)}$		
$\frac{13}{2}$	1	-168	$-86\sqrt{(3)}$	$-42\sqrt{(5)}$	$-3\sqrt{(10)}$	$64\sqrt{(33)}$	$77\sqrt{(6)}$	$66\sqrt{(13)}$	
$\frac{15}{2}$	1	-252	$-87\sqrt{(7)}$	$-43\sqrt{(15)}$	$-9\sqrt{(55)}$	$34\sqrt{(3)}$	$31\sqrt{(39)}$	$117\sqrt{(7)}$	$91\sqrt{(15)}$

O_3^2

S	F	$\langle 1\|-1\rangle$	$\langle 2\|0\rangle$	$\langle 3\|1\rangle$	$\langle 4\|2\rangle$	$\langle 5\|3\rangle$	$\langle 6\|4\rangle$	$\langle 7\|5\rangle$	$\langle 8\|6\rangle$
2	1	0	$\sqrt{(6)}$						
3	1	0	$\sqrt{(30)}$	$2\sqrt{(15)}$					
4	3	0	$\sqrt{(10)}$	$2\sqrt{(7)}$	$2\sqrt{(7)}$				
5	1	0	$\sqrt{(210)}$	$4\sqrt{(42)}$	$18\sqrt{(3)}$	$12\sqrt{(5)}$			
6	1	0	$2\sqrt{(105)}$	$12\sqrt{(10)}$	$9\sqrt{(30)}$	$4\sqrt{(165)}$	$5\sqrt{(66)}$		
7	3	0	$2\sqrt{(21)}$	$10\sqrt{(3)}$	$5\sqrt{(22)}$	$4\sqrt{(11)}$	$5\sqrt{(26)}$	$2\sqrt{(91)}$	
8	1	0	$6\sqrt{(35)}$	$2\sqrt{(1155)}$	$18\sqrt{(110)}$	$8\sqrt{(195)}$	$5\sqrt{(546)}$	$18\sqrt{(35)}$	$14\sqrt{(30)}$

S	F	$\langle \frac{3}{2}\|-\frac{1}{2}\rangle$	$\langle \frac{5}{2}\|\frac{1}{2}\rangle$	$\langle \frac{7}{2}\|\frac{3}{2}\rangle$	$\langle \frac{9}{2}\|\frac{5}{2}\rangle$	$\langle \frac{11}{2}\|\frac{7}{2}\rangle$	$\langle \frac{13}{2}\|\frac{9}{2}\rangle$	$\langle \frac{15}{2}\|\frac{11}{2}\rangle$
$\frac{3}{2}$	$\frac{1}{2}$	$\sqrt{(3)}$						
$\frac{5}{2}$	$\frac{1}{2}$	$\sqrt{(2)}$	$\sqrt{(10)}$					
$\frac{7}{2}$	$\frac{1}{2}$	$2\sqrt{(15)}$	$9\sqrt{(5)}$	$5\sqrt{(21)}$				
$\frac{9}{2}$	$\frac{1}{2}$	$5\sqrt{(6)}$	$9\sqrt{(14)}$	$10\sqrt{(21)}$	42			
$\frac{11}{2}$	$\frac{1}{2}$	$3\sqrt{(35)}$	$6\sqrt{(70)}$	$30\sqrt{(6)}$	$21\sqrt{(15)}$	$9\sqrt{(55)}$		
$\frac{13}{2}$	$\frac{1}{2}$	$14\sqrt{(3)}$	$18\sqrt{(15)}$	$75\sqrt{(2)}$	$7\sqrt{(330)}$	$27\sqrt{(22)}$	$11\sqrt{(78)}$	
$\frac{15}{2}$	$\frac{1}{2}$	$12\sqrt{(7)}$	$9\sqrt{(105)}$	$25\sqrt{(33)}$	$14\sqrt{(165)}$	$54\sqrt{(13)}$	$11\sqrt{(273)}$	$13\sqrt{(105)}$

O_3^3

S	F	$\langle 2 \| -1 \rangle$	$\langle 3 \| 0 \rangle$	$\langle 4 \| 1 \rangle$	$\langle 5 \| 2 \rangle$	$\langle 6 \| 3 \rangle$	$\langle 7 \| 4 \rangle$	$\langle 8 \| 5 \rangle$
2	6	1						
3	6	$\sqrt{(10)}$	$\sqrt{(5)}$					
4	6	$5\sqrt{(2)}$	$\sqrt{(35)}$	$\sqrt{(14)}$				
5	6	$5\sqrt{(7)}$	$2\sqrt{(35)}$	$2\sqrt{(21)}$	$\sqrt{(30)}$			
6	6	$7\sqrt{(10)}$	$2\sqrt{(105)}$	$10\sqrt{(3)}$	$\sqrt{(165)}$	$\sqrt{(55)}$		
7	6	$14\sqrt{(6)}$	$5\sqrt{(42)}$	$5\sqrt{(33)}$	$5\sqrt{(22)}$	$\sqrt{(286)}$	$\sqrt{(91)}$	
8	6	$6\sqrt{(70)}$	$\sqrt{(2310)}$	$5\sqrt{(77)}$	$\sqrt{(1430)}$	$\sqrt{(910)}$	$\sqrt{(455)}$	$2\sqrt{(5)}$

S	F	$\langle \frac{3}{2} \| -\frac{3}{2} \rangle$	$\langle \frac{5}{2} \| -\frac{1}{2} \rangle$	$\langle \frac{7}{2} \| \frac{1}{2} \rangle$	$\langle \frac{9}{2} \| \frac{3}{2} \rangle$	$\langle \frac{11}{2} \| \frac{5}{2} \rangle$	$\langle \frac{13}{2} \| \frac{7}{2} \rangle$	$\langle \frac{15}{2} \| \frac{9}{2} \rangle$
$\frac{3}{2}$	3	1						
$\frac{5}{2}$	3	4	$\sqrt{(10)}$					
$\frac{7}{2}$	3	10	$4\sqrt{(5)}$	$\sqrt{(35)}$				
$\frac{9}{2}$	3	20	$5\sqrt{(14)}$	$4\sqrt{(14)}$	$2\sqrt{(21)}$			
$\frac{11}{2}$	3	35	$4\sqrt{(70)}$	$2\sqrt{(210)}$	$4\sqrt{(30)}$	$\sqrt{(165)}$		
$\frac{13}{2}$	3	56	$14\sqrt{(15)}$	$20\sqrt{(6)}$	$5\sqrt{(66)}$	$4\sqrt{(55)}$	$\sqrt{(286)}$	
$\frac{15}{2}$	3	84	$8\sqrt{(105)}$	$5\sqrt{(231)}$	$20\sqrt{(11)}$	$2\sqrt{(715)}$	$4\sqrt{(91)}$	$\sqrt{(455)}$

O_4^0

S	F	⟨0‖0⟩	⟨1‖1⟩	⟨2‖2⟩	⟨3‖3⟩	⟨4‖4⟩	⟨5‖5⟩	⟨6‖6⟩	⟨7‖7⟩	⟨8‖8⟩
1	0	0	0							
2	12	6	−4	1						
3	60	6	1	−7	3					
4	60	18	9	−11	−21	14				
5	420	6	4	−1	−6	−6	6			
6	60	84	64	11	−54	−96	−66	99		
7	12	756	621	251	−249	−704	−869	−429	1001	
8	420	36	31	17	−3	−24	−39	−39	−13	52

S	F	$\langle \frac{1}{2} \Vert \frac{1}{2} \rangle$	$\langle \frac{3}{2} \Vert \frac{3}{2} \rangle$	$\langle \frac{5}{2} \Vert \frac{5}{2} \rangle$	$\langle \frac{7}{2} \Vert \frac{7}{2} \rangle$	$\langle \frac{9}{2} \Vert \frac{9}{2} \rangle$	$\langle \frac{11}{2} \Vert \frac{11}{2} \rangle$	$\langle \frac{13}{2} \Vert \frac{13}{2} \rangle$	$\langle \frac{15}{2} \Vert \frac{15}{2} \rangle$
$\frac{1}{2}$	0	0							
$\frac{3}{2}$	0	0	0						
$\frac{5}{2}$	60	2	−3	1					
$\frac{7}{2}$	60	9	−3	−13	7				
$\frac{9}{2}$	84	18	3	−17	−22	18			
$\frac{11}{2}$	120	28	12	−13	−33	−27	33		
$\frac{13}{2}$	60	108	63	−13	−92	−132	−77	143	
$\frac{15}{2}$	60	189	129	23	−101	−201	−221	−91	273

O_4^1

S	F	$\langle 1\|0\rangle$	$\langle 2\|1\rangle$	$\langle 3\|2\rangle$	$\langle 4\|3\rangle$	$\langle 5\|4\rangle$	$\langle 6\|5\rangle$	$\langle 7\|6\rangle$	$\langle 8\|7\rangle$
2	3	$-\sqrt{(6)}$	1						
3	3	$-5\sqrt{(3)}$	$-4\sqrt{(10)}$	$5\sqrt{(6)}$					
4	3	$-9\sqrt{(5)}$	$-30\sqrt{(2)}$	$-5\sqrt{(14)}$	$35\sqrt{(2)}$				
5	21	$-\sqrt{(30)}$	$-5\sqrt{(7)}$	$-5\sqrt{(6)}$	0	$6\sqrt{(10)}$			
6	3	$-10\sqrt{(42)}$	$-53\sqrt{(10)}$	-195	$-21\sqrt{(30)}$	$15\sqrt{(22)}$	$165\sqrt{(3)}$		
7	3	$-27\sqrt{(14)}$	$-111\sqrt{(6)}$	$-250\sqrt{(2)}$	$-91\sqrt{(11)}$	-99	$44\sqrt{(26)}$	$143\sqrt{(14)}$	
8	21	$-15\sqrt{(2)}$	$-7\sqrt{(70)}$	$-10\sqrt{(66)}$	$-21\sqrt{(15)}$	$-15\sqrt{(13)}$	0	$13\sqrt{(30)}$	130

S	F	$\langle \frac{1}{2}\|-\frac{1}{2}\rangle$	$\langle \frac{3}{2}\|\frac{1}{2}\rangle$	$\langle \frac{5}{2}\|\frac{3}{2}\rangle$	$\langle \frac{7}{2}\|\frac{5}{2}\rangle$	$\langle \frac{9}{2}\|\frac{7}{2}\rangle$	$\langle \frac{11}{2}\|\frac{9}{2}\rangle$	$\langle \frac{13}{2}\|\frac{11}{2}\rangle$	$\langle \frac{15}{2}\|\frac{13}{2}\rangle$
$\frac{3}{2}$	$\frac{1}{2}$	0	0						
$\frac{5}{2}$	3	0	$-5\sqrt{(2)}$	$2\sqrt{(5)}$					
$\frac{7}{2}$	6	0	$-3\sqrt{(15)}$	$-5\sqrt{(3)}$	$5\sqrt{(7)}$				
$\frac{9}{2}$	21	0	$-3\sqrt{(6)}$	$-2\sqrt{(21)}$	-2	12			
$\frac{11}{2}$	12	0	$-4\sqrt{(35)}$	$-25\sqrt{(2)}$	$-15\sqrt{(3)}$	$3\sqrt{(5)}$	$15\sqrt{(11)}$		
$\frac{13}{2}$	3	0	$-90\sqrt{(3)}$	$-114\sqrt{(5)}$	$-79\sqrt{(10)}$	$-20\sqrt{(33)}$	$55\sqrt{(6)}$	$110\sqrt{(13)}$	
$\frac{15}{2}$	6	0	$-45\sqrt{(7)}$	$-53\sqrt{(15)}$	$-31\sqrt{(55)}$	$-100\sqrt{(3)}$	$-5\sqrt{(39)}$	$65\sqrt{(7)}$	$91\sqrt{(15)}$

O_4^2

S	F	$\langle 1\|-1\rangle$	$\langle 2\|0\rangle$	$\langle 3\|1\rangle$	$\langle 4\|2\rangle$	$\langle 5\|3\rangle$	$\langle 6\|4\rangle$	$\langle 7\|5\rangle$	$\langle 8\|6\rangle$
1	1	0							
2	3	−4	$\sqrt{(6)}$						
3	3	−20	$-\sqrt{(30)}$	$6\sqrt{(15)}$					
4	3	−60	$-11\sqrt{(10)}$	$10\sqrt{(7)}$	$30\sqrt{(7)}$				
5	21	−20	$-\sqrt{(210)}$	0	$10\sqrt{(3)}$	$12\sqrt{(5)}$			
6	3	−280	$-22\sqrt{(105)}$	$-24\sqrt{(10)}$	$23\sqrt{(30)}$	$24\sqrt{(165)}$	$45\sqrt{(66)}$		
7	3	−504	$-94\sqrt{(21)}$	$-130\sqrt{(3)}$	$15\sqrt{(22)}$	$116\sqrt{(11)}$	$121\sqrt{(26)}$	$66\sqrt{(91)}$	
8	7	−360	$-54\sqrt{(35)}$	$-6\sqrt{(1155)}$	$-3\sqrt{(110)}$	$12\sqrt{(195)}$	$15\sqrt{(546)}$	$78\sqrt{(35)}$	$78\sqrt{(30)}$

S	F	$\langle \frac{3}{2}\|-\frac{1}{2}\rangle$	$\langle \frac{5}{2}\|\frac{1}{2}\rangle$	$\langle \frac{7}{2}\|\frac{3}{2}\rangle$	$\langle \frac{9}{2}\|\frac{5}{2}\rangle$	$\langle \frac{11}{2}\|\frac{7}{2}\rangle$	$\langle \frac{13}{2}\|\frac{9}{2}\rangle$	$\langle \frac{15}{2}\|\frac{11}{2}\rangle$
$\frac{5}{2}$	3	$-5\sqrt{(2)}$	$3\sqrt{(10)}$					
$\frac{7}{2}$	6	$-4\sqrt{(15)}$	$\sqrt{(5)}$	$5\sqrt{(21)}$				
$\frac{9}{2}$	21	$-5\sqrt{(6)}$	$-\sqrt{(14)}$	$2\sqrt{(21)}$	18			
$\frac{11}{2}$	12	$-8\sqrt{(35)}$	$-3\sqrt{(70)}$	$5\sqrt{(6)}$	$13\sqrt{(15)}$	$9\sqrt{(55)}$		
$\frac{13}{2}$	3	$-210\sqrt{(3)}$	$-62\sqrt{(15)}$	$-15\sqrt{(2)}$	$13\sqrt{(330)}$	$95\sqrt{(22)}$	$55\sqrt{(78)}$	
$\frac{15}{2}$	6	$-120\sqrt{(7)}$	$-23\sqrt{(105)}$	$-15\sqrt{(33)}$	$8\sqrt{(165)}$	$80\sqrt{(13)}$	$25\sqrt{(273)}$	$39\sqrt{(105)}$

O_4^3

S	F	$\langle 2\|-1\rangle$	$\langle 3\|0\rangle$	$\langle 4\|1\rangle$	$\langle 5\|2\rangle$	$\langle 6\|3\rangle$	$\langle 7\|4\rangle$	$\langle 8\|5\rangle$
2	3	1						
3	3	$\sqrt{(10)}$	$3\sqrt{(5)}$					
4	3	$5\sqrt{(2)}$	$3\sqrt{(35)}$	$5\sqrt{(14)}$				
5	3	$5\sqrt{(7)}$	$6\sqrt{(35)}$	$10\sqrt{(21)}$	$7\sqrt{(30)}$			
6	3	$7\sqrt{(10)}$	$6\sqrt{(105)}$	$50\sqrt{(3)}$	$7\sqrt{(165)}$	$9\sqrt{(55)}$		
7	3	$14\sqrt{(6)}$	$15\sqrt{(42)}$	$25\sqrt{(33)}$	$35\sqrt{(22)}$	$9\sqrt{(286)}$	$11\sqrt{(91)}$	
8	3	$6\sqrt{(70)}$	$3\sqrt{(2310)}$	$25\sqrt{(77)}$	$7\sqrt{(1430)}$	$9\sqrt{(910)}$	$11\sqrt{(455)}$	$26\sqrt{(35)}$

S	F	$\langle \frac{5}{2}\|-\frac{1}{2}\rangle$	$\langle \frac{7}{2}\|\frac{1}{2}\rangle$	$\langle \frac{9}{2}\|\frac{3}{2}\rangle$	$\langle \frac{11}{2}\|\frac{5}{2}\rangle$	$\langle \frac{13}{2}\|\frac{7}{2}\rangle$	$\langle \frac{15}{2}\|\frac{9}{2}\rangle$
$\frac{5}{2}$	3	$\sqrt{(10)}$					
$\frac{7}{2}$	3	$4\sqrt{(5)}$	$2\sqrt{(35)}$				
$\frac{9}{2}$	3	$5\sqrt{(14)}$	$8\sqrt{(14)}$	$6\sqrt{(21)}$			
$\frac{11}{2}$	$12\sqrt{(5)}$	$\sqrt{(14)}$	$\sqrt{(42)}$	$3\sqrt{(6)}$	$\sqrt{(33)}$		
$\frac{13}{2}$	3	$14\sqrt{(15)}$	$40\sqrt{(6)}$	$15\sqrt{(66)}$	$16\sqrt{(55)}$	$5\sqrt{(286)}$	
$\frac{15}{2}$	6	$4\sqrt{(105)}$	$5\sqrt{(231)}$	$30\sqrt{(11)}$	$4\sqrt{(715)}$	$10\sqrt{(91)}$	$3\sqrt{(455)}$

O_4^4

S	F	$\langle 2\|-2\rangle$	$\langle 3\|-1\rangle$	$\langle 4\|0\rangle$	$\langle 5\|1\rangle$	$\langle 6\|2\rangle$	$\langle 7\|3\rangle$	$\langle 8\|4\rangle$
2	12	1						
3	12	5	$\sqrt{(15)}$					
4	12	15	$5\sqrt{(7)}$	$\sqrt{(70)}$				
5	12	35	$5\sqrt{(42)}$	$3\sqrt{(70)}$	$\sqrt{(210)}$			
6	12	70	$21\sqrt{(10)}$	$15\sqrt{(14)}$	$5\sqrt{(66)}$	$3\sqrt{(55)}$		
7	12	126	$70\sqrt{(3)}$	$5\sqrt{(162)}$	$15\sqrt{(33)}$	$5\sqrt{(143)}$	$\sqrt{(1001)}$	
8	12	210	$6\sqrt{(1155)}$	$15\sqrt{(154)}$	$5\sqrt{(1001)}$	$\sqrt{(15015)}$	$5\sqrt{(273)}$	$2\sqrt{(455)}$

S	F	$\langle \frac{5}{2}\|-\frac{3}{2}\rangle$	$\langle \frac{7}{2}\|-\frac{1}{2}\rangle$	$\langle \frac{9}{2}\|\frac{1}{2}\rangle$	$\langle \frac{11}{2}\|\frac{3}{2}\rangle$	$\langle \frac{13}{2}\|\frac{5}{2}\rangle$	$\langle \frac{15}{2}\|\frac{7}{2}\rangle$
$\frac{3}{2}$	12						
$\frac{5}{2}$	12	$\sqrt{(5)}$					
$\frac{7}{2}$	12	$5\sqrt{(3)}$	$\sqrt{(35)}$				
$\frac{9}{2}$	$12\sqrt{(7)}$	$5\sqrt{(3)}$	$5\sqrt{(2)}$	$3\sqrt{(2)}$			
$\frac{11}{2}$	$12\sqrt{(2)}$	35	$3\sqrt{(105)}$	$5\sqrt{(21)}$	$\sqrt{(165)}$		
$\frac{13}{2}$	12	$42\sqrt{(5)}$	$35\sqrt{(6)}$	$15\sqrt{(22)}$	$15\sqrt{(11)}$	$\sqrt{(5 \cdot 143)}$	
$\frac{15}{2}$	12	$42\sqrt{(15)}$	$10\sqrt{(231)}$	$15\sqrt{(77)}$	$5\sqrt{(429)}$	$\sqrt{(5005)}$	$\sqrt{(1365)}$

O_5^0

S	F	⟨0‖0⟩	⟨1‖1⟩	⟨2‖2⟩	⟨3‖3⟩	⟨4‖4⟩	⟨5‖5⟩	⟨6‖6⟩	⟨7‖7⟩	⟨8‖8⟩
3	180	0	5	−4	1					
4	240	0	9	4	−11	4				
5	2520	0	4	4	−1	−6	3			
6	1080	0	−20	26	11	−18	−33	22		
7	60	0	675	1000	751	−44	−979	−1144	1001	
8	1260	0	55	88	83	36	−39	−104	−91	104

S	F	$\langle \frac{1}{2} \Vert \frac{1}{2} \rangle$	$\langle \frac{3}{2} \Vert \frac{3}{2} \rangle$	$\langle \frac{5}{2} \Vert \frac{5}{2} \rangle$	$\langle \frac{7}{2} \Vert \frac{7}{2} \rangle$	$\langle \frac{9}{2} \Vert \frac{9}{2} \rangle$	$\langle \frac{11}{2} \Vert \frac{11}{2} \rangle$	$\langle \frac{13}{2} \Vert \frac{13}{2} \rangle$	$\langle \frac{15}{2} \Vert \frac{15}{2} \rangle$
$\frac{5}{2}$	30	10	−5	1					
$\frac{7}{2}$	90	15	17	−23	7				
$\frac{9}{2}$	630	6	11	1	−14	6			
$\frac{11}{2}$	420	20	44	29	−21	−57	33		
$\frac{13}{2}$	270	60	145	139	28	−132	−187	143	
$\frac{15}{2}$	630	45	105	131	77	−33	−143	−143	143

O_5^1

S	F	$\langle 1\|0\rangle$	$\langle 2\|1\rangle$	$\langle 3\|2\rangle$	$\langle 4\|3\rangle$	$\langle 5\|4\rangle$	$\langle 6\|5\rangle$	$\langle 7\|6\rangle$	$\langle 8\|7\rangle$
3	6	$10\sqrt{(3)}$	$-9\sqrt{(10)}$	$5\sqrt{(6)}$					
4	42	$6\sqrt{(5)}$	$-5\sqrt{(2)}$	$-5\sqrt{(14)}$	$10\sqrt{(2)}$				
5	84	$4\sqrt{(30)}$	0	$-10\sqrt{(6)}$	$-15\sqrt{(2)}$	$\sqrt{(10)}$			
6	36	$20\sqrt{(42)}$	$12\sqrt{(10)}$	90	$-29\sqrt{(30)}$	$-15\sqrt{(22)}$	$110\sqrt{(3)}$		
7	30	$90\sqrt{(14)}$	$65\sqrt{(6)}$	$-83\sqrt{(6)}$	$-106\sqrt{(11)}$	-374	$-11\sqrt{(26)}$	$143\sqrt{(14)}$	
8	42	$330\sqrt{(2)}$	$33\sqrt{(70)}$	$-5\sqrt{(66)}$	$-94\sqrt{(15)}$	$-150\sqrt{(13)}$	$-65\sqrt{(42)}$	$13\sqrt{(30)}$	780

S	F	$\langle \frac{1}{2}\|-\frac{1}{2}\rangle$	$\langle \frac{3}{2}\|\frac{1}{2}\rangle$	$\langle \frac{5}{2}\|\frac{3}{2}\rangle$	$\langle \frac{7}{2}\|\frac{5}{2}\rangle$	$\langle \frac{9}{2}\|\frac{7}{2}\rangle$	$\langle \frac{11}{2}\|\frac{9}{2}\rangle$	$\langle \frac{13}{2}\|\frac{11}{2}\rangle$	$\langle \frac{15}{2}\|\frac{13}{2}\rangle$
$\frac{5}{2}$	6	10	$-10\sqrt{(2)}$	$\sqrt{(7)}$					
$\frac{7}{2}$	6	60	$\sqrt{(15)}$	$-40\sqrt{(3)}$	$15\sqrt{(7)}$				
$\frac{9}{2}$	210	3	$\sqrt{(6)}$	$-\sqrt{(21)}$	-6	6			
$\frac{11}{2}$	84	40	$4\sqrt{(35)}$	$-10\sqrt{(2)}$	$-25\sqrt{(3)}$	$-12\sqrt{(5)}$	$15\sqrt{(11)}$		
$\frac{13}{2}$	18	420	$170\sqrt{(3)}$	$-9\sqrt{(5)}$	$-111\sqrt{(10)}$	$-80\sqrt{(33)}$	$-55\sqrt{(6)}$	$165\sqrt{(13)}$	
$\frac{15}{2}$	42	360	$105\sqrt{(7)}$	$16\sqrt{(15)}$	$-27\sqrt{(55)}$	$-220\sqrt{(3)}$	$-55\sqrt{(39)}$	0	$143\sqrt{(15)}$

O_5^2

S	F	$\langle 1\|-1\rangle$	$\langle 2\|0\rangle$	$\langle 3\|1\rangle$	$\langle 4\|2\rangle$	$\langle 5\|3\rangle$	$\langle 6\|4\rangle$	$\langle 7\|5\rangle$	$\langle 8\|6\rangle$
3	6	0	$-\sqrt{(30)}$	$\sqrt{(15)}$					
4	24	0	$-7\sqrt{(10)}$	$-5\sqrt{(7)}$	$10\sqrt{(7)}$				
5	6	0	$\sqrt{(210)}$	$-10\sqrt{(42)}$	0	$42\sqrt{(5)}$			
6	12	0	$-2\sqrt{(105)}$	$-27\sqrt{(10)}$	$-9\sqrt{(30)}$	$3\sqrt{(165)}$	$15\sqrt{(66)}$		
7	30	0	$-10\sqrt{(21)}$	$-41\sqrt{(3)}$	$-13\sqrt{(22)}$	$-4\sqrt{(11)}$	$11\sqrt{(26)}$	$11\sqrt{(91)}$	
8	6	0	$-66\sqrt{(35)}$	$-19\sqrt{(1155)}$	$-63\sqrt{(110)}$	$-28\sqrt{(195)}$	$5\sqrt{(546)}$	$117\sqrt{(35)}$	$182\sqrt{(30)}$

S	F	$\langle \frac{3}{2}\|-\frac{1}{2}\rangle$	$\langle \frac{5}{2}\|\frac{1}{2}\rangle$	$\langle \frac{7}{2}\|\frac{3}{2}\rangle$	$\langle \frac{9}{2}\|\frac{5}{2}\rangle$	$\langle \frac{11}{2}\|\frac{7}{2}\rangle$	$\langle \frac{13}{2}\|\frac{9}{2}\rangle$	$\langle \frac{15}{2}\|\frac{11}{2}\rangle$
$\frac{5}{2}$	$\frac{1}{2}$	$-15\sqrt{(2)}$	$3\sqrt{(10)}$					
$\frac{7}{2}$	3	$-4\sqrt{(15)}$	$-9\sqrt{(5)}$	$5\sqrt{(21)}$				
$\frac{9}{2}$	$\frac{15}{2}$	$-7\sqrt{(6)}$	$-9\sqrt{(14)}$	$-2\sqrt{(21)}$	42			
$\frac{11}{2}$	6	$-8\sqrt{(35)}$	$-13\sqrt{(70)}$	$-35\sqrt{(6)}$	$7\sqrt{(15)}$	$21\sqrt{(55)}$		
$\frac{13}{2}$	$\frac{9}{2}$	$-70\sqrt{(3)}$	$-78\sqrt{(15)}$	$-225\sqrt{(2)}$	$-7\sqrt{(330)}$	$45\sqrt{(22)}$	$55\sqrt{(78)}$	
$\frac{15}{2}$	3	$-120\sqrt{(7)}$	$-81\sqrt{(105)}$	$-175\sqrt{(33)}$	$-56\sqrt{(165)}$	0	$55\sqrt{(273)}$	$143\sqrt{(105)}$

O_5^3

S	F	$\langle 2\|-1\rangle$	$\langle 3\|0\rangle$	$\langle 4\|1\rangle$	$\langle 5\|2\rangle$	$\langle 6\|3\rangle$	$\langle 7\|4\rangle$	$\langle 8\|5\rangle$
3	36	$-\sqrt{(10)}$	$2\sqrt{(5)}$					
4	12	$-35\sqrt{(2)}$	$2\sqrt{(35)}$	$20\sqrt{(14)}$				
5	72	$-10\sqrt{(7)}$	$-\sqrt{(35)}$	$5\sqrt{(21)}$	$7\sqrt{(30)}$			
6	216	$-7\sqrt{(10)}$	$-\sqrt{(105)}$	$5\sqrt{(3)}$	$2\sqrt{(165)}$	$4\sqrt{(55)}$		
7	60	$-70\sqrt{(6)}$	$-16\sqrt{(42)}$	$2\sqrt{(33)}$	$29\sqrt{(22)}$	$13\sqrt{(286)}$	$22\sqrt{(91)}$	
8	36	$-66\sqrt{(70)}$	$-8\sqrt{(2310)}$	$-10\sqrt{(77)}$	$7\sqrt{(1430)}$	$19\sqrt{(910)}$	$34\sqrt{(455)}$	$104\sqrt{(35)}$

S	F	$\langle \frac{3}{2}\|-\frac{3}{2}\rangle$	$\langle \frac{5}{2}\|-\frac{1}{2}\rangle$	$\langle \frac{7}{2}\|\frac{1}{2}\rangle$	$\langle \frac{9}{2}\|\frac{3}{2}\rangle$	$\langle \frac{11}{2}\|\frac{5}{2}\rangle$	$\langle \frac{13}{2}\|\frac{7}{2}\rangle$	$\langle \frac{15}{2}\|\frac{9}{2}\rangle$
$\frac{5}{2}$	12	-5	$\sqrt{(10)}$					
$\frac{7}{2}$	36	$-2\sqrt{(5)}$	$-\sqrt{(5)}$	$2\sqrt{(35)}$				
$\frac{9}{2}$	180	-7	$-\sqrt{(14)}$	$\sqrt{(14)}$	$2\sqrt{(21)}$			
$\frac{11}{2}$	12	-280	$-23\sqrt{(70)}$	$\sqrt{(210)}$	$49\sqrt{(30)}$	$14\sqrt{(165)}$		
$\frac{13}{2}$	108	-70	$-14\sqrt{(15)}$	$-5\sqrt{(6)}$	$5\sqrt{(66)}$	$11\sqrt{(55)}$	$5\sqrt{(286)}$	
$\frac{15}{2}$	36	-420	$-34\sqrt{(105)}$	$-10\sqrt{(231)}$	$35\sqrt{(11)}$	$14\sqrt{(715)}$	$55\sqrt{(91)}$	$22\sqrt{(455)}$

O_5^4

S	F	$\langle 2\|-2\rangle$	$\langle 3\|-1\rangle$	$\langle 4\|0\rangle$	$\langle 5\|1\rangle$	$\langle 6\|2\rangle$	$\langle 7\|3\rangle$	$\langle 8\|4\rangle$
3	12	0	$\surd(15)$					
4	12	0	$5\surd(7)$	$2\surd(70)$				
5	12	0	$5\surd(42)$	$6\surd(70)$	$3\surd(210)$			
6	36	0	$7\surd(10)$	$10\surd(14)$	$5\surd(66)$	$4\surd(55)$		
7	60	0	$14\surd(3)$	$2\surd(462)$	$9\surd(33)$	$4\surd(143)$	$\surd(1001)$	
8	12	0	$2\surd(1155)$	$30\surd(154)$	$15\surd(1001)$	$4\surd(15015)$	$25\surd(273)$	$2\surd(455)$

S	F	$\langle \frac{5}{2}\|-\frac{3}{2}\rangle$	$\langle \frac{7}{2}\|-\frac{1}{2}\rangle$	$\langle \frac{9}{2}\|\frac{1}{2}\rangle$	$\langle \frac{11}{2}\|\frac{3}{2}\rangle$	$\langle \frac{13}{2}\|\frac{5}{2}\rangle$	$\langle \frac{15}{2}\|\frac{7}{2}\rangle$
$\frac{5}{2}$	6	$\surd(5)$					
$\frac{7}{2}$	6	$5\surd(3)$	$3\surd(35)$				
$\frac{9}{2}$	30	$\surd(21)$	$3\surd(14)$	$3\surd(14)$			
$\frac{11}{2}$	6	$35\surd(2)$	$9\surd(210)$	$25\surd(42)$	$7\surd(330)$		
$\frac{13}{2}$	18	$14\surd(5)$	$35\surd(6)$	$25\surd(22)$	$35\surd(11)$	$3\surd(715)$	
$\frac{15}{2}$	6	$42\surd(15)$	$30\surd(231)$	$75\surd(77)$	$35\surd(429)$	$9\surd(5005)$	$11\surd(1365)$

O_5^5

S	F	$\langle 3\|-2\rangle$	$\langle 4\|-1\rangle$	$\langle 5\|0\rangle$	$\langle 6\|1\rangle$	$\langle 7\|2\rangle$	$\langle 8\|3\rangle$
3	60	$\sqrt{(6)}$					
4	60	$3\sqrt{(14)}$	$\sqrt{(14)}$				
5	120	$7\sqrt{(16)}$	$3\sqrt{(21)}$	$3\sqrt{(7)}$			
6	120	210	$7\sqrt{(3)}$	$3\sqrt{(77)}$	$3\sqrt{(22)}$		
7	60	$126\sqrt{(2)}$	$14\sqrt{(33)}$	$3\sqrt{(462)}$	$3\sqrt{(858)}$	$\sqrt{(2002)}$	
8	60	$42\sqrt{(66)}$	$36\sqrt{(77)}$	$6\sqrt{(2002)}$	$7\sqrt{(858)}$	$3\sqrt{(2002)}$	$4\sqrt{(273)}$

S	F	$\langle \frac{5}{2}\|-\frac{5}{2}\rangle$	$\langle \frac{7}{2}\|-\frac{3}{2}\rangle$	$\langle \frac{9}{2}\|-\frac{1}{2}\rangle$	$\langle \frac{11}{2}\|\frac{1}{2}\rangle$	$\langle \frac{13}{2}\|\frac{3}{2}\rangle$	$\langle \frac{15}{2}\|\frac{5}{2}\rangle$
$\frac{5}{2}$	60	1					
$\frac{7}{2}$	60	6	$\sqrt{(21)}$				
$\frac{9}{2}$	60	21	$4\sqrt{(21)}$	$3\sqrt{(14)}$			
$\frac{11}{2}$	60	56	$21\sqrt{(6)}$	$6\sqrt{(42)}$	$\sqrt{(462)}$		
$\frac{13}{2}$	180	42	$28\sqrt{(2)}$	$7\sqrt{(22)}$	$4\sqrt{(33)}$	$\sqrt{(143)}$	
$\frac{15}{2}$	60	84	$42\sqrt{(33)}$	$24\sqrt{(77)}$	$3\sqrt{(3003)}$	$2\sqrt{(3003)}$	$\sqrt{(3003)}$

O_6^0

S	F	$\langle 0\|0\rangle$	$\langle 1\|1\rangle$	$\langle 2\|2\rangle$	$\langle 3\|3\rangle$	$\langle 4\|4\rangle$	$\langle 5\|5\rangle$	$\langle 6\|6\rangle$	$\langle 7\|7\rangle$	$\langle 8\|8\rangle$
1	0	0	0							
2	0	0	0	0						
3	180	−20	15	−6	1					
4	1260	−20	1	22	−17	4				
5	2520	−40	−12	36	29	−48	15			
6	7560	−40	−20	22	43	8	−55	22		
7	3780	−200	−125	50	197	176	−55	−286	143	
8	13860	−120	−85	2	93	128	65	−78	−169	104

S	F	$\langle \frac{1}{2}\|\frac{1}{2}\rangle$	$\langle \frac{3}{2}\|\frac{3}{2}\rangle$	$\langle \frac{5}{2}\|\frac{5}{2}\rangle$	$\langle \frac{7}{2}\|\frac{7}{2}\rangle$	$\langle \frac{9}{2}\|\frac{9}{2}\rangle$	$\langle \frac{11}{2}\|\frac{11}{2}\rangle$	$\langle \frac{13}{2}\|\frac{13}{2}\rangle$	$\langle \frac{15}{2}\|\frac{15}{2}\rangle$
$\frac{1}{2}$	0	0							
$\frac{3}{2}$	0	0	0						
$\frac{5}{2}$	0	0	0	0					
$\frac{7}{2}$	1260	−5	9	−5	1				
$\frac{9}{2}$	5040	−8	6	10	−11	3			
$\frac{11}{2}$	7560	−20	4	25	11	−31	11		
$\frac{13}{2}$	2160	−200	−25	185	227	−11	−319	143	
$\frac{15}{2}$	13860	−75	−25	45	87	59	−39	−117	65

O_6^1

S	F	⟨1‖0⟩	⟨2‖1⟩	⟨3‖2⟩	⟨4‖3⟩	⟨5‖4⟩	⟨6‖5⟩	⟨7‖6⟩	⟨8‖7⟩
3	15	$10\sqrt{(3)}$	$-3\sqrt{(10)}$	$\sqrt{(6)}$					
4	45	$14\sqrt{(5)}$	$21\sqrt{(2)}$	$-13\sqrt{(14)}$	$14\sqrt{(2)}$				
5	30	$28\sqrt{(30)}$	$96\sqrt{(7)}$	$-14\sqrt{(6)}$	$-231\sqrt{(2)}$	$63\sqrt{(10)}$			
6	90	$20\sqrt{(42)}$	$84\sqrt{(10)}$	126	$-35\sqrt{(30)}$	$-63\sqrt{(22)}$	$154\sqrt{(3)}$		
7	45	$150\sqrt{(14)}$	$525\sqrt{(6)}$	$735\sqrt{(2)}$	$-42\sqrt{(11)}$	−1386	$-231\sqrt{(26)}$	$429\sqrt{(14)}$	
8	115	$210\sqrt{(2)}$	$87\sqrt{(70)}$	$91\sqrt{(66)}$	$70\sqrt{(15)}$	$-126\sqrt{(13)}$	$-143\sqrt{(42)}$	$-91\sqrt{(30)}$	1092

S	F	$\langle\frac{1}{2}\Vert-\frac{1}{2}\rangle$	$\langle\frac{3}{2}\Vert\frac{1}{2}\rangle$	$\langle\frac{5}{2}\Vert\frac{3}{2}\rangle$	$\langle\frac{7}{2}\Vert\frac{5}{2}\rangle$	$\langle\frac{9}{2}\Vert\frac{7}{2}\rangle$	$\langle\frac{11}{2}\Vert\frac{9}{2}\rangle$	$\langle\frac{13}{2}\Vert\frac{11}{2}\rangle$	$\langle\frac{15}{2}\Vert\frac{13}{2}\rangle$
$\frac{7}{2}$	30	0	$7\sqrt{(15)}$	$-14\sqrt{(3)}$	$3\sqrt{(7)}$				
$\frac{9}{2}$	120	0	$14\sqrt{(6)}$	$-2\sqrt{(21)}$	−42	21			
$\frac{11}{2}$	540	0	$4\sqrt{(35)}$	$14\sqrt{(2)}$	$-7\sqrt{(3)}$	$-14\sqrt{(5)}$	$7\sqrt{(11)}$		
$\frac{13}{2}$	360	0	$50\sqrt{(3)}$	$45\sqrt{(5)}$	$6\sqrt{(10)}$	$-17\sqrt{(33)}$	$-44\sqrt{(6)}$	$33\sqrt{(13)}$	
$\frac{15}{2}$	330	0	$75\sqrt{(7)}$	$60\sqrt{(15)}$	$21\sqrt{(55)}$	$-56\sqrt{(3)}$	$-49\sqrt{(39)}$	$-78\sqrt{(7)}$	$91\sqrt{(15)}$

O_6^2

S	F	$\langle 1\|-1\rangle$	$\langle 2\|0\rangle$	$\langle 3\|1\rangle$	$\langle 4\|2\rangle$	$\langle 5\|3\rangle$	$\langle 6\|4\rangle$	$\langle 7\|5\rangle$	$\langle 8\|6\rangle$
3	24	15	$-2\sqrt{(30)}$	$\sqrt{(15)}$					
4	30	84	0	$-36\sqrt{(7)}$	$27\sqrt{(7)}$				
5	120	84	$2\sqrt{(210)}$	$-11\sqrt{(42)}$	$-42\sqrt{(3)}$	$-42\sqrt{(5)}$			
6	72	420	$22\sqrt{(105)}$	$-63\sqrt{(10)}$	$-84\sqrt{(30)}$	$-14\sqrt{(165)}$	$70\sqrt{(66)}$		
7	1080	70	$10\sqrt{(21)}$	$-7\sqrt{(3)}$	$-14\sqrt{(22)}$	$-21\sqrt{(11)}$	0	$11\sqrt{(91)}$	
8	264	630	$78\sqrt{(35)}$	$\sqrt{(1155)}$	$-42\sqrt{(110)}$	$-49\sqrt{(195)}$	$-20\sqrt{(546)}$	$39\sqrt{(35)}$	$182\sqrt{(30)}$

S	F	$\langle \frac{3}{2}\|-\frac{1}{2}\rangle$	$\langle \frac{5}{2}\|\frac{1}{2}\rangle$	$\langle \frac{7}{2}\|\frac{3}{2}\rangle$	$\langle \frac{9}{2}\|\frac{5}{2}\rangle$	$\langle \frac{11}{2}\|\frac{7}{2}\rangle$	$\langle \frac{13}{2}\|\frac{9}{2}\rangle$	$\langle \frac{15}{2}\|\frac{11}{2}\rangle$
$\frac{7}{2}$	24	$7\sqrt{(15)}$	$-21\sqrt{(5)}$	$5\sqrt{(21)}$				
$\frac{9}{2}$	240	$7\sqrt{(6)}$	$-3\sqrt{(14)}$	$-5\sqrt{(21)}$	21			
$\frac{11}{2}$	216	$12\sqrt{(35)}$	$-\sqrt{(70)}$	$-35\sqrt{(6)}$	$-14\sqrt{(15)}$	$14\sqrt{(55)}$		
$\frac{13}{2}$	720	$35\sqrt{(3)}$	$3\sqrt{(15)}$	$-36\sqrt{(2)}$	$-4\sqrt{(330)}$	$-3\sqrt{(22)}$	$11\sqrt{(78)}$	
$\frac{15}{2}$	1320	$30\sqrt{(7)}$	$3\sqrt{(105)}$	$-7\sqrt{(33)}$	$-7\sqrt{(165)}$	$-21\sqrt{(13)}$	$\sqrt{(273)}$	$13\sqrt{(105)}$

O_6^3

S	F	⟨2‖−1⟩	⟨3‖0⟩	⟨4‖1⟩	⟨5‖2⟩	⟨6‖3⟩	⟨7‖4⟩	⟨8‖5⟩
2	0	0						
3	18	$-3\sqrt{(10)}$	$2\sqrt{(5)}$					
4	90	$-7\sqrt{(2)}$	$-2\sqrt{(35)}$	$4\sqrt{(14)}$				
5	180	$-6\sqrt{(7)}$	$-5\sqrt{(35)}$	$-\sqrt{(21)}$	$7\sqrt{(30)}$			
6	36	$-63\sqrt{(10)}$	$-43\sqrt{(105)}$	$-175\sqrt{(3)}$	$14\sqrt{(165)}$	$84\sqrt{(55)}$		
7	90	$-70\sqrt{(6)}$	$-64\sqrt{(42)}$	$-70\sqrt{(33)}$	$-21\sqrt{(22)}$	$21\sqrt{(286)}$	$66\sqrt{(91)}$	
8	198	$-18\sqrt{(70)}$	$-8\sqrt{(2310)}$	$-50\sqrt{(77)}$	$-7\sqrt{(1430)}$	$3\sqrt{(910)}$	$22\sqrt{(455)}$	$104\sqrt{(35)}$

S	F	$\langle \frac{5}{2} \Vert -\frac{1}{2} \rangle$	$\langle \frac{7}{2} \Vert \frac{1}{2} \rangle$	$\langle \frac{9}{2} \Vert \frac{3}{2} \rangle$	$\langle \frac{11}{2} \Vert \frac{5}{2} \rangle$	$\langle \frac{13}{2} \Vert \frac{7}{2} \rangle$	$\langle \frac{15}{2} \Vert \frac{9}{2} \rangle$
$\frac{5}{2}$	0	0					
$\frac{7}{2}$	36	$-7\sqrt{(5)}$	$2\sqrt{(35)}$				
$\frac{9}{2}$	360	$-2\sqrt{(14)}$	$-\sqrt{(14)}$	$2\sqrt{(21)}$			
$\frac{11}{2}$	$36\sqrt{(5)}$	$-27\sqrt{(14)}$	$-16\sqrt{(42)}$	$7\sqrt{(6)}$	$28\sqrt{(33)}$		
$\frac{13}{2}$	360	$-14\sqrt{(15)}$	$-29\sqrt{(6)}$	$-4\sqrt{(66)}$	$6\sqrt{(55)}$	$6\sqrt{(286)}$	
$\frac{15}{2}$	1980	$-2\sqrt{(105)}$	$-2\sqrt{(231)}$	$-7\sqrt{(11)}$	0	$3\sqrt{(91)}$	$2\sqrt{(455)}$

O_6^4

S	F	$\langle 2\|-2\rangle$	$\langle 3\|-1\rangle$	$\langle 4\|0\rangle$	$\langle 5\|1\rangle$	$\langle 6\|2\rangle$	$\langle 7\|3\rangle$	$\langle 8\|4\rangle$
3	60	6	$\surd(15)$					
4	180	−14	$-\surd(7)$	$2\surd(70)$				
5	60	−168	$-13\surd(42)$	$12\surd(70)$	$15\surd(210)$			
6	180	−168	$-35\surd(10)$	$8\surd(14)$	$21\surd(66)$	$28\surd(55)$		
7	60	−1260	$-546\surd(3)$	$-6\surd(4211)$	$143\surd(33)$	$126\surd(143)$	$45\surd(1001)$	
8	660	−252	$-6\surd(1155)$	$-6\surd(154)$	$3\surd(1001)$	$2\surd(15015)$	$19\surd(273)$	$12\surd(455)$

S	F	$\langle \frac{5}{2}\|-\frac{3}{2}\rangle$	$\langle \frac{7}{2}\|-\frac{1}{2}\rangle$	$\langle \frac{9}{2}\|\frac{1}{2}\rangle$	$\langle \frac{11}{2}\|\frac{3}{2}\rangle$	$\langle \frac{13}{2}\|\frac{5}{2}\rangle$	$\langle \frac{15}{2}\|\frac{7}{2}\rangle$
$\frac{5}{2}$	60						
$\frac{7}{2}$	60	$-7\surd(3)$	$3\surd(55)$				
$\frac{9}{2}$	$60\surd(7)$	$-16\surd(3)$	$6\surd(2)$	$3\surd(2)$			
$\frac{11}{2}$	$60\surd(18)$	−63	$-\surd(105)$	$13\surd(21)$	$7\surd(165)$		
$\frac{13}{2}$	360	$-56\surd(5)$	$-21\surd(6)$	$13\surd(22)$	$46\surd(11)$	$6\surd(715)$	
$\frac{15}{2}$	660	$-42\surd(15)$	$-6\surd(231)$	$3\surd(77)$	$7\surd(429)$	$3\surd(5005)$	$5\surd(1365)$

O_6^5

S	F	$\langle 3\|-2\rangle$	$\langle 4\|-1\rangle$	$\langle 5\|0\rangle$	$\langle 6\|1\rangle$	$\langle 7\|2\rangle$	$\langle 8\|3\rangle$
3	30	$\sqrt{(6)}$					
4	90	$\sqrt{(14)}$	$2\sqrt{(14)}$				
5	60	$7\sqrt{(6)}$	$9\sqrt{(21)}$	$15\sqrt{(7)}$			
6	180	14	$21\sqrt{(3)}$	$5\sqrt{(77)}$	$7\sqrt{(22)}$		
7	90	$42\sqrt{(2)}$	$28\sqrt{(33)}$	$10\sqrt{(462)}$	$7\sqrt{(858)}$	$3\sqrt{(2002)}$	
8	30	$42\sqrt{(66)}$	$108\sqrt{(77)}$	$30\sqrt{(2002)}$	$49\sqrt{(858)}$	$27\sqrt{(2002)}$	$41\sqrt{(273)}$

S	F	$\langle \frac{5}{2}\|-\frac{5}{2}\rangle$	$\langle \frac{7}{2}\|-\frac{3}{2}\rangle$	$\langle \frac{9}{2}\|-\frac{1}{2}\rangle$	$\langle \frac{11}{2}\|\frac{1}{2}\rangle$	$\langle \frac{13}{2}\|\frac{3}{2}\rangle$	$\langle \frac{15}{2}\|\frac{5}{2}\rangle$
$\frac{5}{2}$	1	0					
$\frac{7}{2}$	60	0	$\sqrt{(21)}$				
$\frac{9}{2}$	120	0	$2\sqrt{(21)}$	$3\sqrt{(14)}$			
$\frac{11}{2}$	180	0	$7\sqrt{(6)}$	$4\sqrt{(42)}$	$\sqrt{(462)}$		
$\frac{13}{2}$	360	0	$14\sqrt{(2)}$	$7\sqrt{(22)}$	$6\sqrt{(33)}$	$2\sqrt{(143)}$	
$\frac{15}{2}$	60	0	$42\sqrt{(33)}$	$48\sqrt{(77)}$	$9\sqrt{(3003)}$	$8\sqrt{(3003)}$	$5\sqrt{(3003)}$

O_6^6

S	F	$\langle 3\|-3\rangle$	$\langle 4\|-2\rangle$	$\langle 5\|-1\rangle$	$\langle 6\|0\rangle$	$\langle 7\|1\rangle$	$\langle 8\|2\rangle$
3	360	1					
4	360	7	$2\sqrt{(7)}$				
5	360	28	$14\sqrt{(3)}$	$\sqrt{(210)}$			
6	360	84	$14\sqrt{(30)}$	$7\sqrt{(66)}$	$2\sqrt{(231)}$		
7	360	210	$42\sqrt{(22)}$	$28\sqrt{(33)}$	$2\sqrt{(3003)}$	$\sqrt{(3003)}$	
8	360	462	$42\sqrt{(110)}$	$12\sqrt{(1001)}$	$14\sqrt{(429)}$	$7\sqrt{(715)}$	$2\sqrt{(2002)}$

S	F	$\langle \frac{7}{2}\|-\frac{5}{2}\rangle$	$\langle \frac{9}{2}\|-\frac{3}{2}\rangle$	$\langle \frac{11}{2}\|-\frac{1}{2}\rangle$	$\langle \frac{13}{2}\|\frac{1}{2}\rangle$	$\langle \frac{15}{2}\|\frac{3}{2}\rangle$
$\frac{7}{2}$	360	$\sqrt{(7)}$				
$\frac{9}{2}$	360	14	$2\sqrt{(21)}$			
$\frac{11}{2}$	360	$28\sqrt{(3)}$	$7\sqrt{(30)}$	$\sqrt{(462)}$		
$\frac{13}{2}$	720	$21\sqrt{(10)}$	$7\sqrt{(66)}$	$7\sqrt{(33)}$	$\sqrt{(429)}$	
$\frac{15}{2}$	$360\sqrt{(11)}$	$42\sqrt{(5)}$	84	$4\sqrt{(273)}$	$7\sqrt{(39)}$	$\sqrt{(455)}$

APPENDIX II. DESCENT OF SYMMETRY[33]

(a) Descent of symmetry of cubic Hamiltonian along fourfold axis

$$H_D^{\text{cub}} \begin{array}{l} \Rightarrow H_C^{\text{cub}} \\ \\ \Rightarrow H_D^{\text{tetr}} \end{array} \Rightarrow H_D^{\text{orth}} \Rightarrow H_C^{\text{mono}} \Rightarrow H_C^{\text{tric}},$$

$$H_D^{\text{cub}} \Rightarrow H_D^{\text{tetr}} \begin{array}{l} \Rightarrow H_C^{\text{tetr}} \\ \\ \Rightarrow H_D^{\text{orth}} \end{array} \Rightarrow H_C^{\text{mono}} \Rightarrow H_C^{\text{tric}}.$$

(b) Descent of symmetry of cubic Hamiltonian along threefold axis

$$H_D^{\text{cub}} \begin{array}{l} \Rightarrow H_C^{\text{cub}} \\ \\ \Rightarrow H_D^{\text{trig}} \end{array} \Rightarrow H_C^{\text{trig}} \Rightarrow H_C^{\text{tric}}.$$

(c) Descent of symmetry of hexagonal Hamiltonian along threefold axis

$$H_D^{\text{hex}} \Rightarrow H_D^{\text{trig}} \begin{array}{l} \Rightarrow H_C^{\text{trig}} \\ \\ \Rightarrow H_C^{\text{mono}} \end{array} \Rightarrow H_C^{\text{tric}},$$

$$H_D^{\text{hex}} \Rightarrow H_C^{\text{hex}} \begin{array}{l} \Rightarrow H_C^{\text{trig}} \\ \\ \Rightarrow H_C^{\text{mono}} \end{array} \Rightarrow H_C^{\text{tric}}.$$

(d) Descent of symmetry of hexagonal Hamiltonian along twofold axis

$$H_D^{\text{hex}} \Rightarrow H_D^{\text{orth}} \Rightarrow H_C^{\text{mono}} \Rightarrow H_C^{\text{tric}}.$$

APPENDIX III. SITE SYMMETRIES OF HOST CRYSTALS[33]

The following table gives the characteristics, site symmetries, and spin-Hamiltonian types of transition metal ions in commonly studied host lattices. The table contains representatives from a wide range of space group types, site symmetries, and spin Hamiltonians. Of particular interest is the nearest-neighbor coordination complex given in column 5, and in most cases this is tetrahedral, MX_4, or octahedral, MX_6. Ordinarily the symmetry of this complex is lower than cubic, but the deviation from cubic can be small, and in many cases the dominant terms in the spin Hamiltonian are cubic. The transition ion site is noted in column 6, and column 7 gives the site symmetry. Column 8 gives the Hamiltonian, and column 9 gives the Wyckoff reference (volume and page) to the structure description. Both International and Schönflies group notations are listed.

Host lattice	Formula	Crystal system	Space group[1]	Complex	Crystal site	Point symmetry	Hamiltonian	Wyckoff **vol**-page[1]	Comment
α-Alum	$KAl(SO_4)_2 \cdot 12H_2O$	Cubic	T_h^6 (Pa$\bar{3}$)	$Al(H_2O)_6$	4a	C_{3i} ($\bar{3}$)	$\mathscr{H}_C^{trig}$	**3**-872	
				$K(H_2O)_6$	4b	C_{3i} ($\bar{3}$)	$\mathscr{H}_C^{trig}$		
β-Alum	$CsCr(SO_4)_2 \cdot 12H_2O$	Cubic	T_h^6 (Pa$\bar{3}$)				$\mathscr{H}_C^{trig}$	**3**-873	Cs coordination is 12
Calcite	$CaCO_3$	Trigonal	D_{3d}^6 (R$\bar{3}$c)	Ca	2b	C_{3i} ($\bar{3}$)	$\mathscr{H}_C^{trig}$	**2**-359	
Cesium chloride	CsCl	Cubic	O_h^1 (Pm$\bar{3}$m)	$CsCl_8$	1a	O_h (m$\bar{3}$m)	$\mathscr{H}_D^{cube}$	**1**-103	Each atom at center of other atom group
Corundum	Al_2O_3	Trigonal ($\approx$hcp)	D_{3d}^6 (R$\bar{3}$c)	AlO_6	4c	C_3 (3)	$\mathscr{H}_C^{trig}$	**2**-6	Slightly-distorted hcp oxygens
Fluorite	CaF_2	Cubic	O_h^5 (Fm$\bar{3}$m)	CaF_8	4a	O_h (m$\bar{3}$m)	$\mathscr{H}_D^{cube}$	**1**-239	
Lanthanum chloride	$LaCl_3$	Hexagonal	C_{6h}^2 ($P6_3$/m)	$LaCl_9$	2d	C_{3h}($\bar{6}$)	$\mathscr{H}_C^{hexa}$	**2**-77	Two different La–Cl distances
Nickel arsenide	NiAs	Hexagonal	C_{6v}^4 (P6/mmm)	$NiAs_6Ni_2$	2a	C_{3v} (3m)	$\mathscr{H}_D^{trig}$	**1**-122	
				$AsNi_6$	2b	C_{3v} (3m)	$\mathscr{H}_D^{trig}$		
Periclase	MgO	Cubic (fcc)	O_h^5 (Fm$\bar{3}$m)	MgO_4	4a	O_h (m$\bar{3}$m)	$\mathscr{H}_D^{cube}$	**1**-85	
Perovskite (cubic)	$BaTiO_3$	Cubic	O_h^1 (Pm$\bar{3}$m)	BaO_{12}	1a	O_h (m$\bar{3}$m)	$\mathscr{H}_D^{cube}$	**1**-391	O + Ba together form fcc lattice
				TiO_6	1b	O_h (m$\bar{3}$m)	$\mathscr{H}_D^{cube}$		
Peroskite (orthorhombic)	$BaTiO_3$	Orthorhombic	C_{2v}^{14} (Amm2)	BaO_{12}	2a	C_{2v} (mm2)	$\mathscr{H}_D^{orth}$	**2**-405	Distortion is of tetragonal form
				TiO_6	2b	C_{2v} (mm2)	$\mathscr{H}_D^{orth}$		
Rutile	TiO_2	Tetragonal	D_{2h}^{14} (P4/mmm)	TiO_6	2a	D_{2h} (mmm)	$\mathscr{H}_D^{orth}$	**1**-250	Two different Ti–O distances

Host lattice	Formula	Crystal system	Space group	Complex	Crystal site	Point symmetry	Hamiltonian	Wyckoff **vol**-page	Comment
Scheelite	$CaWO_4$	Tetragonal	C_{4h}^6 ($I4_1/a$)	CaO_6	4b	$S_4(\bar{4})$	$\mathscr{H}_C^{tetr}$	**3**-19	
Sphalerite or Zinc blende	ZnS	Cubic	T_d^2 ($F\bar{4}3m$)	ZnS_4	4a	T_d ($\bar{4}3\,m$)	$\mathscr{H}_D^{cube}$	**1**-108	ZnS_4 and SZn_4 tetrahedra
Spinel	$MgAl_2O_4$	Cubic (≈fcc)	O_h^7 ($Fd\bar{3}m$)	MgO_4	8a	T_d ($\bar{4}3\,m$)	$\mathscr{H}_D^{cube}$	**3**-75	Almost perfect fcc oxygen packing
				AlO_6	16b	D_{3d} ($\bar{3}\,m$)	$\mathscr{H}_D^{trig}$		
Spinel, inverse	AB_2O_4	Cubic (≈fcc)	O_h^7 ($Fd\bar{3}m$)	BO_4	8a			**3**-76	All A and half B atom in 16b site
				(A, B)O_6	16b				
Tutton salt	$(NH_4)_2SO_4 \cdot MgSO_4 \cdot 6H_2O$	Monoclinic	C_{2h}^5 ($P2_1/a$)	$Mg(H_2O)_6$	2a	C_i ($\bar{1}$)	$\mathscr{H}_C^{tric}$	**3**-821	All Mg-OH_2 Distances not the same
Zincite (Wurzite ZnS)	ZnO	Hexagonal	C_{6v}^4 (P6/mmm)	ZnO_4	2b	C_{3v} (3 m)	$\mathscr{H}_D^{hexa}$	**1**-111	Also OZn_4 tetrahedra
Zircon	$ZrSiO_4$	Tetragonal	D_{4h}^{19} (I4/amd)	ZrO_4	2a	D_{2d} ($\bar{4}2\,m$)	$\mathscr{H}_D^{tetr}$	**3**-15	
				SiO_4	2b	D_{2d} ($\bar{4}2\,m$)	$\mathscr{H}_D^{tetr}$		

[1] R. W. G. Wyckoff, *Crystal Structures* (Wiley, New York, 2nd ed, Vol. 1, 1964; Vol. 2, 1965; Vol. 3, 1965). Wyckoff differs in some space group notations, from that in the International Tables.

CHAPTER VIII

TRANSITION ION DATA TABULATION

Sushil K. Misra

Physics Department, Concordia University, Montreal, Quebec H3G 1M8, Canada

Tabulations of experimental data of spin-Hamiltonian parameters of the various transition ions.

The published data on spin-Hamiltonian parameters (SHP) are included in the following table. The various ions are organized in the following order: ($3d^n$) iron group: $3d^0$ (Sc^{3+}, Ti^{4+}, V^{5+}), $3d^1$ (VO^{2+}, Ti^{3+}, V^{4+}, Cr^{5+}, Mn^{6+}), $3d^2$ (Ti^{2+}, V^{3+}, Cr^{4+}, Mn^{5+}), $3d^3$ (V^{2+}, Cr^{3+}, Mn^{4+}), $3d^4$ (Cr^{2+}, Mn^{3+}, Fe^{4+}), $3d^5$ (Cr^+, Mn^{2+}, Fe^{3+}, Co^{4+}), $3d^6$ (Mn^+, Fe^{2+}, Co^{3+}), $3d^7$ (Mn^0, Fe^+, Co^{2+}, Ni^{3+}, Cu^{4+}), $3d^8$ (Fe^0, Co^+, Ni^{2+}, Cu^{3+}), $3d^9$ (Ni^+, Cu^{2+}), $3d^{10}$ (Cu^+), $3d^{10}4s^1$ (Cu^0, Zn^+); ($4d^n$) palladium group: $4d^0$ (Zr^{4+}, Mo^{6+}), $4d^1$ (Y^{2+}, Zr^{3+}, Nb^{4+}, Mo^{5+}), $4d^2$ (Nb^{3+}, Ru^{6+}), $4d^3$ (Nb^{2+}, Mo^{3+}, Tc^{4+}), $4d^4$ (Ru^{4+}), $4d^5$ (Ru^{3+}, Rh^{4+}), $4d^6$ (Rh^{3+}), $4d^7$ (Ru^+, Rh^{2+}), $4d^9$ (Rh^0, Pd^+, Ag^{2+}), $4d^{10}$ (Ag^+, Cd^{2+}), $4d^{10}5s^1$ (Ag^0, Cd^+, Sn^{3+}); ($5d^n$) platinum group: $5d^1$ (Ta^{4+}, W^{5+}, Re^{6+}), $5d^2$ (Re^{5+}), $5d^3$ (Re^{4+}), $5d^4$ (Pt^{6+}), $5d^5$ (Ir^{4+}), $5d^7$ (Os^+, Ir^{2+}, Pt^{3+}), $5d^9$ (Tl^{2+}, Pb^{3+}), $5d^{10}$ (Tl^+, Pb^{2+}), $5d^{10}6s^1$ (Pb^+); ($4f^n$) lanthanide group: $4f^0$ (Ce^{4+}), $4f^1$ (La^{2+}, Ce^{3+}), $4f^2$ (Pr^{3+}), $4f^3$ (Nd^{3+}), $4f^5$ (Sm^{3+}), $4f^7$ (Eu^{2+}, Gd^{3+}), $4f^8$ (Tb^{3+}), $4f^9$ (Dy^{3+}), $4f^{10}$ (Ho^{3+}), $4f^{11}$ (Ho^{2+}, Er^{3+}), $4f^{12}$ (Tm^{3+}), $4f^{13}$ (Tm^{2+}, Yb^{3+}); ($5f^n$) actinide group: $5f^1$ (U^{5+}), $5f^2$ ($(PuO_2)^{2+}$, U^{4+}), $5f^3$ (U^{3+}, Np^{4+}), $5f^4$ (U^{2+}), $5f^5$ (Am^{4+}, Pu^{3+}), $5f^7$ (Am^{2+}, Cm^{3+}), $5f^9$ (Cf^{3+}). Figure 1 displays elements detected by EPR that appear in data tabulations. References on specific samples are given, except for the cases where the reference cited is either that of a review article (ART72, BD74, BD80, BM73, BM83, BM86, CJ88, Koh79, Msrev91, MWrev90, VJ90), or a book (LB, PF74), the specific reference for the particular sample is to be found in that source. The temperatures at which the parameters have been determined are given in most cases. However, in some cases no temperature was reported for which the temperature column is either blank or contains the notation "NR." The microwave frequency is not specified, in the majority of cases it is that of X-band (~ 9.5 GHz). In the columns of $\tilde{g}$, $\tilde{A}$, and $\tilde{Q}$ tensors, if three values are specified they indicate the three principle values, while if only one value is given it represents the isotropic value unless otherwise indicated. For the cases of axial

(*continued on page 279*)

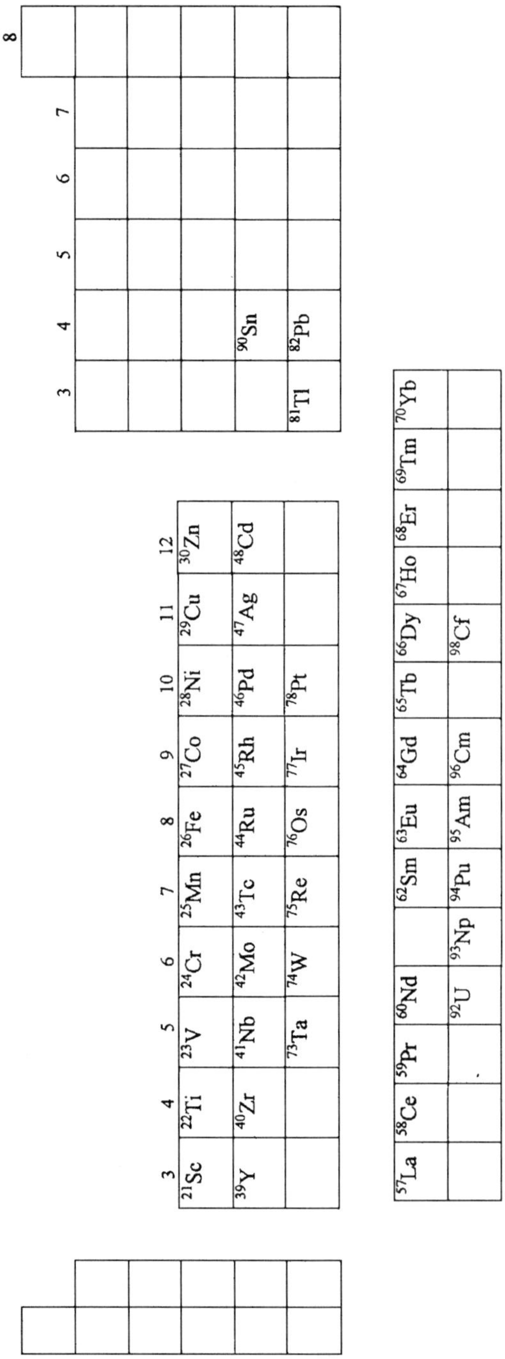

FIG. 1. Elements detected by EPR that appear in the data tabulations.

Sc^{3+} $3d^0$. Data tabulation of spin-Hamiltonian parameters.

Sample	T (°K)	$\tilde{g}$	$\tilde{A}^{45}$ (10^{-4} cm^{-1})	Ref.
BaF	1.14–4.2	$g_1 = 1.956$	$A_1 = -68.3$	BD74
CaF	1.14–4.2	$g_1 = 1.972$	$A_1 = -65.8$	BD74
SrCl	1.14–4.2	$g_1 = 1.953$	$A_1 = -62.7$	BD74
SrF	1.14–4.2	$g_1 = 1.964$	$A_1 = -67.1$	BD74

Ti^{4+} $3d^0$. Data tabulation of spin-Hamiltonian parameters.

Sample	T (°K)	Doping	$\tilde{g}$	D (GHz)	E (GHz)	$\tilde{A}^{47,49}$ (10^{-4} cm^{-1})	Ref.
TiO_2	1.4; 10		$g_c = 1.945$			$A_{110} < 2.5$, $A_{\bar{1}10} < 2.5$, $A_c = 63.9$	PF74
	63		$g_c = 1.446$			$A_{110} = 92.0$, $A_{\bar{1}10} = 40.5$	PF74
Al_2O_3	77	Mg + Ti	$g_x = 2.030$, $g_y = 2.016$, $g_z = 2.003$				BD74
		Li + Ti	$g_x = 2.023$, $g_y = 2.014$, $g_z = 2.003$	±3.05	±0.8		
		Ti + Oxy	$g_x = 2.020$, $g_y = 2.013$, $g_z = 2.003$	±3.44	±0.86		

V^{5+} $3d^0$. Data tabulation of spin-Hamiltonian parameters.

Sample	T (°K)	$\tilde{g}$	Ref.
UCl_5	77–300	1.188	BM86
V_2MoO_8	3.8	$g_x = 1.969$, $g_y = 1.970$, $g_z = 1.874$	BD74

DATA TABULATION (VO^{2+} $3d^1$). Spin-Hamiltonian parameters for VO^{2+} in various host crystals at room temperature, unless otherwise indicated. The values of hyperfine tensors components indicated in ref. Koh79 are absolute values of these components.

Host	T (°K)	Site	$\tilde{g}$	$\tilde{A}$ (10^{-4} cm^{-1})	Ref.
Alums					
$CsAl(SO_4)_2 \cdot 12H_2O$			$g_\parallel = 1.932$, $g_\perp = 1.979$	$\lvert A_\parallel \rvert = 183.4$, $\lvert A_\perp \rvert = 65.7$	Koh79
$KAl(SO_4)_2 \cdot 12H_2O$			$g_\parallel = 1.936$, $g_\perp = 1.977$	$\lvert A_\parallel \rvert = 177.0$, $\lvert A_\perp \rvert = 68.0$	Koh79
$NH_4Al(SO_4)_2 \cdot 12H_2O$			$g_\parallel = 1.940$, $g_\perp = 1.978$	$\lvert A_\parallel \rvert = 176.0$, $\lvert A_\perp \rvert = 67.0$	Koh79
$RbAl(SO_4)_2 \cdot 12H_2O$			$g_\parallel = 1.932$, $g_\perp = 1.975$	$\lvert A_\parallel \rvert = 182.2$, $\lvert A_\perp \rvert = 66.6$	Koh79
Methyl ammonium gallium alum			$g_\parallel = 1.939$, $g_\perp = 1.976$	$\lvert A_\parallel \rvert = 174.2$, $\lvert A_\perp \rvert = 67.0$	Koh79
Formates					
Ba Formate			$g_x = 1.978$, $g_y = 1.981$, $g_z = 1.935$	$A_x = 63.0, 64.0$, $A_z = 173.0$	CJ88
$Sr(HCOO)_2 \cdot 2H_2O$			$g_x = 1.972$, $g_z = 1.944$	$A_x = 66.0$, $A_z = 165.0$	Koh79
$Ba_2Zn(HCOO)_6 \cdot 4H_2O$			$g_x = 1.974$, $g_y = 1.979$, $g_z = 1.943$	$A_x = 60.7$, $A_y = 57.7$, $A_z = 166.5$	CJ88
Glasses					
$\mathbf{R_2O \cdot 4SiO_2}$ (glasses)	293				BM86
R = Na			$g_\parallel = 1.938$, $g_\perp = 1.993$	$A_\parallel = 166$, $A_\perp = 62$	
R = K			$g_\parallel = 1.939$, $g_\perp = 1.995$	$A_\parallel = 167$, $A_\perp = 62$	
R = Cs			$g_\parallel = 1.936$, $g_\perp = 1.995$	$A_\parallel = 166$, $A_\perp = 62$	
$\mathbf{xNa_2O \cdot yZrO_2 \cdot zSiO_2}$ **glasses**					BM86
($x = 20.0$, $y = 5.0$, $z = 75.0$)	77, 293		$g_\parallel = 1.935$, $g_\perp = 1.989$	$A_\parallel = -167.8$, $A_\perp = -62.7$	
($x = 12.5$, $y = 10.0$, $z = 77.5$)	77, 293		$g_\parallel = 1.932$, $g_\perp = 1.993$	$A_\parallel = -172.0$, $A_\perp = 66.0$	
Halides					
CsCl			$g_x = 1.980$, $g_y = 1.979$, $g_z = 1.934$	$A_x = 67.0$, $A_y = 71.0$, $A_z = 180.0$	CJ88
			$g_0 = 1.970$	$A_0 = 103.1$	CJ88
NH_4Br			$g_0 = 1.974$	$A_0 = 308.0$*	CJ88
			$g_x = 1.991$, $g_z = 1.938$	$A_x = 194.0$*, $A_z = 541.0$*	CJ88
NH_4Cl	LT		$g_0 = 1.966$	$A_0 = 103.9$	Koh79
			$g_\parallel = 1.925$, $g_\perp = 1.977$	$\lvert A_\parallel \rvert = 174.0$, $\lvert A_\perp \rvert = 67.8$	Koh79
NH_4I			$g_x = 1.980$, $g_z = 1.948$	$A_x = 188.0$*, $A_z = 517.0$*	CJ88
KBr			$g_0 = 1.965$	$A_0 = 108.8$	Koh79
			$g_\parallel = 1.933$, $g_\perp = 1.986$	$\lvert A_\perp \rvert = 61.8$	Koh79

KCl		$g_0 = 1.969$	$A_0 = 106.8$	Koh79
		$g_\parallel = 1.932$, $g_\perp = 1.988$	$\lvert A_\perp \rvert = 70.0$	Koh79
KI		$g_0 = 1.932$	$A_0 = 109.5$	Koh79
		$g_\parallel = 1.932$, $g_\perp = 1.983$	$\lvert A_\parallel \rvert = 194.5$, $\lvert A_\perp \rvert = 60.2$	Koh79
RbBr		$g_0 = 1.967$	$A_0 = 105.7$	CJ88
		$g_x = 1.981$, $g_z = 1.939$	$A_x = 72.15$, $A_z = 171.4$	CJ88
RbCl		$g_0 = 1.967$	$A_0 = 106.0$	Koh79
		$g_\parallel = 1.939$, $g_\perp = 1.968$	$\lvert A_\perp \rvert = 64.0$	Koh79
RbI		$g_0 = 1.968$	$A_0 = -104.24$	CJ88
BNaCl		$g_0 = 1.971$	$A_0 = 104.0$	Koh79
		$g_\parallel = 1.925$, $g_\perp = 1.996$	$\lvert A_\perp \rvert = 64.2$	Koh79
$BaBr_2 \cdot 2H_2O$		$g_0 = 1.963$	$A_0 = 107.0$	Koh79
		$g_\parallel = 1.928$, $g_\perp = 1.979$	$\lvert A_\parallel \rvert = 183.0$, $\lvert A_\perp \rvert = 73.0$	Koh79
$BaCl_2 \cdot 2H_2O$		$g_0 = 1.966$	$A_0 = 106.0$	Koh79
		$g_\parallel = 1.930$, $g_\perp = 1.977$	$\lvert A_\parallel \rvert = 177.0$, $\lvert A_\perp \rvert = 74.0$	Koh79
$CsGaBr_4$		$g_0 = 1.967$	$A_0 = 108.0$	CJ88
		$g_x = 1.984$, $g_z = 1.929$	$A_x = 71.0$*, $A_z = 182.0$	CJ88
$CsGaCl_4$		$g_0 = 1.964$	$A_0 = 107.0$	CJ88
$(NH_4)_2AlCl_5$	IV	$g_x = 1.979$, $g_z = 1.948$	$A_x = 62.8$, $A_z = 168.8$	Koh79
$(NH_4)_2SbCl_5$	IV	$g_x = 1.979$, $g_z = 1.948$	$A_x = 62.8$, $A_z = 168.8$	Koh79
$(NH_4)_3AlF_6$	IV	$g_x = 1.977$, $g_z = 1.937$	$A_x = 64.0$, $A_y = 64.0$, $A_z = 178.0$	Koh79
$(NH_4)_3VO_2F_4$	I	$g_x = 1.962$, $g_y = 1.962$, $g_z = 1.927$	$A_x = 61.0$, $A_y = 61.0$, $A_z = 174.0$	CJ88
	II	$g_x = 1.961$, $g_y = 1.961$, $g_z = 1.927$	$A_x = 58.0$, $A_y = 58.0$, $A_z = 202.0$	CJ88
$AlCl_6 \cdot 6H_2O$		$g_x = 1.974$, $g_z = 1.929$	$A_x = 72.0$, $A_z = 187.0$	Koh79
$(NH_4)_3InCl_6 \cdot H_2O$	IV	$g_x = 1.985$, $g_z = 1.945$	$A_x = 63.8$, $A_z = 173.0$	Koh79
$Cs_3Sb_2Cl_9$		$g_0 = 1.966$	$A_0 = 106.0$	CJ88
Bis (glycine) $CdCl_2$	I	$g_x = 1.996$, $g_y = 1.971$, $g_z = 1.922$	$A_x = 88.0$, $A_y = 110.0$, $A_z = 198.0$	CJ88
	II	$g_x = 2.003$, $g_y = 1.976$, $g_z = 1.922$	$A_x = 110.0$, $A_y = 145.0$, $A_z = 198.0$	CJ88
Langbeinites				
$K_2Mg_2(SO_4)_3$	I, III	$g_x = 1.980$, $g_y = 1.990$, $g_z = 1.936$	$A_x = 69.7$, $A_y = 96.9$, $A_z = 181.2$	Koh79
	II, IV	$g_x = 1.980$, $g_y = 1.988$, $g_z = 1.936$	$A_x = 69.7$, $A_y = 107.4$, $A_z = 181.2$	Koh79
$(NH_4)_2Mg_2(SO_4)_3$	I, III	$g_x = 1.979$, $g_y = 1.992$, $g_z = 1.932$	$A_x = 72.6$, $A_y = 97.6$, $A_z = 182.7$	Koh79
	II, IV	$g_x = 1.979$, $g_y = 1.988$, $g_z = 1.932$	$A_x = 72.6$, $A_y = 103.3$, $A_z = 182.7$	Koh79
$(NH_4)_2Zn_2(SO_4)_3$	I, III	$g_x = 1.979$, $g_y = 1.993$, $g_z = 1.934$	$A_x = 72.5$, $A_y = 97.0$, $A_z = 183.0$	Koh79
	II, IV	$g_x = 1.979$, $g_y = 1.990$, $g_z = 1.933$	$A_x = 72.5$, $A_y = 107.8$, $A_z = 183.0$	Koh79

DATA TABULATION (VO^{2+} $3d^1$) *(continued)*

Host	T (°K)	Site	$\tilde{g}$	$\tilde{A}$ (10^{-4} cm^{-1})	Ref.
Nitrates					
$CsNO_3$			$g_0 = 1.964$	$A_0 = 105.5$	Koh79
			$g_\parallel = 1.919$, $g_\perp = 1.984$	$\lvert A_\parallel\rvert = 195.4$, $\lvert A_\perp\rvert = 62.1$	Koh79
KNO_3			$g_0 = 1.986$	$A_0 = 106.5$	Koh79
			$g_\parallel = 1.935$, $g_\perp = 1.974$	$\lvert A_\parallel\rvert = 179.8$, $\lvert A_\perp\rvert = 71.9$	Koh79
$NaNO_3$			$g_0 = 1.965$	$A_0 = 105.5$	Koh79
			$g_\parallel = 1.947$, $g_\perp = 1.974$	$\lvert A_\parallel\rvert = 198.2$, $\lvert A_\perp\rvert = 61.8$	Koh79
NH_4NO_3			$g_0 = 1.963$	$A_0 = 106.3$	Koh79
			$g_\parallel = 1.939$, $g_\perp = 1.975$	$\lvert A_\parallel\rvert = 186.5$, $\lvert A_\perp\rvert = 66.4$	Koh79
$Ba(NO_3)_2$			$g_0 = 1.964$	$A_0 = 106.4$	Koh79
			$g_\parallel = 1.919$, $g_\perp = 1.984$	$\lvert A_\parallel\rvert = 178.3$, $\lvert A_\perp\rvert = 72.3$	Koh79
$UO_2(NO_3)_2 \cdot 6H_2O$			$g_0 = 1.964$	$A_0 = 106.8$	Koh79
			$g_\parallel = 1.930$, $g_\perp = 1.981$	$\lvert A_\parallel\rvert = 178.0$, $\lvert A_\perp\rvert = 71.0$	Koh79
Zn (antipyrine)$_2$ $(NO_3)_2$			$g_x = 1.976$, $g_z = 1.934$	$A_x = 65.9$, $A_z = 171.0$	CJ88
Oxalates					
$Cd(COO)_2 \cdot 3H_2O$	295		1.984, 1.984, 1.918	−0.219*, −0.219*, −0.555*	MW89C
	80		1.984, 1.984, 1.917	−0.215*, −0.215*, −0.560*	
	4.2		1.982, 1.982, 1.916	−0.217*, −0.217*, −0.560*	
$Cs_2C_2O_4 \cdot H_2O$			$g_x = 1.977$, $g_y = 1.978$, $g_z = 1.934$	$A_x = 68.5$, $A_y = 67.1$, $A_z = 175.2$	CJ88
$K_2C_2O_4 \cdot H_2O$			$g_x = 1.984$, $g_y = 1.987$, $g_z = 1.945$	$A_x = 65.4$, $A_y = 53.4$, $A_z = 161.7$	CJ88, Koh79
		Powder	$g_x = 1.993$, $g_z = 1.940$	$A_x = 58.5$, $A_z = 177.5$	CJ88
$K_2C_2O_4 \cdot H_2O$	295	I	1.981, 1.983, 1.944	0.182*, 0.149*, 0.483* ($\tilde{Q} = -0.001$*, -0.002*, -0.002*)	MW89A
		II	1.984, 1.978, 1.943	0.181*, 0.152*, 0.499*	
	80	I	1.980, 1.991, 1.943	0.182*, 0.154*, 0.489* ($\tilde{Q} = -0.001$*, -0.001*, -0.002*)	MW89A
		II	1.988, 1.982, 1.940	0.182*, 0.147*, 0.499*	
	4.2	I	1.977, 1.986, 1.943	0.185*, 0.153*, 0.488* ($\tilde{Q} = -0.000$*, -0.002*, 0.002*)	MW89A
		II	1.984, 1.981, 1.934	0.191*, 0.145*, 0.506*	
KHC_2O_4		I	$g_x = 1.993$, $g_y = 1.990$, $g_z = 1.935$	$A_x = 88.0$, $A_y = 74.0$, $A_z = 187.0$	CJ88, Koh79
		II	$g_x = 1.994$, $g_y = 1.989$, $g_z = 1.938$	$A_x = 89.0$, $A_y = 73.0$, $A_z = 188.0$	CJ88, Koh79
		I	$g_x = 1.976$, $g_y = 1.979$, $g_z = 1.930$	$A_x = 67.6$, $A_y = 65.6$, $A_z = 176.3$	CJ88
		II	$g_x = 1.977$, $g_z = 1.934$	$A_x = 69.2$, $A_z = 175.2$	CJ88
		Powder	$g_x = 1.992$, $g_z = 1.936$	$A_x = 81.0$, $A_z = 190.5$	CJ88

$(NH_4)_2C_2O_4 \cdot H_2O$		I	$g_x = 1.982, g_y = 1.980, g_z = 1.935$	$A_x = 70.8, A_y = 69.6, A_z = 178.8$	CJ88
		II	$g_x = 1.985, g_y = 1.992, g_z = 1.936$	$A_x = 65.0, A_y = 62.0, A_z = 179.2$	CJ88
		Powder	$g_z = 1.930$	$A_x = 73.0, A_z = 196.0$	CJ88
$NH_4HC_2O_4 \cdot \frac{1}{2}H_2O$			$g_0 = 1.967$	$A_0 = 96.6$	CJ88
$Rb_2C_2O_4 \cdot H_2O$			$g_x = 1.975, g_y = 1.984, g_z = 1.943$	$A_x = 58.2, A_y = 50.3, A_z = 157.9$	CJ88
$(NH_4)_3VO_2(C_2O_4) \cdot 2H_2O$			$g_x = 1.978, g_z = 1.946$	$A_x = 55.5, A_z = 153.2$	Koh79
$K_2(TiO)(C_2O_4) \cdot 2H_2O$			$g_x = 1.972, g_z = 1.940$	$A_x = 60.0, A_z = 163.0$	Koh79
$K_2(TiO)_2(C_2O_4)_2 \cdot 4H_2O$			$g_x = 1.967, g_y = 1.985, g_z = 1.954$	$A_x = 100$	Koh79
$(tpmp)_2[TiO(C_2O_4)_2]$: $[VO(mnt)_2]$		VIII	$g_x = 1.981, g_y = 1.988, g_z = 1.946$	$A_x = 48.1, A_y = 38.1, A_z = 145.1$	Koh79
Phosphates					
α-$VOPO_4$			$g = 1.965$		CJ88
α-$VOPO_4 \cdot 2H_2O$			$g_x = 1.982, g_z = 1.938$	$A_x = 69.0, A_z = 192.5$	CJ88
KH_2PO_4		Powder	$g_x = 1.976, g_z = 1.923$	$A_x = 77.0, A_z = 204.0$	CJ88
KH_2PO_4		I	$g_x = 1.975, g_z = 1.924$	$A_x = 70.9, A_z = 182.0$	Koh79
		II	$g_x = 1.982, g_z = 1.929$	$A_x = 68.1, A_z = 177.1$	Koh79
		III	$g_x = 1.985, g_z = 1.933$	$A_x = 69.0, A_z = 180.8$	Koh79
$NH_4H_2PO_4$		I	$g_x = 1.985, g_z = 1.929$	$A_x = 72.0, A_z = 181.0$	CJ88, Koh79
		II	$g_x = 1.985, g_z = 1.925$	$A_x = 68.6, A_z = 172.0$	CJ88, Koh79
$(NH_4)H_2PO_4$	77–300	I	$g_\parallel = 1.929, g_\perp = 1.985$	$A_\parallel = 181.0, A_\perp = 72.0$	BM86
		II	$g_\parallel = 1.925, g_\perp = 1.982$	$A_\parallel = 172.0, A_\perp = 68.0$	
$(NH_4)MgPO_4 \cdot 6H_2O$		I	$g_x = 1.978, g_z = 1.929$	$A_x = 76.0, A_z = 190.0$	CJ88
		II	$g_x = 1.980, g_z = 1.932$	$A_x = 77.0, A_z = 189.0$	CJ88
		III	$g_x = 1.979, g_z = 1.929$	$A_x = 76.0, A_z = 189.0$	CJ88
Selenates					
$MgSeO_4 \cdot 6H_2O$			$g_x = 1.979, g_y = 1.981, g_z = 1.934$	$A_x = 72.8, A_y = 71.6, A_z = 179.0$	CJ88
$(NH_4)_2SeO_4$		I	$g_x = 1.985, g_y = 1.989, g_z = 1.920$	$A_x = 73.6, A_y = 77.5, A_z = 188.6$	CJ88, Koh79
		II	$g_x = 1.982, g_z = 1.922$	$A_x = 70.9, A_z = 183.3$	CJ88, Koh79
$(NH_4)_2SeO_4$	298	I	$g_1 = 1.985, g_2 = 1.989, g_3 = 1.920$	$A_1 = 73.6, A_2 = 77.5, A_3 = 188.0$	BM86
		II	$g_\parallel = 1.922, g_\perp = 1.982$	$A_\parallel = 183.3, A_\perp = 70.9$	
$Zn(SeO_4)_2 \cdot 6H_2O$			$g_x = 1.974, g_z = 1.917$	$A_x = 72.8, A_z = 182.0$	CJ88
$K_2CO(SeO_4)_2 \cdot 6H_2O$			$g_x = 1.979, g_y = 1.979, g_z = 1.938$	$A_x = 207.8^*, A_y = 206.7^*, A_z = 528.0^*$	CJ88
Selenites					
$LiH_3(SeO_3)_2$		I	$g_x = 1.982, g_y = 1.985, g_z = 1.932$	$A_x = 63.6, A_y = 59.2, A_z = 171.6$	CJ88, Koh79
		II	$g_x = 1.979, g_y = 1.986, g_z = 1.926$	$A_x = 69.6, A_y = 61.0, A_z = 174.2$	CJ88, Koh79
		III	$g_x = 1.982, g_y = 1.992, g_z = 1.936$	$A_x = 66.7, A_y = 63.2, A_z = 166.1$	CJ88, Koh79

DATA TABULATION (VO^{2+} $3d^1$) *(continued)*

Host	T (°K)	Site	$\tilde{g}$	$\tilde{A}$ (10^{-4} cm^{-1})	Ref.
$LiH_3(SeO_3)_2$	298	I	$g_x = 1.982$, $g_y = 1.932$, $g_z = 1.985$	$A_x = 63.6$, $A_y = 171.6$, $A_z = 59.2$	BM86
		II	$g_x = 1.926$, $g_y = 1.986$, $g_z = 1.979$	$A_x = 174.2$, $A_y = 61.0$, $A_z = 69.6$	
		III	$g_x = 1.992$, $g_y = 1.982$, $g_z = 1.936$	$A_x = 63.2$, $A_y = 66.7$, $A_z = 166.1$	
$NaH_3(SeO_3)_2$			$g_x = 1.988$, $g_y = 1.972$, $g_z = 1.920$	$A_x = 66.0$, $A_y = 61.2$, $A_z = 171.5$	Koh79
$KH_3(SeO_3)_2$			$g_x = 1.992$, $g_y = 1.998$, $g_z = 1.925$	$A_x = 71.3$, $A_y = 65.4$, $A_z = 1174.8$	Koh79
$RbH_3(SeO_3)_2$	298	I	$g_\parallel = 1.920$, $g_\perp = 1.974$	$A_\parallel = 177.0$, $A_\perp = 70.3$	BM86
		II	$g_\parallel = 1.926$, $g_\perp = 1.988$	$A_\parallel = 174.7$, $A_\perp = 71$	
		III	$g_\parallel = 1.923$, $g_\perp = 1.990$	$A_\parallel = 173$, $A_\perp = 78$	
$RbH_3(SeO_3)_2$		I	$g_x = 1.974$, $g_z = 1.920$	$A_x = 70.3$, $A_z = 177.0$	CJ88, Koh79
		II	$g_x = 1.988$, $g_z = 1.926$	$A_x = 71.0$, $A_z = 174.7$	CJ88, Koh79
		III	$g_x = 1.990$, $g_z = 1.923$	$A_x = 78.0$, $A_z = 173.0$	CJ88, Koh79
Sulfates					
$3CdSO_4 \cdot 8H_2O$		I	$g_x = 1.987$, $g_y = 1.999$, $g_z = 1.922$	$A_x = 79.0$, $A_y = 74.0$, $A_z = 199.0$	CJ88
		II	$g_x = 1.995$, $g_y = 1.993$, $g_z = 1.922$	$A_x = 80.0$, $A_y = 75.0$, $A_z = 200.0$	CJ88
		Powder	$g_x = 1.987$, $g_z = 1.923$	$A_x = 78.0$, $A_z = 199.0$	CJ88
$MgSO_4 \cdot 7H_2O$		I	$g_x = 1.987$, $g_y = 1.989$, $g_z = 1.939$	$A_x = 75.5$, $A_y = 72.7$, $A_z = 180.3$	CJ88
		II	$g_z = 1.940$	$A_z = 179.3$	CJ88
$ZnSO_4 \cdot 7H_2O$		I	$g_z = 1.935$	$A_z = -711.0$*	CJ88
		II	$g_z = 1.938$	$A_z = -689.0$*	CJ88
				$A_x = -212.5$*, $A_y = -218.1$*, $A_z = -545.2$*	CJ88
$ZnSO_4 \cdot 7H_2O$		I	$g_x = 1.986$, $g_y = 1.985$, $g_z = 1.945$	$A_x = 71.3$, $A_y = 73.6$, $A_z = 187.0$	Koh79
		II	$g_x = 1.993$, $g_y = 1.985$, $g_z = 1.945$	$A_x = 71.4$, $A_y = 73.2$, $A_z = 177.2$	Koh79
Cs_2SO_4		I	$g_x = 1.995$, $g_y = 1.996$, $g_z = 1.938$	$A_x = 75.9$, $A_y = 72.3$, $A_z = 178.3$	Koh79
		II	$g_x = 1.996$, $g_y = 1.999$, $g_z = 1.967$	$A_x = 72.3$, $A_y = 61.2$, $A_z = 131.4$	Koh79
		III	$g_x = 1.994$, $g_y = 2.001$, $g_z = 1.950$	$A_x = 76.4$, $A_y = 71.0$, $A_z = 163.0$	Koh79
$(NH_4)_2SO_4$		I	$g_x = 1.981$, $g_y = 1.990$, $g_z = 1.930$	$A_x = 74.8$, $A_y = 74.3$, $A_z = 199.1$	CJ88, Koh79
		II	$g_x = 1.979$, $g_y = 1.987$, $g_z = 1.933$	$A_x = 74.2$, $A_y = 74.1$, $A_z = 197.4$	CJ88, Koh79
		III	$g_x = 1.984$, $g_y = 1.993$, $g_z = 1.927$	$A_x = 79.2$, $A_y = 68.6$, $A_z = 200.8$	CJ88, Koh79
$(NH_4)_2SO_4$	110–300	I	$g_x = 1.990$, $g_y = 1.981$, $g_z = 1.930$	$A_x = 74.3$ G, $A_y = 74.8$ G, $A_z = 199.1$ G	BM86
		II	$g_x = 1.987$, $g_y = 1.979$, $g_z = 1.933$	$A_x = 74.1$ G, $A_y = 74.2$ G, $A_z = 197.4$ G	
		III	$g_x = 1.993$, $g_y = 1.984$, $g_z = 1.927$	$A_x = 68.6$ G, $A_y = 79.2$ G, $A_z = 200.8$ G	

Rb_2SO_4		I	$g_x = 1.976, g_y = 1.975, g_z = 1.931$	$A_x = 68.1, A_y = 67.0, A_z = 176.7$	CJ88, Koh79
		II	$g_x = 1.970, g_z = 1.933$	$A_x = 69.3, A_z = 177.2$	CJ88, Koh79
		III	$g_x = 1.972, g_z = 1.931$	$A_x = 66.0, A_z = 173.8$	CJ88, Koh79
		IV	$g_x = 1.974, g_z = 1.933$	$A_z = 177.2$	CJ88, Koh79
		Powder	$g_x = 1.978, g_z = 1.931$	$A_x = 68.0, A_z = 178.0$	CJ88
$LiKSO_4$		Powder	$g_x = 1.986, g_z = 1.928$	$A_x = 74.0, A_z = 187.0$	CJ88
$LiNH_4SO_4$		I	$g_x = 2.001, g_y = 1.981, g_z = 1.931$	$A_x = 77.0, A_y = 69.0, A_z = 206.0$	CJ88
		II	$g_x = 2.007, g_y = 1.977, g_z = 1.932$	$A_x = 83.0, A_y = 72.0, A_z = 205.0$	CJ88
NH_4HSO_4		I		$A_x = 75.0, A_y = 71.4, A_z = 204.0$	CJ88
		II		$A_x = 75.7, A_y = 70.0, A_z = 203.8$	CJ88
	295.5	I		$A_x = 69.3, A_y = 66.0, A_z = 184.4$	Koh79
		II		$A_x = 70.0, A_y = 64.7, A_z = 184.2$	Koh79
	353.2	I		$A_x = 70.4, A_y = 63.9, A_z = 183.9$	Koh79
		II		$A_x = 70.5, A_y = 63.6, A_z = 183.7$	Koh79
$Na_2Zn(SO_4)_2 \cdot 4H_2O$		I	$g_x = 1.985, g_z = 1.927$	$A_x = 72.0, A_z = 184.9$	CJ88, Koh79
		II	$g_x = 1.988, g_z = 1.930$	$A_x = 73.0, A_z = 183.4$	CJ88, Koh79
		III	$g_x = 1.987, g_z = 1.932$	$A_x = 70.0, A_z = 181.0$	CJ88, Koh79
$Na_2Zn(SO_4)_2 \cdot 4H_2O$	305	I	$g_\parallel = 1.927, g_\perp = 1.985$	$A_\parallel = 184.9, A_\perp = 72.0$	BM86
		II	$g_\parallel = 1.930, g_\perp = 1.988$	$A_\parallel = 183.4, A_\perp = 73$	
		III	$g_\parallel = 1.932, g_\perp = 1.987$	$A_\parallel = 181.0, A_\perp = 70$	
$K_3Cd(SO_4)_2 \cdot 6H_2O$		I	$g_x = 1.978, g_y = 1.991, g_z = 1.938$	$A_x = 72.0, A_y = 72.0, A_z = 181.0$	CJ88
		II	$g_x = 1.989, g_y = 1.986, g_z = 1.939$	$A_x = 72.0, A_y = 71.0, A_z = 181.0$	CJ88
			$g_x = 1.986, g_y = 1.994, g_z = 1.938$	$A_x = 82.0, A_y = 69.0, A_z = 188.0$	CJ88
		Powder	$g_x = 1.986, g_z = 1.930$	$A_x = 74.0, A_z = 196.0$	CJ88
		I		$A_x = -218.3^*, A_y = -211.7^*, A_z = -559.1^*$	CJ88
		II		$A_x = -212.3^*, A_y = -205.7^*, A_z = -548.0^*$	CJ88
$(NH_4)_3H(SO_4)_2$		I		$A_x = 87.0, A_y = 84.0, A_z = 234.0$	CJ88
		II		$A_x = 86.0, A_y = 83.0, A_z = 234.0$	CJ88
			$g_x = 1.976, g_y = 1.973, g_z = 1.917$	$A_x = 73.3, A_y = 75.7, A_z = 204.1$	CJ88
$(ND_4)_3D(SO_4)_2$			$g_0 = 1.955$	$A_0 = 117.7$ G	CJ88
$(NH_4)_6Sb(SO_4)_3F_{12}$			$g_0 = 1.970$	$A_0 = 110.8$	CJ88
			$g_x = 1.981, g_z = 1.942$	$A_x = 71.0, A_z = 192.3$	CJ88
		I	$g_x = 1.997, g_y = 1.993, g_z = 1.927$	$A_x = 82.0, A_y = 76.0, A_z = 200.0$	CJ88
$HONH_3LiSO_4 \cdot 6H_2O$					
		II	$g_x = 1.996, g_y = 1.993, g_z = 1.929$	$A_x = 80.0, A_y = 74.0, A_z = 198.0$	CJ88
$(EDTA)_2Mg(SO_4)_2 \cdot 6H_2O$			$g_x = 1.987, g_y = 1.983, g_z = 1.951$	$A_x = 80.7, A_y = 64.8, A_z = 151.3$	Koh79

DATA TABULATION (VO^{2+} $3d^1$) *(continued)*

Host	T (°K)	Site	$\tilde{g}$	$\tilde{A}$ (10^{-4} cm^{-1})	Ref.
Tutton salts					
$Cs_2Cd(SO_4)_2 \cdot 6H_2O$		Powder	$g_x = 1.987, g_z = 1.921$	$A_x = 70.0, A_z = 177.0$	CJ88
$Cs_2Co(SeO_4)_2 \cdot 6H_2O$			$g_x = 1.979, g_y = 1.979, g_z = 1.932$	$A_x = 71.0, A_y = 69.7, A_z = 178.5$	CJ88
$Cs_2Fe(SO_4)_2 \cdot 6H_2O$		I	$g_x = 1.989, g_y = 1.990, g_z = 1.938$	$A_x = 73.2, A_y = 71.4, A_z = 181.3$	Koh79
		II	$g_x = 1.988, g_y = 1.992, g_z = 1.941$	$A_x = 72.1, A_y = 71, A_z = 178.9$	Koh79
$Cs_2Mg(SO_4)_2 \cdot 6H_2O$			$g_x = 1.979, g_y = 1.979, g_z = 1.933$	$A_x = 73.93, A_y = 73.94, A_z = 183.26$	Koh79
$Cs_2Zn(SO_4)_2 \cdot 6H_2O$		I	$g_x = 1.985, g_y = 1.983, g_z = 1.936$	$A_x = 71.6, A_y = 70.0, A_z = 181.0$	Koh79
		II	$g_x = 1.986, g_y = 1.984, g_z = 1.934$	$A_x = 72.0, A_y = 69.0, A_z = 180.0$	Koh79
$K_2Cd(SO_4)_2 \cdot 6H_2O$		I	$g_x = 1.978, g_y = 1.991, g_z = 1.938$	$A_x = 72, A_y = 77, A_z = 181$	Koh79
		II	$g_x = 1.989, g_y = 1.986, g_z = 1.939$	$A_x = 72, A_y = 71, A_z = 181$	Koh79
$K_2Cd(SO_4)_2 \cdot 6H_2O$	295		1.986, 1.982, 1.917	$-0.222^*, -0.209^*, -0.549^*$ ($\tilde{Q} = -0.001^*, -0.004^*, 0.005^*$)	MW89E
$K_2Co(SO_4)_2 \cdot 6H_2O$		I	$g_x = 1.996, g_y = 1.995, g_z = 1.940$	$A_x = 71, A_y = 68, A_z = 177$	Koh79
		II	$g_x = 1.986, g_z = 1.939$	$A_x = 72, A_z = 181$	Koh79
$K_2Mg(SO_4)_2 \cdot 6H_2O$		I		$A_x = -212.0^*, A_y = -207.0^*, A_z = -543.5^*$	CJ88
		II		$(A_x + A_y)/2 = 201.0^*, A_z = -535.0^*$	CJ88
		III		$(A_x + A_y)/2 = -196.0^*, A_z = -528.0^*$	CJ88
		I	$g_x = 2.005, g_y = 1.997, g_z = 1.930$	$A_x = 67, A_y = 77, A_z = 178$	Koh79
		II	$g_x = 2.002, g_y = 1.998, g_z = 1.931$	$A_x = 63, A_y = 69, A_z = 181$	Koh79
$K_2Zn(SO_4)_2 \cdot 6H_2O$				$A_x = -211.7^*, A_y = -206.7^*, A_z = -543.4^*$	CJ88
		I	$g_x = 2.004, g_y = 1.996, g_z = 1.933$	$A_x = 62, A_y = 75, A_z = 179$	Koh79
		II	$g_x = 2.005, g_y = 1.998, g_z = 1.930$	$A_x = 68, A_y = 69, A_z = 184$	Koh79
$(NH_4)_2Cd(SO_4)_2 \cdot 6H_2O$		I	$g_x = 1.986, g_y = 1.988, g_z = 1.935$	$A_x = 71.3, A_y = 69.0, A_z = 180.0$	CJ88, Koh79
		II	$g_x = 1.982, g_y = 1.984, g_z = 1.931$	$A_x = 71.4, A_y = 69.5, A_z = 183.9$	CJ88, Koh79
		I	$g_x = 1.999, g_y = 1.992, g_z = 1.947$	$A_x = 206.8^*, A_y = 200.7^*, A_z = 517.8^*$	CJ88
		II	$g_x = 1.998, g_y = 1.993, g_z = 1.945$	$A_x = 206.9^*, A_y = 200.8^*, A_z = 517.2^*$	CJ88
		Powder	$g_x = 1.994, g_z = 1.946$	$A_x = 74.0, A_z = 190.0$	CJ88
$(NH_4)_2Cd(SO_4)_2 \cdot 6H_2O$	295	I	$g_1 = 1.936, g_2 = 1.981, g_3 = 1.985$	$A_1 = -0.534^*, A_2 = -0.216^*, A_3 = -0.200^*$	MS90A
		II	$g_1 = 1.924, g_2 = 1.980, g_3 = 1.985$	$A_1 = -0.536^*, A_2 = -0.219^*, A_3 = -0.203^*$	
	80	I	$g_1 = 1.948, g_2 = 1.996, g_3 = 1.994$	$A_1 = -0.533^*, A_2 = -0.222^*, A_3 = -0.202^*$	
		II	$g_1 = 1.949, g_2 = 1.995, g_3 = 1.994$	$A_1 = -0.561^*, A_2 = -0.219^*, A_3 = -0.198^*$	
	4.2	I	$g_1 = 1.938, g_2 = 1.999, g_3 = 1.989$	$A_1 = -0.551^*, A_2 = -0.234^*, A_3 = -0.194^*$	
		II	$g_1 = 1.942, g_2 = 1.990, g_3 = 1.992$	$A_1 = -0.526^*, A_2 = -0.240^*, A_3 = -0.195^*$	

$(NH_4)_2Co(SO_4)_2 \cdot 6H_2O$		I	$g_x = 2.001$, $g_y = 1.990$, $g_z = 1.938$	$A_x = 77.0$, $A_y = 82.0$, $A_z = 201.0$	CJ88, Koh79
		II	$g_x = 2.001$, $g_y = 1.991$, $g_z = 1.936$	$A_x = 77.0$, $A_y = 81.0$, $A_z = 202.0$	CJ88, Koh79
			$g_x = 1.982$, $g_z = 1.932$	$A_x = 72.0$, $A_z = 174.0$	CJ88
$(NH_4)_2Co(SO_4)_2 \cdot 6H_2O$	295	I	$g_1 = 1.960$, $g_2 = 1.996$, $g_3 = 1.997$	$A_1 = -0.533^*$, $A_2 = -0.195^*$, $A_3 = -0.178^*$	MSL91
		II	$g_1 = 1.953$, $g_2 = 1.985$, $g_3 = 1.999$	$A_1 = -0.511^*$, $A_2 = -0.253^*$, $A_3 = -0.166^*$	
$(NH_4)_2Fe(SO_4)_2 \cdot 6H_2O$			$g_x = 1.990$, $g_z = 1.941$	$A_x = 73.0$, $A_z = 180.0$	CJ88, Koh79
			$g_0 = 1.970$	$A_0 = 302.0$ MHz	CJ88, Koh79
$(NH_4)_2Fe(SO_4)_2 \cdot 6H_2O$	295	I	$g_1 = 1.936$, $g_2 = 1.981$, $g_3 = 1.985$	$A_1 = -0.534^*$, $A_2 = -0.214^*$, $A_3 = -0.181^*$	MSO90
		II	$g_1 = 1.937$, $g_2 = 1.980$, $g_3 = 1.974$	$A_1 = -0.541^*$, $A_2 = -0.258^*$, $A_3 = -0.178^*$	
	80	I	$g_1 = 1.936$, $g_2 = 1.991$, $g_3 = 1.987$	$A_1 = -0.532^*$, $A_2 = -0.217^*$, $A_3 = -0.184^*$	
		II	$g_1 = 1.940$, $g_2 = 1.983$, $g_3 = 1.966$	$A_1 = -0.563^*$, $A_2 = -0.257^*$, $A_3 = -0.201^*$	
	4.2	I	$g_1 = 1.936$, $g_2 = 1.981$, $g_3 = 1.985$	$A_1 = -0.534^*$, $A_2 = -0.214^*$, $A_3 = -0.181^*$	
		II	$g_1 = 1.924$, $g_2 = 1.977$, $g_3 = 1.975$	$A_1 = -0.524^*$, $A_2 = -0.237^*$, $A_3 = -0.157^*$	
$(NH_4)_2Mg(SO_4)_2 \cdot 6H_2O$		I	$g_x = 2.006$, $g_y = 1.988$, $g_z = 1.942$	$A_x = 79.0$, $A_y = 81.0$, $A_z = 202.0$	CJ88, Koh79
		II	$g_x = 2.003$, $g_y = 1.997$, $g_z = 1.9372$	$A_x = 76.0$, $A_y = 89.0$, $A_z = 199.0$	CJ88, Koh79
		Powder	$g_x = 1.988$, $g_z = 1.937$	$A_x = 74.0$, $A_z = 190.0$	CJ88
$(NH_4)_2Mg(SO_4)_2 \cdot 6H_2O$	295	I	$g_1 = 1.942$, $g_2 = 1.983$, $g_3 = 1.986$	$A_1 = -0.523^*$, $A_2 = -0.222^*$, $A_3 = -0.207^*$	MS90A
		II	$g_1 = 1.938$, $g_2 = 1.983$, $g_3 = 1.986$	$A_1 = -0.532^*$, $A_2 = -0.223^*$, $A_3 = -0.194^*$	
	80	I	$g_1 = 1.951$, $g_2 = 1.996$, $g_3 = 1.994$	$A_1 = -0.522^*$, $A_2 = -0.259^*$, $A_3 = -0.181^*$	
		II	$g_1 = 1.951$, $g_2 = 1.994$, $g_3 = 1.994$	$A_1 = -0.536^*$, $A_2 = -0.235^*$, $A_3 = -0.183^*$	
	4.2	I	$g_1 = 1.939$, $g_2 = 1.992$, $g_3 = 1.987$	$A_1 = -0.523^*$, $A_2 = -0.261^*$, $A_3 = -0.186^*$	
		II	$g_1 = 1.947$, $g_2 = 1.992$, $g_3 = 1.988$	$A_1 = -0.553^*$, $A_2 = -0.235^*$, $A_3 = -0.197^*$	
$(NH_4)_2Zn(SO_4)_2 \cdot 6H_2O$				$A_x = -261.9^*$, $A_y = -212.2^*$, $A_z = -547.9^*$	CJ88
		I	$g_x = 1.981$, $g_y = 1.980$, $g_z = 1.933$	$A_x = 71.4$, $A_y = 72.6$, $A_z = 182.8$	Koh79
		II	$g_x = 1.981$, $g_y = 1.980$, $g_z = 1.932$	$A_x = 71.0$, $A_y = 72.6$, $A_z = 182.8$	Koh79
		III	$g_x = 1.981$, $g_z = 1.930$	$A_x = 72.5$, $A_z = 184.4$	Koh79
$(NH_4)_2Zn(SO_4)_2 \cdot 6H_2O$	295	I	$g_1 = 1.938$, $g_2 = 1.985$, $g_3 = 1.974$	$A_1 = -0.565^*$, $A_2 = -0.288^*$, $A_3 = -0.172^*$	MS90
		II	$g_1 = 1.932$, $g_2 = 1.982$, $g_3 = 1.988$	$A_1 = -0.534^*$, $A_2 = -0.259^*$, $A_3 = -0.168^*$	
$(NH_4)_2Zn(SO_4)_2 \cdot 6H_2O$	room		$g_z = 1.972$, $g_y = 1.973$	$A_\parallel = 82.63$, $A_\perp = 82.46$	LB
$Rb_2Fe(SO_4)_2 \cdot 6H_2O$		I	$g_x = 1.989$, $g_y = 1.992$, $g_z = 1.938$	$A_x = 74.5$, $A_y = 72.5$, $A_z = 183.9$	CJ88
		II	$g_x = 1.984$, $g_y = 1.990$, $g_z = 1.936$	$A_x = 76.3$, $A_y = 72.0$, $A_z = 183.6$	CJ88
$Rb_2Mg(SO_4)_2 \cdot 6H_2O$		I	$g_x = 1.984$, $g_y = 1.986$, $g_z = 1.936$	$A_x = 72.0$, $A_y = 71.0$, $A_z = 181.7$	CJ88
		II	$g_x = 1.982$, $g_y = 1.984$, $g_z = 1.936$	$A_x = 71.5$, $A_y = 70.0$, $A_z = 179.8$	CJ88
		Powder	$g_x = 1.986$, $g_z = 1.938$	$A_x = 72.0$, $A_z = 190.0$	CJ88

DATA TABULATION (VO^{2+} $3d^1$) *(continued)*

Host	T (°K)	Site	$\tilde{g}$	$\tilde{A}$ (10^{-4} cm^{-1})	Ref.
$Rb_2Zn(SO_4)_2 \cdot 6H_2O$		I	$g_x = 1.978$, $g_y = 1.982$, $g_z = 1.931$	$A_x = 70.3$, $A_y = 69.0$, $A_z = 180.0$	CJ88, Koh79
		II	$g_x = 1.985$, $g_z = 1.937$	$A_x = 71.0$, $A_y = 70.0$, $A_z = 179.6$	CJ88, Koh79
		Powder	$g_x = 1.980$, $g_z = 1.934$	$A_x = 78.0$, $A_y = 69.0$, $A_z = 190.0$	CJ88
				$A_x = 71.0$, $A_z = 180.0$	
				$A_x = 71.5$, $A_z = 179.6$	
				$A_x = 71.0$, $A_z = 190.0$	
		Powder	$g_x = 1.980$, $g_z = 1.934$	$A_x = 71.0$, $A_z = 190.0$	CJ88
Tutton salts (with Selenates)					
$Cs_2Zn(SeO_4)_2 \cdot 6H_2O$			$g_x = 1.975$, $g_y = 1.978$, $g_z = 1.938$	$A_x = 71.3$, $A_y = 70.4$, $A_z = 178.0$	CJ88
$K_2Mg(SeO_4)_2 \cdot 6H_2O$			$g_x = 1.980$, $g_y = 1.982$, $g_z = 1.937$	$A_x = 210.0$*, $A_y = 209.0$*, $A_z = 533.0$*	CJ88
$(NH_4)_2Co(SeO_4)_2 \cdot 6H_2O$			$g_x = 1.979$, $g_z = 1.935$	$A_x = 70.4$, $A_z = 177.6$	CJ88
$Rb_2Co(SeO_4)_2 \cdot 6H_2O$			$g_x = 1.979$, $g_y = 1.980$, $g_z = 1.932$	$A_x = 71.5$, $A_y = 70.0$, $A_z = 177.9$	CJ88
$(NH_4)_3Zn(SeO_4)_2 \cdot 6H_2O$		I	$g_x = 1.983$, $g_y = 1.985$, $g_z = 1.935$	$A_x = 73.0$, $A_y = 72.0$, $A_z = 182.0$	CJ88
		II	$g_x = 1.986$, $g_y = 1.984$, $g_z = 1.935$	$A_x = 71.0$, $A_y = 70.0$, $A_z = 181.0$	CJ88
Other inorganic compounds					
SiO_2, α-Cristobalite			$g_x = 1.977$, $g_z = 1.940$	$A_x = 61.0$, $A_z = 166.0$	CJ88
$Mg(ClO_4)_2 \cdot 6H_2O$			$g_0 = 1.9664$	$A_0 = 107.7$	CJ88
			$g_x = 1.979$, $g_y = 1.981$, $g_z = 1.934$	$A_x = 72.8$, $A_y = 71.6$, $A_z = 179.0$	CJ88
$K_4[Fe(CN)_6] \cdot 3H_2O$	90		1.992	-55.5	LB
	room			$A^{51} = -55.5$, $A^{50} = -21.1$	LB
$Mg(CH_3COO)_2 \cdot 4H_2O$			$g_x = 1.981$, $g_z = 1.922$	$A_x = 66.7$, $A_z = 182.8$	Koh79
Organic compounds					
α-Ca Formate		I	$g_x = 1.982$, $g_z = 1.939$	$A_x = 67.0$, $A_z = 178.0$	CJ88
		II	$g_x = 1.981$, $g_z = 1.938$	$A_x = 68.0$, $A_z = 179.6$	CJ88
		III	$g_x = 1.982$, $g_z = 1.939$	$A_x = 67.0$, $A_z = 178.0$	CJ88
Al (urea) $_6Cl_3 \cdot 6H_2O$			$g_x = 1.987$, $g_z = 1.936$	$A_x = 64.0$, $A_z = 179.0$	Koh79
Al (urea) $_6(ClO_4)_3 \cdot 6H_2O$			$g_x = 1.987$, $g_z = 1.932$	$A_x = 64.0$, $A_z = 175.0$	Koh79
Al (urea) $_6I_3 \cdot 6H_2O$			$g_x = 1.987$, $g_z = 1.934$	$A_x = 64.0$, $A_z = 182.0$	Koh79
β-Ni(mesalim)$_2$			$g_x = 1.984$, $g_y = 1.980$, $g_z = 1.954$	$A_x = 59.7$, $A_y = 48.4$, $A_z = 156.3$	Koh79
bis-Zn(benzac)$_2 \cdot C_2H_2OH$			$g_x = 1.981$, $g_y = 1.976$, $g_z = 1.941$	$A_x = 60.5$, $A_y = 55.3$, $A_z = 164.8$	Koh79
$Ca(d \cdot C_4H_4O_6) \cdot 4H_2O$		I	$g_x = 1.978$, $g_y = 1.984$, $g_z = 1.913$	$A_x = 70.2$, $A_y = 66.7$, $A_z = 172.4$	Koh79
		II	$g_x = 1.978$, $g_y = 1.981$, $g_z = 1.934$	$A_x = 64.7$, $A_y = 63.8$, $A_z = 168.9$	Koh79

$NaKC_4H_4O_6 \cdot 4H_2O$		I	$g_x = 1.979$, $g_y = 1.987$, $g_z = 1.958$	$A_x = 53.0$, $A_y = 42.0$, $A_z = 150.4$	CJ88
		II	$g_x = 1.989$, $g_y = 1.992$, $g_z = 1.959$	$A_x = 37.8$, $A_y = 366.8$, $A_z = 148.5$	CJ88
		III	$g_x = 1.996$, $g_z = 1.956$	$A_x = 40.0$, $A_z = 159.0$	CJ88
		IV	$g_z = 1.947$	$A_z = 162.0$	CJ88
			$g_x = 1.973$, $g_y = 1.980$, $g_z = 1.947$	$A_x = 55.0$, $A_y = 33.6$, $A_z = 152.2$	CJ88
$SrC_4H_4O_6 \cdot 4H_2O$		I	$g_x = 1.996$, $g_y = 2.009$, $g_z = 1.947$	$A_x = 68.0$, $A_y = 72.0$, $A_z = 186.0$	CJ88
		II	$g_x = 2.000$, $g_y = 2.002$, $g_z = 1.942$	$A_x = 69.0$, $A_y = 70.3$, $A_z = 186.0$	CJ88
		Powder	$g_x = 1.990$, $g_z = 1.943$	$A_x = 68.5$, $A_z = 185.4$	CJ88
$(CH_3NH_3)_2SbCl_6$			$g_0 = 1.965$	$A_0 = 107.0$	CJ88
Dimethylalanine		I	$g_x = 1.994$, $g_z = 1.925$	$A_x = 77.0$, $A_y = 59.0$, $A_z = 182.0$	Koh79
		II	$g_x = 1.991$, $g_z = 1.939$	$A_x = 82.0$, $A_y = 46.0$, $A_z = 181.0$	Koh79
Glycine			$g_x = 1.986$, $g_z = 1.943$	$A_x = 85.0$, $A_y = 59.0$, $A_z = 156.0$	Koh79
l-Alanine			$g_x = 2.006$, $g_z = 1.955$	$A_x = 57.0$, $A_y = 66.0$, $A_z = 171.0$	Koh79
$Na_3C_6H_5O_7 \cdot 5H_2O$			$g_x = 1.998$, $g_y = 1.992$, $g_z = 1.934$	$A_x = 70.0$, $A_y = 66.6$, $A_z = 175.4$	CJ88
Potassium succinate		I	$g_x = 1.978$, $g_z = 1.926$	$A_x = 69.4$, $A_z = 178.2$	CJ88
		II	$g_x = 1.997$, $g_z = 1.936$	$A_x = 65.8$, $A_z = 170.3$	CJ88
Rochelle salt			$g_x = 1.979$, $g_z = 1.955$	$A_x = 53.8$, $A_z = 150.8$	Koh79
Thiourea			$g_0 = 1.968$	$A_0 = 104.1$	CJ88
			$g_x = 1.985$, $g_z = 1.936$	$A_x = 70.4$, $A_z = 169.0$	CJ88
trans · Pd (benzac)$_2$			$g_x = 1.981$, $g_y = 1.976$, $g_z = 1.943$	$A_x = 61.4$, $A_y = 62.1$, $A_z = 164.3$	Koh79
Tri Sarcosine calcium		I		$A_x = 70.0$, $A_y = 68.5$, $A_z = 185.0$	CJ88
chloride		II		$A_x = 76.2$, $A_y = 68.4$, $A_z = 182.5$	CJ88
		III		$A_x = 70.4$, $A_y = 67.9$, $A_z = 185.2$	CJ88
		IV		$A_x = 74.6$, $A_y = 68.1$, $A_z = 190.0$	CJ88
		V		$A_x = 74.6$, $A_y = 70.0$, $A_z = 190.2$	CJ88
Tris (sarcosine) $CaCl_2$		I	$g_x = 1.987$, $g_z = 1.911$	$A_x = 79.0$, $A_z = 174.0$	Koh79
		II	$g_x = 1.991$, $g_z = 1.930$	$A_x = 72.0$, $A_z = 178.0$	Koh79
Urea oxalate			$g_0 = 1.967$	$A_0 = 96.06$	CJ88
			$g_x = 1.984$, $g_z = 1.931$	$A_x = 66.90$, $A_z = 155.3$	CJ88
$[Mg(H_2O)_6H_2$ EDTA]			$g_x = 1.983$, $g_y = 1.987$, $g_z = 1.952$	$A_x = 64.8$, $A_y = 80.7$, $A_z = 151.3$	CJ88
		I	$g_x = 1.977$, $g_y = 1.991$, $g_z = 1.939$	$A_x = 60.0$, $A_y = 67.3$, $A_z = 175.4$	CJ88
		II	$g_x = 1.979$, $g_y = 1.985$, $g_z = 1.936$	$A_x = 62.0$, $A_y = 65.6$, $A_z = 173.5$	CJ88
Triglycine Sulphate		I	$g_x = 1.985$, $g_y = 1.998$, $g_z = 1.929$	$A_x = 73.3$, $A_y = 71.8$, $A_z = 178.2$	Koh79
		II	$g_x = 1.985$, $g_y = 1.998$, $g_z = 1.929$	$A_x = 73.5$, $A_y = 70.6$, $A_z = 178.4$	Koh79
$NaKC_4H_4O_6 \cdot 6H_2O$	150–300		$g_x = 1.974$, $g_y = 1.973$, $g_z = 1.980$	$A_x = -152.2$, $A_y = 55.0$, $A_z = 33.6$	BM86

Ti^{3+} $3d^1$. Data tabulation of spin-Hamiltonian parameters.

Sample	T (°K)	Site	$\tilde{g}$	$\tilde{A}^{47,49}$ (10^{-4} cm^{-1})	Ref.
CdS	27		$g_x = 1.961$, $g_y = 1.950$, $g_z = 1.865$		BD74
CsAl	1.3–4.2		$g_\parallel = 1.194$, $g_\perp = 0.6674$	$D = 39.3$ MHz	BD80
MgO	20		$g_x = 1.849$, $g_y = 1.946$, $g_z = 1.767$	$A_z = 12.5$	ART 72
B–Ti–Li	20–77		1.918		BM73
B–Ti–H	20–77		1.908		
CaF_2	300		$g_\parallel = 1.762$, $g_\perp = 1.849$	$A_\perp = 33.7$ G	BD80
TiO_2	4.2		$g_{110} = 1.974$, $g_{\bar{1}10} = 1.977$, $g_c = 1.941$		PF74
TiF_3 (p)	77		1.91		LB
$KTaO_2$	77, 300		$g_\parallel = 1.997$, $g_\perp = 1.904$	$A^{47} = 10.8$, $A^{49} = 11.4$	BM83
Al_2O_3	4.2		$g_\parallel = 1.067$, $g_\perp < 0.1$		PF74
Ti_2O_3	4.2		$g_\parallel = 2.003$, $g_\perp = 1.904$		BD80
$AlCl_3 \cdot 6H_2O$ (p)	77; 4		1.93		LB
$AlCl_3 \cdot 6H_2O$ (p)	77		$g_{max} = 1.91$, $g_{min} = 1.84$		LB
	200		1.91		LB
$TiBr_3 \cdot 6H_2O$ (p)	77		1.90		LB
$TiI_3 \cdot 6H_2O$ (p)	77		1.91		LB
$CuAlS_2$	21, 80	Tetragonal	$g_\parallel = 1.902$, $g_\perp = 1.978$	$A_\parallel^{47,49} = 26.3$	BD80
	4.2		$g_1 = 1.526$, $g_2 = 1.300$, $g_3 = 1.430$	$A_\perp^{47,49} = 18.3$	BD80
$BaBSiO_4$	15–300				PF74
$ZnWO_4$	77		1.91		LB
$CsTi(SO_4)_2 \cdot 12H_2O$	4.2–2.5		$g_\parallel = 1.25$, $g_\perp = 1.14$		LB
$CsTi(SO_4)_2 \cdot 12H_2O$ (p)	7.88		1.53		LB
	6.33		1.35		LB
$CsTi(SO_4)_2 \cdot 12H_2O$	295		$g_x = 1.14$, $g_y = 1.14$, $g_z = 1.25$		WM68
$CsAl(SO_4)_2 \cdot 12H_2O$	295	I	$g_x = 0.23$, $g_y = 0.23$, $g_z = 1.166$		WM68
		II	$g_x = 0.696$, $g_y = 0.696$, $g_z = 1.192$		
		III	$g_x = 0.931$, $g_y = 0.931$, $g_z = 1.241$		
$KAl(SO_4)_2 \cdot 12H_2O$	4.2, 295		$g_x = 1.828$, $g_y = 1.898$, $g_z = 1.979$		MS75, BD80
$NaAl(SO_4)_2 \cdot 12H_2O$	295	I	~ 1.93		WM68
$RbAl(SO_4)_2 \cdot 12H_2O$	4.2		$g_x = 1.715$, $g_y = 1.767$, $g_z = 1.895$		LB
$RbAl(SO_4)_2 \cdot 12H_2O$	295		$g_x = 1.715$, $g_y = 1.767$, $g_z = 1.895$		WM68
$TlAl(SO_4)_2 \cdot 12H_2O$	295		$g_x = 1.790$, $g_y = 1.834$, $g_z = 1.938$		WM68

$K_3[Ti(CN)_6] \cdot 2KCN$ (*p*)	room	1.99	LB
$(.7 - x)B_2O_3 \cdot 3CaO \cdot xTiO_2$		1.942	ART 72
$xP_2O_5(1 - x)Al_2O_3$ glasses			BM83
($x = 0.725$)	77	$g_1 = 1.910$, $g_2 = 1.870$, $g_3 = 1.805$	
	143	$g_1 = 1.924$, $g_2 = 1.894$, $g_3 = 1.845$	
($x = 0.700$)	108	$g_1 = 1.919$, $g_2 = 1.870$, $g_3 = 1.810$	
	289	$g_1 = 1.930$, $g_2 = 1.915$, $g_3 = 1.880$	
($x = 0.625$)	77	$g_1 = 1.912$, $g_2 = 1.865$, $g_3 = 1.780$	
($x = 0.600$)	108	$g_1 = 1.918$, $g_2 = 1.876$, $g_3 = 1.806$	
	328	$g_1 = 1.932$, $g_2 = 1.917$, $g_3 = 1.884$	
($x = 0.300$)	108	$g_1 = 1.935$, $g_2 = 1.918$, $g_3 = 1.868$	
	332	$g_1 = 1.952$, $g_2 = 1.930$, $g_3 = 1.885$	
	474	$g_1 = 1.960$, $g_2 = 1.940$, $g_3 = 1.900$	
$Na_2O–TiO_2–SiO_2$ glass		1.93	ART 72
$30CaO \cdot 69B_2O_3 \cdot 1TiO_2$ (oxide glass)	NR	1.95	BD74
K borate glasses	113, 293	$g_{\parallel} = 1.939$, $g_{\perp} = 1.975$	BD74

V^{4+} **$3d^1$**. Data tabulation of spin-Hamiltonian parameters.

Sample	T (°K)	Site	$\tilde{g}$	$\tilde{A}^{47,49}$ (10^{-4} cm^{-1})	Ref.
GeO_2	77		$g_x = 1.947$, $g_y = 1.921$, $g_z = 1.964$	$A_x = 18.5$, $A_y = 37.3$, $A_z = 134.6$	
	NR		$g_x = 1.921$, $g_y = 1.919$, $g_z = 1.953$		BD80
SnO_2	NR		$g_x = 1.934$, $g_y = 1.928$, $g_z = 1.971$		BD80
TiO_2	77		$g_c = 1.912$, $D = -0.68$ cm^{-1}, $E = -0.68$ cm^{-1}	$A_{110} = 30.9$, $A_{\bar{1}10} = 141.5$, $A_c = 15.0$	PF74
	4.2, 77, 300		$g_x = 1.987$, $g_y = 1.993$, $g_z = 1.941$	$\lvert A_x\rvert = 45.4$, $\lvert A_y\rvert = 60.5$, $\lvert A_z\rvert = 111.4$	BD74
	77	Interstitial	$g_1 = 1.940$, $g_2 = 1.986$, $g_3 = 1.993$	$A_1 = 111$, $A_2 = 45$, $A_3 = 60$	BM83
	77	Substitution	$g_1 = 1.955$, $g_2 = 1.913$, $g_3 = 1.913$	$A_1 = 142$, $A_2 = 42$, $A_3 = 31$	BM83
	77	Vanadyl	$g_1 = 1.906$, $g_2 = 1.967$, $g_3 = 1.967$	$A_1 = 192$, $A_2 = 75$, $A_3 = 75$	BM83
	NR		$g_x = 1.915$, $g_y = 1.908$, $g_z = 1.963$		BD80
MoO_3	295	I	$g_\parallel = 1.934$, $g_\perp = 1.987$	$A_\parallel = 109$, $A_\perp = 45$	ART 72
		II	$g_\parallel = 1.942$, $g_\perp = 1.989$	$A_\parallel = 98$, $A_\perp = 42$	
Al_2O_3	300		$g_\parallel = 1.97$, $g_\perp = 1.97$	$A = 11.321$, $B = 11.321$	PF74
Al_2O_3 (γ-irrad.)	2.3–4.2		$g_\parallel = 1.38$, $g_\perp < 0.1$	$\lvert A_\parallel\rvert = 2$	BD74
Nb_2O_5	~300		$g_1 = 1.918$, $g_2 = 1.965$, $g_3 = 1.971$	$A_x = 56.9$, $A_y = 52.5$, $A_z = 160.4$	BD80
V_2O_5	NR		$g_\parallel = 1.911$, $g_\perp = 1.983$		
V_2O_5	296		$g_x = 1.983$, $g_y = 1.980$, $g_z = 1.928$		BM86
$BaVS_3$	$T > 30$		$g = 1.90$		BM83
	$T < 30$		$g < 1.90$		BM83
$SrTiO_3$	4.2		$g_\parallel = 1.942$, $g_\perp = 1.895$	$\lvert A_\parallel\rvert = 146.78$, $\lvert A_\perp\rvert = 44.04$	BM86

$mV_2O_5 \cdot (100-m)P_2O_5$ glasses					BM83
($m = 90$)	77, 295		$g_\parallel = 1.916$, $g_\perp = 1.992$	$A_\parallel = 152$, $A_\perp = 64$	
($m = 10$)	77, 295		$g_\parallel = 1.933$, $g_\perp = 1.990$	$A_\parallel = 194$, $A_\perp = 77$	
($m = 2$)	77, 295		$g_\parallel = 1.933$, $g_\perp = 1.990$	$A_\parallel = 196$, $A_\perp = 78$	
$24CaO \cdot V_2O_5 \cdot 25P_2O_5$ glasses	77, 295		$g_\parallel = 1.925$, $g_\perp = 1.982$	$A_\parallel = 186.6$, $A_\perp = 69.4$	BM83
$24BaO \cdot V_2O_5 \cdot 25P_2O_5$ glasses	77, 295		$g_\parallel = 1.931$, $g_\perp = 1.980$	$A_\parallel = 181.8$, $A_\perp = 70$	BM83
$xPbO \cdot (1-x)SiO_2$ ($x = 0.22$–0.80) glasses	295		$g_\parallel = 1.906$, $g_\perp = 1.969$	$A_\parallel = 160$, $A_\perp = 60$	BM83
Li_2O–Al_2O_3–P_2O_5 glasses	295		$g_\parallel = 1.92$, $g_\perp = 1.98$	$A_\parallel = 187$, $A_\perp = 71$	BM83
Li_2O–P_2O_5 glasses	295		$g_\parallel = 1.92$, $g_\perp = 1.98$	$A_\parallel = 187$, $A_\perp = 71$	BM83
Al_2O_3–P_2O_5 glasses	295		$g_\parallel = 1.92$, $g_\perp = 1.98$	$A_\parallel = 187$, $A_\perp = 71$	BM83
$Na_2O \cdot 2B_2O_3$ glasses	300		$g_\parallel = 1.946$, $g_\perp = 1.975$	$A_\parallel = 167.1$, $A_\perp = 54.8$	BM83
$Na_2O \cdot P_2O_3$ glasses	300		$g_\parallel = 1.934$, $g_\perp = 1.975$	$A_\parallel = 178.6$, $A_\perp = 63.8$	BM83
$Na_2O \cdot 2SiO_2$ glasses	300		$g_\parallel = 1.939$, $g_\perp = 1.973$	$A_\parallel = 171.8$, $A_\perp = 57.9$	BM83
CaV_2O_6	NR		$g_x = 1.975$, $g_y = 1.975$, $g_z = 1.930$	$A_x = 162$ MHz, $A_y = 162$ MHz, $A_z = 501$ MHz	BD80
CaV_2O_7	NR		$g_x = 1.966$, $g_y = 1.975$, $g_z = 1.948$	$A_x = 150$ MHz, $A_y = 138$ MHz, $A_z = 175$ MHz	BD80
$Cu_xV_2O_5$ ($0.26 < x < 0.64$)	3.8–300		$g_1 = 1.941$, $g_2 = 1.95$, $g_3 = 1.18$		BD80
$Ca_2Al_3Si_3O_{12}(OH)$	77	III	$g_y = 1.981$, $g_z = 1.876$	$\|A_y\| = 6.1$, $\|A_z\| = 155.5$	BM73
$Ca_3In_2Ge_3O_{12}$	77		$g_\parallel = 1.874$, $g_\perp = 1.983$	$\|A_\parallel\| = 150.4$, $\|A_\perp\| = 36.7$, $\|a\| = 22.1$	BD80
$BeAlSiO_4OH$	295		$g_x = 1.968$, $g_y = 1.960$, $g_z = 1.944$	$A_{xx} = 140$ MHz, $A_{yy} = 155$ MHz, $A_{zz} = 490$ MHz	BD80
$CaNaMg_2V_3O_{12}$	77		$g_\parallel = 1.856$, $g_\perp = 1.979$	$A_\parallel = -152$, $A_\perp = -31$	BM73
BaO–V_2O_5–CuO glasses					BM86
$V_2O_5(60\%)$–$BaO(35\%)$–$CuO(5\%)$	77, 300		2.082		
$V_2O_5(60\%)$–$BaO(30\%)$–$CuO(10\%)$	77, 300		2.084		
$V_2O_5(60\%)$–$BaO(25\%)$–$CuO(15\%)$	77, 300		2.084		
$V_2O_5(60\%)$–$BaO(20\%)$–$CuO(20\%)$	77, 300		2.08		

V^{4+} $3d^1$ *(continued)*

Sample	T (°K)	Site	$\tilde{g}$	$\tilde{A}^{47,49}$ (10^{-4} cm^{-1})	Ref.
(25%)P_2O_5–(55%)P_2O_5–(y%)CaO–(x%)Cu glasses					BM86
$y = 20$, $x = 0$	77, 300		1.964		
$y = 20$, $x = 1$	77, 300		2.021		
$y = 20$, $x = 1$	77, 300		2.045		
$y = 20$, $x = 3$	77, 300		2.047		
$y = 20$, $x = 5$	77, 300		2.055		
$y = 20$, $x = 10$	77, 300		2.00		
$y = 5$, $x = 15$	77, 300		2.016		
(x%)CuO–(y%)BaO–(z%)V_2O_5 glasses					BM86
$x = 5$, $y = 35$, $z = 60$	77		2.082		
$x = 10$, $y = 30$, $z = 10$	77		2.084		
$x = 15$, $y = 25$, $z = 15$	77		2.084		
$x = 20$, $y = 20$, $z = 20$	77		2.08		
$x = 5$, $y = 30$, $z = 5$	77		2.085		
$x = 10$, $y = 25$, $z = 10$	77		2.088		
$x = 5$, $y = 25$, $z = 5$	77		2.078		
$x = 10$, $y = 20$, $z = 10$	77		2.087		
xZnO · $(1 - x)P_2O_5$ glasses					BD80
(P_2O_5) $x = 44.9$% (ZnO) $x = 55.1$%	295		$g_{\parallel} = 1.925$, $g_{\perp} = 1.988$	$A_{\parallel} = 178.5$, $A_{\perp} = 71$	
(P_2O_5) $x = 39.8$% (ZnO) $x = 60.2$%	295		$g_{\parallel} = 1.925$, $g_{\perp} = 1.989$	$A_{\parallel} = 177.5$, $A_{\perp} = 71$	
(P_2O_5) $x = 35$%, (ZnO) $x = 65$%	295		$g_{\parallel} = 1.925$, $g_{\perp} = 1.987$	$A_{\parallel} = 176$, $A_{\perp} = 70$	
(P_2O_5) $x = 33$%, (ZnO) $x = 67$%	295		$g_{\parallel} = 1.928$, $g_{\perp} = 1.986$	$A_{\parallel} = 175$, $A_{\perp} = 69$	

$mM_2O_x = 15 \cdot (90 - m)V_2O_5 \cdot 10P_2O_5$					BM86
M = Li ($m = 3\%$)	77, 295	I	$g_\parallel = 1.917$, $g_\perp = 1.992$	$A_\parallel = 154.5$, $A_\perp = 64.3$	
M = Li ($m = 6\%$)	77, 295	I	$g_\parallel = 1.918$, $g_\perp = 1.992$	$A_\parallel = 156.9$, $A_\perp = 64.6$	
M = Li ($m = 20\%$)	77, 295	I	$g_\parallel = 1.923$, $g_\perp = 1.990$	$A_\parallel = 165.2$, $A_\perp = 65.4$	
M = Li ($m = 30\%$)	77, 295	I	$g_\parallel = 1.927$, $g_\perp = 1.987$	$A_\parallel = 169.1$, $A_\perp = 65.8$	
M = LiO_2 ($m = 40\%$)	77, 295	I	$g_\parallel = 1.932$, $g_\perp = 1.985$	$A_\parallel = 171.3$, $A_\perp = 66$	
M = Na ($m = 5\%$)	77, 295	I	$g_\parallel = 1.918$, $g_\perp = 1.992$	$A_\parallel = 155.8$, $A_\perp = 64.6$	
M = Na ($m = 10\%$)	77, 295	I	$g_\parallel = 1.920$, $g_\perp = 1.991$	$A_\parallel = 159$, $A_\perp = 65$	
M = Na ($m = 30\%$)	77, 295	I	$g_\parallel = 1.928$, $g_\perp = 1.987$	$A_\parallel = 168.1$, $A_\perp = 66.2$	
M = Na ($m = 40\%$)	77, 295	I	$g_\parallel = 1.933$, $g_\perp = 1.985$	$A_\parallel = 170.1$, $A_\perp = 66.4$	
M = K ($m = 3\%$)	77, 295	I	$g_\parallel = 1.917$, $g_\perp = 1.992$	$A_\parallel = 154.2$, $A_\perp = 64.4$	
M = K ($m = 6\%$)	77, 295	I	$g_\parallel = 1.918$, $g_\perp = 1.992$	$A_\parallel = 156.2$, $A_\perp = 64.8$	
M = K_2O ($m = 20\%$)	77, 295	I	$g_\parallel = 1.924$, $g_\perp = 1.989$	$A_\parallel = 163.5$, $A_\perp = 66$	
M = K_2O ($m = 30\%$)	77, 295	I	$g_\parallel = 1.929$, $g_\perp = 1.987$	$A_\parallel = 167$, $A_\perp = 66.6$	
M = K_2O ($m = 40\%$)	77, 295	I	$g_\parallel = 1.934$, $g_\perp = 1.984$	$A_\parallel = 168.9$, $A_\perp = 66.8$	
M = Rb ($m = 0\%$)	77, 295	I	$g_\parallel = 1.916$, $g_\perp = 1.992$	$A_\parallel = 152$, $A_\perp = 64$	
M = Rb ($m = 20\%$)	77, 295	I	$g_\parallel = 1.926$, $g_\perp = 1.989$	$A_\parallel = 161.8$, $A_\perp = 66.6$	
M = Rb ($m = 30\%$)	77, 295	I	$g_\parallel = 1.931$, $g_\perp = 1.985$	$A_\parallel = 164.8$, $A_\perp = 67.3$	
M = Rb ($m = 40\%$)	77, 295	I	$g_\parallel = 1.937$, $g_\perp = 1.982$	$A_\parallel = 166.5$, $A_\perp = 67.7$	
M = Cs ($m = 3\%$)	77, 295	I	$g_\parallel = 1.918$, $g_\perp = 1.992$	$A_\parallel = 153.5$, $A_\perp = 64.6$	
M = Cs ($m = 6\%$)	77, 295	I	$g_\parallel = 1.919$, $g_\perp = 1.991$	$A_\parallel = 155$, $A_\perp = 62.5$	
M = Cs ($m = 30\%$)	77, 295	I	$g_\parallel = 1.933$, $g_\perp = 1.986$	$A_\parallel = 162.6$, $A_\perp = 61.8$	
M = Cs ($m = 40\%$)	77, 295	I	$g_\parallel = 1.940$, $g_\perp = 1.983$	$A_\parallel = 164$, $A_\perp = 68.5$	
$mM_2O \cdot (50 - m)V_2O_5 \cdot 50P_2O_5$ glasses					BM86
M = Li ($m = 48\%$)	77, 295	II	$g_\parallel = 1.935$, $g_\perp = 1.989$	$A_\parallel = 185.7$, $A_\perp = 70.1$	
M = N ($m = 48\%$)	77, 295	II	$g_\parallel = 1.936$, $g_\perp = 1.989$	$A_\parallel = 184.5$, $A_\perp = 70.2$	
M = K ($m = 48\%$)	77, 295	II	$g_\parallel = 1.938$, $g_\perp = 1.989$	$A_\parallel = 183.4$, $A_\perp = 70.4$	
M = Rb_2O ($m = 40\%$)	77, 295	II	$g_\parallel = 1.940$, $g_\perp = 1.988$	$A_\parallel = 178.5$, $A_\perp = 70.3$	
M = Cs ($m = 40\%$)	77, 295	II	$g_\parallel = 1.942$, $g_\perp = 1.988$	$A_\parallel = 176.5$, $A_\perp = 70.6$	
M = Cs ($m = 48\%$)	77, 295	II	$g_\parallel = 1.944$, $g_\perp = 1.987$	$A_\parallel = 179$, $A_\perp = 71.3$	

Cr^{5+} $3d^1$. Data tabulation of spin-Hamiltonian parameters ($S = \frac{5}{2}$; $I = \frac{3}{2}$).

Sample	T (°K)	Site	$\tilde{g}$	$\tilde{A}^{53}$ (10^{-4} cm^{-1})	Ref.
TiO_2 (rutile)	4.2		$g_x = 1.928$, $g_y = 1.976$, $g_z = 2.105$	$A_x = 27.4$, $A_y = 24.3$, $A_z = 37.6$	BM73
	77	Orthorhombic	$g_x = 1.949$, $g_y = 1.968$, $g_z = 1.935$	$A_x = 9.7$, $A_y = 9.1$, $A_z = 32.0$	BM83
	200	Axial	$g_\parallel = 1.944$, $g_\perp = 1.950$		BM83
MVO_3					
M = Li	77–300		$g_\parallel = 1.896$, $g_\perp = 1.984$	$A_\parallel = 37$, $A_\perp = 7$	BM86
M = Na	77–300		$g_\parallel = 1.896$, $g_\perp = 1.982$	$A_\parallel = 36.5$, $A_\perp = 7.0$	BM86
M = K	77–300		$g_\parallel = 1.903$, $g_\perp = 1.981$	$A_\parallel = 34.0$, $A_\perp = 6.8$	BM86
M = Rb	77–300		$g_\parallel = 1.903$, $g_\perp = 1.981$	$A_\parallel = 33.5$, $A_\perp = 6.5$	BM86
$SrTiO_3$	77–140	Orthorhombic	$g_x = 1.981$, $g_y = 1.956$, $g_z = 1.945$	$\|A_x\| < 3$, $\|A_y\| = 16$, $\|A_z\| = 28$	BM86
$SrTiO_3$	4.2–300		$g_\parallel = 1.961$, $g_\perp = 1.920$	$\|A_\parallel\| = 34.7$, $\|A_\perp\| = 10.4$	BD74
	77		$g_1 = 1.981$, $g_2 = 1.956$, $g_3 = 1.945$	$A_1 < 3$, $A_2 = 16$, $A_3 = 28$	BM83
KD_2AsO_4	300		$g_\parallel = 1.955$, $g_\perp = 1.974$		BM83
	140		$g_\parallel = 1.956$, $g_\perp = 1.975$		
KD_2AsO_4	300		$g_\parallel = 1.955$, $g_\perp = 1.974$		BM83
	77		$g_x = 1.984$, $g_y = 1.978$, $g_z = 1.945$		
KD_2PO_4	300		$g_\parallel = 1.955$, $g_\perp = 1.974$		BM83
	140		$g_\parallel = 1.956$, $g_\perp = 1.975$		
Be_2SiO_4 (X-irrad.)	4.2, 77		$g_x = 1.951$, $g_y = 1.958$, $g_z = 1.993$	$A_c = \pm 25.2$ G	BD74
$K_2Cr_2O_7$ (X-irrad.)–CrO_4^{3-}	77	I	$g_x = 1.959$, $g_y = 1.985$, $g_z = 1.984$		BM83
		II	$g_x = 1.959$, $g_y = 1.968$, $g_z = 1.985$		
$K_2Cr_2O_7$ (X-irrad.)–CrO^{3-}	77	I	$g_x = 1.989$, $g_y = 1.967$, $g_z = 1.915$	$\|A_x\| = 10.62$, $\|A_y\| = 6.62$, $\|A_z\| = 30.53$	BM83
		II	$g_x = 1.988$, $g_y = 1.970$, $g_z = 1.914$	$\|A_x\| = 10.62$, $\|A_y\| = 6.62$, $\|A_z\| = 30.53$	
K_3CrO_4 (alkali glass)	20		$g_\parallel = 1.98$, $g_\perp = 1.97$		LB
$K_3[CrO_8]$ (p)	room		$g_\parallel = 1.944$, $g_\perp = 1.986$		LB
$K_3[NbO_8]$	room		$g_\parallel = 1.944$, $g_\perp = 1.983$		LB
Li_3AsO_4	4.2, 77, 195		$g_x = 1.938$, $g_y = 1.936$, $g_z = 1.973$	$A_x = 22.3$, $A_y = 24.5$, $A_z = 2.0$	BM86
Li_3PO_4	4.2, 77, 195		$g_x = 1.741$, $g_y = 1.772$, $g_z = 1.940$	$A_x = 30$, $A_y \le 19$, $A_z = 11.5$	BM86
Li_3VO_4	4.2, 77, 195		$g_x = 1.915$, $g_y = 1.898$, $g_z = 1.940$	$A_x = 21.9$, $A_y = 25.1$, $A_z = 4.5$	BM86
$Rb_2[CrOCl_5]$ (p)	292		1.986		LB
$[(CH_3)_4N][CrOCl_4]$ (p)	room		$g_\parallel = 1.989$, $g_\perp = 1.989$		LB
$x Na_2O \cdot y ZrO_2 \cdot z SiO_2$ glasses	77–293	PCI	$g_\parallel = 1.913$, $g_f = 1.985$		BM86
		PCII	$g_\parallel = 1.959$, $g_\perp = 1.980$		
$Ba_5(PO_4)_3Cl$ (CrO_4^{3-} complex)	4.2		$g_x = 1.922$, $g_y = 1.932$, $g_z = 1.974$	$A_x = 26.2$, $A_y = 22.3$, $A_z = 6.5$	BM86

$Sr_5(PO_4)_3Cl$ (CrO_4^{3-} complex)	4.2		$g_x = 1.935$, $g_y = 1.943$, $g_z = 1.982$	$\lvert A_x\rvert = 26.5$, $\lvert A_y\rvert = 22.1$, $\lvert A_z\rvert = 6.7$	BM86
$Ca_5(PO_4)_3F$	4.2–240	I	$g_x = 1.949$, $g_y = 1.950$, $g_z = 1.980$		BM83
		II	$g_x = 1.943$, $g_y = 1.959$, $g_z = 1.993$		
$(NH_4)_2SO_4$ (X-irrad.) (CrO_4^{2-} complex)	300		$g_x = 1.983$, $g_y = 1.986$, $g_z = 1.947$		BM83
	208		$g_x = 1.971$, $g_y = 1.988$, $g_z = 1.947$		
$NH_4H_2AsO_4$	300		$g_\parallel = 1.955$, $g_\perp = 1.974$		BM83
	77		$g_x = 1.983$, $g_y = 1.977$, $g_z = 1.946$		
$ND_4D_2AsO_4$	300		$g_\parallel = 1.955$, $g_\perp = 1.974$		BM83
	77		$g_x = 1.983$, $g_y = 1.977$, $g_z = 1.946$		

Mn^{6+} $3d^1$. Data tabulation of spin-Hamiltonian parameters.

Sample	T (°K)	$\tilde{g}$	$\tilde{A}^{55}$ (10^{-4} cm^{-1})	Ref.
MnO_3 (p)	room	1.96		LB
$BaSO_4$	4.2	$g_x = 1.960$, $g_y = 1.968$, $g_z = 1.980$	$A_x = -128$, $A_y = -79$, $A_z = 48$	BD74
K_2CrO_4	20	$g_x = 1.970$, $g_y = 1.966$, $g_z = 1.938$	$A_x = 25$	LB

Ti^{2+} $3d^2$. Data tabulation of spin-Hamiltonian parameters.

Sample	T (°K)	$\tilde{g}$	D (10^{-4} cm^{-1})	$\tilde{A}^{47,49}$ (10^{-4} cm^{-1})	Ref.
CdS	27	$g_\parallel = 1.916$, $g_\perp = 1.914$	475	$A = 12.9$	BM73
BaF_2	4.2–295	1.927			ART 72
CaF_2	4.2–295	1.942			ART 72
$SrCl_2$	1.2	1.907		$A^{47} = A^{49} = 18.25$	BD74
SrF_2	4.2–295	1.930			ART 72

V^{3+} $3d^2$. Data tabulation of spin-Hamiltonian parameters.

Sample	T (°K)	$\tilde{g}$	D (10^{-4} cm^{-1})	$\tilde{A}^{47,49}$ (10^{-4} cm^{-1})	Ref.
ZnO	NR	$g_\parallel = 1.945$, $g_\perp = 1.937$	7467	$\lvert A\rvert = 68$, $\lvert B\rvert = 93$, $\lvert A - P\rvert = 65$	ART 72
	4.2	$g_\parallel = 1.945$, $g_\perp = 1.933$	7460	$A = 66.0$, $B = 77.1$	BM73
α-Al_2O_3	NR	$g_\parallel = 1.910$, $g_\perp = 1.72$	8.29 cm^{-1}		BD80
Al_2O_3	NR	$g_\parallel = 1.910$, $g_\perp = 1.72$	248.7 GHz	$E = 1.1$ GHz	BD80
Al_2O_3	NR		8.45 cm^{-1}		BD80
xZnO–$y$$P_2O_5$					BD80
($x = 46.5\%$, $y = 53.5\%$)	295	$g_\parallel = 1.919$, $g_\perp = 1.980$		$A_\parallel = 187$, $A_\perp = 75$	
($x = 50.6\%$, $y = 49.4\%$)	295	$g_\parallel = 1.919$, $g_\perp = 1.980$		$A_\parallel = 187$, $A_\perp = 75$	
($x = 55.1\%$, $y = 44.9\%$)	295	$g_\parallel = 1.920$, $g_\perp = 1.982$		$A_\parallel = 186$, $A_\perp = 74$	
($x = 60.2\%$, $y = 39.8\%$)	295	$g_\parallel = 1.920$, $g_\perp = 1.982$		$A_\parallel = 187$, $A_\perp = 75$	
$CsAl(SO_4)_2 \cdot 12H_2O$	4.2	$g_\parallel = 1.975$	6.43 K	$\lvert A_\parallel\rvert = 1{,}019$	BM73, CM71
$[C(NH_2)_3] \cdot V(SO_4)_2 \cdot 6H_2O$		$g_\parallel = 2.32$			BM86
$[C(NH_2)_3] \cdot V(SO_4)_2 \cdot 6H_2O$	4.2	$g_\parallel = 2.00$, $g_\perp = 1.91$	38, 500, $E \leq -200$	$A = 101$, $B = 72$	BM83
$[C(NH_2)_3] \cdot Al(SO_4)_2 \cdot 6H_2O$	4.2	$g_\parallel = 2.01$, $g_\perp = 1.92$	37, 500	$A = 101$, $B = 72$	BM83

Cr^{4+} $3d^2$. Data tabulation of spin-Hamiltonian parameters.

Sample	T (°K)	$\tilde{g}$	D (10^{-4} cm^{-1})	A	Ref.
α-Al_2O_3	4.2	$g_\parallel = 1.913$	226.7 GHz	$\lvert A\rvert = 64.8$ MHz	BD80
α-Al_2O_3	NR	$g_\parallel = 1.913$, $g_\perp = 1.80$	7.55 cm^{-1}		BD80
Al_2O_3 (γ-irrad.)	1.3–4.2	$g = 1.97$	70,000		BM83

Mn^{5+} $3d^2$. Data tabulation of spin-Hamiltonian parameters (* indicates unit of 10^{-4} cm^{-1}).

Sample	T (°K)	$\tilde{g}$	D (GHz) ($= b_2^0$)	E (GHz)	$\tilde{A}^{55}$ (GHz)	Ref.
$CaWO_4$	4.2	$g_\parallel = 1.9790$	4.874		$A = 0.194$	BM83
Ca_2PO_4Cl	295	$g_\parallel = 1.980$, $g_\perp = 1.972$	13.194	−1.816	$A = 0.207$, $B = 0.201$	MMCB94
	190	$g_\parallel = 1.974$, $g_\perp = 1.972$	14.072	−1.844	$A = 0.213$, $B = 0.200$	
	120	$g_\parallel = 1.980$, $g_\perp = 1.973$	14.520	−1.878	$A = 0.188$, $B = 0.194$	
$Na_3VO_4 \cdot 12H_2O$	90; 20	2.0			62.5*	LB
$Sr_5(PO_4)_3Cl$	296	$g_\parallel = 1.966$, $g_\perp = 1.977$	11.79	−1.305	$A = 0.203$, $B = 0.201$	MMC94
	190	$g_\parallel = 1.963$, $g_\perp = 1.979$	12.41	−1.498	$A = 0.196$, $B = 0.201$	
	120	$g_\parallel = 1.961$, $g_\perp = 1.977$	12.77	−1.599	$A = 0.197$, $B = 0.200$	

Fe^{6+} $3d^2$. Data tabulation of spin-Hamiltonian parameters.

Sample	T (°K)	$\tilde{g}$	D (10^{-4} cm^{-1})	Ref.
K_2CrO_4	90	2.00	$D_x = 504$, $D_y = 180$, $D_z = 684$	LB
K_2FeO_4 (p)	193	2.00		LB
	20	2.14		

V^{2+} $3d^3$. Data tabulation of spin-Hamiltonian parameters.

Sample	T (°K)	$\tilde{g}$	D (10^{-4} cm^{-1})	E (10^{-4} cm^{-1})	$\tilde{A}^{47,49}$ (10^{-4} cm^{-1})	Ref.
CaO	290; 77; 20	1.968			$A = 76.15$	PF74
MgO	290	1.980			$A = 74.24$	PF74
MgO	290	$g_x = g_y = g_z = 1.991$	± 1819	± 419	$A_x = A_y = A_z = 72$	BM73
	300	$g_x = 1.981$, $g_y = 1.979$, $g_z = 1.981$	-580	1,126	$A_x = -73.4$, $A_y = -73.3$, $A_z = -74.6$	BD80
NaCl (irradiated)	293 Cubic	1.971			79.1	BM83
NaCl (non-irradiated)	240 Cubic	1.968			79.6	BM83
ZnO	1.3	$g_\parallel = 1.977$, $g_\perp \sim 2$			$A = 46.7$	PF74
CdI_2	300	$g_\parallel = 2.01$, $g_\perp = 2.01$	-985			BM83
	77	$g_\parallel = 2.01$, $g_\perp = 2.01$	-877		$A_\parallel = -64$, $A_\perp = -64$	
	4.2	$g_\parallel = 2.01$, $g_\perp = 2.01$	-870			
	4.2, 300	$g_\parallel = 2.01$, $g_\perp = 2.01$	-985		$A_\parallel = -64$, $A_\perp = -64$	BD80
PbI_2	300	$g_\parallel = 2.015$, $g_\perp = 2.00$	± 4190		$A_\parallel = -70$, $A_\perp = -66$	BM83
	77	$g_\parallel = 2.015$, $g_\perp = 2.00$	± 4350			
Al_2O_3	300	$g_\parallel = 1.991$, $g_\perp = 1.991$	-1601.2		$A = 73.538$, $B = -74.267$	PF74
$CsMgCl_3$	77	$g_\parallel = 1.974$, $g_\perp = 1.976$	931 G		$A = 82.3$ G, $B = 79.8$ G	ART 72
$KMgF_3$	77	1.972			$A = 86.2$, $A' = 7.1$, $B' = 2.0$	PF74
	77				$A_\parallel = -6.5$, $A_\perp = -1.7$	BD80
	77	1.972			$\lvert A\rvert = 21.3$ MHz, $\lvert B\rvert = 6.0$ MHz	BD74
	4.2	1.972			$A = -258.337$ MHz, $U = 0.110$ MHz	BM86
$KZnF_3$	4.2	1.969			$A = -259.157$ MHz, $U = 0.110$ MHz	BM86
	4.2	1.969			$\lvert A_{av}\rvert = 85.95$ MHz	BD74
$ZnSiF_6 \cdot 6H_2O$	4.2	$g_\parallel = g_\perp = 1.980$	810		$A = B = 85.2$	BD80
$Ca_2Al_3Si_3O_{12}(OH)$	295	$g_{iso} = 1.940$	$\lvert D\rvert < 140$	$\lvert E\rvert < 50$	$A_{ave} = \pm 94$	BM73

Data tabulation of Cr^{3+} $3d^3$. spin-Hamiltonian parameters. The g-values are dimensionless, while the units for b_2^0, b_2^2 are cm^{-1}, unless indicated otherwise (* indicates units of MHz).

Host	T (°K)	Site	$\tilde{g}$	b_2^0 (= D) (cm^{-1})	b_2^2 (= E) (cm^{-1})	$\tilde{A}^{53}$ (10^{-4} cm^{-1})	Ref.
Alums							
Cs Ga Alum (deuterated)	300		$g_\parallel = 1.975$, $g_\perp = 1.966$	−0.075			BD74
	4.2		$g_\parallel = 1.975$, $g_\perp = 1.970$	−0.070		51.7*	
Cs Ga Alum (hydrated)	300		$g_\parallel = 1.972$, $g_\perp = 1.969$	−0.073			BD74
	4.2		$g_\parallel = 1.972$, $g_\perp = 1.972$	−0.067		51.6*	
$KAl(SO_4)_2 \cdot 12H_2O$				0.116	0.008		VJ90
$NH_4Al(SO_4)_2 \cdot 12H_2O$		I	$g_y = 1.978$, $g_z = 1.977$	0.013	0.076		BD74
		II	$g_y = 1.978$, $g_z = 1.978$	0.013	0.284		BD74
		III	$g_y = 1.978$, $g_z = 1.977$	0.012	0.217		BD74
$NH_4Al(SO_4)_2 \cdot 12H_2O$			$g_x = 1.968$, $g_z = 1.975$	0.048			VJ90
Rb Ga Alum (deuterated)	300		$g_\parallel = 1.975$, $g_\perp = 1.965$	0.087			BD74
	4.2		$g_\parallel = 1.980$, $g_\perp = 1.969$	0.053		52.1*	
Rb Ga Alum (hydrated)	300		$g_\parallel = 1.975$, $g_\perp = 1.968$	0.086			BD74
	4.2		$g_\parallel = 1.975$, $g_\perp = 1.973$	0.053		52.1*	
$TlAl(SO_4)_2 \cdot 12H_2O$			$g_x = 1.972$, $g_z = 1.976$	0.061			VJ90
Garnets							
YGaG (garnet)	290		$g_\parallel = 1.977$, $g_\perp = 1.976$				VJ90
	300		$g_\perp = 1.976$	0.350			VJ90
	77		$g = 1.98$	0.349			VJ90
			$g = 1.98$				VJ90
YAlG (garnet)	300		$g = 1.98$	0.255			PF74
			$g = 1.98$	0.262			VJ90
Yttrium–Aluminum garnet			$g_z = 1.953$	0.256			VJ90
Halides							
CsBr			$g_x = 1.986$, $g_z = 1.970$	0.339	0.044		VJ90
CsCl			$g_x = 1.983$, $g_z = 1.972$	0.217			VJ90
NH_4Cl	77–330		$g_x = 1.980$, $g_y = 1.980$, $g_z = 1.986$	3.189*	0.247*		ART 72
CdF_2	77		$g_\parallel = 1.94$, $g_\perp = 1.97$	$b_2^0 = 16{,}700$			ART 72
$AlCl_3 \cdot 6H_2O$			$g = 1.977$	−0.033			VJ90

Data tabulation of Cr^{3+} $3d^3$ (*continued*)

Host	T (°K)	Site	$\tilde{g}$	b_2^0 (= D) (cm^{-1})	b_2^2 (= E) (cm^{-1})	$\tilde{A}^{53}$ (10^{-4} cm^{-1})	Ref.
$CsCdBr_3$			$g_x = 2.010$, $g_z = 2.015$	0.246			VJ90
$CsCdF_3$			$g_x = 1.973$, $g_z = 1.970$	−0.074			VJ90
$CsMgBr_3$			$g_x = 2.012$, $g_y = 0.228$, $g_z = 2.012$				VJ90
$CsMgCl_3$			$g_x = 1.984$, $g_z = 1.984$	0.222			VJ90
$KCdF_3$			$g_x = 1.972$, $g_z = 1.971$	±0.051			VJ90
$KCdF_3$	488		$g_\parallel = 1.971$, $g_\perp = 1.972$	$\lvert D \rvert = -512$			BM86
$KMgF_3$	77		1.973			$A = 86.2$, $A' = 9.4$, $B' = 2.3$	PF74
$KZnF_3$			$g_x = 1.974$, $g_z = 1.972$	−0.054			VJ90
$LiBaF_3$ (X-irrad.)	77	Tetragonal	$g_\parallel = 1.957$, $g_\perp = 1.970$				BM86
$RbCdF_3$			$g_x = 1.973$, $g_z = 1.971$	−0.060			VJ90
$CsCdCl_4$			$g_z = 1.99$	0.076	−0.004		VJ90
Cs_2CdF_4			$g_x = 1.973$, $g_z = 1.971$	−0.070			VJ90
K_2MgF_4			$g_x = 1.974$, $g_z = 1.973$	−0.042			VJ90
K_2ZnF_4			$g_x = 1.975$, $g_z = 1.972$	−0.187	0.010		VJ90
Rb_2CdF_4			$g_x = 1.973$, $g_y = 1.972$, $g_z = 1.975$	0.055	0.011		VJ90
Rb_2ZnF_4			$g_x = 1.973$, $g_y = 1.973$, $g_z = 1.976$	0.040	−0.001		VJ90
$Na_5Al_3F_{14}$			$g_x = 1.990$, $g_y = 1.960$, $g_z = 1.975$	0.417	0.018		VJ90
$Cs_2NaAl_3F_{12}$			$g_x = 1.972$, $g_z = 1.962$	0.153	−0.092		VJ90
Zinc hexaimidazole dichloride			$g_z = 1.968$	0.183	−0.057		VJ90
Molybdenates							
$KSc(MoO_4)_2$			$g_x = 1.969$, $g_y = 1.969$, $g_z = 1.964$	0.479	0.008		VJ90
$RbIn(MoO_4)_2$			$g_x = 1.968$, $g_z = 1.964$	0.540			VJ90
$RbSc(MoO_4)_2$			$g_x = 1.968$, $g_z = 1.965$	0.482			VJ90

Nitrates							
$Al(NO_3)_3 \cdot 9H_2O$			$g_x = 1.964$, $g_y = 1.986$, $g_z = 1.976$	± 0.091	± 0.013		VJ90
Oxides							
CaO	290, 77		1.973			17.0	PF74
MgO	290		1.980			16.0	PF74
MgO			$g_x = 1.979$, $g_z = 1.979$	± 0.015			VJ90
SrO	20		1.968			17.2	PF74
SnO_2	4.2	Interstitial	$g_x = g_y = 1.975$, $g_z = 1.971$	−0.211	$E = 0.043$		BD74
	4.2	Substitutional	$g_x = 1.973$, $g_y = 1.974$, $g_z = 1.971$	−0.197	$E = 0.053$		
TiO_2	4.2; 350		$g_c = 1.97$	0.400	$3E = 0.390$	$A_c = 72.7$	PF74
MoO_3	NR		$g_x = 1.965$	± 0.762	± 0.111		ART 72
			$g_y = 1.964$				
			$g_z = 1.952$				
WO_3			$g_z = 1.980$	2.802	0.087		VJ90
Al_2O_3	4.2		$g_\parallel = 1.984$, $g_\perp = 1.984$	−0.191		$A = 16.2$, $B = 16.2$	PF74
	300		$g_\parallel = 2.003$, $g_\perp = 2.002$	−0.193			
Naβ-alumina			$g_x = 1.94$, $g_y = 2.03$, $g_z = 1.93$	0.5	0.065		VJ90
$BaTiO_3$			$g_z = 1.975$	0.012	0.003		VJ90
$KNbO_3$			$g_z = 1.969$	0.396			VJ90
$LaAlO_3$	291		1.983	0.045		19.0	PF74
	80		1.983	0.059			
$LiNbO_3$			$g_z = 1.969$	0.396			VJ90
$LiNbO_3$	295		1.960	0.39			YCRC93
$SrTiO_3$	295		1.978			15.8	PF74
$BeAl_2O_4$	290	B	$g_\parallel = 1.986$, $g_\perp = 1.989$	$b_2^0 = -9200$			PF74
$MgCr_2O_4$			$g_z = 1.977$				VJ90
$ZnAl_2O_4$	300	B	$g_\parallel = 1.984$, $g_\perp = 1.980$	0.930			PF74, ART 72
			$g_x = 1.979$, $g_y = 1.982$, $g_z = 1.979$	± 0.245	± 0.058		VJ90

Data tabulation of Cr^{3+} $3d^3$ (*continued*)

Host	T (°K)	Site	$\tilde{g}$	$b_2^0\ (= D)$ (cm^{-1})	$b_2^2\ (= E)$ (cm^{-1})	$\tilde{A}^{53}$ (10^{-4} cm^{-1})	Ref.
$ZnGa_2O_4$	NR		$g_\parallel = 1.978$, $g_\perp = 1.987$	5230			ART 72
CaV_2O_6			$g_z = 1.967$				VJ90
Y_4TiO_7	NR		$g_\parallel = 1.974$, $g_\perp = 1.977$	± 9540			ART 72
$SrAl_{12}O_{19}$			$g_x = 1.981$, $g_y = 1.978$, $g_z = 1.982$	13.15 Hz	1.84 Hz		VJ90
$Ca_3Ga_2Ge_3O_{12}$			$g_x = 1.973$, $g_z = 1.964$	± 0.553			VJ90
$(ZrO_2)_{0.9}(Y_2O_3)_{0.1}$			$g_x = 1.98$, $g_z = 1.96$	1.4	0.0003		VJ90
$xNa_2O \cdot yZrO_2 \cdot zSiO_2$	77–293		$g_{eff} = 4.90$	$3100 < D < 11{,}800$	$E/D = 0.18$		BM86
Sulphates							
$LiSO_4 \cdot H_2O$			$g_x = 1.985$, $g_y = 1.984$, $g_z = 1.988$	0.130	0.016		VJ90
$Li_2SO_4 \cdot H_2O$	77		$g_x = 1.985$, $g_y = 1.984$, $g_z = 1.988$	$D = 1300$	$E = 160$	$\lvert A \rvert = 17.8$	BM86
$CsAl(SO_4)_2 \cdot 12H_2O$	295		1.975	3			MS74
$NaAl(SO_4)_2 \cdot 12H_2O$	NR		$g_x = 1.980$, $g_y = 1.966$, $g_z = 1.975$	−942	−322		ART 72
$Al_2(SO_4)_3 \cdot 18H_2O$			$g = 1.975$	0.052			VJ90
$(NH_4)_xRb_{1-x}Al(SO_4)_2 \cdot 12H_2O$							VJ90
($x = 0, 0.2, 0.4, 0.6, 0.8, 1.0$)	305		1.978				BM86
Guanidinium aluminum	296	I	$g_\parallel = g_\perp = 1.980$	$b_2^0 = -0.732$*			MGS77
Sulphate hexahydrate		II	$g_\parallel = g_\perp = 1.976$	$b_2^0 = -0.585$*			MGS77
	78	I	$g_\parallel = 1.976$, $g_\perp = 1.975$	$b_2^0 = -1.087$*			MGS77
		II	$g_\parallel = 1.976$, $g_\perp = 1.976$	$b_2^0 = -0.837$*			MGS77
	1.6	I	$g_\parallel = 1.984$, $g_\perp = 1.985$	$b_2^0 = -1.164$*			MGS77
		II	$g_\parallel = 1.987$, $g_\perp = 1.986$	$b_2^0 = -0.892$*			MGS77
$CH_3NH_3Al(SO_4)_2 \cdot 12H_2O$			$g_z = 1.98$	−0.094			VJ90
Sulphides							
$GAS \cdot 6H_2O$			$g_z = 1.997$	−0.060	0.002		VJ90
$AlGaS_2$			$g = 1.988$	0.022			VJ90
$Cu_{0.5}Cr_{0.5}PS_3$			$g_x = 1.991$, $g_z = 1.988$				VJ90
$CdI_{2(1-X)}Cr_XS_4$				−0.185			VJ90

Sample	T (°K)	$\tilde{g}$	D	E	$\tilde{A}$	Ref.
Other						
$GASe \cdot 6H_2O$			−0.108			VJ90
$NH_4H_2PO_4$	151	$g_\parallel = 1.98$, $g_\perp = 1.98$	±1000	±220		ART 72
$MgIn_2S_4$	77	$g_\parallel = 1.98$, $g_\perp = 1.89$			$A_\parallel = 16.9$	BM86
NH_4Al_4		$g_x = 1.974$, $g_z = 1.970$	−0.103	0.009		VJ90
$KH_3(SeO_3)_2$		$g_z = 1.98$	0.222	0.037		VJ90
$RbH_3(SeO_3)_2$		$g_z = 1.970$	0.540	0.177		VJ90
$KH_3(SeO_4)_3$	300	1.98	$D = 2400$	$E = 400$		BM86
$FeCr_2S_{4-x}Se_x$ ($x = 0, 0.25, 0.5, 1.0$)	113	1.810				BM86
	173	1.885				
	213	1.902				
	300	1.970				
Ga tris acetylacetonate		$g_z = 1.997$	0.593	0.01		VJ90

Mn^{4+} $3d^3$. Data tabulation of spin-Hamiltonian parameters (* indicates units of GHz).

Sample	T (°K)	$\tilde{g}$	D (10^{-4} cm^{-1})	E (10^{-4} cm^{-1})	$\tilde{A}^{55}$ (10^{-4} cm^{-1})	Ref.
TiO_2	4.2; 300	$g_c = 1.990$			$A_c = 0.93$	PF74
	78–470	$g = 1.99$				BM83
	300	$g_x = 1.995$	$\lvert D\rvert = 12.23$*	$\lvert E\rvert = 3.918$*	$\lvert A_x\rvert = 0.217$*	BM86
		$g_y = 1.991$			$\lvert A_y\rvert = 0.211$*	
		$g_z = 1.990$			$\lvert A_z\rvert = 0.218$*	
	4.2	$g_x = 1.988$	11.618*	3.703*	$\lvert A_x\rvert = 0.211$*	BM86
		$g_y = 2.005$			$\lvert A_y\rvert = 0.209$*	
		$g_z = 1.996$			$\lvert A_z\rvert = 0.210$*	
Al_2O_3	1.6–295	$g_\parallel = 1.994$, $g_\perp = 1.994$	−1957.0		$A = 170.01$, $B = 170.01$	PF74
$SrTiO_3$	295	1.994			75.0	PF74
$Ca_2Mn_3O_8$	78–600	1.979				BM86
$Cd_2Mn_3O_8$	78–600	1.982				BM86
$Cs_2[GeF_6]$	77				72.0	LB
Mg_6MnO_8	77–550	$g = 1.994$				BM83
$(NH_4)_6MnMo_9O_{32} \cdot 8H_2O$	NR	$g_\parallel = 1.992$, $g_\perp = 1.988$	±8610		$\lVert A_\parallel\rvert = 76$, $\lvert A_\perp\rvert = 68.4$	ART 72
Aluminosilicate glasses	295	$g = 1.993$	$Dq = 25{,}600$ cm^{-1}		$A = 62$ G	BM83

Cr^{2+} $3d^4$. Data tabulation of spin-Hamiltonian parameters.

Sample	T (°K)	$\tilde{g}$	D (10^{-4} cm^{-1})	E (10^{-4} cm^{-1})	a (10^{-4} cm^{-1})	Ref.
CdS	4.2	$g_{\parallel} = 1.934,\ g_{\perp} = 1.970$	−18,050	225	1500	BM73
	NR	$g_{\parallel} = 1.935,\ g_{\perp} = 1.968$	−1.802 cm^{-1}	$F = -0.247$ cm^{-1}	0.148 cm^{-1}	BD80
CdTe	4.2–20	$g_{\parallel} = 1.980,\ g_{\perp} = 1.980$	2600		500	BM73
ZnS	4.2–20	$g_{\parallel} = 1.94,\ g_{\perp} = 1.98$	−17,700		1930	BM73
	NR	$g_{\parallel} = 1.935,\ g_{\perp} = 1.978$	−1.828 cm^{-1}	$F = -0.19$ cm^{-1}	0.185 cm^{-1}	BD80
ZnSe	4.2–20	$g_{\parallel} = 1.961,\ g_{\perp} = 1.98$	−24,800		240	BM73
	NR	$g_{\parallel} = 1.957,\ g_{\perp} = 2.001$	−2.458 cm^{-1}	$F = -0.142$ cm^{-1}	0.021 cm^{-1}	BD80
ZnTe	4.2–20	$g_{\parallel} = 1.982,\ g_{\perp} = 1.982$	23,000		1400	BM73
Al_2O_3	2.0, 4.2	1.980		2		BM86
$CrSO_4{\cdot}5H_2O$ (sc)	Room	$g_{\parallel} = 1.95,\ g_{\perp} = 1.99$	22,400	1000		LB
$CdCr_3Se_4$	100–300	$g = 1.99$				BD80
$Cd_{1-x}In_xCr_2S_4$ ($0 \le x \le 0.4$)	77–300	$g = 1.99$				BM83
$Cd_{1-x}In_xCr_2S_{4-x}P_x$ ($0 \le x \le 0.5$)	77–300	$g = 1.99$				BM83

Mn^{3+} $3d^4$. Data tabulation of spin-Hamiltonian parameters.

Sample	T (°K)	$\tilde{g}$	D (10^{-4} cm^{-1})	$3E$ (10^{-4} cm^{-1})	$\frac{1}{3}F$	$\tilde{A}^{47,49}$ (10^{-4} cm^{-1})	Ref.
TiO_2	4.2; 77	$g_c = 1.99$	−34,000	3480	650.0	$A_{110} = 184.51$ $A_{\bar{1}10} = 152.81$ $A_c = 80.61$	PF74
$SrTiO_3$	2–300	$g_{\parallel} \approx 8,\ g_{\perp} < 0.4$					BD80
γ-MnOOH	296	1.997					BM86

Fe^{4+} $3d^4$. Data tabulation of spin-Hamiltonian parameters.

Sample	T (°K)	$\tilde{g}$	D (10^{-4} cm^{-1})	E (10^{-4} cm^{-1})	a (10^{-4} cm^{-1})	Ref.
$SrTiO_3$	4.2–77	$g_{\parallel} = 2.007$	−18,050	225	1504	BD80

Cr^{+} $3d^5$. Data tabulation of spin-Hamiltonian parameters.

Sample	T (°K)	Site	$\tilde{g}$	D (10^{-4} cm^{-1})	$\frac{1}{3}F$ $(= b_4^0)$ (10^{-4} cm^{-1})	$\tilde{A}^{53}$ (10^{-4} cm^{-1})	Ref.
CaS	77		$g = 1.999$			$A = 17.4$	BM83
CdTe	100–300		$g = 2.000$				BD80
GaAS	80		$g_{iso} = 1.999$				BM83
MgO (γ-irrad.)	77		1.996			17.6	BD74
MgTe	4.2, 77–300		$g = 2.004$	$\lvert D\rvert = 28.1$	$\lvert a - F\rvert = 11.0$, $\lvert a\rvert = 9.2$	$\lvert A\rvert = 13.4$	BM83
NaF (X-irrad.)	4.2–300		$g = 2.001$				BD80
ZnS	77	Site 1	$g_1 = 2.003$				BD80
			$g_2 = 2.046$				
			$g_3 = 2.050$				
		Site 2	$g = 2.004$				BD80
$KMgF_3$	77		2.001		2.25	$A' = 23.0$, $B' = 17.5$	PF74

DATA TABULATION (Mn^{2+} $3d^5$). Spin-Hamiltonian parameters for Mn^{2+} in various host crystals at different temperatures. The g-values are dimensionless, while the A-values are in 10^{-4} cm^{-1}, unless otherwise indicated (* indicates GHz, G indicates gauss).

Host (site)	T (°K)	Site	$\tilde{g}$ (g_z, g_x, g_y)	b_2^0 $(= D)$, b_2^2 $(= E)$ (10^{-4} cm^{-1})	b_4^0, b_4^2, b_4^4 (10^{-4} cm^{-1})	$\tilde{A}$ (A_z, A_x, A_y), Q', Q'' (10^{-4} cm^{-1})	Ref.
Acetates							
$CaCd(CH_3COO)_4 \cdot 6H_2O$	300		$g_x = 2.02, 2.005, 2.004$	509.4 G, 468 G	−0.7 G, —, —	−91 G, −93 G, −97 G	MSrev91
	295		$g_\parallel = 2.005$, $g_\perp = 2.005$	−1.427*, 1.486*	0.001*, 0.043*, 0.070*	$A' = -0.269$*, $B' = -0.269$*	MKK92
	148.5		$g_\parallel = 2.003$, $g_\perp = 2.002$	−1.437*, 1.450*	0.003*, 0.046*, 0.024*	$A' = -0.270$*, $B' = -0.267$*	
	4.2		$g_\parallel = 2.005$, $g_\perp = 2.005$	−1.499*, 1.435*	−0.001*, 0.233*, 0.250*	$A' = -0.275$*, $B' = -0.260$*	
$Ni(CH_3COO)_2 \cdot 4H_2O$	RT		$g_\parallel = 1.984$, $g_\perp = 2.000$	1.271*, −0.254*	−0.023*, —, —	$A_\parallel = -0.270$*, $A_\perp = -0.270$*	MSrev91
	62		$g_\parallel = 1.952$, $g_\perp = 1.966$	1.435*, −0.396*	0.019*, —, —	$A_\parallel = -0.295$*, $A_\perp = -0.205$*	
	3.2		$g_\parallel = 1.953$, $g_\perp = 1.989$	1.454*, −0.413*	0.022*, —, —	$A_\parallel = -0.254$*, $A_\perp = -0.214$*	
$Pb(CH_3COO)_2 \cdot 3H_2O$	223–333		$g = 2.02$			$A = -98$ G	MSrev91
$Sr(CH_3OO)_2 \cdot \frac{1}{2}H_2O$	300		2.012, 2.005, 2.008	328.6 G, 66.0 G	−1.3 G, —, —	−93.0 G, −93.5 G, −91.8 G	MSrev91
	295		$g_\parallel = 2.002$, $g_\perp = 2.000$	0.917*, −0.192*	0.005*, −0.017*, 0.024*		MK91
	77		$g_\parallel = 2.012$, $g_\perp = 2.001$	0.669*, −0.501*	0.012*, 0.244*, 0.125*		
	4.2		$g_\parallel = 2.015$, $g_\perp = 2.002$	0.593*, −0.645*	0.002*, 0.599*, 0.623*		
$B_2Zn(CH_3COO)_6 \cdot 4H_2O$	298		1.999, 2.000, 2.003	294.6, −180.0	−2.7, —, —	−83.3 G, −85.2 G, −84.4 G	MSrev91
Fluosilicates							
$Co(H_2O)_6SiF_6$	288		$g = 2.001$	−191 G, 0 G	2.6 G, —, —	$A_\parallel = -99.5$ G, $A_\perp = -98.6$ G	MSrev91
	250		$g = 2.006$	−295 G, 120 G	3 G, —, —	$A_\parallel = -99$ G, $A_\perp = -98$ G	

$Mg(H_2O)_6SiF_6$	290			−274, 180	3.5, —, —	$A_{\parallel} = -92$, $A_{\perp} = -92$	MSrev91
	293		$g = 2.001$	−273, 183	3.5, —, —	$A_{\parallel} = -93.3$, $A_{\perp} = -93.3$	
	308	I	$g = 2.001$	−277, 0	3.5, —, —	$A_{\parallel} = -92$, $A_{\perp} = -92$	
	308	II	$g = 2.001$	−286, 0	3.5, —, —	$A_{\parallel} = -92$, $A_{\perp} = -92$	MSrev91
$CoSiF_6 \cdot 6H_2O$	RT		$g = 1.998$	−183.1 G, —	3.8 G, —, —	$A = -95$ G	MSrev91
$FeSiF_6 \cdot 6D_2O$	320		$g = 2.003$	269, —			MSrev91
	314		$g = 2.002$	270, —			
	200		$g = 2.003$	263, —			
	143		$g = 2.003$	261, —			
$FeSiF_6 \cdot 6H_2O$	4.2		$g = 2.001$	−254, −120	8, —, —	$A = -91$ G	MSrev91
	4–60			270, —			
	314		$g = 2.002$	270, —			
	300		$g = 2.003$	263, —			
	222		$g = 2.001$				
	4		$g = 2.001$				
$MgSiF_6 \cdot 6H_2O$	295		$g_{\parallel} = 1.962$, $g_{\perp} = 1.959$	0.836*, −0.235*	0.008*, —, —	$A_{\parallel} = -0.278$*, $A_{\perp} = -0.257$*	MSrev91
	85		$g_{\parallel} = 1.992$, $g_{\perp} = 1.963$	0.787*, −0.350*	0.010*, —, —	$A_{\parallel} = -0.284$*, $A_{\perp} = -0.249$*	
	5		$g_{\parallel} = 2.000$, $g_{\perp} = 1.962$	0.788*, −0.401*	0.015*, —, —	$A_{\parallel} = -0.276$*, $A_{\perp} = -0.251$*	
$ZnSiF_6 \cdot 6H_2O$	15		$g = 2.000$	−130.5, —			MSrev91
	297		$g = 2.000$	−170.2, —			
	50		$g_{\parallel} = 2.001$	−132.2, —		$A_{\parallel} = -90.1$, $A_{\perp} = -90.2$	
			$g_{\parallel} = 2.000$	9.3, —		$A_{\parallel} = -89.6$, $A_{\perp} = -89.7$	
	300		$g_{\perp} = 2.001$	−170.4, —			
	NR						
			$g = 2.000$	7.9, −130	11, —, —		

DATA TABULATION (Mn^{2+} $3d^5$) *(continued)*

Host (site)	T (°K)	Site	$\tilde{g}$ (g_z, g_x, g_y)	b_2^0 $(= D)$, b_2^2 $(= E)$ $(10^{-4}$ cm$^{-1})$	b_4^0, b_4^2, b_4^4 $(10^{-4}$ cm$^{-1})$	$\tilde{A}$ (A_z, A_x, A_y), Q', Q'' $(10^{-4}$ cm$^{-1})$	Ref.
Halides							
NaCl	NR		$g = 1.991$	−655, 228	≤1, —, —		MSrev91
	RT		$g_\parallel = 2.005$, $g_\perp = 2.003$	159.9, —	−0.5, —, —	$A = -82.1$	
NH_4Br	77–573		$g = 2.006$			$A = 95$ G	MSrev91
NH_4Cl	RT			4.51*, 0.74*	0.007*, —, —	$A_\parallel = -0.245$, $A_\perp = 0.253$	MSrev91
NH_4I	300		$g_\parallel = 2.002$, $g_\perp = 2.003$	−1611 G, —		$A_\parallel = 90$ G, $A_\perp = 90$ G	MSrev91
				6 G, —			
	77		$g_\parallel = 2.001$, $g_\perp = 2.001$	−4.77*, —		$A_\parallel = 0.254$*, $A_\perp = 0.254$*	
	40		$g_\parallel = 2.001$, $g_\perp = 2.001$	−4.85*, —		$A_\parallel = 0.254$*, $A_\perp = 0.254$*	
	5		$g_\parallel = 2.001$, $g_\perp = 2.001$	−5.01*, —		$A_\parallel = 0.254$*, $A_\perp = 0.254$*	
	77–573		$g = 2.002$			$A = 84.5$ G	
BaF_2	RT		$g = 2.00$			$A_\parallel = -0.241$, $A_\perp = -0.236$*	MSrev91
			$g = 2.00$			$A_\parallel = -0.234$*, $A_\perp = -0.266$*	
CaF_2	77		$g = 2.001$			$A = 195$ G	MSrev91
$CdCl_2$	4.2–543		$g = 2.002$			$A = -86.$ G	MSrev91
CdI_2	300		$g = 2.004$			$A = 70$	MSrev91
	77		$g = 2.004$			$A = 70$	
	4.2		$g = 2.004$			$A = 70$	
$FeBr_2$	LHT		$g_\parallel = 1.94$, $g_\perp = 1.51$				MSrev91
MgF_2	300					$A = -95.8$	MSrev91
						$A = -93.2$	
	20					$A = -96.1$	
						$A = -94.2$	MSrev91

PbI_2	300		$g = 2.005$			$A = 72$	
	77		$g = 2.005$			$A = 72$	
$SrCl_2$	4.2–300		$g = 2.006$			$A = 82.5$	MSrev91
SrF_2	77–300		$g = 2.004$			$A = 0.265*$	MSrev91
$CsCaCl_3$	4.2–450		$g = 2.001$			$A = 89.5$ G	MSrev91
				2 G, —			
$CsCdCl_3$	RT	I	$g = 2.002$	6.8 G, —		$A = 86$ G	MSrev91
				0.5 G, —			
		II	$g = 2.002$	18 G, —		$A = 86$ G	
				0.5 G, —			
$CsSrCl_3$	293		$g = 2.007$	13, —		$A = 83$	MSrev91
				0.4, —			
	393		$g = 2.001$	0, —		$A = 82.7$	
$LiBaF_3$	77		$g = 2.001$		2, —, —	$A = -88.3$	MSrev91
NH_4MnF_3	80–300		$g = 2.001$				MSrev91
$RbMgBr_3$	449–738	I	$g = 2.003$	979, —	4.5, —, —		MSrev91
	235–449	II	$g = 2.003$	983, —	4.5, —, —		
	77–228	III	$g = 2.003$	983, —			
$BaMgF_4$	93			−230 G, —			MSrev91
	183			−226 G, —			
	293			−220 G, —		−0.270*, −0.268*, −0.272*	
	368			−195 G, —			
	403			−193 G, —			
	538			−188 G, —			
$BaZnF_4$	RT			−901, —		$A = -88.1$	MSrev91
				−555, —			
Cs_2CdCl_4	4.2		$g = 2.003$			$A_{\parallel} = -86$ G, $A_{\perp} = -85.5$ G	MSrev91
	300		$g = 2.001$	196 G, —		$A_{\parallel} = -84.7$ G, $A_{\perp} = -85.5$ G	
				4.8 G, —			
Cs_2ZnBr_4	NR					$A_{\parallel} = -68.6$, $A_{\perp} = -70.6$	MSrev91
$Cd(BF_4)_2 \cdot 6H_2O$	325	I	2.001, 2.001, 2.001	76.3 G, —	1.8 G, —, —	−95.0, −95.0, −95.0	MSrev91

DATA TABULATION (Mn^{2+} $3d^5$) *(continued)*

Host (site)	T (°K)	Site	$\tilde{g}$ (g_z, g_x, g_y)	b_2^0 (= D), b_2^2 (= E) (10^{-4} cm^{-1})	b_4^0, b_4^2, b_4^4 (10^{-4} cm^{-1})	$\tilde{A}$ (A_z, A_x, A_y), Q', Q'' (10^{-4} cm^{-1})	Ref.
	323	II	2.001, 2.002, 2.002	57.0 G, —	4.9 G, —, —	−95.0, −95.0, −95.0	
	297	II	2.001, 2.002, 2.002	46.8 G, —	4.10 G, —, —	−95.0, −96.0, −96.0	
	187	II	2.001, 2.002, 2.002	12.0 G, —	4.8 G, —, —	−96.0, −96.0, −96.0	
$Co(BF_4)_2 \cdot 6H_2O$	LT, RT		2.000, 2.000, 2.000	100.5 G, 5.4 G	4 G, —, —	−96 G, −96 G, −96 G	MSrev91
$Fe(BF_4)_2 \cdot 6H_2O$	LT, RT		2.000, 2.000, 2.000	104.2 G, 11.4 G	4 G, —, —	−96 G, −96 G, −96 G	MSrev91
$Ni(BF_4)_2 \cdot 6H_2O$	LT, RT		1.999, 1.999, 1.999	108.9 G, 15 G	3.5 G, —, —	−96 G, −96 G, −96 G	MSrev91
$Zn(BF_4)_2 \cdot 6H_2O$	94–353		$g = 2.000$	—, —	4.5 G, -, —	$A_\parallel = 97.9$ G, $A_\perp = 97.0$ G	MSrev91
	LT, RT		2.002, 2.002, 2.002	107.5 G, 6.9 G	4.2 G, —, —	−96 G, −96 G, −96 G	
$Na_2ZnCl_4 \cdot 3H_2O$	4.2		$g_\parallel = 2.001$, $g_\perp = 2.008$	−943, —	4, —, —	$A_\parallel = 74.5$, $A_\perp = 73$	MSrev91
Rb_2CdCl_4	4.2		$g = 1.988$	247 G, —	0.5 G, —, —	$A_\parallel = -86.4$ G, $A_\perp = -86.8$ G	MSrev91
	300		$g = 1.998$	245 G, —	1.6 G, —, —	$A_\parallel = -82.8$ G, $A_\perp = -86.8$ G	
	4.2–300		$g_\parallel = 1.998$, $g_\perp = 2.000$	245 G, 0	1 G, —, —	$A_\parallel = -82.8$ G, $A_\perp = -86.4$ G	
Rb_2ZnCl_4	HT			480 G, 81 G	—, —, —		MSrev91
$(NH_4)_2SbF_5$	NR			35, —		$A = -95$	MSrev91
$ZnNbOF_5 \cdot 6H_2O$	LT, RT		$g_\parallel = 2.004$, $g_\perp = 2.003$	−168.8, 0	4.5, —, —	$A_\parallel = -90.9$, $A_\perp = -87.9$	MSrev91
$CoNbOF_5 \cdot 6H_2O$	LT, RT	I	$g_\parallel = 2.0095$, $g_\perp = -2000$	−176.8, 0	7.0, —, —	$A_\parallel = -90.7$, $A_\perp = -86.0$	MSrev91
		II	$g_\parallel = 2.013$, $g_\perp = 2.002$	−270.0, 84	7.1, —, —	$A_\parallel = -91.0$, $A_\perp = -86.7$	

$Ca(H_2O)_6SnCl_6$	293	$g = 2.001$	−176.5, —	−2.4, —, —	$A_\parallel = -90.6$, $A_\perp = -89.6$	MSrev91
	77	$g = 2.001$	−188.8, —	−2.9, —, —	$A_\parallel = -91.3$, $A_\perp = -90.5$	
$Co(H_2O)_6PtCl_6$	291	$g = 2.001$	−167 G, 0	3.0 G, —, —	$A_\parallel = -97.5$ G, $A_\perp = -95.1$ G	MSrev91
$Co(H_2O)_6SnF_6$	293	$g = 2.001$	−205 G, —	2.6 G, —, —	$A_\parallel = -98.5$, $A_\perp = -96.5$	MSrev91
$Zn(H_2O)_6SnF_6$	293	$g = 2.001$	−200 G, —	2.6 G, —, —	$A_\parallel = -97.5$, $A_\perp = -95.6$	MSrev91
	77	$g = 2.001$	−186 G, —	3.0 G, —, —	$A_\parallel = -98.5$, $A_\perp = -96.4$	
Cs_2HfCl_6	12–400	$g = 2.004$	470 G, 9 G		$A = 77$ G	MSrev91
Cs_2NaYCl_6	RT	$g_\parallel = 1.99$, $g_\perp = 2.006$	857.6, —	−2.6, —, —	$A = 82$	MSrev91
Ba_2ZnF_7	RT		−297, —	−3.2, —, —	$A = -93.5$	MSrev91
$[N(CH_3)_4]_2CoCl_4$	LT–RT		89 G, —			MSrev91
$[N(CH_3)_4]_2ZnCl_4$	6, 80, 300	$g = 2.002$			$A = -74.8$	MSrev91
$(CH_3NH_3)_2ZnCl_4$	RT	2.002	−174, —			MSrev91
		2.004	−150, —			
		2.003	9, —			
$[(en)NH_2]_2ZnCl_4$	RT	2.002	861.6, —			MSrev91
		2.002	725.4, —			
		2.002				
$(CH_3)_3NHCoCl_3 \cdot 2H_2O$	1.8–300	$g = 2$				MSrev91
Formates						
α-$Ca(HCOO)_2$	RT	2.021, —, —	−367.5 G, 287.9 G	−0.8 G, —, —	$A = -92.0$ G	MSrev91
	LNT	2.018, —, —	−420.2 G, 300 G	−1.4 G, —, —	$A = -91.7$	
$Ba_2Co(HCOO)_6 \cdot 4H_2O$	90, 290	2.002, 2.003, 2.003	300.0, −182.0	−1.8, —, —	83.5, 86.0, 86.0	MSrev91
Langbeinites						
$Cd_2(NH_4)_2(SO_4)_3$	77–300	$g_\parallel = 2.004$, $g_\perp = 2.013$	0.485*, —	0.005*, —, —	$A_\parallel = -0.278$*, $A_\perp = -0.246$*	MSrev91
$Tl_2Cd_2(SO_4)_3$	77–300	$g_\parallel = 2.005$, $g_\perp = 2.003$	0.454*, —	0.005*, —, —	$A_\parallel = -0.274$*, $A_\perp = -0.235$*	MSrev91

DATA TABULATION (Mn^{2+} $3d^5$) *(continued)*

Host (site)	T (°K)	Site	$\tilde{g}$ (g_z, g_x, g_y)	b_2^0 (= D), b_2^2 (= E) (10^{-4} cm^{-1})	b_4^0, b_4^2, b_4^4 (10^{-4} cm^{-1})	$\tilde{A}$ (A_z, A_x, A_y), Q', Q'' (10^{-4} cm^{-1})	Ref.
Molybdenates							
$CdMoO_4$	4.2–350			57.7, —	−0.6, —, —	$A_\parallel = -86.8$ G, $A_\perp = -88.0$ G	MSrev91
$CeMoO_4$	4.2–350			33.0, —	−0.5, —, —	$A_\parallel = -88.0$ G, $A_\perp = -88.9$ G	MSrev91
$PbMoO_4$	4.2–350			72.3, —	−0.4, —, —	$A_\parallel = -87.9$ G, $A_\perp = -89.1$ G	MSrev91
$SrMoO_4$	4.2–350			24.3, —	−0.4, —, —	$A_\parallel = -89.2$ G, $A_\perp = -89.2$ G	MSrev91
Niobates							
$LiNbO_3$	4.2–1080			783, —	−1.2, —, —	$A = 87$	MSrev91
$NaNbO_3$	LT, RT	P phase		4, —			MSrev91
		N phase		10, —			
	4-1023	Q phase	$g = 2.003$			$A = 84$	
	HT	T_2 phase	$g = 2.003$	180, —	—, —, —	$A = 84$	
	653		$g = 2.002$	275, —	—, —, —	$A = -78.6$	
	300		$g = 2.002$	—, 120		$A = -83$	
	77		$g = 2.000$	227, —	—, —, —	$A = -83$	
Nitrates							
$Cd(NO_3)_2 \cdot 4H_2O$	300		2.004, 2.004, 2.007	420 G, −48 G	1.1 G, —, —	−95.5 G, −95.7 G, −93.2 G	MSrev91
$Mg(OH_2)_6(NO_3)_2$	RT		2.002, 2.003, —	−359, 306	3.3, —, —	$A_\parallel = -89.7$, $A_\perp = -87.0$	MSrev91
$Ni(OH_2)_6(NO_3)_2$	RT		1.997, 1.998, —	−384.2, 327	3.8, —, —	$A_\parallel = -88.6$, $A_\perp = -86.7$	MSrev91
$Pr_2Mg_3(NO_3)_{12} \cdot 24H_2O$	298	I	$g_\parallel = 2.005$, $g_\perp = 2.007$	−187.6, —	3.6, —, —	$A_\parallel = -89.9$, $A_\perp = -89.6$	MSrev91
		II	$g_\parallel = 2.008$, $g_\perp = 2.009$	20.63, —	4.7, —, —	$A_\parallel = -89.4$, $A_\perp = -90.0$	
	77	I	$g_\parallel = 2.004$, $g_\perp = 2.007$	−220, —	3.7, —, —	$A_\parallel = -89.9$, $A_\perp = -89.7$	

		II	$g_{\parallel} = 2.008$, $g_{\perp} = 2.009$	43.0, —	4.8, —, —	$A_{\parallel} = -89.5$, $A_{\perp} = -90.0$	
$Ce_2Mg_3(NO_3)_{12} \cdot 24H_2O$	290	I	$g_{\parallel} = 2.003$, $g_{\perp} = 2.001$	−188.7, —	4.3, —, —	$A_{\parallel} = -90.4$, $A_{\perp} = -89.6$	MSrev91
		II	$g_{\parallel} = 2.002$, $g_{\perp} = 2.003$	−19.4, —	4.4, —, —	$A_{\parallel} = -90.5$, $A_{\perp} = -89.5$	
	77	I	$g_{\parallel} = 2.003$, $g_{\perp} = 2.004$	−214.2, —	4.5, —, —	$A_{\parallel} = -90.9$, $A_{\perp} = -89.5$	
		II	$g_{\parallel} = 2.003$, $g_{\perp} = 2.003$	−38.6, —	4.5, —, —	$A_{\parallel} = -89.9$, $A_{\perp} = -89.1$	
Oxalates							
$CdC_2O_4 \cdot 3H_2O$	RT		$g_z = 2.000$	577, −219	0.34, —, —	$A_z = -90$	MSrev91
			$g_x = 2.001$			$A_x = -89$	
			$g_y = 2.001$			$A_y = -88$	
$K_2C_2O_4 \cdot H_2O$	29		$g_z = 2.007$	−383, 225		$A_z = -82.8$ G	MSrev91
			$g_x = 2.004$			$A_x = -84$ G	
			$g_y = 2.004$	0.34, —		$A_y = -84$ G	
$NH_4C_2O_4 \cdot H_2O$	77–345		$g_z = 1.999$	565 G, 68 G		$A_z = 91.5$ G	MSrev91
			$g_x = 1.996$			$A_x = 93.0$ G	
			$g_y = 1.999$			$A_y = 91.5$ G	
Oxides							
BeO (powder)	150, 300		$g = 2.001$			$A = -79.2$	MSrev91
CaO	100		$g_{\parallel} = 2.002$			$A = -67.2$	MSrev91
	295		$g = 2.001$			$A = -80.8$	
	100		$g = 2.001$			$A = -80.6$	
CaO	290		2.001			80.8	PF74
	77		2.001			81.6	MSrev91
CaO (powder)	295		$g = 2.001$			$A = -80.6$	MSrev91
	100		$g = 2.001$			$A = -87.4$	
MgO	295		$g = 2.001$		9.1, —, —	$A = -81.4$	MSrev91
	100		$g = 2.001$		9.4, —, —	$A = -81.6$	
MgO (powder)	295		$g = 2.001$		9.2, —, —	$A = -81.2$	MSrev91
	100		$g = 2.001$		9.4, —, —	$A = -81.4$	
MgO	290		2.001		9.33, —, —	−81.0	PF74
SrO	290		2.001		2.15, —, —	−78.7	PF74

DATA TABULATION (Mn^{2+} $3d^5$) *(continued)*

Host (site)	T (°K)	Site	$\tilde{g}$ (g_z, g_x, g_y)	b_2^0 (= D), b_2^2 (= E) (10^{-4} cm^{-1})	b_4^0, b_4^2, b_4^4 (10^{-4} cm^{-1})	$\tilde{A}$ (A_z, A_x, A_y), Q', Q'' (10^{-4} cm^{-1})	Ref.
SrO	290		$g = 2.001$			$A = -78.3$	MSrev91
	100		$g = 2.001$			$A = -80.2$	
	1.5		$g_\parallel = 2.001$, $g_\perp = 2.002$			$A = -80.2$	
	295		$g = 2.001$			$A = -78.2$	
	100					$A = -78.3$	
SrO (powder)	295		$g = 2.001$			$A = -80.2$	MSrev91
	100		$g = 2.001$			$A = -90.3$	
ZnO	300		2.001	236.2, —	$a - F = 5.23$	$B = 74.1$	PF74
IO_2	300		$g_\parallel = 2.001$, $g_\perp = 1.997$			$A = -78.7$	MSrev91
	77		$g_\parallel = 2.000$, $g_\perp = 1.991$			$A = -78.7$	
ZrO_2	77, 300					$A = 82$ G	MSrev91
Al_2O_3	NR					$A_\parallel = -85.1$ G, $A_\perp = -83.7$ G	MSrev91
Al_2O_3	300		$g_\parallel = 2.002$, $g_\perp = 2.000$	194.2, —		$A = -79.6$, $B = -78.8$	PF74
α-Al_2O_3	292					$A_\parallel = -85.1$ G, $A_\perp = -83.7$ G	MSrev91
β-Al_2O_3	NR		$g_\parallel = 2.000$, $g_\perp = 2.002$			$A_\parallel = -83.9$ G, $A_\perp = -83.3$ G	MSrev91
Na–βAl_2O_3	77–300	I	$g_\perp = 2.001$,	−310 G, —		$A_\perp = -83.5$ G,	MSrev91
		II	$g_\perp = 2.001$	−250 G, —		$A_\perp = -83.0$ G	
		III	$g_\perp = 2.002$	−160 G, —		$A_\perp = -82.1$ G	
	NR		$g_\parallel = 2.001$, $g_\perp = 2.002$	−270 G, —		$A_\parallel = -82.0$ G, $A_\perp = -82.8$ G	
$KMgF_3$	290	A	2.002		3.24, —, —	$A = 91.0$	PF74
$MgAl_2O_4$	290	A	2.000			$A = 74.9$	PF74
$Li_{0.5}Al_{2.5}O_4$	290	B?	2.002			$A = 77.2$	PF74
$Li_{0.5}Ga_{2.5}O_4$	290	A	2.008	±0.080, —			PF74

La_2O_3	NR	$g_\parallel = 2.001$, $g_\perp = 2.003$			$A_\parallel = -70.8$, $A_\perp = -67.67$	MSrev91
$(La_2O_3)_{0.95}(CeO_2)_{0.05}$	RT	$g_\parallel = 2.004$, $g_\perp = 2.007$	−0.600*, —	−0.0004*, —, —	$A_\parallel = -0.187$*, $A_\perp = 1.183$*	MSrev91
CaO–SiO_2–Al_2O_3 (glass)	NR	$g = 2.002$			$A = -77.1$	MSrev91
$(CaO)_{40-45}(P_2O_5)_{20-25}$	NR	$g = 4.29$	2400 G, —		$A = -88$ G	MSrev91
$(WO_3)_{20-40}$ (glass)			2400 G, —			MSrev91
$KTaO_3$	77	$g_\parallel = 1.998$, $g_\perp = 2.004$	1480, —	−1.2, —, —	$A_\parallel = 85.9$, $A_\perp = 82.7$	MSrev91
	4.2		>0, —			
	65–900	$g = 2.000$	1470, —		$A_\parallel = 86$, $A_\perp = 80$	
$KTa_xNb_{1-x}O_3$	110–560	$g = 2.000$			$A = 84$	MSrev91
$LiCsO_4$	152	$g_\perp = 2.005$	2.237*, −0.806*	0.14*, 0.26*, 0.08*	$A_\parallel = A_\perp = -0.267$*, $Q' = -0.012$*, $Q'' = -0.001$*	MM93
	90	$g_\perp = 2.007$	2.238*, −0.844*	0.25*, 0.71*, 0.20*	$A_\parallel = A_\perp = -0.264$*, $Q' = -0.015$*, $Q'' = 0.004$*	MSrev91
$Li(NH_4)TaO_3 \cdot H_2O$	77–300	2.007, 2.010, 2.005	609 G, −327 G	1.8 G, —, —	−88 G, −95 G, −90 G	MSrev91
$ZnGa_2O_4$	RT	$g = 2.001$			$A = -76.2$	MSrev91
$Cd_2Nb_2O_7$	200	$g = 2.001$	434, —		$A = -66$	MSrev91
	77	$g = 2.000$	341, —		$A = -66$	
	4.2–300	$g = 2.000$	486, —		$A = -66$	
$Pb_5Ge_3O_{11}$	RT	$g = 2.00$			$A = 0.26$*	MSrev91
	4.2	$g_\parallel = 2.006$, $g_\perp = 1.999$	−8.97*, —	−0.004*, —, —	$A_\parallel = -0.246$*, $A_\perp = -0.252$*	
	77	$g_\parallel = 2.003$, $g_\perp = 2.000$	−8.92*, —	−0.003*, —, —	$A_\parallel = -0.245$*, $A_\perp = -0.251$*	
	296	$g_\parallel = 2.004$, $g_\perp = 2.000$	−8.67*, —	−0.003*, —, —	$A_\parallel = -0.244$*, $A_\perp = -0.249$*	
Perchlorates						
$Cd(ClO_4)_2 \cdot 6D_2O$	77–380	$g_\parallel = 2.001$, $g_\perp = 2.001$	48.9 G, —		$A_\parallel = -93.0$ G, $A_\perp = -94.0$ G	MSrev91

DATA TABULATION (Mn^{2+} $3d^5$) *(continued)*

Host (site)	T (°K)	Site	$\tilde{g}$ (g_z, g_x, g_y)	b_2^0 (= D), b_2^2 (= E) (10^{-4} cm^{-1})	b_4^0, b_4^2, b_4^4 (10^{-4} cm^{-1})	$\tilde{A}$ (A_z, A_x, A_y), Q', Q'' (10^{-4} cm^{-1})	Ref.
$Cd(ClO_4)_2 \cdot 6H_2O$	77–380		$g_\parallel = 2.001$, $g_\perp = 2.000$	48.0 G, —	3 G, —, —	$A_\parallel = -93.0$ G, $A_\perp = -94.0$ G	MSrev91
$Co(ClO_4)_2 \cdot 6H_2O$	77-HT		$g_\parallel = 2.002$, $g_\perp = 2.00$	116 G, —	11 G, —, —	$A_\parallel = 92.0$ G, $A_\perp = 92.0$ G	MSrev91
$Mg(ClO_4)_2 \cdot 6H_2O$	RT–HT		2.002, 2.001, 2.001	146 G, −2.7 G	8 G, —, —	−93 G, −94 G, −94 G	MSrev91
$Ni(ClO_4)_2 \cdot 6H_2O$	77-HT		$g_\parallel = 2.000$, $g_\perp = 2.000$	118 G, —	10 G, —, —	$A_\parallel = 92.0$ G, $A_\perp = 92.0$ G	MSrev91
$Zn(ClO_4)_2 \cdot 6H_2O$	300–380		$g_\parallel = 2.002$, $g_\perp = 2.001$	125.7 G, —	12 G, —, —	$A_\parallel = 93.0$ G, $A_\perp = 93.0$ G	MSrev91
	LT, RT, HT		$g_\parallel = 2.002$, $g_\perp = 2.001$	126 G, 1.5 G	12 G, —, —	$A_\parallel = -93.0$ G, $A_\perp = -94.0$ G	
$Zn(NH_3)_4(ClO_4)_2$	77–300			5, —			MSrev91
Phosphates							
$AlPO_4$ (powder)	RT		$g = 2.000$	1630, 150	—, —, —	$A = 77$	MSrev91
	473		$g = 1.999$				
$LuPO_4$	RT	I	$g = 2.001$	218.5, —	0.4, —, —	$A = -90.1$	MSrev91
		II	$g = 2.004$	−177.3, −25	0.7, —, —	$A = -88.2$	
$ScPO_4$	RT	I	$g = 2.002$	525.4, —	0.7, —, —	$A = -86.7$	MSrev91
		II	$g = 2.001$	−254.3, −138	1.2, —, —	$A = -88.7$	
YPO_4	RT		$g = 2.005$	207.3, —		$A = -88.6$	MSrev91
$Mg(PO_3)_2$ (powder)	NR		$g = 2.000$	220 G, —	60 G, —, —	$A = -93$ G	MSrev91
$Pb_3(PO_4)_2$	298		2.003, 2.009, 1.996	−425 G, 280.5 G	2 G, —, —	−95.5 G	MSrev91
	523		2.000, 2.001, 2.001	−222.5 G, 0 G	−2.4 G, —, —	−92.7 G	
	298			−425 G, 279 G	—, —, —	−91.0 G	
	423			−288 G, 96 G	—, —, —	−92.5 G	
	448			−298 G, 27 G	—, —, —	−92.5 G	
	473			−222 G, 0 G	—, —, —	−92.5 G	
Propionates							
$Ca_2Ba(C_2H_5COO)_6$	LT, RT, HT		$g_\parallel = 2.009$, $g_\perp = 2.022$	0.391*, —	−0.001*, —, —	$A_\parallel = -0.269$*, $A_\perp = -0.1285$*	MSrev91
	113–493		$g = 2.002$			$A = 95.2$ G	

$Ca_2Pb(C_2H_5COO)_6$	RT–HT		$g_\parallel = 2.006$, $g_\perp = 2.004$	0.697*, −0.644*	−0.009*, —, —	$A_\parallel = -0.270$*, $A_\perp = -0.269$*	MSrev91
$Ca_2Sr(C_2H_5COO)_6$	LT, RT, HT		$g_\parallel = 2.100$, $g_\perp = 2.087$	0.141*, −0.082*	0.006*, —, —	$A_\parallel = -0.278$*, $A_\perp = -0.267$*	MSrev91
Sulfates							
$CoSO_4 \cdot 7H_2O$	77–300	I	$g = 2.001$	459, 402	—, —, —	$A_\parallel = -86$, $A_\perp = -87$	MSrev91
		II	$g = 1.996$	234, 51	—, —, —	$A_\parallel = -86$, $A_\perp = -87$	
$MgSO_4 \cdot 7H_2O$				1.18*, 0.58*	0.01*, —, —	$A_\parallel = -0.269$*, $A_\perp = -0.266$*	MSrev91
	295		$g_\parallel = 2.002$, $g_\perp = 1.968$	−1.118*, 0.367*	0.004*, —, —	$A_\parallel = -0.268$*, $A_\perp = -0.267$*	
	85		$g_\parallel = 1.996$, $g_\perp = 1.998$	−1.241*, 0.254*	0.006*, —, —	$A_\parallel = -0.276$*, $A_\perp = -0.258$*	
	3		$g_\parallel = 2.006$, $g_\perp = 1.963$	−1.292*, 0.670*	0.007*, —, —	$A_\parallel = -0.277$*, $A_\perp = -0.271$*	
$NiSO_4 \cdot 7H_2O$	295		$g_\parallel = 1.970$, $g_\perp = 1.979$	−1.257*, 0.383*	−0.003*, —, —	$A_\parallel = -0.265$, $A_\perp = -0.262$*	MSrev91
	85		$g_\parallel = 2.015$, $g_\perp = 1.985$	−1.535*, 0.520*	−0.009*, —, —	$A_\parallel = -0.272$*, $A_\perp = -0.254$*	
	2		$g_\parallel = 1.995$, $g_\perp = 1.999$	−1.595*, 0.321*	0.004*, —, —	$A_\parallel = -0.264$*, $A_\perp = -0.265$*	
	RT		$g_\parallel = 2.000$, $g_\perp = 2.000$	1.298*, −0.200*	−0.001*, —, —	$A_\parallel = -0.258$*, $A_\perp = -0.249$*	
	LNT		$g_\parallel = 2.012$, $g_\perp = 2.001$	1.554*, −0.252*	−0.003*, —, —	$A_\parallel = -0.260$*, $A_\perp = -0.246$*	
	LHT		$g_\parallel = 2.014$, $g_\perp = 2.001$	1.589*, −0.259*	−0.07*, —, —	$A_\parallel = -0.256$*, $A_\perp = -0.249$*	
$Co_{0.5}Zn_{0.5}SO_4 \cdot 7H_2O$	LNT–RT		2.009, 1.997, —	−1.3833*, 1.1915*	0.0015*, —, —	−0.2679*, 0.2514*, —	MSrev91
$SrSO_4$	4.2–350			−108.9, —	−1.2, —, —	$A_\parallel = -90.5$ G, $A_\perp = -90.1$ G	MSrev91
$(NH_4)_2SO_4$	113–403		2.001, 2.018, —	2.153*, −0.715*	—, —, —	$A_\parallel = -0.258$, $A_\perp = -0281$	MSrev91

DATA TABULATION (Mn^{2+} $3d^5$) (*continued*)

Host (site)	T (°K)	Site	$\tilde{g}$ (g_z, g_x, g_y)	b_2^0 (= D), b_2^2 (= E) (10^{-4} cm^{-1})	b_4^0, b_4^2, b_4^4 (10^{-4} cm^{-1})	$\tilde{A}$ (A_z, A_x, A_y), Q', Q'' (10^{-4} cm^{-1})	Ref.
$Pb_{1-x}Mn_xTe$	4.2		1.996				KM88
($x = 0.0004$)						$A = 64.7$ G	
($x = 0.0012$)						$A = 65.2$ G	
($x = 0.002$)						$A = 65.5$ G	
$Na_2Cd(SO_4)_2 \cdot 4H_2O$	RT		2.004, 2.008, 2.008	550, −282	4, —, —	−84.5, −84.3, −84.3	MSrev91
$Na_2Co(SO_4)_2 \cdot 4H_2O$	90–290		2.000, 2.003, —	−328, 231	0.3, —, —	−88.0, −87.0, —	MSrev91
$Na_2Zn(SO_4)_2 \cdot 4H_2O$	RT		2.011, 2.012, 2.012	−344, 249	1.4, —, —	−87.4, −87.3, −87.3	MSrev91
$Tl_2Cd_2(SO_4)_3$	295		$g_\parallel = 2.005$, $g_\perp = 2.003$	0.454*, —	$b_4^0 = 0.005$* $b_4^3 = -0.265$*	$A = -0.274$*, $B = -0.235$*	MK87
	RT, HT		$g = 2.004$	453 G, 177 G	16 G, —, —	$A = -92$ G	MSrev91
$(NH_4)_3H(SO_4)_2$							
$MgSeSO_4 \cdot 6H_2O$	298	I	$g_\parallel = 2.004$, $g_\perp = 2.005$	238.6, 165	3.7, —, —	$A_\parallel = 90$, $A_\perp = -89.6$	MSrev91
		II	$g_\parallel = 2.008$, $g_\perp = 2.009$	404.0, 321	−0.5, —, —	$A_\parallel = -90.4$, $A_\perp = -89$	
$C(NH_2)_3Al(SO_4)_2 \cdot 6H_2O$	RT	I	$g_\parallel = 2.004$, $g_\perp = 2.004$	370, —	5, —, —	$A_\parallel = -87.5$, $A_\perp = -87.5$	MSrev91
		II	$g_\parallel = 2.004$, $g_\perp = 2.004$	344, —	5, —, —	$A_\parallel = -87.5$, $A_\perp = -87.5$	
Sulphides							
MgS	6, 80, 300		$g = 2.002$			$A = -74.8$	MSrev91
ZnS	4.2						MSrev91
	20–300	I			—, —, —	$A = -64.0$	
		II	$g = 2.002$	−107.5, —	—, —, —	$A = -64.5$	
		III	$g = 2.003$	−131.0, —	—, —, —	$A = -64.5$	
			$g = 2.003$	−132.9, —	—, —, —		
	RT, LHT	I		35.9, —	—, —, —		
		II		−99.9, —	—, —, —		
		III					

Na_2S	296		$g = 2.003$			$A_{\parallel} = -65.2$	MSrev91
	77		$g = 2.003$			$A_{\parallel} = -65.7$	
FeS_2 (powder)	5–295		$g = 2.000$			$A = -95$ G	MSrev91
RuS_2	RT		$g = 2.135$			$A = 81$	MSrev91
$CdPS_3$	RT		2.006, 2.007, 2.001	365, −18	0.35, —, —	−69.2, −69.4, −70.5	MSrev91
	360		2.006, 2.007, 2.001	−365, −16	0.35, —, —	−70.5, −69.2, −69.2	
	143		2.005, 2.002, 2.002	−353, 0	0.15, —, —	−69, −69, −70	
$CdScS_4$	4.2–330		$g = 2.002$			$A = 65.9$	MSrev91
$CdLu_2S_4$	4.2–330		$g = 2.003$			$A = 67.6$	MSrev91
CdY_2S_4	4.2–330		$g = 2.002$			$A = 68.2$	MSrev91
Titanates							
$CdTiO_3$	4.5–600		$g_{\parallel} = 2.008$, $g_{\perp} = 2.004$	−120, −26	—, —, —	$A = -82.7$	MSrev91
$MgTiO_3$	300		$g_{\parallel} = 2.000$, $g_{\perp} = 2.000$	164, —	7, —, —	$A = -79.3$	MSrev91
$SrTiO_3$	4.2–RT	I	$g = 2.000$			$A = 82.6$	MSrev91
		II	$g_{\parallel} = 200$, $g_{\perp} = 5.9$				
		III	$g_{\parallel} = 2.003$, $g_{\perp} = 2.008$			$A_{\parallel} = 76$, $A_{\perp} = 65$	
$SrTiO_3 \cdot 3H_2O$	RT		2.001	483 G, —		−96.2 G	MSrev91
			1.978	303 G, —		−94.9 G	
			1.978	0.7 G, —		−94.9 G	
Tungstate							
$CaWO_4$ (powder)	NR		$g = 2.000$	1473 G, —		$A = -95$ G	MSrev91
$CsWO_4$	4.2–350			−28.9, —	−1.3, —, —	$A_{\parallel} = -88.9$ G, $A_{\perp} = -89.4$ G	MSrev91
$PbWO_4$	4.2–350			−134, —	−1.5, —, —	$A_{\parallel} = -89.6$ G, $A_{\perp} = -89.1$ G	MSrev91
Tutton salts							
$CdCs_2(SO_4)_2 \cdot 6H_2O$	RT			−0.803*, 0.334*	0.006*, —, —	−0.263*, −0.267*, −0.264*	MSrev91
$Cd(NH_4)_2(SO_4)_2 \cdot 6H_2O$	295		2.003, 2.001, —	−0.709*, 0.642*	0.008*, —, —	$A_{\parallel} = -0.276$*, $A_{\perp} = -0.251$*	MSrev91
	85		2.009, 2.003, —	−0.697*, 0.875*	0.002*, —, —	$A_{\parallel} = -0.276$*, $A_{\perp} = -0.251$*	

DATA TABULATION (Mn^{2+} $3d^5$) *(continued)*

Host (site)	T (°K)	Site	$\tilde{g}$ (g_z, g_x, g_y)	b_2^0 $(= D)$, b_2^2 $(= E)$ $(10^{-4}$ $cm^{-1})$	b_4^0, b_4^2, b_4^4 $(10^{-4}$ $cm^{-1})$	$\tilde{A}$ (A_z, A_x, A_y), Q', Q'' $(10^{-4}$ $cm^{-1})$	Ref.
	5		2.000, 2.000, —	−0.715*, 0.863*	0.002*, —, —	$A_\parallel = -0.273$*, $A_\perp = -0.262$*	
$CdRb_2(SO_4)_2 \cdot 6H_2O$	298		2.007, 2.010, —	−314, 168	2.8, —, —	$A_\parallel = -90.0$, $A_\perp = -89.0$	MSrev91
$CoCs_2(SO_4)_2 \cdot 6H_2O$	RT		2.003, 2.000, 2.000	−260.6, 60	1.9, —, —	−88.0, −87.8, −87.8	MSrev91
$Co(NH_4)_2(SO_4)_2 \cdot 6H_2O$	295		2.002, 2.004, —	−0.712*, 0.630*	0.010*, —, —	$A_\parallel = -0.270$*, $A_\perp = -0.270$*	MSrev91
$CoRb_2(SO_4)_2 \cdot 6H_2O$	RT		2.006, 2.007, 2.007	−303, 114	3.5, —, —	−89.2, −88.6, −88.6	MSrev91
$CoTl_2(SO_4)_2 \cdot 6H_2O$	RT		2.006, 2.007, 2.007	−254, 135	2.6, —, —	−89.5, −89.4, −89.4	MSrev91
$FeCs_2(SO_4)_2 \cdot 6H_2O$	298		2.003, 2.007, —	−265.8, 72	2.8, —, —	$A_\parallel = -88.9$, $A_\perp = -88.6$	MSrev91
$FeRb_2(SO_4)_2 \cdot 6H_2O$	298		2.010, 2.014, —	−289.2, 120	3.6, —, —	$A_\parallel = -89.5$, $A_\perp = -88.0$	MSrev91
$MgCs_2(SO_4)_2 \cdot 6H_2O$	298		2.006, 2.011, —	−257.8, 96	2.1, —, —	$A_\parallel = -89.2$, $A_\perp = -88.0$	MSrev91
$Mg(NH_4)_2(SO_4)_2 \cdot 6H_2O$	RT			−1.067*, 0.514*	0.007*, —, —	−0.265*, −0.264*, −0.261*	MSrev91
$MgRb_2(SO_4)_2 \cdot 6H_2O$	298		2.005, 2.011, —	−309, 225	2.0, —, —	$A_\parallel = -90.5$, $A_\perp = -90.0$	MSrev91
$MgTl_2(SO_4)_2 \cdot 6H_2O$	RT		2.007, 2.005, 2.005	−251, 126	2.7, —, —	−89.7, −89.5, −89.5*	MSrev91
$NiK_2(SO_4)_2 \cdot 6H_2O$	295		2.058, 2.025, —	−0.750*, 0.564*	0.300*, —, —	$A_\parallel = -0.275$*, $A_\perp = -0.232$*	MSrev91
	5		2.017, 1.991, —	−1.198*, 0.432*	0.125*, —, —	$A_\parallel = -0.115$*, $A_\perp = -0.228$*	
	1.8		2.024, 1.974, —	−1.260*, 0.395*	0.111*, —, —	$A_\parallel = -0.169$*, $A_\perp = -0.228$*	
$NiK_2(SO_4)_2 \cdot 6H_2O$	295		2.009, 1.994, —	−0.807*, 0.292*	−0.188*, —, —	$A_\parallel = -0.249$*, $A_\perp = -0.264$*	MSrev91

	60	2.000, 2.000, —	−1.000*, −0.300*	−0.320*, —, —	$A_{\parallel} = -0.200$*, $A_{\perp} = -0.260$*	
	3.2	2.068, 1.997, —	−1.022*, 0.304*	−0.322*, —, —	$A_{\parallel} = -0.202$*, $A_{\perp} = -0.256$*	
$Ni(NH_4)_2(SO_4)_2 \cdot 6H_2O$	295	2.061, 2.002, —	−0.690*, 0.651*	−0.118*, —, —	$A_{\parallel} = -0.295$*, $A_{\perp} = -0.249$*	MSrev91
	85	2.022, 2.000, —	—, 0.855*	−0.112*, —, —	$A_{\parallel} = -0.298$*, $A_{\perp} = -0.242$*	
	5	1.931, 1.992, —	−0.377*, 1.242*	−0.300*, —, —	$A_{\parallel} = -0.279$*, $A_{\perp} = -0.255$*	
$ZnCs_2(SO_4)_2 \cdot 6H_2O$	298	2.008, 2.011, —	−266, 96	2.3, —, —	$A_{\parallel} = -89.5$, $A_{\perp} = -88.6$	MSrev91
$ZnK_2(SO_4)_2 \cdot 6H_2O$	290	2.007, 2.006, —	−376 G, 18 G	0.26 G, —, —	$A_{\parallel} = -93$ G, $A_{\perp} = -92$ G	MSrev91
	295	1.996, 1.999, —	−1.068*, 0.504*	0.018*, —, —	$A_{\parallel} = -0.265$*, $A_{\perp} = -0.269$*	
	85	2.001, 2.021, —	−1.315*, 0.528*	0.019*, —, —	$A_{\parallel} = -0.259$*, $A_{\perp} = -0.264$*	
	5	2.002, 2.013, —	−1.359*, 0.489*	0.021*, —, —	$A_{\parallel} = -0.258$*, $A_{\perp} = -0.260$*	
	RT		−0.723*, 0.566*	0.007*, —, —	−0.264*, −0.266*, −0.266*	
$Zn(NH_4)_2(SO_4)_2 \cdot 6H_2O$	RT		−0.721*, 0.677*	0.005*, —, —	−0.264*, −0.265*, −0.264*	MSrev91
$ZnRb_2(SO_4)_2 \cdot 6H_2O$	298	2.010, 2.011, —	−300, 165	4.7, —, —	$A_{\parallel} = -89.0$, $A_{\perp} = -88.0$	MSrev91
$ZnTl_2(SO_4)_2 \cdot 6H_2O$	RT	2.009, 2.011, 2.011	−261, 114	3.7, —, —	−89.3, −88.8, −88.8	MSrev91
$ZnTiF_6 \cdot 6H_2O$	300	$g_{\parallel} = 2.001$, $g_{\perp} = 2.001$	−182.2, 0	$a - F = 8$	−90.7	RDH89
	115	$g_{\parallel} = 2.001$	−218, $E \sim 40$	$a - F = 8$		MSrev91
Tutton salts (with selenates)						
$CoK_2(SeO_4)_2 \cdot 6H_2O$	RT	1.999, 1.999, —	287 G, 180 G	2.5 G, —, —	−94.0 G, −95.5 G, —	MSrev91
$Co(NH_4)_2(SeO_4)_2 \cdot 6H_2O$	90–300	1.999, 1.999, —	287 G, 180 G	0.01 G, —, —	−94.0 G, −95.5 G, —	MSrev91

DATA TABULATION (Mn^{2+} $3d^5$) *(continued)*

Host (site)	T (°K)	Site	$\tilde{g}$ (g_z, g_x, g_y)	b_2^0 (= D), b_2^2 (= E) (10^{-4} cm^{-1})	b_4^0, b_4^2, b_4^4 (10^{-4} cm^{-1})	$\tilde{A}$ (A_z, A_x, A_y), Q', Q'' (10^{-4} cm^{-1})	Ref.
$MgCs_2(SeO_4)_2 \cdot 6H_2O$	NR		2.005, 2.004, —	−272.3 G, −151 G	3.2 G, —, —	−95.1 G, −94.8 G, —	MSrev91
$MgK_2(SeO_4)_2 \cdot 6H_2O$	NR		2.004, 2.004, —	−374.4 G, −296 G	6.3 G, —, —	−93.7 G, −94.0 G, —	MSrev91
$NiK_2(SeO_4)_2 \cdot 6H_2O$	294		1.998, 1.993, —	−1.046*, 0.281*	0.014*, —, —	−0.258*, −0.253*, —	MSrev91
$Ni(NH_4)_2(SeO_4)_2 \cdot 6H_2O$	RT		1.996, —, —	274, −126	2.8, —, —	$A_\parallel = -86$, $A_\perp = -87$	MSrev91
$ZnK_2(SeO_4)_2 \cdot 6H_2O$	290		2.005, 2.006, —	−388 G, 30 G	0.23 G, —, —	−94 G, −92 G, —	MSrev91
Other							
Mg	2–300			126, —	—, —, —		MSrev91
LiH	50		$g = 1.991$	450 G, —		$A = 71.4$	MSrev91
	20		2.057, 2.001, 4.56				
	293		$g = 1.98$	312, —, 0.01		$A = 65$	
	40		$g = 1.991$	450 G, —		$A = 71.5$	
GaSe	4.2–500		$g = 2.004$				MSrev91
GeTe	140–300					$A = 56.2$	MSrev91
$Pb_{1-x}Mn_xTe$	4.2		1.996				ZM88
$RuSe_2$	RT		$g = 2.149$			$A = 80.4$	MSrev91
NaN_3	308		$g = 2.007$			−88 G	MSrev91
	473		$g = 2.007$			−88 G	
$CaCO_3$	NR		2.002, 2.004, 1.991			−90.2, −88.7, −90.2	MSrev91
			$g_\parallel = 2.002$, $g_\perp = 2.002$	74.97, —	2.382, —, —	$A = 87.01$, $Q' = 0.17$, $B = 87.26$	SJ66

$NaNO_2$	LT, RT, HT	2.003				MSrev91
$BiVO_4$	295		27.6, 82.2	−0.031, −0.28, 0.32		YRC94
$CdSc_2Se_4$	4.2–330	$g = 2.004$			$A = 62.6$	MSrev91
CdY_2Se_4	4.2–330	$g = 2.004$			$A = 64.4$	MSrev91
$CdLu_2Se_4$	4.2–330	$g = 2.003$			$A = 64.3$	MSrev91
$CdC_4H_4O_6 \cdot 5H_2O$	4.2–300	$g_\parallel = 2.005$, $g_\perp = 1.999$	−0.526*, —	−0.005*, —, —	$A_\parallel = -0.277$*, $A_\perp = -0.251$*	MSrev91
$CdC_4H_4O_6 \cdot 5H_2O$	RT	2.002, 1.980, 1.992	705 G, 117 G	2 G, —, —	−96 G, −97 G, −96.5 G	MSrev91
$LiKC_2H_4O_6 \cdot H_2O$	77–300	2.005, 2.013, 2.005	672 G, −33 G	2.4 G, —, —	−88 G, −90 G, −91 G	MSrev91
$LiNH_4C_2H_4O_6 \cdot H_2O$	77–300	2.007, 2.010, 2.005	609 G, −327 G	1.8 G, —, —	−88 G, −90 G, −91 G	MSrev91
$Na_8Al_6Si_6O_{24}Cl_2$	NR	$g = 2.003$			$A = 77$	MSrev91

Fe^{3+} $3d^5$. Data tabulation of spin-Hamiltonian parameters (* indicates units of GHz).

Sample	T (°K)	Site	$\tilde{g}$	$D\,(=b_2^0)$ $(10^{-4}\ cm^{-1})$	$E\,(=b_2^2)$ $(10^{-4}\ cm^{-1})$	a $(10^{-4}\ cm^{-1})$	$\frac{1}{3}F(=b_4^0)$ $(10^{-4}\ cm^{-1})$	$\tilde{A}^{57}$ $(10^{-4}\ cm^{-1})$	Ref.
Alums									
$KAl(SO_4)_2 \cdot 12H_2O$	room		2.00	160					LB
$NH_4Al(SO_4)_2 \cdot 12H_2O$	293–77		2.00	160					LB
	4		2.00	180					
$RbAl(SO_4)_2 \cdot 12H_2O$	90		2.003	22		−134	−3		LB
	20		2.003	31		−134	−3.3		
$RbAl(SO_4)_3 \cdot 12H_2O$	4.2		2.003	−31		133.4	$F=4$	$A=-10.77$, $B=-10.74$	BM83
$RbGa(SO_4)_3 \cdot 12H_2O$	4.2		2.003	−84		133.4	$F=4$	$A=-10.76$, $B=-10.74$	BM83
Halides									
NaCl	300		2.001						BM73
NH_4Cl	295			−1720		$a-F=9.0$			ART 72
CaF_2	300		$g=2.005$			$\|a\|=37.3$			BM73
$SrCl_2$	4.2		2.014				$b_4=35.5$		BM83
$AlCl_3$	NR		$g_x=2.001$ $g_y=2.002$ $g_z=1.995$	−1890	16.1	−2.2	$a+\frac{2}{3}F=-16$		ART 72
$AlCl_3 \cdot 6H_2O$	77		2.002	1500		160	310		LB
$FeCl_3(p)$	298, 77		2.013						LB
$FeF_3 \cdot 4.5H_2O(p)$	room		1.99						LB
$FeF_3 \cdot 9H_2O(p)$	room		2.02						LB
$LiBaF_3$ (X-irrad.)	4.2, 77		2.003				$A=120$		BM86
$KMgF_3$	77		2.003				25.75	$A'=360$, $B'=18.0$	PF74
	300		$g_\parallel=2.008$ $g_\perp=2.008$	−3572	75, 269		$b_4^0=75$	$b_4^4=269$	BM86
$KZnF_3$	293			−89.9		6.3		$A_\parallel=23.6$ $A_\perp=14.3$	BD74
	4.2		2.0021	−90.80		6.9		$A_\parallel=24.19$ $A_\perp=15.054$	BD74

$KZnF_3$	77		2.003	107.9		49.7	$F=-2.7$		BD74
$(NH_4)_2InCl_5 \cdot H_2O$	295		$g_x=2.016$ $g_y=2.014$ $g_z=2.014$	-4.570*	2.545*		$b_4^0=0.027$* $b_4^2=0.032$* $b_4^4=0.137$*		MS76
	295		$g_\parallel=2.012$ $g_\perp=2.015$	-4.563*	2.551*		$b_4^0=0.032$* $b_4^2=0.021$* $b_4^4=0.217$*		ML92C
	77		$g_\parallel=2.017$ $g_\perp=2.019$	-5.182*	2.576*		$b_4^0=0.039$* $b_4^2=0.014$* $b_4^4=0.208$*		ML92C
	20		$g_\parallel=2.019$ $g_\perp=2.020$	-5.201*	2.587*		$b_4^0=0.037$* $b_4^2=0.018$* $b_4^4=0.213$*		ML92C
	4		$g_\parallel=2.020$ $g_\perp=2.021$	5.205*	2.590*		$b_4^0=0.038$* $b_4^2=0.019$* $b_4^4=0.211$*		ML92C
$(NH_4)_2SbF_5$	77–300		1.996	$\lvert D\rvert=690$	$\lvert E\rvert=15.2$	$\lvert a\rvert=32.8$	$\lvert F\rvert=6$		BD80
Oxides									
CaO	77		2.005				$b_4^0=31.9$		PF74
	20		2.006				$b_4^0=32.6$	$A=10.5$	PF74
MgO	290		2.004					11.4	PF74
	300	Cubic				205			BM83
MgO ($S'=\frac{1}{2}$)	1.5–?		$g_\parallel=2.004$ $g_\perp=2.003$	4700		26.2		$A^{57}=10.9$	BM73
SrO	90		$g_x=g_y=g_z$ $=2.006$	294.9	140.7	$B_4^0=0.031$, $B_4^2=-1.00$, $B_4^4=0.98$			BM83
ZnO	290		2.006	-593.7		$(\frac{1}{2}a)=19.95$	$(a-F)=4.0$	$B=9.02$	PF74
	298		$g_\parallel=2.005$ $g_\perp=2.004$	-593.77		46.36	$F=10.72$		BD80
α-LiIO	295		$g_\parallel=g_\perp=2.00$	≈ 7.5*			$b_4^0\approx -0.1$* $b_4^3\approx -0.1$*		BD80
SnO_2	76, 293		2.00	$\lvert D\rvert=15$*	$E/D=0.065$				BD74
TiO_2	78	I	$g_\parallel=2.004$ $g_\perp=2.004$	447			$b_4^0=55.9$ $b_4^4=267$		BM73
	300	I	$g_\parallel=2.005$ $g_\perp=2.005$	308.7			$b_4^0=53.5$ $b_4^4=257$		BM73

Fe^{3+} $3d^5$ (*continued*)

Sample	T (°K)	Site	$\tilde{g}$	$D\,(=b_2^0)$ $(10^{-4}$ cm$^{-1})$	$E\,(=b_2^2)$ $(10^{-4}$ cm$^{-1})$	a $(10^{-4}$ cm$^{-1})$	$\frac{1}{3}F(=b_4^0)$ $(10^{-4}$ cm$^{-1})$	$\tilde{A}^{57}$ $(10^{-4}$ cm$^{-1})$	Ref.
	78	II	$g=2.002$	−5110	−3540		$b_4^0=-26$		BM73
							$b_4^2=60$		
							$b_4^4=100$		
	300	II	$g=2.002$	−4970	−3720		$b_4^0=-14$		BM73
							$b_4^2=87$		
							$b_4^4=70$		
TiO_2	1.4, 78		$g_c=2.000$	6780	2210		183.0		PF74
$Ca(OH)_2$	4.2, 77	I	$g_x=g_y=2.008$	$B_2^0=1799.4$ G			$B_4^0=9.42$ G		BD80
			$g_z=2.001$				$B_4^3=-17.29$ G		
		II	$g_x=g_y=2.009$	$B_2^0=1796.6$ G			$B_4^0=9.44$ G		
			$g_z=1.998$				$B_4^3=19.24$ G		
MoO_3	NR			12, 400		3000			ART 72
SnO_4	300		4.27		$E/D=0.333$				BM73
Al_2O_3	290		$g_\perp=2.003$	1684	131.0				PF74
Quartz (Al_2O_3)			$g\sim4.3$	$D\sim3000$	$E/D\sim0.300$				BM73
La_2O_3	4.2, 293		$g=2.002$	$B_2^0=3430$ G			$B_4^0=0.24$ G		BD80
Y_2O_3–Ga_2O_3	293			$\lvert D\rvert=0.133$ cm^{-1}		$a=0.020$ cm^{-1}	$\lvert F\rvert=0.003$ cm^{-1}		BD80
Y_2O_6	77		2.006	128.4		$\lvert a\rvert=77$	$\lvert a-F\rvert=77.5$		BD80
α-$LiIO_3$	4.2	Li site A					$A_x=A_y=0.236$ MHz, $A_z=0.475$ MHz		BD80
		Li site B					$A_x=A_y=-0.291$ MHz, $A_z=0.586$ MHz		
	295		2.007	$B_2^0=879.8$			$B_4^0=-0.69$		
							$B_4^3=\pm6.5$		
			$g_\parallel=2.001$	7.831*			$b_4^0=-0.133$*		
			$g_\perp=2.008$				$b_4^3=-0.101$*		
Al_2SiO_5	4.2, 98, 293	6	$g_x=2.017$	1.94	0.43		$B_4^0=-0.137$ kG		BM83
			$g_y=2.066$				$B_4^2=-0.112$ kG		
			$g_z=1.990$						

		7	$g_x = 2.012$	1.02	0.08		$B_4^0 = 0.003$ kG	
			$g_y = 2.018$				$B_4^2 = 0.018$ kG	
			$g_z = 1.999$					
$PbO \cdot 3B_2O_3$ glasses	78–300		2.0, 4.20, 6.0 (transitions)					BM83
$Li_{0.5}Al_{2.5}O_4$								
Ordered	290	A	2.006	± 0.104			± 0.01	PF74
Disordered	290	B		0.13				
$Pb_5Ge_3O_{11}$	1.4–5		$g_\parallel = 2.004$	−25.2	−0.2, −0.8			
			$g_\perp = 1.992$					
$Mg_xFe_{3-x}O_{4-\delta}$								BM83
$(x = 1.41)$	77–520		1.996					
$(x = 1.39)$	77–520		1.996					
$(x = 1.35)$	77–520		1.998					
$(x = 1.31)$	77–520		2.001					
$(x = 1.18)$	77–520		2.003					
$NaAlSi_3O_8$	293		$g_{zz} = 2.001$	200.5			$B_4^0 = 10.6$	BD74
			$g_{xx} = 2.002$	39.8			$B_4^2 = 10.3$	
			$g_{yy} = 2.002$				$B_4^4 = -59.0$	
$LiAlSi_4O_{10}$	77		$g_x = 2.016$	± 7.317*	± 0.506*			ART 72
			$g_y = 2.008$					
			$g_z = 2.007$					
$Ca_2Al_3Si_3O_{12}(OH)$	4.2, 300		$g_\parallel = 2.000$	7870	150			BM73
			$g_\perp = 2.000$					
$Ca_2Al_3O_{12}(OH)$	295		$g_z = 1.96$	± 0.14				BM73
$K_5BW_{12}O_{40} \cdot nH_2O$	295		$g_x = 2.010$	−96	−15	80	$F = -17$	BM73
			$g_y = 1.992$					
			$g_z = 2.000$					
Perovskites								
$BaTiO_3$	425		$g_\parallel = 2.003$				17.0	PF74
	300			930.0			51.0	
$MgTiO_3$	300		$g_\parallel = 2.003$	844		$(a - F) = 118$, $a = -69$	$F/a = 2.7$, $F = -189$	BM86
			$g_\perp = 2.005$					
$SrTiO_3$	300		2.004				98.8	PF74
	80		2.004	7.7				

Fe^{3+} $3d^5$ (*continued*)

Sample	T (°K)	Site	$\tilde{g}$	$D\,(=b_2^0)$ $(10^{-4}\ cm^{-1})$	$E\,(=b_2^2)$ $(10^{-4}\ cm^{-1})$	a $(10^{-4}\ cm^{-1})$	$\frac{1}{3}F(=b_4^0)$ $(10^{-4}\ cm^{-1})$	$\tilde{A}^{57}$ $(10^{-4}\ cm^{-1})$	Ref.
	77		2.004	7.3			110.0		
	4.2		2.00	16.1			112.8		
	1.9		2.004	17.9			115.0		
	77–295	Rhombic	2.002	2700	4900				BM83
$LiNbO_3$	313–573		$g_\parallel = 1.995$	153.3			$B_4^0 = 16.1$		BD74
	293		1.99	198.7			$B_4^0 = -0.85$ G		
							$B_4^3 = 10$ G		
$LiNbO_3$	293	Axial	$g_\parallel = 2.002$	1106.7		160	$a - F = 130$		BD74
Spinels									
$BeAl_2O_4$	NR		2.003	8.572*	2.020*		$b_4^0 = -0.311$*		ART 72
							$b_4^2 = 0.014$*		
							$b_4^4 = 1.461$*		
$MgAl_2O_4$	290	B	2.001	± 0.247			± 0.024		PF74
$ZnAl_2O_4$	290	B	$g_\parallel = 2.000$	−3401.7		575	$a - F = \pm 470.7$		BM73
			$g_\perp = 2.002$						
$ZnGa_2O_4$	298		$g_\parallel = 2.005$	−0.248		0.0444	$(a - F) = 0.0353$		BM86
			$g_\perp = 2.008$						
	4.5		$g_\parallel = 2.006$	−0.244		0.0465	$(a - F) = 0.0372$		BM86
			$g_\perp = 2.008$						
$Li_{0.5}Ga_{2.5}O_4$	290	A	2.008	0.080					PF74
$ZnNa_2GeO_4$	77	I	$g_x = 2.003$	2180	−2000		$b_4^0 = 0$		BD80
			$g_y = 2.003$				$b_4^2 = -40$		
			$g_z = 2.004$				$b_4^4 = 40$		
		II	$g_x = 2.006$	1775	−540		$b_4^0 = -5$		BD80
			$g_y = 2.005$				$b_4^2 = -60$		
			$g_z = 2.007$				$b_4^4 = 60$		
Sulphates									
$Fe_2(SO_4)_3 \cdot 3H_2O$ (p)	room		2.01						LB
$Fe_2(SO_4)_3 \cdot 9H_2O$ (p)	room		2.01						LB
Other									
CdS	77	Orthorhombic	$g_0 = 2.018$	5200	$E/D = 0.17$				BD74

ZnS (pre-annealing)	77		$g_{\parallel}=2.024$						BD80
			$g_{\perp}=2.018$						
ZnS (after annealing)	77		$g_1=2.003$						BD80
			$g_2=2.046$						
			$g_3=2.050$						
EDTA	77, 90		$g_{\mathrm{eff}}\sim 4.3$	7000	$E/D=0.327$			$A^{57}=9.5$	BM73
GaAs	4.2, 77		2.042				$B_4=2.75$		BM86
GaAs	NR		$g_{\parallel}=2.04,$						BD80
			$g_{\perp}=2.23$						
GaP	4.2, 77		2.045				$B_4=3.25$		BM86
TGS	77, 300		$g_{\parallel}=g_{\perp}=2.030$	2.73*	0.26*	<0.01*			BM86
YAlG (garnet)	77	Octahedral	2.003	−705.8			$B_4^0=-23.4$		BM83
	77	Cubic	2.004	−655.3			$B_4^0=7.35$		
							$\lvert B_4^4\rvert=3.85$		
	295	C_{3v}		$B_2^0=-350$					BD80
	295	S_3		$B_2^0=-329$					BD80
YGaG (garnet)	295	Octahedral	2.003	−1295.0		186.0	2.89		PF74
	4.2	Octahedral	2.003	−1320.0		190.0	3.78		
	295	Tetrahedral	2.005	−885.0		62.0	4.11		
	4.2	Tetrahedral	2.005	−880.0		62.0	4.22		
$AlPO_4$	295		2.002	3.33*	1.14*	0.67*			BD80
$GaPO_4$	295		2.002	5.89*	1.17*	1.04*			BD80
AS_2Se_3	77		~ 4.3						BD74
α-$K_3(C_2O_4)_3$	90		$g_x=2.012$	0.199*	0.133*		$b_4^0=0.0009$*		BD80
			$g_y=2.012$				$b_4^2=0$*		
			$g_z=2.012$				$b_4^4=0$*		
β-$K_3(C_2O_4)_3$			$g_x=2.000$	0.159*	0.147*		$b_4^0=0.0003$*		BD80
			$g_y=2.000$				$b_4^2=0$*		
			$g_z=2.000$				$b_4^4=0$*		
α-$NaMg(C_2O_4)_3$			$g_x=2.002$	0.030*	0.017*		$b_4^0=-0.002$*		BD80
			$g_y=2.002$				$b_4^2=-0.011$*		
			$g_z=2.006$				$b_4^4=0.004$*		
β-$NaMg(C_2O_4)_3$			$g_x=2.003$	0.044*	0*		$b_4^0=-0.001$*		BD80
			$g_y=2.003$				$b_4^2=0$*		
			$g_z=2.002$				$b_4^4=0$*		

Fe^{3+} $3d^5$ (*continued*)

Sample	T (°K)	Site	$\tilde{g}$	$D\,(=b_2^0)$ $(10^{-4}\ cm^{-1})$	$E\,(=b_2^2)$ $(10^{-4}\ cm^{-1})$	a $(10^{-4}\ cm^{-1})$	$\frac{1}{3}F(=b_4^0)$ $(10^{-4}\ cm^{-1})$	$\tilde{A}^{57}$ $(10^{-4}\ cm^{-1})$	Ref.
$AgGaS_2$	1.7–4		$g_\parallel=2.032$						BM83
			$g_\perp=1.973$						
$TlAsSe_2$	77		~ 4.3						BD74
$KNbO_3$	300	I	2.005	∓ 5338 MHz	± 879 MHz		$b_4^0=\pm 170$ MHz		BD80
							$b_4^2=\mp 69$ MHz		
							$b_4^4=\mp 730$ MHz		
		II		± 22*	± 4.0*		$b_4^0=\mp 4$*		
$KTaO_3$	4.2		1.99				172.5		PF74
	4.2		$g_\parallel=1.99$, $g_\perp=6.0$						PF74
α-$LiIO_3$	4.2–450	A	$g_\parallel=2.007$	293.3			$B_4^0=-0.69$		BM83
			$g_\perp=2.007$				$B_4^2=\pm 14.5$		
		B	$g_\parallel=2.007$	313			$B_4^0=-0.80$		
			$g_\perp=2.007$						
$PbTiO_3$	290		$g_\parallel=2.00$, $g_\perp=5.97$						PF74
$CaWO_4$	293	I	$g_x=2.001$	21.472	7.039		$B_4^0=0.002$		BM83
			$g_y=2.011$				$B_4^2=-0.002$		
			$g_z=2.003$				$B_4^4=-0.008$		
		II	$g_x=1.998$	21.709	7.133		$B_4^0=-0.003$		
			$g_y=2.002$				$B_4^2=-0.001$		
			$g_z=2.015$				$B_4^4=-0.001$		
$MgWO_4$	77		$g_x=g_y=g_z$ $=2.00$	20.6*	14.4*		$b_4^0=-0.21$*		BM83
							$b_4^2=-2.4$*		
							$b_4^4=2.0$*		
$PbMoO_4$, $CaMoO_4$ (X-irrad.)	77–500	I		$\|D\|=0.69$	$\|E\|=0.12$				BM86
		II	$g_1=3.875$	$\|D\|=0.44$	$\|E\|=0.12$				
			$g_2=3.316$						
			$g_3=4.962$						
		III	4.28						BM83

$AlAsPO_4$	295		2.002	6.57*	1.11*	0.82*		BD80
$GaAsPO_4$	295		2.002	6.77*	1.11*	1.53*		BD80
FeP_3O_9(*p*)	300		2.025					BD80
$FeH_3P_sO_6 \cdot 3H_2O$	295		2.007					BM83
$Fe_2(HPO_3)_3 \cdot 9H_2O$	295		2.007					BM83
$Fe_4H_{33}P_{15}O_{45} \cdot 6H_2O$	295		2.012					BM83
$(ZnOH) \cdot (ZnAsO_4)$	4.2, 77, 290	I (cubic)	2.004	5.03*	$b_2^2/b_2^0=0.088$			BD74
		II	2.004	7.92*	$b_2^2/b_2^0=0.486$			BD74
		III	2.004	4.1*	$b_2^2/b_2^0=0.34$			BD74
$(CH_3NH_4)Al(SO_4)_3 \cdot 12H_2O$	90			1880		100		LB
$(CH_3NH_4)Al(SO_4)_3 \cdot 12H_2O$	90			1893		≈ -130	≈ -40	LB
$[(NH_2)_3ClAl(SO_4)_3] \cdot 6H_2O$	293	I	2.003	−1850.8		219.7	1053	LB
		II	2.003	−1760.1		205.9	1005	
GASH	295	I	$g_x=2.004$	−8.662*			$b_4^0=0.181$*	MS76
			$g_z=2.001$				$b_4^3=0.513$*	
		II	$g_x=2.002$	−8.499*			$b_4^0=0.181$*	MS76
			$g_z=2.002$				$b_4^3=0.184$*	
	79	I	$g_x=2.004$	−8.888*			$b_4^0=0.173$*	MS76
			$g_z=1.995$				$b_4^3=2.963$*	
		II	$g_x=2.005$	−8.765*			$b_4^0=0.173$*	MS76
			$g_z=1.997$				$b_4^3=2.763$*	
	1.6	I	$g_x=1.992$	−8.931*			$b_4^0=0.193$*	MS76
			$g_z=1.994$				$b_4^3=2.998$*	
		II	$g_x=1.997$	−8.808*			$b_4^0=0.199$*	MS76
			$g_z=1.996$				$b_4^3=3.271$*	
$KAl(SeO_4)_2 \cdot 12H_2O$	90		2.003	−103		−127	-2 ± 2	LB
	20			−115		−127	-2 ± 1	
$K_3[Co(CN)_6]$	20		$g_x=2.35$					LB
			$g_y=2.10$					
			$g_z=0.915$					
$Fe_4[Fe(CN)_6]_3$ (*p*)	room		2.06					LB
$(NH_4)_2Al(SO_4)_2$	290		1.99	10–44 G		± 127 G		BM73

Fe^{3+} $3d^5$ (*continued*)

Sample	T (°K)	Site	$\tilde{g}$	$D(=b_2^0)$ $(10^{-4}$ cm$^{-1})$	$E(=b_2^2)$ $(10^{-4}$ cm$^{-1})$	a $(10^{-4}$ cm$^{-1})$	$\frac{1}{3}F(=b_4^0)$ $(10^{-4}$ cm$^{-1})$	$\tilde{A}^{57}$ $(10^{-4}$ cm$^{-1})$	Ref.
$(Ph_4P)_2[Fe(S_2C_2O_2)_2X]$									BM83
(X = I)	78		4	42,000		0			
(X = Br)	78		4	42,000		6000			
Aluminosilicate glasses	295		$g=4.28$						BM83
Chalcogenide glasses			4.28						BD74
Cristobalite	300	I	$g_x=4.05$	2800	$E/D=0.290$				BM73
			$g_y=3.87$						
			$g_z=4.44$						
	77	II	$g_x=4.10$	3000	$E/D=0.300$				BM73
			$g_y=3.96$						
			$g_z=4.38$						
	77	III	$g_x=4.11$	2900	$E/D=0.305$				BM73
			$g_y=3.95$						
			$g_z=4.34$						
Phlogopite (mica)	300	I	$g_0=2.0$		$E/D=0.09$				BD74
	300	II	$g_0=2.0$	2500	$E/D=0.058$				BD74
	300	III	$g_0=2.0$	3000	$E/D=0.275$				BD74
	300	IV	$g_0=2.0$		$E/D=0.33$				BD74
Sillimanite	300	II	$g_x=g_y=g_z$ $=2.004$	0.58*	0.54*				BM73
	300	III	$g=2.004$	0.39*	1.35*		$B_4^0=0.37$*		BM73
							$B_4^2=0.09$*		
Triphenylene	4.2–77		$g_x=5.719$		$E/D=0.011$				BM83
			$g_y=6.266$						
			$g_z=2.001$						

Mn^+ $3d^6$. Data tabulation of spin-Hamiltonian parameters.

Sample	T (°K)	$\tilde{g}$	$\tilde{A}^{55}$ (10^{-4} cm^{-1})	Ref.
NaCl (X-irrad.)	77	$g_{iso} = 1.999$	231	LB, BM73

Fe^{2+} $3d^6$. Data tabulation of spin-Hamiltonian parameters ($S = 1$). (Here Δ is the separation within the lowest doublet.)

Sample	T (°K)	$\tilde{g}$	D (10^{-4} cm^{-1})	E (10^{-4} cm^{-1})	$\tilde{A}^{57}$ (10^{-4} cm^{-1})	Δ (10^{-4} cm^{-1})	Ref.
CaO	4.2, 2	3.30					PF74
MgO	4.2	3.428					PF74
	1.3–4.2	6.836 (single quantum)					BD80
	1.3–4.2	3.428 (double quantum)					BD80
	1.3–4.2	2.279 (triple quantum)					BD80
SrO	8	$g_\parallel = 3.992$			$A_\parallel^{57} = 302{,}000$		BM83
$CdCl_2$	2	$g_\perp = 3.5$	29,000				BM73
	4.2	$g_\parallel = 3.68$, $g_\perp = 3.40$	39,000				
ZnF_2	90	$g_\parallel = 8.97$, $g_\perp = 0$				2240	LB
	20	$g_\parallel = 8.97$, $g_\perp = 0$				2030	
Al_2O_3	4.2	$g_\parallel = 3.428$, $g_\perp < 0.1$					BD74
$SrTiO_3$	77–295	2.00	95,800				BM83
$MgWO_4$	78	2.00	−6860	1740			LB
$FeSiF_6 \cdot 6H_2O$	4.2	$g_\parallel = 1.96$	118,800	6700			BM86
$FeZrF_6$	298	2.15					BM83
$FeHPO_3 \cdot 3H_2O$	295	2.003					BM83
$FeH_2P_2O_5$	295	2.003					BM83
$FeH_4P_2O_6 \cdot \frac{1}{2}H_2O$	295	2.009					BM83
Cs_2ZnCl_5	NR	2.53	45,900	8000			ART 72
$Fe(C_{17}H_{35}COO)_2$ (p)	room	2.01					LB
$Na_3[Fe(CN)_5NO]$	77	$g_\parallel = 2.007$, $g_\perp = 2.037$					LB

Co^{3+} ... Co^{4+} $3d^5$ ... $3d^6$. Data tabulation of spin-Hamiltonian parameters.

Sample	T (°K)	$\tilde{g}$	Ref.
$[(NH_3)_5Co \cdot O_2 \cdot Co(NH_3)_5]$	room	$g_\parallel = 2.12$, $g_\perp = 1.99$	LB
$(SO_4)_2HSO_4 \cdot 3H_2O(p)$	room	2.023	LB

Mn^0 $3d^7$. Data tabulation of spin-Hamiltonian parameters.

Sample	T (°K)	$\tilde{g}$	D (10^{-4} cm^{-1})	$\tilde{A}^{55}$ (10^{-4} cm^{-1})	Ref.
NaCl (X-irrad.)	293	$g = 2.002$	160	−19.5	BD74

Fe^+ $3d^7$. Data tabulation of spin-Hamiltonian parameters.

Sample	T (°K)	$\tilde{g}$	D (10^{-4} cm^{-1})	$\tilde{A}^{57}$ (10^{-4} cm^{-1})	Ref.
CaO	4.2, 2	4.158		33.9	PF74
GaAs	4.2, 77	2.122		$\|a/h\| = 10.705$	BM86
GaP	4.2, 77	2.122		$\|A/h\| = 126$	BM86
KCl	93–573	$g_\parallel = 2.007$, $g_\perp = 2.100$			BD80
MgO	20	4.15			PF74
NaCl	93–573	$g_\parallel = 1.996$, $g_\perp = 2.159$		$A = 11.9$, $B = 8.1$	BD80
NaF (X-irrad.)	4.2–20	$g_x = 2.027$, $g_y = 5.797$, $g_z = 4.588$			BM86
ThO_2	77	$g'_\parallel = 1.989$, $g'_\perp = 1.069$	$\|D\| = 29{,}500$	$A^{57}_\parallel = 11.9\,G$, $A^{57}_\perp = 8.8\,G$	BD74
$SrTiO_3$ (irradiated)	77–295	$g_\parallel = 1.999$, $g_\perp = 2.058$			BM83

Co^{2+} $3d^7$. Data tabulation of spin-Hamiltonian parameters (* indicates units of GHz).

Sample	T (°K)	Site	$\tilde{g}$	D $(= b_2^0)$ (cm^{-1})	$\tilde{A}^{59}$ (10^{-4} cm^{-1})	Ref.
Diamond (synthesized type ASO, ASR, ASM, SAM)	77		$g_\parallel = 4.117, g_\perp = 4.43$		$A_\parallel = 2.45$ MHz, $A_\perp \approx 260$ MHz	BD80
$Cd_{1-x}Co_xSe$ ($x = 0.4\%$)	4.2		$g_\parallel = 2.295, g_\perp = 2.294$	0.468		IA95
$Cd_{1-x}Co_xS$ ($x = 1.6\%$)	4.2		$g_\parallel = 2.26$	0.640		BA94
CaO	20, 4.2		$g = 4.375$		$A = 131.5$	PF74
CdTe	4.2		$g = 2.30$		$\|A\| = 23.2$	BM73
GaAs	4.2		2.189			BM83
	20	Tetra-hedral	2.356			BM83
GaP	4.2; 77		2.122		$A' = 0.126$	BM86
	20	Tetra-hedral	2.164			BM83
KCl (X, γ-irrad.)	93–573		$g_\parallel = 1.998, g_\perp = 2.306$		$A = 12.1, B = 5.4$	BD80
	77		$g_\parallel = 2.008, g_\perp = 2.102$		$A_\parallel = 69.96, A_\perp = -61.85$	
MgO	20, 4.2		$g = 4.279$		$A = 97.79$	PF74
NaCl	93–573		$g_\parallel = 1.992, g_\perp = 2.194$		$A = 15.3, B = 7.4$	BD80
	77		$g_\parallel = 2.007, g_\perp = 2.080$		$A_\parallel = 61.10, A_\perp = -76.74$	
NH_4I	4.2		$g_\parallel = 6.69, g_\perp = 2.01$		$\|A\| = 0.015$*, $B = 0.005$*	CU85
SrO	19		$g_\parallel = 2.174, g_\perp = 5.478$		$A = 25, B = 295$	BM83
ZnO	1.3		$g_\parallel = 2.243, g_\perp = 2.279$	2.75	$A = 16.11, B = 3.0$	PF74
ZnS	1.4–7		$g_\parallel = 2.253, g_\perp = 2.28$	19.5*		BM83
$MgSO_4 \cdot 7H_2O$	4.2		$g_{x'x'} = 2.928, g_{y'y'} = 2.198, g_{z'z'} = 7.008$		$A_{x'x'} = 0.246$*, $A_{y'y'} = 0.000$*, $A_{z'z'} = 0.765$*	MW87
$NiSO_4 \cdot 7H_2O$	4.2		$g_{x'x'} = 2.936, g_{y'y'} = 2.320, g_{z'z'} = 7.100$		$A_{x'x'} = 0.303$*, $A_{y'y'} = 0.000$*, $A_{z'z'} = 0.806$*	MW87
$ZnSO_4 \cdot 7H_2O$	20		$g_x = 2.30, g_y = 3.30, g_z = 6.90$		$A_x = 28, A_z = 254$	LB
$CoCl_2 \cdot 6H_2O$	1.5		$g_a = 2.8, g_b = 4.6, g_c = 4.8$			BM86
	4.2		$g_b = 4.25, g_c = 4.0$			
$Ca(OH)_2$	1.6, 4.2		$g_x = 3.938, g_y = 3.939, g_z = 5.025$		$A_x = 190.71, A_y = 191.54, A_z = 279.96$	BD80

Co^{2+} $3d^7$ *(continued)*

Sample	T (°K)	Site	$\tilde{g}$	D $(= b_2^0)$ (cm^{-1})	$\tilde{A}^{59}$ $(10^{-4}$ $cm^{-1})$	Ref.
CdF_2	1.5–20		$g = 2.288$		$\|A\| = 22.3, f = -0.011, \|F\| = 2.1$	BD80
$SrCl_2$	1.7–4.2		$g = 2.392$		$A = 22\,G$	BM83
TiO_2	4.2		$g_{x'} = 2.13, g_{y'} = 5.36, g_{z'} = 4.36$		$A_{x'} = 54.8, A_{y'} = 124.2, A_{z'} = 67.5$	BD74
	4.2	Regular	$g_x = 2.079, g_y = 5.885, g_z = 3.735$		$A_x = 38.28, A_y = 124.7, A_z = 27.0$	BD74
	4.2	GI	$g_x = 2.21, g_y = 6.03, g_z = 3.56$		$A_x = 34.2, A_y = 150.0, A_z = 28.2$	BD74
	4.2	GIII	$g_x = 2.17, g_y = 5.40, g_z = 4.81$		$A_x = 71.9, A_y = 136.4, A_z = 82.9$	BD74
	4.2	GIV	$g_x = 2.44, g_y = 5.20, g_z = 4.96$		$A_x = 36.3, A_y = 152.5, A_z = 126.0$	BD74
ZnF_2	20		$g_x = 2.6, g_y = 6.05, g_z = 4.1$		$A_x = -43, A_y = 217, A_z = 67$	LB
Cu_2O	4.2		$g_1 = 7.53, g_2 = 1.695, g_3 = 1.274$			BM83
Al_2O_3	4.2		$g_\parallel = 2.292, g_\perp = 4.497$		$A = 32.4, B = 97.2$	PF74
Y_xCo_{1-x}	300		2.11			BM83
Zr_xCo_{1-x}	300		2.11			BM83
$ZnWO_4$	4.2		$g_x = 2.390, g_y = 3.41, g_z = 6.77$		$A_x = 0.070^*$, $A_y = 0.210^*$, $A_z = 0.900^*$, $\|B\| = 0.030^*$	BD74
$CsCaCl_3$	4.2		$g_\parallel = 4.48, g_\perp = 4.36$		$A_\parallel = 69.7\,G, A_\perp = 63.9\,G$	BM86
$CsMgCl_3$	NR		$g_\parallel = 7.37, g_\perp = 2.51$		$A_\parallel = 339, A_\perp = 21$	ART 72
$KMgF_3$	4.2		4.28			PF74
$KTaO_3$	4.2		$g_\parallel = 2.061, g_\perp = 4.933$		$\|A\| = 57, \|B\| = 74$	BM73
$KZnF_3$	4.2		$g = 4.376$	0.011		BD74
$LiBaF_3$	4.2	Cubic	$g = 4.316$		$A = 94.7$	BM86
$ZnMoO_4$	4.2, 18	I	$g_x = 4.021, g_y = 3.495, g_z = 5.468$	$\Delta E = 196$	$A = 0.610^*$, $B = 0.267^*$, $C = 0.153^*$	BM73
	33					
		II	$g_x = 3.778, g_y = 3.350, g_z = 5.780$	$\Delta E = 206$	$A = 0.723^*$, $B = 0.217^*$, $C = 0.115^*$	BM73
$CdCl_2L_2$						BM86
L = Pyrazole	4.2		$g_1 = 6.96, g_2 = 2.66, g_3 = 1.99$		$A_1 = 134\,G, A_2 = 60\,G, A_3 = 55\,G$	
L = Indazole	4.2		$g_1 = 5.97, g_2 = 4.06, g_3 = 2.28$		$A_1 = 110\,G, A_2 = 60\,G, A_3 = 45\,G$	
$CdIn_2S_4$	1.1, 4.2		$g_\parallel = 6.00, g_\perp = 4.07$		$A_\parallel = 85.0\,G, A_\perp = 101\,G$	BD74
Cs_3CoBr_5	77, 4		$g_\parallel = 2.6, g_\perp = 2.4$			LB
Cs_3CoCl_5	77, 4		$g_\parallel = 2.41, g_\perp = 2.33$			LB

$K_3Co(CN)_6$ (*p*) (γ-irrad.)	77		$g_{xx} = 2.091, g_{yy} = 2.090, g_{zz} = 2.008$	$A_{xx} = -70.07, A_{yy} = -60.60, A_{zz} = 68.08$	BD80
$K_3Co(CN)_6$ (γ-irrad.)	77		$g_{xx} = 2.093, g_{yy} = 2.091, g_{zz} = 2.012$	$A_{xx} = -60.35, A_{zz} = 68.17$	BD80
$Mg(CH_3COO)_2 \cdot 4H_2O$	8–12	I	$g_x = 2.546, g_y = 3.935, g_z = 6.032$	$A_x = 36.8, A_y = 73.4, A_z = 193$	BM86
	8–12	II	$g_x = 2.532, g_y = 3.895, g_z = 6.030$	$A_x = 28, A_y = 88, A_z = 190$	BM86
$K_2Zn(SO_4)_2 \cdot 6H_2O$	20		$g_z = 6.56, g_{min} = 2.50, g_2 = 3.35$	$A_z = 286, A_{min} = 65, A_2 = 80$	LB
$(NH_4)_2Co(SO_4)_2 \cdot 6H_2O$	77		$g_{\parallel} = 6.45, g_{\perp} = 3.06$		BM86, BM83
$(NH_4)_2Zn(SO_4)_2 \cdot 6H_2O$	20		$g_z = 6.45, g_{min} = 3.06, g_2 = 3.06$	$A_z = 245, A_{min} = 20, A_2 = 20$	LB
$(NH_4)_2Zn(SO_4)_2 \cdot 6D_2O$	20		$g_z = 6.41$	$A_z = 252$	LB
$Rb_2Zn(SO_4)_2 \cdot 6H_2O$	20		$g_z = 6.65, g_{min} = 2.7, g_2 = 3.3$	$A_z = 293, A_{min} = 49$	LB
$La_2Mg_3(NO_3)_{12} \cdot 24H_2O$	4.2	I	$g_{\parallel} = 7.23, g_{\perp} = 2.310$	$A = 278.8, B \leq 1$	LB
		II	$g_{\parallel} = 4.050, g_{\perp} = 4.430$	$A = 80.7, B = 103.3$	
$Mg_3Bi_2(NO_3)_{12} \cdot 24H_2O$	20	I	$g_{\parallel} = 7.29, g_{\perp} = 2.338$	$A = 283, B \leq 1$	LB
		II	$g_{\parallel} = 4.108, g_{\perp} = 4.385$	$A = 85, B = 103$	
$Mg_3(Bi + Ce)_2(NO_3)_{12} \cdot 24H_2O$	4				LB
Ce = 0%		I	$g_{\parallel} = 7.20, g_{\perp} = 2.39$	$A = 302, B < 1$	
		II	$g_{\parallel} = 4.145, g_{\perp} = 4.415$	$A = 95, B = 103$	
Ce = 10%		I	$g_{\perp} = 2.37$		
		II	$g_{\parallel} = 4.112, g_{\perp} = 4.45$	$A = 82, B = 114$	
Ce = 20%		I	$g_{\parallel} = 7.41, g_{\perp} = 2.36$		
		II	$g_{\parallel} = 4.22, g_{\perp} = 4.22$	$A = 90, B = 90$	
Ce = 50%		I	$g_{\parallel} = 10.55$		
		II	$g_{\parallel} = 4.30, g_{\perp} = 4.31$	$A = 89.7, B = 107$	
Ce = 80%		I	$g_{\parallel} = 7.33, g_{\perp} = 2.36$		
		II	$g_{\parallel} = 4.02, g_{\perp} = 4.45$		
$ZnSiF_6 \cdot 6H_2O$	20		$g_{\parallel} = 5.82, g_{\perp} = 3.44$	$A = 184, B = 47$	LB
$B_{12r}O_7$ (vitamin)	83, 223		$g_1 = 1.994, g_2 = 2.012, g_3 = 2.079$	$A_1 = -0.021^*, A_2 = -0.027^*, A_3 = -0.063^*$	BM86
Co pyridine chloride	4.2		$g_x = 1.77, g_y = 2.62, g_z = 5.82$		BM83

Ni^{3+} $3d^7$. Data tabulation of spin-Hamiltonian parameters ($S = \frac{1}{2}$).

Sample	T (°K)	$\tilde{g}$	$\tilde{A}^{61}$ (10^{-4} cm^{-1})	Ref.
GaP	20	2.089		BM83
GdAs	20	2.114		BM83
InP	20	2.098		BM83
SrO	4.2	$g_\parallel = 4.36$, $g_\perp = 4.647$		PF74
ZnO	4.2	$g_\parallel = 2.143$, $g_\perp = 4.318$		PF74
SnO_2	95, 300	$g_x = 1.96$, $g_y = 6.84$, $g_z = 5.10$		ART72
TiO_2	4.2, 78	$g_c = 2.237$		PF74
Al_2O_3	1.6	$g_\parallel = 2.316$, $g_\perp = 4.98$	33.8	PF74
	10	$g = 2.133$		BM86
	1.2–20	$g_1 = 2.142$, $g_2 = 0.048$		BM86
	1.2, 4.2	$g_1 = 2.140$, $g_2 = 0.052$		BM86
$AgGaS_2$	1.7–4.2	$g_\parallel = 2.14$, $g_\perp = 4.52$		BM83
$SrTiO_3$	203	$g = 2.180$		PF74
	80	$g_\parallel = 2.172$, $g_\perp = 2.184$		
	20	$g_\parallel = 2.136$, $g_\perp = 2.202$		
	4.2	$g_\parallel = 2.110$, $g_\perp = 2.213$		
$Na_2Zn(SO_4)_2 \cdot 4\,H_2O$ (X-irrad.)	300	$g_x = 2.079$, $g_y = 2.100$, $g_z = 2.451$	$\lvert A_x\rvert = \lvert A_y\rvert = 27$, $\lvert A_z\rvert = 11$	BM83
Na_2O-Al_2O_3-SiO_2-TiO_2 glasses(γ-irrad.)	77	$g_\parallel \approx 2.25$, $g_\perp \approx 2.085$		BM83

Fe^0 $3d^8$. Data tabulation of spin-Hamiltonian parameters.

Sample	T (°K)	$\tilde{g}$	$\tilde{A}^{57}$ (10^{-4} cm^{-1})	Ref.
SiO_2	298	2.081		BM86

Co^+ $3d^8$. Data tabulation of spin-Hamiltonian parameters.

Sample	T (°K)	Site	$\tilde{g}$	$\tilde{A}$ (10^{-4} cm^{-1})	Ref.
KCl	295	I	2.14		BD80
		II	1.99		
MgO	77		2.173	54.0	PF74

Ni^{2+} $3d^8$. Data tabulation of spin-Hamiltonian parameters (* indicates units of GHz).

Sample	T (°K)	Site	$\tilde{g}$	D (10^{-4} cm^{-1})	E (10^{-4} cm^{-1})	$\tilde{A}^{61}$ (10^{-4} cm^{-1})	Ref.
AgCl	30		$g_\parallel = 2.714$, $g_\perp = 2.138$				BM38
AgCl	4.2		2.276			$A = 7.35$ G, $B = 2.41$ G	BM86
CaO	20, 4.4		2.327				PF74
LiCl	77	I	$g_x = 2.27$, $g_y = 2.26$, $g_z = 2.31$	$\|D\| = 5.640$	$\|E\| = 800$		BM86
		II	$g_\parallel = 2.370$, $g_\perp = 2.30$	$\|D\| = 31{,}000$	$\|E\| = 0.0$		
MgO	290		2.225				PF74
	77		2.215			8.3	
	4.2		2.234				
NaCl (X, γ-irrad.)	300		$g_\parallel = 2.143$, $g_\perp = 2.051$				BD80
	93		$g_\parallel = 2.110$, $g_\perp = 2.040$				
$CdBr_2$		Tetragonal	$g_\parallel = 2.19$, $g_\perp = 2.18$	22,000			ART 72
		Cubic	$g = 2.24$				
$NiSO_4 \cdot 7H_2O$	room		2.20	−35,600	−15,000		LB
$NiBr_2$ (p)	room		2.27				LB
$NiCl_2$	290		2.25				LB
TiO_2	4.2		$g_c = 2.1$			$A_c = -29.0$	PF74
NiO–MgO solid solutions	77		2.22				BM83
$GdCO_2$	100–300	Cubic	$g = 2.07$				BM83
$Ni[BrO_3]_2 \cdot 6H_2O$	290		2.29	19,300	0		LB
$Ni(BF_4)_2 \cdot 6H_2O$	<173		$g_\parallel = 2.28$, $g_\perp = 2.31$	210			BM86
$Cd(NO_3)_2 \cdot 6NH_3$	100–300		2.17	~0.3 cm^{-1}			BD80
$Ni(NO_3)_2 \cdot 6NH_3$	100–300		2.18	0.425 cm^{-1}			BD80
$Zn(NO_3)_2 \cdot 6NH_3$	100–300		2.17	0.606 cm^{-1}			BD80
$CsMgCl_3$			$g_\parallel = 2.257$, $g_\perp = 2.241$	18,980 G			ART 72
$KMgF_3$	20, 77		2.280				PF74
$KZnF_3$	4.2		2.287				BD74
$SrTiO_3$	80		2.204				PF74
	90		2.24				
	20.4		2.30				
$NiSnCl_6 \cdot 6H_2O$	298		$g_\parallel = 2.28$, $g_\perp = 2.30$	0.078 cm^{-1}			BD80
	273		$g_\parallel = 2.33$, $g_\perp = 2.35$	0.252 cm^{-1}			
	195		$g_\parallel = 2.35$, $g_\perp = 2.38$	0.252 cm^{-1}			
	77		$g_\parallel = 2.27$, $g_\perp = 2.17$	0.403 cm^{-1}			
	4.2		$g_\parallel = 2.26$, $g_\perp = 2.19$	0.458 cm^{-1}			

Ni^{2+} $3d^8$ *(continued)*

Sample	T (°K)	Site	$\tilde{g}$	D (10^{-4} cm^{-1})	E (10^{-4} cm^{-1})	$\tilde{A}^{61}$ (10^{-4} cm^{-1})	Ref.
$ZnSiF_6 \cdot 6H_2O$	4.2, 77, 300		$g_\parallel = g_\perp = 2.23$				BD80
Al_2O_3	290–4.2		$g_\parallel = 2.196$, $g_\perp = 2.186$				PF74
$K_2Ni(SO_4)_2 \cdot 6H_2O$	290		2.25	−33,000	−5170		LB
$K_2Ni(SeO_4)_2 \cdot 6H_2O$	290		2.25	−30,000	−10,000		LB
$(NH_4)_2Ni(SO_4)_2 \cdot 6H_2O$	290		2.25	−22,400	−3870		LB
	90		2.25	−19,900	−4860		
$(NH_4)_2Ni(SeO_4)_2 \cdot 6H_2O$	290		2.25	−18,900	−7900		LB
	90		2.25	−17,300	−8200		
$Tl_2Ni(SO_4)_2 \cdot 6H_2O$	290		2.25	−26,000	−1000		LB
$Mg_3La_2(NO_3)_{12} \cdot 24H_2O$			2.24	−1770	–		LB
$Ni(CN)_2 \cdot NH_3$ (p)	290, 77		2.38				LB
$Ni(CN)_2 \cdot NH_3 \cdot C_6H_6$ (p)	room		2.19				LB
$Ni(NH_3)_6Br_2$	300		2.178				LB
	90		2.173				
$Ni(NH_3)_6Cl_2$	300		2.175				LB
	90		2.169				
$Ni(NH_3)_6I_2$	300		2.191				LB
	90		2.177				
$Ni(ND_3)_6I_2$	300		2.191				LB
$Ni(NH_3)_6ClO_4$ (p)	room		2.17				LB
$NiSiF_6 \cdot 6H_2O$	290		2.3	−5000	–		LB
	195		2.29	−3200	–		
	90		2.26	−1700	–		
	20		2.29	−1200	–		
	4.2–0.4		2.24	−1130	–		
$NiSnCl_6 \cdot 6H_2O$	0.083–4.2		$g_\parallel = 2.247$	4470			BM83
$Ni_2CdCl_6 \cdot 12H_2O$	295	I	$g_\parallel = 2.274$, $g_\perp = 2.323$	−6.813*			MMC94
		II	$g_\parallel = 2.236$, $g_\perp = 2.291$	−31.12*			MMC94
	120	I	$g_\parallel = 2.250$, $g_\perp = 2.320$	−5.920*			MMC94
		II	$g_\parallel = 2.285$, $g_\perp = 2.295$	−34.26*			MMC94
	3.8	I	$g_\parallel = 2.293$, $g_\perp = 2.287$	−5.216*			MMC94
		II	$g_\parallel = 2.098$, $g_\perp = 2.288$	−35.43*			MMC94
$[TMPD^+]_2[Ni(mnt)_2]$ (p)	80–300			$\lvert D \rvert = 216$ G	$\lvert E \rvert = 29$ G		BM86
$Na_2O–Al_2O_3–SiO_2–TiO_2$ glasses (γ-irrad.)	77, 298		$g_\parallel = 2.004$, $g_\perp \approx 2.085$				BM83

Cu^{3+} $3d^8$. Data tabulation of spin-Hamiltonian parameters.

Sample	T (°K)	$\tilde{g}$	$\tilde{A}$	Ref.
Al_2O_3	1.4	$g_\parallel = 2.078$, $g_\perp = 2.077$	$A^{63} = -64.316$, $A^{65} = -68.893$	PF74

Ni^{+} $3d^9$. Data tabulation of spin-Hamiltonian parameters.

Sample	T (°K)	Site	$\tilde{g}$	$\tilde{A}^{61}$	Ref.
Diamond (synthetic)	77		$g = 2.032$	$\|A\| = 6.5$ G	BD74
$AgCl[(NiCl_6)^{5-}]$	1.8		$g_\parallel = 2.737$, $g_\perp = 2.150$		BM86
CaO	77		$g_\parallel = 2.067$, $g_\perp = 2.383$		PF74
CdS	4.2		$g_\parallel = 1.229$, $g_\perp = 1.512$		BD74
CdS (irradiated)	4.2		$g_\parallel = 1.230$, $g_\perp = 1.510$	$A = 120.9$ G, $B = 133.5$ G	BD74
GaP (epitaxial layers)	4.2		$\|g\| = 0.934$		BM86
MgO	77		2.169		PF74
NaCl (X-irrad.)	77	I	$g_\parallel = 2.048$, $g_\perp = 2.270$		BM73
		II	$g_\parallel = 2.047$, $g_\perp = 2.299$		
SrO (γ-irrad.)	<90	I	$g_x = 2.408$, $g_y = 2.378$, $g_z = 2.030$		BM86
	$100 < T < 200$	II	$g_x = g_y = g_z = 2.033$		BM86
	$T > 220$	III	$g_x = 2.396$, $g_y = 2.390$, $g_z = 2.033$		BM86
ZnS	4.2		$g = 1.399$		BD74
NiCaX zeolite (p)	77		$g_\parallel = 2.6$–2.8, $g_\perp = 2.096$		BM83
$CuGaS_2$	5		$g_\parallel = 1.918$, $g_\perp = 2.324$		BD80
	4.2		$g_\parallel = 1.918$, $g_\perp = 2.328$	$A_\parallel < 12$, $A_\perp = 65$	
$SrTiO_3$	203, 80, 20		$g_\parallel = 2.029$, $g_\perp = 2.352$		PF74
	77		$g_\parallel = 2.375$, $g_\perp = 2.084$		
$Ca(OH)_3$ (X-irrad.)	77–293		$g_x = 2.085$, $g_y = 2.004$, $g_z = 1.996$	$A_x = 39$, $A_y = 52$, $A_z = 28$	BM86
$MgO \cdot Al_2O_3 \cdot 2.5SiO_2$ glasses (γ-irrad.)	295		$g_\parallel = 2.38$, $g_\perp = 2.08$		BM83

DATA TABULATION (Cu^{2+}) $3d^9$. Spin-Hamiltonian parameters for Cu^{2+} in various host crystals at different temperatures. The g-values are dimensionless, while the A-values are in $10^{-4}\,cm^{-1}$, unless otherwise indicated (* indicates GHz).

Host	T (°K)	Site	$\tilde{g}$	$\tilde{A}^{63}$, $\tilde{A}^{65}$, $\tilde{Q}$	Ref.
Acetates					
$Ca(CH_3COO)_2 \cdot 8H_2O$	300		$g_x = 2.06$, $g_y = 2.06$, $g_z = 2.36$	$A_x = 16$ G, $A_y = 16$ G, $A_z = 130$ G	MWrev90
$Cd(CH_3COO)_2 \cdot 2H_2O$			$g_x = 2.038$, $g_y = 2.164$, $g_z = 2.414$	$A_x = 50.1$, $A_y = 25.3$, $A_z = 101.6$	MWrev90
$Cu(CH_3COO)_2 \cdot H_2O$ (γ-irrad.)			$g_x = 2.079$, $g_y = 2.079$, $g_z = 2.377$	$A_x = 6$, $A_y = 6$, $A_z = 121$	MWrev90
$Mg(CH_3COO)_2 \cdot 4H_2O$	295		$g_\parallel = 2.374$, $g_\perp = 2.096$	$A_\parallel^{63,65} = 0.011^*$, $A_\perp^{63,65} = 0.003^*$	MW89B
	77		$g_\parallel = 2.388$, $g_\perp = 2.096$	$A_\parallel^{63} = 0.012^*$, $A_\perp^{63} = 0.003^*$	
				$A_\parallel^{65} = 0.013^*$, $A_\perp^{65} = 0.003^*$	
	4.2		$g_\parallel = 2.388$, $g_\perp = 2.096$	$A_\parallel^{63} = 0.012^*$, $A_\perp^{63} = 0.003^*$	
				$A_\parallel^{65} = 0.013^*$, $A_\perp^{65} = 0.003^*$	
$Sr(CH_3COO)_2 \cdot \frac{1}{2}H_2O$	303		$g_x = 2.077$, $g_y = 2.077$, $g_z = 2.373$	$A_x = 0.011^*$, $A_y = 0.011^*$, $A_z = 0.292^*$	MWrev90
	85		$g_x = 2.067$, $g_y = 2.063$, $g_z = 2.382$	$A_x = 0.069^*$, $A_y = 0.058^*$, $A_z = 0.410^*$	
	5		$g_x = 2.087$, $g_y = 2.081$, $g_z = 2.387$	$A_x = 0.071^*$, $A_y = 0.067^*$, $A_z = 0.417^*$	
$Sr(CH_3COO)_2 \cdot 4H_2O$	300		$g_x = 2.075$, $g_y = 2.075$, $g_z = 2.373$	$A_x = 5$, $A_y = 5$, $A_z = 135$	MWrev90
$Zn(CH_3COO)_4 \cdot 2H_2O$	4.2		$g_x = 2.080$, $g_y = 2.080$, $g_z = 2.369$	$A_x = 10$, $A_y = 10$, $A_z = 124$	MWrev90
$CaCd(CH_3COO)_4 \cdot 4H_2O$	77, 300		$g_x = 2.060$, $g_y = 2.060$, $g_z = 2.360$	$A_x = 13$, $A_y = 13$, $A_z = 119$	MWrev90
$CaCd(CH_3COO)_4 \cdot 4D_2O$	295		$g_1 = 2.365$, $g_2 = 2.054$	$A_1 = 0.410^*$, $A_2 = 0.030^*$	MKK92
	77		$g_1 = 2.370$, $g_2 = 2.060$	$A_1 = 0.410^*$, $A_2 = 0.055^*$	MWrev90
$CaCu_x(CH_3COO)_4 \cdot 6H_2O$	RT		$g_x = 2.061$, $g_x = 2.061$, $g_y = 2.061$,	$A_x = 15$, $A_y = 15$, $A_z = 143$	MWrev90
	LNT		$g_y = 2.060$, $g_z = 2.365$, $g_z = 2.355$	$A_x = 26$, $A_y = 24$, $A_z = 136$	
$Cu(NH_3)_2(CH_3COO)_2$	301		$g_x = 2.107$, $g_y = 2.211$, $g_z = 2.054$		MWrev90
$Cd(NH_3)_4(CH_3COO)_2$		I	$g_0 = 2.125$		MWrev90
		II	$g_0 = 2.135$	$A_0 = 69$	
	77	I	$g_x = 2.065$, $g_y = 2.065$, $g_z = 2.236$	$A_x = 11$, $A_y = 11$, $A_z = 185$	
		II	$g_x = 2.071$, $g_y = 2.071$, $g_z = 2.271$	$A_x = 20$, $A_y = 20$, $A_z = 185$	
Aluminates					
Na β-aluminate	490	B	$g_x = 2.08$, $g_y = 98$, $g_z = 2.452$		MWrev90
	300	$B^{63}Cu$	$g_x = 2.08$, $g_y = 109$, $g_z = 2.443$	$A_x < 10$, $A_z < 10$	
		$B^{65}Cu$	$g_x = 2.08$, $g_y = 117$, $g_z = 2.443$	$A_x < 10$, $A_z < 10$	

	150	$B^{63}Cu$	$g_x = 2.08,\ g_y = 116,\ g_z = 2.429$		
		$B^{65}Cu$	$g_x = 2.08,\ g_y = 125,\ g_z = 2.429$		
	490	B1	$g_x = 2.07,\ g_y = 90,\ g_z = 2.473$		
	300	B1	$g_x = 2.07,\ g_y = 104,\ g_z = 2.467$	$A_x < 10,\ A_z < 10$	
	150	B1	$g_x = 2.07,\ g_y = 114,\ g_z = 2.450$		
T1 β-aluminate	300	B	$g_x = 2.08,\ g_y = 2.08,\ g_z = 2.450$	$A_x < 10,\ A_y < 10,\ A_z = 90$	MWrev90
	140	B	$g_x = 2.08,\ g_y = 2.08,\ g_z = 2.444$	$A_x < 10,\ A_y < 10,\ A_z = 113$	
	300	B1	$g_x = 2.08,\ g_y = 2.08,\ g_z = 2.484$	$A_x < 10,\ A_y < 10,\ A_z = 91$	
	140	B1	$g_x = 2.08,\ g_y = 2.08,\ g_z = 2.476$	$A_x < 10,\ A_y < 10,\ A_z = 97$	
Ag β-aluminate	300		$g_x = 2.076,\ g_y = 2.076,\ g_z = 2.453$	$A_x < 7,\ A_y < 7,\ A_z = 106$	MWrev90
	140		$g_x = 2.071,\ g_y = 2.071,\ g_z = 2.445$	$A_x < 7,\ A_y < 7,\ A_z = 115$	
Fluosilicates (titanates, zircanates)					
$FeSiF_6 \cdot 6H_2O$	295		$g_x = 2.00,\ g_y = 2.00,\ g_z = 2.90$		MWrev90
	77		$g_x = 2.160,\ g_y = 2.160,\ g_z = 2.335$		
$MgSiF_6 \cdot 6H_2O$	4.2		$g_x = 2.10,\ g_y = 2.10,\ g_z = 2.47$	$A_z = 110$	MWrev90
	270		$g_x = 2.25,\ g_y = 2.25,\ g_z = 2.23$		
$ZnSiF_6 \cdot 6H_2O$	4.2		$g_x = 2.100,\ g_y = 2.100,\ g_z = 2.460$	$A_x = 14,\ A_y = 14,\ A_z = 107$	MWrev90
	77		$g_x = 2.23,\ g_y = 2.23,\ g_z = 2.22$	$A_x = 29,\ A_y = 29,\ A_z = 21$	
$ZnTiF_6 \cdot 6H_2O$	180		$g_x = 2.223,\ g_y = 2.223,\ g_z = 2.226$		MWrev90
	4.2		$g_x = 2.100,\ g_y = 2.100,\ g_z = 2.470$	$A_x = 3,\ A_y = 30,\ A_z = 106$	
	158		$g_x = 2.173,\ g_y = 2.173,\ g_z = 2.331$	$A_z = 53$	
$ZnGeF_6 \cdot 6H_2O$	194		$g_x = 2.225,\ g_y = 2.225,\ g_z = 2.219$	$A_x = 25$ G, $A_y = 25$ G, $A_z = 19$ G	MWrev90
	40		$g_x = 2.404,\ g_y = 2.404,\ g_z = 2.971$	$A_x = 16$ G, $A_y = 16$ G, $A_z = 93$ G	
$ZnZrF_6 \cdot 6H_2O$	4.2	^{63}Cu	$g_z = 2.470$		MWrev90
		^{65}Cu	$g_z = 2.470$		
$K_2Zn(ZrF_6)_2 \cdot 6H_2O$	300		$g_x = 2.39,\ g_y = 2.109,\ g_z = 2.07$	$A_x = 106,\ A_y < 11,\ A_z < 30$	MWrev90
	77		$g_x = 2.466,\ g_y = 2.106,\ g_z = 2.061$		
	4.2		$g_x = 2.469,\ g_z = 2.060$		
Formates					
$Ba(HCOO)_2$	RT	I	$g_x = 2.076,\ g_y = 2.071,\ g_z = 2.362$	$A_x = -7,\ A_y = -11,\ A_z = -151$	MWrev90
	LNT	I	$g_x = 2.062,\ g_y = 2.056,\ g_z = 2.366$	$A_x = -6,\ A_y = -8,\ A_z = -155$	
$Sr(HCOO)_2 \cdot 2H_2O$	RT		$g_x = 2.387,\ g_y = 2.086,\ g_z = 2.136$	$A_x = 116$ G, $A_y = 0$ G, $A_z = 58$ G	MWrev90
	RT	I	$g_x = 2.100,\ g_y = 2.077,\ g_z = 2.399$	$A_x = 25,\ A_y = 15,\ A_z = 119$	NWrev90
Halides					
$AgBr_{1-x}Cl_x$	NR		$g_x = 2.065,\ g_y = 2.065,\ g_z = 2.078$		MWrev90
AgCl	4.2		$g_x = 2.067,\ g_y = 2.067,\ g_z = 2.302$	$A_x = 43$ G, $A_y = 43$ G, $A_z = 107$ G	MWrev90

DATA TABULATION (Cu^{2+}) $3d^9$ *(continued)*

Host	T (°K)	Site	$\tilde{g}$	$\tilde{A}^{63}$, $\tilde{A}^{65}$, $\tilde{Q}$	Ref.
CsCl	NR		$g_x = 2.033$, $g_y = 2.033$, $g_z = 2.224$	$A_x = 54.7$, $A_y = 52.7$, $A_z = 62.9$	MWrev90
NaCl	4.2–77		$g_x = 2.07$, $g_y = 2.07$, $g_z = 2.373$	$A_x = 40$ G, $A_y = 40$ G, $A_z = 95$ G	MWrev90
NaCl	80		$g_x = 2.077$, $g_y = 2.077$, $g_z = 2.232$	$A_x = 36.9$, $A_y = 36.9$, $A_z = 111.3$	MWrev90
LiCl	80	I	$g_x = 2.161$, $g_y = 2.161$, $g_z = 2.375$	$A_x = 30.6$, $A_y = 30.6$, $A_z = 81$ G	MWrev90
		II	$g_x = 2.161$, $g_y = 2.161$, $g_z = 2.342$	$A_x = 30.6$, $A_y = 30.6$, $A_z = 87.5$ G	
		III	$g_x = 2.061$, $g_y = 2.061$, $g_z = 2.302$	$A_x = 30.6$, $A_y = 30.6$, $A_z = 77.0$	
NaF	80		$g_x = 2.085$, $g_y = 2.085$, $g_z = 2.413$	$A_x = 39.4$, $A_y = 39.4$, $A_z = 146.2$	MWrev90
NH_4Br,ND_4Br	300	I	$g_x = 2.205$, $g_y = 2.205$, $g_z = 2.035$	$A_x = 66$, $A_y = 66$, $A_z = 72$	MWrev90
	77	I	$g_x = 2.259$, $g_y = 2.175$, $g_z = 2.027$	$A_x = 81.0$, $A_y = 0$, $A_z = 114.05$	
	300	II	$g_x = 2.188$, $g_y = 2.195$, $g_z = 2.030$	$A_x = 0$, $A_y = 0$, $A_z = 189$	
	77	II	$g_x = 2.195$, $g_y = 2.195$, $g_z = 2.077$	$A_x = 50$, $A_y = 50$, $A_z = 240$	
NH_4Cl,ND_4Cl	77		$g_x = 2.292$, $g_y = 2.292$, $g_z = 2.021$	$A_x = 61.0$, $A_y = 61.0$, $A_z = 44.3$	MWrev90
	20		$g_x = 2.302$, $g_y = 2.302$, $g_z = 2.019$	$A_x = 60.2$, $A_y = 60.2$, $A_z = 47.2$	
	4.2		$g_x = 2.412$, $g_y = 2.189$, $g_z = 2.019$	$A_x = 97.3$, $A_y = 39.8$, $A_z = 47.6$	
	290		$g_x = 2.275$, $g_y = 2.275$, $g_z = 2.054$	$A_x = 81.3$, $A_y = 81.3$, $A_z = 15$	
	298	I	$g_x = 2.256$, $g_y = 2.256$, $g_z = 2.045$	$A_x = 79$, $A_y = 79$, $A_z = 15$	
	4	I	$g_x = 2.220$, $g_y = 2.220$, $g_z = 1.996$	$A_x = 73$, $A_y = 73$, $A_z = 240$	
	298	I	$g_x = 2.209$, $g_y = 2.209$, $g_z = 2.010$	$A_x = 25$, $A_y = 25$, $A_z = 177.9$	
	4	II	$g_x = 2.237$, $g_y = 2.237$, $g_z = 2.010$	$A_x = 55$, $A_y = 55$, $A_z = 108$	
	130	II	$g_x = 2.239$, $g_y = 2.239$, $g_z = 2.010$	$A_x = 54$, $A_y = 54$, $A_z = 116$	
	298	III	$g_x = 2.223$, $g_y = 2.223$, $g_z = 2.021$	$A_x = 57$, $A_y = 57$, $A_z = 71$	
NaF	78		$g_x = 2.105$, $g_y = 2.105$, $g_z = 2.458$		MWrev90
	1.6		$g_x = 2.133$, $g_y = 2.012$, $g_z = 2.514$		
NH_4F	77		$g_x = 2.077$, $g_y = 2.104$, $g_z = 2.470$	$A_z < 4$	MWrev90
NH_4I	338	I	$g_x = 2.216$, $g_y = 2.216$, $g_z = 2.011$	$A_x = 30$, $A_y = 30$, $A_z = 90$	MWrev90
	293	II	$g_x = 2.216$, $g_y = 2.216$, $g_z = 2.019$	$A_x = 56$, $A_y = 56$, $A_z = 80$	
	103		$g_x = 2.272$, $g_y = 2.272$, $g_z = 2.002$	$A_x = 0.006$, $A_y = 0.006$, $A_z = 0.012$	
	LHT		$g_x = 2.222$, $g_y = 2.222$, $g_z = 1.990$	$A_x = 0.007$, $A_y = 0.007$, $A_z = 0.012$	
$SrCl_2$	4.2–290		$g_x = 2.040$, $g_y = 2.040$, $g_z = 2.236$	$A_x = 40.0$, $A_y = 40.0$, $A_z = 157.8$	MWrev90
CaF_2	4.2		$g_x = 2.103$, $g_y = 2.147$, $g_z = 2.802$	$A_x = 97$, $A_y = 97$, $A_z = 76$	MWrev90
$CuCl_2 \cdot 2H_2O$	300		$g_x = 2.047$, $g_y = 2.102$, $g_z = 2.331$	$A_z = 130$	MWrev90
$CuCl_2 \cdot 2NH_3$	NR		$g_x = 2.3$, $g_y = 2.3$, $g_z = 2.0$		MWrev90

$CuF_2 \cdot 2H_2O$	300		$g_x = 2.08$, $g_y = 2.07$, $g_z = 2.42$	$A_z = 135$	MWrev90
ZnF_2	77–4.2	^{63}Cu	$g_x = 2.427$, $g_y = 2.354$, $g_z = 2.061$	$A_x = 54.9$, $A_y = 23.1$, $A_z = 149.7$	MWrev90
		^{65}Cu	$g_x = 2.427$, $g_y = 2.354$, $g_z = 2.061$	$A_x = 54.9$, $A_y = 23.1$, $A_z = 160.5$	
$Cd(NH_3)_4Cl_2$	100		$g_x = 2.022$, $g_y = 2.018$, $g_z = 2.087$	$A_x = -126$, $A_y = -117$, $A_z = -492$	MWrev90
$Pd(NH_3)_4Cl_2 \cdot H_2O$	300		$g_x = 2.038$, $g_y = 2.038$, $g_z = 2.33$	$A_x = 37.9$, $A_y = 37.9$, $A_z = 201$	MWrev90
$\varnothing_4AsCuCl_3$	RT		$g_x = 2.045$, $g_y = 2.045$, $g_z = 2.245$	$A_x = 57.0$, $A_y = 57.0$, $A_z = 155$	MWrev90
K_2PdCl_4	RT		$g_x = 2.049$, $g_y = 2.049$, $g_z = 2.233$	$A_x = 34.5$, $A_y = 34.5$, $A_z = 163.6$	MWrev90
K_2PdBr_4	RT		$g_x = 2.043$, $g_y = 2.043$, $g_z = 2.143$	$A_x = 45.8$, $A_y = 45.8$, $A_z = 189.5$	MWrev90
$AsCdCl_3$	1.6		$g_x = 2.09$, $g_y = 2.09$, $g_z = 2.38$	$A_x < +18$ G, $A_y > +18$ G, $A_z = -93$ G	MWrev90
	RT		$g_z = 2.18$	$A_z \geq -18$ G	
	351		$g_x = 2.17$, $g_y = 2.17$, $g_z = 1.93$		
	376		$g_x = 2.04$, $g_y = 2.04$, $g_z = 2.11$		
$RbCdCl_3$	4.2		$g_x = 2.085$, $g_y = 2.085$, $g_z = 2.389$	$A_x < 15$ Oe, $A_y < 15$ Oe, $A_z = 76$ Oe	MWrev90
$(RNH_3)_2CuBr_4$, $(NH_3RNH_3)CuBr_4$	300		$g_x = 2.047$, $g_y = 2.047$, $g_z = 2.153$		
Rb_2CdCl_4	$T > 143$		$g_x = 2.245$, $g_y = 2.245$, $g_z = 2.030$		MWrev90
	4.2	I	$g_x = 2.355$, $g_y = 2.133$, $g_z = 2.030$	$A_x = 74.8$, $A_y < 18$, $A_z = 50.2$	
		II	$g_x = 2.133$, $g_y = 2.355$, $g_z = 2.030$	$A_x < 18$, $A_y = 74.8$, $A_z = 50.2$	
	300		$g_x = 1.985$, $g_y = 1.985$, $g_z = 1.980$		MWrev90
Cs_2CuCl_4	4.2–16		$g_x = 2.068$, $g_y = 2.053$, $g_z = 2.220$		MWrev90
Cs_2ZnCl_4	NR	I	$g_x = 2.100$, $g_y = 2.084$, $g_z = 2.445$	$A_x = 4.6$, $A_y = 5.1$, $A_z = 2.5$	MWrev90
		II	$g_x = 2.101$, $g_y = 2.082$, $g_z = 2.447$		MWrev90
$K_2CuCl_4 \cdot 2H_2O$	RT		$g_x = 2.12$, $g_y = 2.12$, $g_z = 2.35$		MWrev90
K_2ZnF_4	$T > 60$	I	$g_x = 2.372$, $g_y = 2.372$, $g_z = 2.04$		MWrev90
	$T < 60$	II	$g_x = 2.46$, $g_y = 2.23$, $g_z = 2.10$		
	$T < 60$	III	$g_x = 2.23$, $g_y = 2.46$, $g_z = 2.10$		
	4.2		$g_x = 2.381$, $g_y = 72.2$, $g_z = 2.014$	$A_z = 42.3$	
	4.2		$g_x = 2.381$, $g_y = 72.2$, $g_z = 2.014$	$A_z = 42.3$	
High T_c superconductors					MWrev90
$BiCaSrCu_2O_x$	77–110		$g_x = 2.056$, $g_z = 2.24$		MWrev90
Bi–Ca–Sr–Cu oxide	77–RT		$g_z = 2.244$		MWrev90
$BaCuO_2$	290		$g_z = 2.09$		MWrev90
	100		$g_z = 2.10$		
$BaCuO_2$ (with exposure)	RT		$g_x = 2.093$, $g_y = 2.039$, $g_z = 2.184$		MWrev90

DATA TABULATION (Cu^{2+}) $3d^9$ *(continued)*

Host	T (°K)	Site	$\tilde{g}$	$\tilde{A}^{63}$, $\tilde{A}^{65}$, $\tilde{Q}$	Ref.
La_2CuO_4	RT		$g_z = 2.08$		MWrev90
Pr_2CuO_4	NR		$g_x = 2.27$, $g_y = 2.27$, $g_z = 3.00$		MWrev90
$Tl_3Ba_2Ca_3Cu_4O_x$	$>T_c$		$g_x = 2.065$, $g_y = 2.065$, $g_z = 2.25$		MWrev90
$YBa_2Cu_3O_x$	1.5–300		$g_x = 2.060$, $g_y = 2.060$, $g_z = 2.200$		MWrev90
$YBa_2Cu_3O_7$	NR		$g_x = 2.160$, $g_y = 2.160$, $g_z = 2.161$	$A_x = 760$, $A_y = 760$, $A_z = 700$	MWrev90
$YBa_2Cu_3O_{7-x}$	300		$g_x = 2.047$, $g_y = 2.048$, $g_z = 2.217$		MWrev90
	90		$g_x = 2.086$, $g_y = 2.086$, $g_z = 2.240$	$A_x = 75$ G, $A_y = 75$ G, $A_z = 90$ G	
$YBa_2Cu_3O_{7-x}(1)$	77		$g_x = 2.055$, $g_y = 2.120$, $g_z = 2.109$		MWrev90
$YBa_2Cu_3O_{7-x}(2)$	77		$g_x = 2.050$, $g_y = 2.050$, $g_z = 2.120$		MWrev90
Y_2BaCuO_5	300		$g_x = 2.107$, $g_y = 2.061$, $g_z = 2.240$		MWrev90
	77		$g_x = 2.054$, $g_y = 2.054$, $g_z = 2.100$		MWrev90
$YBa_3Cu_2O_y$	4–300		$g_x = 2.09$, $g_y = 2.09$, $g_z = 2.23$		MWrev90
$Y_{1-x}Gd_xBa_2Cu_3O_7$	$>T_c$		$g_z = 2.03$		MWrev90
$Y_2Ba_2CuO_5$	77–RT		$g_x = 2.6$, $g_y = 2.06$, $g_z = 2.24$		MWrev90
$Y_2Cu_2O_5$	RT		$g_x = 2.04$, $g_y = 2.04$, $g_z = 2.39$		MWrev90
$Y_{0.2}Ba_{0.5}CuO_x$	LT		$g_x = 2.033$, $g_z = 2.193$		MWrev90
Hydroxide					
$Ca(OH)_2$	300		$g_x = 2.172$, $g_y = 2.172$, $g_z = 2.241$	$A_x = 37$ G, $A_y = 37$ G, $A_z = 48$ G	MWrev90
		$I^{63}Cu$	$g_x = 2.079$, $g_y = 2.079$, $g_z = 2.421$	$A_z = 132$ G	MWrev90
		$I^{65}Cu$	$g_x = 2.079$, $g_y = 2.079$, $g_z = 2.421$	$A_z = 142$ G	MWrev90
		$II^{63}Cu$	$g_x = 2.081$, $g_y = 2.081$, $g_z = 2.401$	$A_z = 140$ G	MWrev90
		$II^{65}Cu$	$g_x = 2.081$, $g_y = 2.081$, $g_z = 2.401$	$A_z = 151$ G	MWrev90
Langbeinite					
$(NH_4)_2Cd(SO_4)_3$	300		$g_x = 2.092$, $g_y = 2.088$, $g_z = 2.329$	$A_x = 55$, $A_y = 39$, $A_z = 172$	MWrev90
Oxides					
CaO	77		2.220	$A^{63} = 21.6$	PF74
	4.2		2.222	$A^{63} = 29.1$	MWrev90
CaO	NR		$g_1 = 2.22$, $g_2 = 0.22$	$A_1 = -21.6$, $A_2 = -10.8$	MWrev90
	77		$g_1 = 2.192$, $g_2 = +0.108$	$A_1 = \pm 37.2$, $A_2 = \pm 11$	MWrev90
	93		$g_0 = 2.21$		MWrev90
MgO	77		2.190	19.0	PF74
MgO	77		$g_1 = 2.22$, $g_2 = 0.22$	$A_1 = \pm 31.2$, $A_2 = \pm 10.8$	MWrev90

SrO	NR	^{63}Cu	$g_x = 2.337,\ g_y = 2.423,\ g_z = 1.993$	$A_x = 2,\ A_y = 22,\ A_z = 79.5$	MWrev90
		^{65}Cu	$g_x = 2.337,\ g_y = 2.423,\ g_z = 1.993$	$A_x = 22,\ A_y = 22,\ A_z = 85$	MWrev90
	77	^{63}Cu	$g_x = 2.082,\ g_y = 2.050,\ g_z = 2.361$	$A_x = 14.5,\ A_y = -8.5,\ A_z = 111$	MWrev90
		^{65}Cu	$g_x = 2.082,\ g_y = 2.050,\ g_z = 2.361$	$A_x = 14.5,\ A_y = -8.5,\ A_z = 119$	
ZnO	NR	^{63}Cu	$g_x = 1.524,\ g_y = 1.524,\ g_z = 0.736$	$A_x = 224.7,\ A_y = 224.7,\ A_z = 198.3$	MWrev90
		^{65}Cu	$g_x = 1.524,\ g_y = 1.524,\ g_z = 0.736$	$A_x = 240.7,\ A_y = 240.7,\ A_z = 212.5$	
		I	$g_x = 1.910,\ g_y = 1.887,\ g_z = 2.848$	$A_x = 0.081^*,\ A_y = 0.073^*,\ A_z = 0.671^*$	
		II	$g_x = 2.56,\ g_y = 2.56,\ g_z = 2.082$	$A_x = 0.060^*,\ A_y = 0.060^*,\ A_z = 0.080^*$	MWrev90
		III	$g_x = 2.120,\ g_y = 2.030,\ g_z = 2.490$	$A_x = 0.047^*,\ A_y = 0.111^*,\ A_z = 0.069^*$	
	4.2–20			$A_y = 0.040^*$	
	4.2–25			$A_y = 0.710^*$	
ZnO	1.2		$g_\parallel = 0.74,\ g_i = 1.50$	$A = 219.0,\ B = 235.0$	PF74
$Cd_3(PO_4)_2$	RT		$g_x = 2.085,\ g_y = 2.085,\ g_z = 2.479$	$A_x = 20.5,\ A_y = 20.5,\ A_z = 69.0$	MWrev90
$Ca_3(PO_4)_2 \cdot 4H_2O$	295–340	I		$A_z = 105$	MWrev90
$(Ca_3(PO_4)_2 \cdot 2H_2O)$		II		$A_z = 86$	
	360–550		$g_x = 2.107,\ g_y = 2.107,\ g_z = 2.539$	$A_x = 27,\ A_y = 27,\ A_z = 53$	
$Ca_{0.5}Ti_2(PO_4)_3$	300		$g_x = 2.060,\ g_y = 2.060,\ g_z = 2.307$		MWrev90
	4.2		$g_x = 2.061,\ g_y = 2.061,\ g_z = 2.325$		
		θ state		$A_x = 0,\ A_y = 0,\ A_z = 72$	MWrev90
		ε state		$A_x = -41,\ A_y = -41,\ A_z = -144$	
Al_2O_3	NR		$g_x = 2.058,\ g_y = 2.058,\ g_z = 2.337$	$A_x = 24,\ A_y = 24,\ A_z = 163$	MWrev90
β-Al_2O_3	4–77, 300		$g_x = 2.032,\ g_y = 2.130,\ g_z = 2.576$	$A_x = 175,\ A_y = 118,\ A_z = 31$	MWrev90
Nb_2O_5	300		$g_x = 2.046,\ g_y = 2.046,\ g_z = 2.174$	$A_x = 24,\ A_y = 24,\ A_z = 195$	MWrev90
$Zn_3B_7O_{13}Cl$	77		$g_x = 2.09,\ g_y = 2.09,\ g_z = 2.417$	$A_x = 15$ G, $A_y = 15$ G, $A_z = 113$ G	MWrev90
Perovskites					
$KTaO_3$	77	I	$g_x = 2.045,\ g_y = 2.045,\ g_z = 2.238$	$A_x = 172,\ A_y = 172,\ A_z = 172$	MWrev90
		II	$g_x = 2.045,\ g_y = 2.045,\ g_z = 2.194$	$A_x \le 30,\ A_y \le 30,\ A_z = 193$	
$LiTaO_3$	77	I	$g_x = 2.050,\ g_y = 2.099,\ g_z = 2.396$	$A_x = 56,\ A_y = 50,\ A_z = 70$	MWrev90
		II	$g_x = 2.167,\ g_y = 2.167,\ g_z = 2.197$		
Phosphates					
β-$Ca_3(PO_4)_2$ (unstressed)	NR		$g_x = 2.061,\ g_y = 2.063,\ g_z = 2.346$	$A_x = 16.8,\ A_y = 15.9,\ A_z = 112.2$	MWrev90
β-$Ca_3(PO_4)_2$ (stressed)			$g_x = 2.055,\ g_y = 2.055,\ g_z = 2.348$	$A_x = 18.2,\ A_y = 18.2,\ A_z = 136.5$	
	120	^{63}Cu	$g_x = 2.055,\ g_y = 2.055,\ g_z = 2.348$	$A_x = -18.2,\ A_y = -18.2,\ A_z = -136.5$	MWrev90

DATA TABULATION (Cu^{2+}) $3d^9$ (*continued*)

Host	T (°K)	Site	$\tilde{g}$	$\tilde{A}^{63}$, $\tilde{A}^{65}$, $\tilde{Q}$	Ref.
		^{65}Cu	$g_x = 2.054$, $g_y = 2.054$, $g_z = 2.350$	$A_x = -19.6$, $A_y = -19.6$, $A_z = -137.4$	
CsH_2PO_4	RT		$g_x = 2.187$, $g_y = 2.187$, $g_z = 2.258$	$A_x = 27$ G, $A_y = 27$ G, $A_z = 30$ G	MWrev90
Selenates					
Rb_2SeO_4	77–293		$g_x = 2.035$, $g_y = 2.165$, $g_z = 2.391$	$A_x = 52.50$, $A_y = 18.79$, $A_z = 114.67$	MWrev90
$ZnSeO_4 \cdot 6H_2O$	77		$g_x = 2.094$, $g_y = 2.094$, $g_z = 2.427$	$A_x = 20.4$ G, $A_y = 20.4$ G, $A_z = 114.6$ G	MWrev90
Spinel					
$ZnAl_2O_4$	300		$g_x = 2.07$, $g_z = 2.32$	$A_x = 4$, $A_z = 15$	MWrev90
	77		$g_z = 2.33$	$A_z = 13.5$	
Sulfates					
$CuSO_4 \cdot 5H_2O$	300		$g_x = 2.083$, $g_y = 2.083$, $g_z = 2.276$		MWrev90
$MgSO_4 \cdot 7H_2O$	300		$g_x = 2.127$, $g_y = 2.127$, $g_z = 2.398$	$A_y = 18$, $A_z = 72$	MWrev90
	77		$g_x = 2.089$, $g_y = 2.098$, $g_z = 2.441$	$A_x = 22$, $A_y = 25$, $A_z = 116$	
	4.2		$g_x = 2.095$, $g_y = 2.097$, $g_z = 2.441$	$A_x = 21$, $A_z = 123$	
$ZnSO_4 \cdot 7H_2O$	77, RT		$g_x = 2.134$, $g_y = 2.101$, $g_z = 2.485$	$A_x = 42.9$ G, $A_y = 13.9$ G, $A_z = 116.2$ G	MWrev90
	4.2		$g_x = 2.096$, $g_y = 2.098$, $g_z = 2.437$	$A_x = 19$, $A_y = 25$, $A_z = 125$	
K_2SO_4	77	I	$g_x = 2.023$, $g_y = 2.174$, $g_z = 2.385$	$A_x = 68.9$, $A_y = 27.8$, $A_z = 102.5$	MWrev90
		II	$g_x = 1.999$, $g_y = 1.999$, $g_z = 2.366$	$A_x = 85.6$, $A_y = 85.6$, $A_z = 70.7$	
		III	$g_x = 2.034$, $g_y = 2.159$, $g_z = 2.405$	$A_x = 48.1$, $A_y = 34.0$, $A_z = 115.7$	
	5–320		$g_x = 2.173$, $g_y = 2.390$, $g_z = 2.022$	$A_x = 25$, $A_y = 102$, $A_z = 68$	
	130–450	I	$g_x = 2.152$, $g_y = 2.031$, $g_z = 2.336$	$A_x = -31$, $A_y = +61$, $A_z = -95.3$	MWrev90
		II	$g_x = 2.365$, $g_y = 2.190$, $g_z = 2.038$	$A_x = -69$, $A_y = +25$, $A_z = +105$	
		III	$g_x = 2.25$, $g_y = 2.25$, $g_z = 1.996$	$A_x = -75$, $A_y = -75$, $A_z = +175$	
$(NH_4)_2SO_4$	246		$g_x = 2.087$, $g_y = 2.087$, $g_z = 2.343$	$A_x = 93.0$ G, $A_z = 58.0$ G	MWrev90
$(ND_2)_2SO_4$	300		$g_x = 2.289$, $g_y = 2.282$, $g_z = 2.045$	$A_x = -74.3$ G, $A_z = 198.3$ G	
	77		$g_x = 2.252$, $g_y = 2.278$, $g_z = 2.025$	$A_x = -81.0$ G, $A_y = -52.9$ G, $A_z = 178.4$ G	

	298		$g_x = 2.249, g_y = 2.183, g_z = 2.011$	$A_x = -24, A_y = -54.3$ G, $A_z = 172$	
		I	$g_x = 2.271, g_y = 2.147, g_z = 2.011$	$A_x = -30, A_y = -48, A_z = 135$	
	77–300	II	$g_x = 2.326, g_y = 2.169, g_z = 2.034$	$A_x = -1, A_y = -55, A_z = -120$	
		III	$g_x = 2.355, g_y = 2.207, g_z = 2.034$	$A_x = 78, A_y = 82, A_z = 90$	
		IV		$A_y = 35$	
Na_2SO_4		I	$g_x = 2.059, g_y = 2.089, g_z = 2.350$	$A_x = 18.76$ G, $A_y = 49.55$ G, $A_z = 149.7$ G	MWrev90
		II	$g_x = 2.041, g_y = 2.109, g_z = 2.347$	$A_x = 18.74$ G, $A_y = 49.10$ G, $A_z = 149.9$ G	
		III	$g_x = 2.063, g_y = 2.094, g_z = 2.344$	$A_x = 19.93$ G, $A_y = 41.71$ G, $A_z = 148.4$ G	
		IV	$g_x = 2.042, g_y = 2.327, g_z = 2.327$	$A_x = 9.29$ G, $A_y = 49.35$ G, $A_z = 143.5$ G	
Rb_2SO_4	77, 290		$g_x = 2.040, g_y = 2.040, g_z = 2.408$	$A_x = 48.1, A_y = 48.1, A_z = 92.2$	MWrev90
$KHSO_4$	300	I	$g_x = 2.201, g_y = 2.017, g_z = 2.385$	$A_x = 57, A_y = 32, A_z = 93$	MWrev90
		II	$g_x = 2.186, g_y = 2.018, g_z = 2.387$	$A_x = 66, A_y = 22, A_z = 85$	
		III	$g_x = 2.204, g_y = 2.023, g_z = 2.385$	$A_x = 66, A_y = 21, A_z = 87$	
		IV	$g_x = 2.201, g_y = 2.014, g_z = 2.380$	$A_x = 69, A_y = 41, A_z = 91$	
	77	I	$g_x = 2.179, g_y = 2.104, g_z = 2.386$	$A_x = 78, A_y = 31, A_z = 96$	
		II	$g_x = 2.214, g_y = 2.012, g_z = 2.379$	$A_x = 73, A_y = 26, A_z = 94$	
		III	$g_x = 2.231, g_y = 2.018, g_z = 2.380$	$A_x = 72, A_y = 25, A_z = 92$	
		IV	$g_x = 2.187, g_y = 2.028, g_z = 2.352$	$A_x = 82, A_y = 18, A_z = 93$	
$LiCsO_4$	295	I	$g_1 = 2.508, g_2 = 2.141, g_3 = 2.141$		MM93
		II	$g_1 = 2.318, g_2 = 2.094, g_3 = 1.917$		MWrev90
$LiKSO_4$	120		$g_x = 2.175, g_y = 2.034, g_z = 2.381$	$A_x = 21, A_y = 63, A_z = 95$	MWrev90
	77		$g_x = 2.181, g_y = 2.044, g_z = 2.400$	$A_x = 24$ G, $A_y = 82$ G, $A_z = 94$ G	
$LiAlSiO_4$	77		$g_x = 2.116, g_y = 2.099, g_z = 2.327$	$A_x = 18, A_y = 27, A_z = 81$	MWrev90
$Cu(NH_3)_4SO_4 \cdot H_2O$	NR		$g_x = 2.075, g_y = 2.075, g_z = 2.212$		MWrev90
$(NH_4)_3H(SO_4)_2$	RT		$g_x = 2.080, g_y = 2.080, g_z = 2.320$	$A_x = 20, A_y = 20, A_z = 96$	MWrev90
	413		$g_0 = 2.240$		
$CoNa_2(SO_4)_2 \cdot 4H_2O$	295		$g_1 = 2.402, g_2 = 2.193, g_3 = 2.026$	$A_1^{63,65} = 0.368^*$, $A_2^{63,65} = 0.236^*$, $A_3^{63,65} = 0.168^*$	MS91
$MgNa_2(SO_4)_2 \cdot 4H_2O$	113		$g_1 = 2.399, g_2 = 2.198, g_3 = 2.030$	$A_1^{65} = 0.443^*$, $A_2^{65} = 0.251^*$, $A_3^{65} = 0.206^*$	MS91
				$A_1^{63} = 0.420^*$, $A_2^{63} = 0.243^*$, $A_3^{63} = 0.199^*$	MWrev90

DATA TABULATION (Cu^{2+}) $3d^9$ *(continued)*

Host	T (°K)	Site	$\tilde{g}$	$\tilde{A}^{63}$, $\tilde{A}^{65}$, $\tilde{Q}$	Ref.
$ZnNa_2SO_4 \cdot 4H_2O$	77		$g_x = 2.391$, $g_y = 2.260$, $g_z = 2.018$	$A_x = 58$, $A_y = 43.6$, $A_z = 73.5$	MWrev90
$(NH_3)_2Zn(SO_4)_2 \cdot 6H_2O$	295		$g_x = 2.26$, $g_y = 2.02$, $g_z = 2.31$		MWrev90
	77		$g_x = 2.20$, $g_y = 2.03$, $g_z = 2.36$	$A_x = 36$, $A_y = 60$, $A_z = 76$	
	20		$g_x = 2.150$, $g_y = 2.030$, $g_z = 2.420$	$A_x = 20$, $A_y = 60$, $A_z = 96$	
Tutton salts [$M'M''(SO_4)_2 \cdot 6H_2O$]					
$M' = Cs$, $M'' = Zn$	300		$g_x = 2.050$, $g_y = 2.164$, $g_z = 2.384$		MWrev90
	120		$g_x = 2.050$, $g_y = 2.127$, $g_z = 2.402$		
$M' = K$, $M'' = Cd$	77		$g_x = 2.172$, $g_y = 2.054$, $g_z = 2.355$	$A_x = 25.6$ G, $A_y = 54.77$ G, $A_z = 101$ G	MWrev90
	77		$g_x = 2.197$, $g_y = 2.128$, $g_z = 2.374$	$A_x = 080^*$, $A_y = 0.137^*$, $A_z = 0.282^*$	
$M' = K$, $M'' = Co$	210		$g_x = 2.227$, $g_y = 2.053$, $g_z = 2.330$	$A_x = 58.5$, $A_y = 52$, $A_z = 60.5$	MWrev90
$M' = K$, $M'' = Zn$	295		$g_x = 2.02$, $g_y = 2.26$, $g_z = 2.31$		MWrev90
	77		$g_x = 2.03$, $g_y = 2.20$, $g_z = 2.36$	$A_x = 60$, $A_y = 36$, $A_z = 76$	
	4.2		$g_x = 2.03$, $g_y = 2.20$, $g_z = 2.42$	$A_x = 60$, $A_y < 20$, $A_z = 96$	
$M' = Na$, $M'' = Zn$	NR		$g_x = 2.035$, $g_y = 2.142$, $g_z = 2.350$	$A_x = 87.7$, $A_y = 6.5$, $A_z = 65.3$	MWrev90
	77–302		$g_x = 2.029$, $g_y = 2.135$, $g_z = 2.349$	$A_x = 8.7$, $A_y = 6.5$, $A_z = 65$	
$M' = NH_4$, $M'' = Cd$	300		$g_x = 2.202$, $g_y = 2.073$, $g_z = 2.331$		MWrev90
$M' = NH_4$, $M'' = Cd$	295		$g_x = 2.071$, $g_y = 2.211$, $g_z = 2.337$		MW89D
	77		$g_x = 2.052$, $g_y = 2.172$, $g_z = 2.361$	$A_x = 0.151^*$, $A_y = 0.074^*$, $A_z = 0.333^*$	
	4.2		$g_x = 2.051$, $g_y = 2.159$, $g_z = 2.429$	$A_x = 0.156^*$, $A_y = 0.077^*$, $A_z = 0.335^*$; $Q_x = -0.001^*$, $Q_y = -0.012^*$, $Q_z = 0.013^*$	
$M' = NH_4$, $M'' = Co$	210		$g_x = 2.310$, $g_y = 2.088$, $g_z = 2.355$	$A_x = 76$, $A_y = 0$, $A_z = 74$	MWrev90
$M' = NH_4$, $M'' = Cu$	293		$g_x = 2.071$, $g_y = 2.218$, $g_z = 2.360$		MWrev90
	160		$g_x = 2.070$, $g_y = 2.131$, $g_z = 2.423$		
	295		$g_x = 2.348$, $g_y = 2.207$, $g_z = 2.060$		
	300		$g_x = 2.06$, $g_y = 2.029$, $g_z = 2.36$		
	77		$g_x = 2.067$, $g_y = 2.11$, $g_z = 2.424$		
	4.2		$g_x = 2.065$, $g_y = 2.11$, $g_z = 2.427$		

$M' = NH_4$, $M'' = Zn$	300		$g_x = 2.06$, $g_y = 2.19$, $g_z = 2.34$		MWrev90
	77		$g_x = 2.066$, $g_y = 2.125$, $g_z = 2.427$	$A_x \leq 35$, $A_y < 35$, $A_z = 120$	
	4.2		$g_x = 2.046$, $g_y = 2.132$, $g_z = 2.434$	$A_x = 49$, $A_y < 22$, $A_z = 108$	MWrev90
$M' = ND_4$, $M'' = Zn$	77		$g_x = 2.283$, $g_y = 2.283$, $g_z = 2.027$	$A_x = 31$, $A_y = 31$, $A_z = 95$	MWrev90
$M' = Rb$, $M'' = Zn$	300		$g_x = 2.04$, $g_y = 2.23$, $g_z = 2.34$		MWrev90
	77		$g_x = 2.044$, $g_y = 2.152$, $g_z = 2.421$	$A_x = 49$, $A_z = 102$	MWrev90
	4.2		$g_x = 2.046$, $g_y = 2.132$, $g_z = 2.434$	$A_x = 49$, $A_z = 108$	MWrev90
$M' = Rb$, $M'' = Cu$	300		$g_x = 2.056$, $g_y = 2.136$, $g_z = 2.42$		MWrev90
	77		$g_x = 2.059$, $g_y = 2.12$, $g_z = 2.425$		
	4.2		$g_x = 2.056$, $g_y = 2.12$, $g_z = 2.421$		
$M' = Rb$, $M'' = Mg$	LNT		$g_x = 2.027$, $g_y = 2.118$, $g_z = 2.374$	$A_x = 57$, $A_y = 0$, $A_z = 100$	MWrev90
$M' = Tl$, $M'' = Mg$	300		$g_x = 2.094$, $g_y = 2.094$, $g_z = 2.395$	$A_x = 34$, $A_y = 34$, $A_z = 106$	MWrev90
	80		$g_x = 2.094$, $g_y = 2.094$, $g_z = 2.390$	$A_x = 26$, $A_y = 26$, $A_z = 100$	
Zeolites					
Ca X-zeolite	4.2, 77, RT		$g_x = 2.300$, $g_y = 2.300$, $g_z = 2.001$	$A_x = 59$, $A_y = 59$, $A_z = 93$	MWrev90
chabazite zeolite	295	I	$g_0 = 2.070$	$A_0 = 42$	MWrev90
		II	$g_x = 2.028$, $g_y = 2.028$, $g_z = 2.324$	$A_x = 18$, $A_y = 18$, $A_z = 159$	MWrev90
Na Y-zeolite	4.2, 77–400				MWrev90
	300				
	77		$g_0 = 2.183$		
			$g_x = 2.088$, $g_y = 2.088$, $g_z = 2.408$	$A_x = 1.5$, $A_y = 1.5$, $A_z = 14.6$	
(Sample 1)	NR		$g_x = 1.74$, $g_y = 1.74$, $g_z = 2.019$	$A_x = 175$ G	MWrev90
(Sample 2)	NR		$g_x = 1.996$, $g_y = 1.996$, $g_z = 1.848$		
(Single Cu^{2+})	77, RT	I	$g_x = 2.066$, $g_y = 2.066$, $g_z = 2.320$	$A_x = 24$, $A_y = 24$, $A_z = 176$	MWrev90
		II	$g_x = 2.066$, $g_y = 2.066$, $g_z = 2.383$	$A_x = 24$, $A_y = 24$, $A_z = 133$	
(Cu^{2+})	77, RT		$g_x = 2.068$, $g_y = 2.068$, $g_z = 2.345$	$A_z = 150$	
Other					
Zn–Cl boracite	293	^{63}Cu	$g_x = 2.080$, $g_y = 2.084$, $g_z = 2.416$	$A_x > 10$ G, $A_y < 20$ G, $A_z = 113.4$ G	MWrev90
		^{65}Cu	$g_x = 2.080$, $g_y = 2.084$, $g_z = 2.416$	$A_x > 10$ G, $A_y < 20$ G, $A_z = 120.0$ G	
α-Quartz	77	^{63}Cu	$g_x = 1.998$, $g_y = 2.000$, $g_z = 2.000$	$A_x = 3.186$*, $A_y = 3.191$*, $A_z = 3.364$*	MWrev90
		^{65}Cu	$g_x = 1.998$, $g_y = 2.000$, $g_z = 2.000$	$A_x = 3.414$*, $A_y = 3.419$*, $A_z = 3.604$*	
	100, 293	^{63}Cu	$g_x = 2.000$, $g_y = 2.000$, $g_z = 1.998$	$A_x = 3.290$*, $A_y = 3.290$*, $A_z = 3.118$*	MWrev90

DATA TABULATION (Cu^{2+}) $3d^9$ *(continued)*

Host	T (°K)	Site	$\tilde{g}$	$\tilde{A}^{63}$, $\tilde{A}^{65}$, $\tilde{Q}$	Ref.
		^{65}Cu	$g_x = 2.000$, $g_y = 2.000$, $g_z = 1.998$	$A_x = 3.520$*, $A_y = 3.520$*, $A_z = 3.311$*	
CdS	NR		$g_x = 1.924$, $g_y = 1.924$, $g_z = 2.221$	$A_x = 23$, $A_y = 23$, $A_z = 77.2$	MWrev90
$CaCO_3$	77	^{63}Cu	$g_x = 2.097$, $g_y = 2.097$, $g_z = 2.520$	$A_x = 17.0$, $A_y = 17.0$, $A_z = 63.5$	MWrev90
		^{65}Cu	$g_x = 2.097$, $g_y = 2.097$, $g_z = 2.520$	$A_x = 17.0$, $A_y = 17.0$, $A_z = 69.0$	
NH_4NO_3	NR	I	$g_x = 2.109$, $g_y = 2.055$, $g_z = 2.185$		
		II	$g_x = 2.264$, $g_z = 2.315$		
$LiNbO_3$	77		$g_x = 2.076$, $g_y = 2.106$, $g_z = 2.318$	$A_x = 30.2$, $A_z = 50.5$	MWrev90
	60–200		$g_x = 2.095$, $g_y = 2.121$, $g_z = 2.403$	$A_x = 29$, $A_z = 40$	
	> 200		$g_x = g_y = 2.17$, $g_z = 2.22$		
$Li_{0.5}Ga_{0.25}O_4$ (ordered)	4.2		$g_x = 2.082$, $g_y = 2.082$, $g_z = 2.386$	$A_x = 0$, $A_y = 0$, $A_z = 100$ G	MWrev90
$Zn_2(OH)PO_4$	290		$g_x = 2.137$, $g_y = 2.065$, $g_z = 2.430$	$A_x \leq 8.5$, $A_y = 40.0$, $A_z = -88.8$	MWrev90
	77		$g_x = 2.134$, $g_y = 2.059$, $g_z = 2.435$	$A_x \leq 4.6$, $A_y = 36.0$, $A_z = -97.3$	
$ZnWO_4$	4.2–300		$g_z = 2.001$	$A_z = 36$–40	MWrev90
$Zn(BrO_3)_2 \cdot 6H_2O$	4.2		$g_x = 2.109$, $g_y = 2.109$, $g_z = 2.451$	$A_x = 13.3$, $A_y = 13.3$, $A_z = 117.8$	MWrev90
$Na_2HAsO_4 \cdot 7H_2O$	77	I	$g_x = 2.492$, $g_y = 2.100$, $g_z = 2.084$	$A_x = -124$ G, $A_y = 13.8$ G, $A_z = 17.4$	MWrev90
		II	$g_x = 2.461$, $g_y = 2.027$, $g_z = 2.179$	$A_x = -88.5$ G, $A_y = -73.4$ G, $A_z = 2.5$	
	298		$g_x = 2.029$, $g_y = 2.160$, $g_z = 2.417$	$A_x = 48.0$, $A_y = 35.0$, $A_z = 91.0$	
$LiH_3(SeO_3)_2$	140		$g_x = 2.362$, $g_y = 2.340$, $g_z = 1.990$	$A_x = 85$, $A_y = 71$, $A_z = 203$	MWrev90
$Ti_{1-x}Cu_x3(Nb_{2x}O_2)$ (powder)	300		$g_x = 2.033$, $g_y = 2.130$, $g_z = 2.376$	$A_x = 175$, $A_y = 118$, $A_z = 31$	MWrev90
$11Ga_2O_3 \cdot Na_2O$	20–77		$g_0 = 2.19$		
$(HfO_2)_{0.9}(Y_2O_3)_{0.1}$	295		$g_x = 2.07$, $g_y = 2.07$, $g_z = 2.40$	$A_x = 12$, $A_y = 12$, $A_z = 114$	MWrev90
ν-$Pb_3V_2O_8$	298		$g_x = 2.076$, $g_y = 2.076$, $g_z = 2.233$		MWrev90
	373		$g_y = 2.18$, $g_0 = 2.063$	$A_x = 105$, $A_y < 9$, $A_z < 30$	
$Be_3A_{12}Si_6O_{18}$	300	^{63}Cu	$g_x = 2.017$, $g_y = 2.092$, $g_z = 2.359$	$A_x = 5.0$, $A_y = 15.6$, $A_z = 114.5$	MWrev90
	77	^{65}Cu	$g_x = 2.017$, $g_y = 2.092$, $g_z = 2.359$	$A_x = 5.0$, $A_y = 15.6$, $A_z = 134.9$	
			$g_x = 2.071$, $g_y = 2.091$, $g_z = 2.359$	$A_x = 5.0$, $A_y = 15.7$, $A_z = 125.6$	
$CuA_{16}(PO_4)_4(OH)_8 \cdot 5H_2O$	300		$g_x = 2.135$, $g_y = 2.373$, $g_z = 2.043$		MWrev90
$Na_{10}Fe_4Cu_4W_{18}O_{70}H_6 \cdot 29H_2O$	77		$g_x = 2.116$, $g_y = 2.099$, $g_z = 2.327$	$A_x = 18$, $A_y = 27$, $A_z = 81$	MWrev90

$(ZnOH)\cdot(ZnAsO_4)$	290	$I^{63}Cu$	$g_x = 2.080,\ g_y = 2.080,\ g_z = 2.338$	$A_x = 68,\ A_y = 68,\ A_z = 117$	MWrev90
		$I^{65}Cu$	$g_x = 2.080,\ g_y = 2.080,\ g_z = 2.338$	$A_x = 68,\ A_y = 68,\ A_z = 125$	
		II	$g_x = 2.253,\ g_y = 2.019,\ g_z = 2.343$	$A_x = 21,\ A_y = 71,\ A_z = 82$	
$Ba_2Co(HCOO)_6\cdot 4H_2O$	300		$g_x = 2.098,\ g_y = 2.098,\ g_z = 2.355$	$A_x = 30,\ A_y = 30,\ A_z = 95$	MWrev90
	77		$g_x = 2.078,\ g_y = 2.078,\ g_z = 2.405$	$A_x = 22,\ A_y = 22,\ A_z = 138$	
$Ba_2Zn(HCOO)_6\cdot 4H_2O$	15		$g_x = 2.090,\ g_y = 2.065,\ g_z = 2.400$	$A_z = 131$	MWrev90
$(ZnS(CH_2)COO))_2\cdot 3H_2O$	120		$g_x = 2.041,\ g_y = 2.174,\ g_z = 2.382$	$A_x = 49.3,\ A_y = 26.0,\ A_z = 97.6$	MWrev90
$Cu(C_6H_5COO)_2\cdot 3H_2O$	0.4–1.0		$g_x = 2.08,\ g_y = 2.05,\ g_z = 2.36$		MWrev90
$Cu(C_6H_5SO_3)_3\cdot 6H_2O$	20	I	$g_x = 2.072,\ g_y = 2.132,\ g_z = 2.437$		MWrev90
		II	$g_x = 2.087,\ g_y = 2.115,\ g_z = 2.428$		
		III	$g_x = 2.070,\ g_y = 2.119,\ g_z = 2.435$		
		IV	$g_x = 2.069,\ g_y = 2.125,\ g_z = 2.447$		
$CaCd(OAc)_4\cdot 6H_2O$	77		$g_x = 2.054,\ g_y = 2.057,\ g_z = 2.356$		MWrev90
Zinc 1-maleate trihydrate	150		$g_x = 2.089,\ g_y = 2.088,\ g_z = 2.425$	$A_x = -12.0,\ A_y = -8.7,$ $A_z = -120.1$	MWrev90
Potassium hydrogen maleate	300		$g_x = 2.062,\ g_y = 2.062,\ g_z = 2.285$	$A_x = 28,\ A_y = 28,\ A_z = 182$	MWrev90
Zinc maleate tetrahydrate			$g_x = 2.095,\ g_y = 2.095,\ g_z = 2.172$		MWrev90
$NaNH_4$ tartrate	77, RT		$g_x = 2.064,\ g_y = 2.095,\ g_z = 2.316$	$A_x = 60.1$ G, $A_y = 29.0,$ $A_z = 118.8$ G	MWrev90
$CaC_4H_4O_6\cdot 4H_2O$	RT		$g_x = 2.117,\ g_y = 2.068,\ g_z = 2.415$	$A_x = 39$ G, $A_y = 35$ G, $A_z = 109$ G	MWrev90
(Weaker species)	LNT		$g_x = 2.122,\ g_y = 2.050,\ g_z = 2.395$	$A_x = 41$ G, $A_y = 1$ G, $A_z = 117$ G	
(Stronger species)	LNT		$g_x = 2.108,\ g_y = 2.063,\ g_z = 2.141$	$A_x = 45$ G, $A_y = 39$ G, $A_z = 114$ G	
	300	I	$g_x = 2.068,\ g_y = 2.117,\ g_z = 2.415$	$A_x = +36,\ A_y = +32,\ A_z = -101$	MWrev90
	77	I	$g_x = 2.063,\ g_y = 2.108,\ g_z = 2.414$	$A_x = +42,\ A_y = +36,\ A_z = -106$	
	77	II	$g_x = 2.05,\ g_y = 2.122,\ g_z = 2.395$	$A_x = +1,\ A_y = +38,\ A_z = -109$	
Cadmium tartrate pentahydrate	77		$g_x = 2.283,\ g_y = 2.020,\ g_z = 2.441$		MWrev90
Anhydrous lead tartrate	RT	I	$g_x = 2.011,\ g_y = 2.146,\ g_z = 2.379$	$A_x = +65.4$ G, $A_y = -82.8$ G, $A_z = -81.5$ G	MWrev90
		II	$g_x = 2.080,\ g_y = 2.183,\ g_z = 2.426$	$A_x = +69.7$ G, $A_y = -27.2$ G, $A_z = -82.8$ G	
		III	$g_x = 1.991,\ g_y = 2.154,\ g_z = 2.395$	$A_x = +56.6$ G, $A_y = +35.6$ G, $A_z = -87.4$ G	
		IV	$g_x = 2.106,\ g_y = 2.152,\ g_z = 2.424$	$A_x = +67.1$ G, $A_y = +55.6$ G, $A_z = -84.9$ G	
$SrC_4H_4O_6\cdot 3H_2O$	77	I	$g_x = 2.038,\ g_y = 2.132,\ g_z = 2.392$	$A_x = 26.3$ G, $A_y = 56.3$ G, $A_z = 110.8$ G	MWrev90

DATA TABULATION (Cu^{2+}) $3d^9$ *(continued)*

Host	T (°K)	Site	$\tilde{g}$	$\tilde{A}^{63}$, $\tilde{A}^{65}$, $\tilde{Q}$	Ref.
		II	$g_x = 2.050$, $g_y = 2.130$, $g_z = 2.371$	$A_x = 19.2$ G, $A_y = 61.4$ G, $A_z = 107.2$ G	
$C(NH_2)_3Al(SO_4)_2 \cdot 6H_2O$	RT	I	$g_x = 2.034$, $g_y = 2.034$, $g_z = 2.380$	$A_x = 20$, $A_y = 20$, $A_z = 108$	MWrev90
		II	$g_x = 2.08$, $g_y = 2.08$, $g_z = 2.422$	$A_x = 30$, $A_y = 30$, $A_z = 103$	
$C(NH_2)_3Al(SO_4)_2 \cdot 6H_2O$(GASH)		I	$g_\parallel = 2.422$, $g_\perp = 2.08$	$A_\parallel = 0.32$*, $A_\perp = 0.09$*	MCK95
	423	II	$g_\parallel = 2.17$, $g_\perp = 2.17$		MWrev90
	373		$g_\parallel = 2.371$, $g_\perp = 2.078$	$A_\parallel = 0.386$*, $A_\perp = 0.121$*	MWrev90
	293		$g_\parallel = 2.373$, $g_\perp = 2.077$	$A_\parallel = 0.382$*, $A_\perp = 0.125$*	MWrev90
	123		$g_\parallel = 2.377$, $g_\perp = 2.075$	$A_\parallel = 0.375$*, $A_\perp = 0.128$*	MWrev90
	82		$g_\parallel = 2.381$, $g_\perp = 2.073$	$A_\parallel = 0.369$*, $A_\perp = 0.135$*	MWrev90
	3.9		$g_\parallel = 2.38$, $g_\perp = 2.07$	$A_\parallel = 0.37$*, $A_\perp = 0.14$*	MWrev90
$Cd(NH_3)_2 \cdot Ni(CN)_4 \cdot 2C_6H_6$	77		$g_x = 2.225$, $g_y = 2.225$, $g_z = 2.020$	$A_x = 53.0$, $A_y = 53.0$, $A_z = 94.2$	MWrev90
	1.2		$g_x = 2.326$, $g_y = 2.160$, $g_z = 2.055$	$A_x = 114.0$, $A_y = 30.2$, $A_z = 93.2$	
$K_2Pt(CN)_4$	4.2–80		$g_x = 2.079$, $g_y = 2.079$, $g_z = 2.231$	$A_x = 0.070$*, $A_y = 0.070$*, $A_z = 0.467$*	MWrev90
$Pt(NH_3)_4CuCl_4$	20–300		$g_x = 2.156$, $g_y = 2.156$, $g_z = 2.055$		MWrev90
$[(C_2H_5)_2NH_2]CuCl_4$	77, 295		$g_x = 2.05$, $g_y = 2.05$, $g_z = 2.25$		MWrev90
$(C_nH_{2n+1}NH_3)_2$			$g_x = 2.05$, $g_y = 2.05$, $g_z = 2.27$		MWrev90
$[N(CH_3)_4]_2CuCl_4$	LT		$g_z = 2.08$		MWrev90
$Cu(NH_3)_4(ClO_4)_2$	RT		$g_x = 2.051$, $g_y = 2.051$, $g_z = 2.230$		MWrev90
	77		$g_x = 2.049$, $g_y = 2.049$, $g_z = 2.228$		
$Cu(NH_3)_5(ClO_4)_2$	RT		$g_0 = 2.136$		MWrev90
	77		$g_0 = 2.130$		
$Cu(ClO_4)_2 \cdot 4H_2O$	RT		$g_x = 2.091$, $g_y = 2.091$, $g_z = 2.429$		MWrev90
$K_2Cu(CN)_4$	100	I	$g_x = 2.065$, $g_y = 2.053$, $g_z = 2.240$	$A_x = 0$, $A_y = 0$, $A_z = 0.260$*	MWrev90
		II	$g_x = 2.005$, $g_y = 2.005$, $g_z = 2.000$	$A_x = 0.074$*, $A_y = 0.074$*, $A_z = 0.262$*	
$K_2C_2O_4 \cdot H_2O$	295		$g_1 = 2.333$, $g_2 = 2.072$, $g_3 = 2.066$	$A_1 = 0.483$*, $A_2 = 0.049$*, $A_3 = 0.050$*	MLW91
	77		$g_1 = 2.331$, $g_2 = 2.083$, $g_3 = 2.067$	$A_1 = 0.438$*, $A_2 = 0.045$*, $A_3 = 0.045$*	MWrev90
	4.2		$g_1 = 2.341$, $g_2 = 2.093$, $g_3 = 2.080$	$A_1 = 0.442$*, $A_2 = 0.043$*, $A_3 = 0.047$*	MWrev90

$K_2PbCu(NO_2)_6$	295	$g_0 = 2.120$		MWrev90
$Rb_2PbCu(NO_2)_6$	295	$g_x = 2.155$, $g_y = 2.155$, $g_z = 2.073$		MWrev90
	183	$g_x = 2.150$, $g_y = 2.150$, $g_z = 2.065$		
$Cs_2BaCu(NO_2)_6$	298	$g_0 = 2.158$		MWrev90
$Cs_2CaCu(NO_2)_6$	298	$g_0 = 2.172$		MWrev90
	77	$g_z = 2.262$		
		$g_x = 2.059$, $g_z = 2.075$		
$Cs_2PbCu(NO_2)_6$	298	$g_0 = 2.160$		MWrev90
	77	$g_x = 2.156$, $g_y = 2.156$, $g_z = 2.066$		
$Cs_2SrCu(NO_2)_6$	298	$g_0 = 2.164$		MWrev90
	77	$g_x = 2.070$, $g_z = 2.265$		
$K_2BaCu(NO_2)_6$	298	$g_0 = 2.142$		MWrev90
	77	$g_x = 2.066$, $g_z = 2.242$		
$K_2CaCu(NO_2)_6$	298	$g_0 = 2.146$		MWrev90
	77	$g_x = 2.066$, $g_z = 2.238$		
$K_2PbCu(NO_2)_6$	298	$g_0 = 2.139$		MWrev90
	77	$g_x = 2.150$, $g_z = 2.060$		
$K_2SrCu(NO_2)_6$	298	$g_0 = 2.143$		MWrev90
	77	$g_x = 2.070$, $g_z = 2.250$		
$Rb_2PbCu(NO_2)_6$	77	$g_x = 2.150$, $g_z = 2.064$		MWrev90
$Rb_2SrCu(NO_2)_6$	298	$g_0 = 2.150$		MWrev90
	77	$g_x = 2.073$, $g_y = 2.073$, $g_z = 2.257$		
$Tl_2CaCu(NO_2)_6$	77	$g_x = 2.071$, $g_y = 2.056$, $g_z = 2.252$		MWrev90
$Tl_2PbCu(NO_2)_6$	77	$g_x = 2.151$, $g_y = 2.151$, $g_z = 2.062$		MWrev90
$Tl_2PbCu(NO_2)_6$	>291	$g_x = 2.072$, $g_y = 2.072$, $g_z = 2.122$		MWrev90
	245–291	$g_x = 2.146$, $g_y = 2.146$, $g_z = 2.072$		
	245	$g_x = 2.146$, $g_y = 2.146$, $g_z = 2.066$		
	<245	$g_x = 2.150$, $g_y = 2.150$, $g_z = 2.062$		
$Tl_2SrCu(NO_2)_6$	298	$g_0 = 2.149$		MWrev90
	77	$g_z = 2.151$		
$Cu_5(SiO_3)OH_2$	120	$g_x = 2.038$, $g_y = 2.038$, $g_z = 2.402$	$A_x = 49.5$, $A_y = 49.5$, $A_z = 122$	MWrev90
$CuAl_6(PO_4)_4(OH)_8 \cdot 4H_2O$	393	$g_x = 2.112$, $g_y = 2.034$, $g_z = 2.305$		MWrev90
	353–173	$g_x = 2.119$, $g_y = 2.034$, $g_z = 2.131$		
	153	$g_x = 2.119$, $g_y = 2.034$, $g_z = 2.320$		
	80	$g_x = 2.118$, $g_y = 2.035$, $g_z = 2.339$		

Cu^+ $3d^{10}$. Data tabulation of spin-Hamiltonian parameters.

Sample	T (°K)	$\tilde{g}$	Ref.
CuCl	4.2, 77	2.11	BD74
NaCl	77, 293	$g_{\parallel} = 2.298$, $g_{=}2.058$	BD74

Zn^+ $3d^{10}$ $4s^1$. Data tabulation of spin-Hamiltonian parameters.

Sample	T (°K)	$\tilde{g}$	$\tilde{A}^{67}$ (GHz)	Ref.
NaCl (X-irrad.)	77	$g = 1.000$	$A = 2.030$	BM83

Cu^0 $3d^{10}$ $4s^1$. Data tabulation of spin-Hamiltonian parameters ($S = \frac{1}{2}$, $I = \frac{3}{2}$).

Sample	T (°K)	$\tilde{g}$	$\tilde{A}^{63,65}$ (GHz)	Ref.
NaCl (X-irrad.)	100	$g_J = 1.995$	$A^{63} = 2.830$, $A^{65} = 3.040$	BM86
ZnO	4.2–25	$g_{xx} = 1.576$, $g_{yy} = 1.576$, $g_{zz} = 0.736$	$A_{xx} = A_{yy} = A_{zz} = 0.710$	BM86

Mo^{6+} $4d^0$. Data tabulation of spin-Hamiltonian parameters.

Sample	T (°K)	$\tilde{g}$	Ref.
V_2O_5	50	$g_\parallel = 1.908$, $g_\perp = 1.991$	BM86

Zr^{4+} $4d^0$. Data tabulation of spin-Hamiltonian parameters ($S = \frac{1}{2}$, $I = \frac{1}{2}$).

Sample	T (°K)	$\tilde{g}$	$\tilde{A}^{103}$ (10^{-4} cm^{-1})	Ref.
$0.056Na_2O$–$0.195CaO$–$0.225B_2O_3$–$0.80Al_2O_3$–$0.444SiO_2$ (γ-irrad.)	77, 300	$g_\parallel = 2.010$, $g_\perp = 2.006$	$A_\parallel = 14.6$ G, $A_\perp = 13.5$ G	BM86

Y^{2+} $4d^1$. Data tabulation of spin-Hamiltonian parameters.

Sample	T (°K)	$\tilde{g}$	$\tilde{A}$	Ref.
ZrO_2	77–293	$g = 1.986$–1.993		BD74

Zr^{3+} $4d^1$. Data tabulation of spin-Hamiltonian parameters ($S = \frac{1}{2}$).

Sample	T (°K)	$\tilde{g}$	$\tilde{A}^{103}$ (10^{-4} cm^{-1})	Ref.
Y_2O_3 (optical and X-ray excitation)	77	$g_\parallel = 1.994$, $g_\perp = 1.907$	$A_\parallel = 29$, $A_\perp = 16$	BM86, BM83

Nb^{4+} $4d^1$. Data tabulation of spin-Hamiltonian parameters.

Sample	T (°K)	Site	$\tilde{g}$	$\tilde{A}^{47,49}$ (10^{-4} cm^{-1})	Ref.
TiO_2	4.2, 25		$g_c = 1.948$	$A_{110} = 8.04$, $A_{\bar{1}10} = 1.75$, $A_c = 30.5$	PF74
$LiNbO_3$ (X-irrad.)	20		$g_\parallel = 1.90$, $g_\perp = 1.72$	$A_\parallel = 110$, $A_\perp = 230$	BM83
$ZrSiO_4$ (X-irrad.)	77–300	Axial	$g_\parallel = 1.862$, $g_\perp = 1.908$	$A_\parallel = 268$, $A_\perp = 127$	BD80
		Monoclinic	$g_x = 1.920$, $g_y = 1.899$, $g_z = 1.872$	$A_x = 110.4$, $A_y = 134.4$, $A_z = 262.2$	

Mo^{5+} $4d^1$. Data tabulation of spin-Hamiltonian parameters (* indicates units of MHz).

Sample	T (°K)	Site	$\tilde{g}$	$\tilde{A}^{95,97}$ (10^{-4} cm^{-1})	Ref.
MgO			1.928		ART 72
GeO_2	293		$g(001) = 1.857$, $g(110) = 1.910$, $g(\bar{1}10) = 1.951$	$A(001) = 24$, $A(110) = 18$, $A(\bar{1}10) = 59$	BD74
SnO_2	77, 300		$g_1 = 1.891$, $g_2 = 1.835$, $g_3 = 1.923$	$A_1 = 24$ G, $A_2 = 30$ G, $A_3 = 70$ G	BD80
TiO_2	4.2, 77		$g_c = 1.788$	$A_{110} = 2.474$, $A_{\bar{1}10} = 65.85$	PF74
	77	Orthorhombic	$g_x = 1.834$, $g_y = 1.759$, $g_z = 1.842$	$A_x = 32.2$, $A_y = 36.8$, $A_z = 74.1$	BM83
TiO_2 (p)	77, 293		$g_1 = 1.919$, $g_2 = 1.822$, $g_z = 1.799$	$A_1 = 65.0$, $A_2 = 25.0$	BD74
MoO_3 (amorphous thin film)	2–4.2		$g_\parallel = 1.89$, $g_\perp = 1.938$		BM86
	77	Surface	$g_\parallel = 1.867$, $g_\perp = 1.958$	$A_\perp \approx 51$*	
$K_{0.33}MoO_3$	1.8–4, 4.2–300		$g_\parallel = 1.900$, $g_\perp = 1.925$, $D = 0.396$ cm^{-1}, $E = 0.004$ cm^{-1}		BD80
$MoO_3 \cdot Bi_2O_3$	77	Bulk	$g_\parallel = 1.890$, $g_\perp = 1.944$	$A_\perp \approx 60$*	BM83
	77	Surface	$g_\parallel = 1.867$, $g_\perp = 1.958$	$A_\perp \approx 51$*	
$MoO_3 \cdot 10Bi_2O_3$	77	Bulk	$g_\parallel = 1.890$, $g_\perp = 1.944$	$A_\perp \approx 60$*	BM83
	77	Surface	$g_\parallel = 1.867$, $g_\perp = 1.958$	$A_\perp \approx 51$*	
$2MoO_3 \cdot Bi_2O_3$	77	Bulk	$g_\parallel = 1.890$, $g_\perp = 1.944$	$A_\perp \approx 60$*	BM83
	77	Surface	$g_\parallel = 1.867$, $g_\perp = 1.958$	$A_\perp \approx 51$*	
$3MoO_3 \cdot Bi_2O_3$	77	Bulk	$g_\parallel = 1.890$, $g_\perp = 1.944$	$A_\perp \approx 60$*	BM83
MoO_8	77, 293	A	$g_x = 1.975$, $g_y = 1.975$, $g_z = 1.945$	$A_x = 21.0$*, $A_y = 7.2$*, $A_z = 8.3$*	BD74
	77, 293	B	$g_x = 1.943$, $g_y = 1.953$, $g_z = 1.875$		
NH_4Cl	77		$g_\parallel = 1.964$, $g_\perp = 1.945$	$A_\parallel = 75.53$, $A_\perp = 38.42$	BD80
Al_2O_3			1.929		ART 72
$SrTiO_3$	1.65		$g_\parallel = 1.72$, $g_\perp = 0$		BD74
$PbWO_4$ (X-irrad.)	4.2, 90		$g_\parallel = 1.893$, $g_\perp = 1.785$		BM83
	1.6–4.2		$g_\parallel = 1.898$, $g_\perp = 1.789$	$\|A_\parallel^{95,97}\| = 4.9$, $\|A_\perp^{95,97}\| = 36.1$	BM83
	85–300				
$MgAl_2O_4$			1.927		ART 72
B_i/Mo catalyst (partially reduced)	77, 900		$g_\parallel = 1.872$, $g_\perp = 1.935$		BM73
β-Nb2O5	NR		$g_x = 1.900$, $g_y = 1.914$, $g_z = 1.836$	$A_x = 30$, $A_y = 25.6$, $A_z = 56.0$	BM86
$[Mo_6O_{19}]$ $(NCl_6H_{36})_2$ (frozen solution, $Mo_5O_{19}^{3-}$ complex)	77		$g_\parallel = 1.919$, $g_\perp = 1.930$	$A_\parallel = 80.5$ G, $A_\perp = 34.5$ G	BM86
$TPA_2NbO(SCN)_5$	77		$g_1 = 1.944$, $g_2 = 1.939$, $g_3 = 1.929$	$A_1 = 94$*, $A_2 = 95$*, $A_3 = 26$*	BM86
	298		$g_1 = 1.944$, $g_2 = 1.939$, $g_3 = 1.929$		
$(Et_4N)_2Zr(SCN)_5$	77	a	$g_1 = 1.942$, $g_2 = 1.940$, $g_3 = 1.930$		BM86
	77	b	$g_1 = 1.942$, $g_2 = 1.940$, $g_3 = 1.930$		

Sample	T (°K)	$\tilde{g}$	Ref.
	298	$g_1 = 1.939,\ g_2 = 1.939,\ g_3 = 1.929$	
$TPA_2NbO(SCN)_5 \cdot n$TPASCN	77	$g_1 = 1.941,\ g_2 = 1.940,\ g_3 = 1.930$	BM86
$TPA_2MoO(SCN)_5$ (in acetone)	77	$g_1 = 1.943,\ g_2 = 1.941,\ g_3 = 1.929$	BM86
$(Et_4N)_2MoO(SCN)_5$ (in acetone)	77	$g_1 = 1.943,\ g_2 = 1.941,\ g_3 = 1.929$	BM86
$TPA_2MoO(SCN)_5$	77	$g_1 = 1.941,\ g_2 = 1.939,\ g_3 = 1.931$	BM86

Ru^{6+} $4d^2$. Data tabulation of spin-Hamiltonian parameters.

Sample	T (°K)	$\tilde{g}$	D (10^{-4} cm^{-1})	Ref.
Na_2RuO_4 (alkali)	20	2.00	≈ 100	LB

Nb^{2+} $4d^3$. Data tabulation of spin-Hamiltonian parameters.

Sample	T (°K)	$\tilde{g}$	$\tilde{A}^{47,49}$ (10^{-4} cm^{-1})	Ref.
ZnO	1.5–25	$g_\parallel = 1.959,\ g_\perp = 1.957$	$\lvert A\rvert = 33.75,\ \lvert B\rvert = 33.25$	BD74

Mo^{3+} $4d^3$. Data tabulation of spin-Hamiltonian parameters (* indicates units of GHz).

Sample	T (°K)	Site	$\tilde{g}$	b_2^0 (= D) (10^{-4} cm^{-1})	b_2^2 (= E) (10^{-4} cm^{-1})	$\tilde{A}^{95,97}$ (10^{-4} cm^{-1})	Ref.
$KMoF_4$ (p)	14		$g = 1.95$				LB
α-Al_2O_3	300			-81.9			BM86
	77			-82.4			
$(NH_4)_2[InCl_5(H_2O)]$	room		$g_\parallel = 1.963,\ g_\perp = 1.940$	$A = 74.7,\ B = 32.6$			LB
$K_2[MoCl_5(H_2O)]$ (p)	300–14		$g = 1.96$				LB
$K_3[InCl_6] \cdot 2H_2O$	90, 20	I	$g = 1.93$	$D \gg 10^4$	$E/D = 0.08$	$A_z = 39$	LB
		II	$g = 1.93$	$D \gg 10^4$	$E/D = 0.15$	$A_z = 39$	
$K_3[InCl_6] \cdot 2H_2O$	20	I, II	$g_\parallel = 1.95,\ g_\perp = 1.939$	$A = 77,\ B = 38.4$			LB
$K_3[MoCl_6]$ (p)	300–14		$g = 1.96$				LB
$Ag_3[Mo(CN)_8]$ (p)	room		1.993				LB
$K_3[Mo(CN)_8]$ (p)	room		1.993				LB
$Y_3Al_5O_{12}$	77, 290		$g_\parallel = g_\perp = 1.967$	26.0*	$\lvert A\rvert = 0.131$*, $\lvert B\rvert = 0.139$*		BM73

Tc^{4+} $4d^3$. Data tabulation of spin-Hamiltonian parameters.

Sample	T (°K)	$\tilde{g}$	D (GHz)	E (GHz)	$\tilde{A}^{99}$ (10^{-4} cm^{-1})	Ref.
SnO_2	4.2	$g_x = 1.972$, $g_y = 1.962$, $g_z = 1.975$	$\|D\| = 74$	$\|E\| = 6$	$A_x = 152$ G, $A_y = 164$ G, $A_z = 166$ G	BD80
$K_2[PtCl_6]$	4.2	1.990			$A_{max} = 138.4$, $A_{min} = 134.0$	LB

Ru^{4+} $4d^4$. Data tabulation of spin-Hamiltonian parameters.

Sample	T (°K)	$\tilde{g}$	Ref.
$K_2[PtCl_6]$	20	1.92	LB
$0.58P_2O_5–0.22Al_2O_3–0.18B_2O_3–0.024Na_2O$ (LP_5) glass	298	2.05 (triplet)	BD80
$0.70SiO_2–0.18Na_2O–0.01B_2O_3$($LP_6$) glass	298	1.98 (singlet)	BD80
$0.51SiO_2–0.26Na_2O–0.23B_2O_3$($LP_4$) glass	298	1.98 (singlet)	BD80

Ru^{3+} $4d^5$. Data tabulation of spin-Hamiltonian parameters.

Sample	T (°K)	Site	$\tilde{g}$	$\tilde{A}^{99,101}$ (10^{-4} cm^{-1})	Ref.
AgBr	4.3–293	A	$g_1 = 2.307$, $g_2 = 2.120$, $g_3 = 0.832$		BM86
		B	$g_1 = 2.492$, $g_2 = 2.492$, $g_3 = 0.3$		
		C	$g_1 = 1.701$, $g_2 = 1.701$, $g_3 = 0.050$		
		X	$g_1 = 1.486$, $g_2 = 4.86$		
AgCl	4.3–293	A′	$g_1 = 2.343$, $g_2 = 2.220$, $g_3 = 0.910$		BM86
		B′	$g_1 = 2.558$, $g_2 = 2.486$, $g_3 = 0.29$		
		C′	$g_1 = 1.767$, $g_2 = 1.767$, $g_3 = 2.074$		
		X′	$g_1 = 1.532$, $g_2 = 1.553$, $g_3 = 2.516$		
CaO	4.2		$g_{iso} = 1.831$	$A^{99} = 44.4$, $A^{101} = 49.7$	BD80
MgO			1.829	$A^{99} = 44.7$, $A^{101} = 49.8$	BM73
Al_2O_3	20		$g_\parallel < 0.06$, $g_\perp = 2.430$ $D = 0.103$ cm^{-1}, $b_4^0 = 0.0026$ cm^{-1}, $b_6^0 = 0.0001$ cm^{-1}		PF74
$K_3[InCl_6]\cdot 2H_2O$	20		$g_x = 1.00$, $g_y = 1.22$, $g_z = 3.24$		LB
$[Co(NH_3)_6]Cl_3$	20	I	$g_x = 2.06$, $g_y = 2.02$, $g_z = 1.72$	$A_x = 48$, $A_y = 48$, $A_z = 49$	LB
		II	$g_x = 1.80$, $g_y = 1.90$, $g_z = 2.06$	$A_x = 48$, $A_y = 48$, $A_z = 50$	
		III	$g_x = 1.15$, $g_y = 1.84$, $g_z = 2.66$	$A_x = 45$, $A_y = 41$, $A_z = 54$	
$[Ru(NH_3)_6]Cl_3\cdot 3HgCl_2$ (sc)	20		$g_x = 2.21$, $g_y = 2.05$, $g_z = 1.5$		LB
$0.577P_2O_5$–$0.22Al_2O_3$–$0.176B_2O_3$–$0.0236Na_2O$ (LP_5) glass	298		4.30		BD80
$0.76SiO_2$–$0.19Na_2O$–$0.005B_2O_3$(LP_8) glass	298		4.30		BD80
$0.80SiO_2$–$0.20Na_2O$ (NS) glass	298		4.30		BD80

Rh^{4+} $4d^5$. Data tabulation of spin-Hamiltonian parameters ($S = \frac{1}{2}$, $I = \frac{1}{2}$).

Sample	T (°K)	$\tilde{g}$	$\tilde{A}^{103}$ (10^{-4} cm^{-1})	Ref.
MgO (γ-irrad.)	20	1.668	32.5	ART 72

Rh^{3+} $4d^6$. Data tabulation of spin-Hamiltonian parameters.

Sample	T (°K)	$\tilde{g}$	$\tilde{A}^{35}$ (10^{-4} cm^{-1})	Ref.
NaCl (γ-irrad.)	77	$g_\parallel = 2.453$, $g_\perp = 2.019$	$A_\parallel^{35} = 24.8$, $A_\perp^{35} = 12.0$	BD80
$Na_3RhCl_6 \cdot 12H_2O$	77	$g_\perp = 2.479$		BD80

Ru^+ $4d^7$. Data tabulation of spin-Hamiltonian parameters ($S = \frac{1}{2}$, $I = \frac{1}{2}$).

Sample	T (°K)	$\tilde{g}$	Ref.
MgO	77	2.170	PF74

Rh^{2+} $4d^7$. Data tabulation of spin-Hamiltonian parameters ($S = \frac{1}{2}$, $I = \frac{1}{2}$).

Sample	T (°K)	$\tilde{g}$	$\tilde{A}^{103}$ (10^{-4} cm^{-1})	Ref.
AgCl		$g_\parallel = 2.011$, $g_\perp = 2.422$		BD80
CaO	93	2.287		BD80
	4.2	$g_\parallel = 2.039$, $g_\perp = 2.405$	$A_\parallel < A_\perp$, $A_\perp = 4.3$	
MgO (γ-irrad.)	20	2.172	12.3	ART 72
LiD-{A}	210	$g_x = 2.038$, $g_y = 2.038$, $g_z = 2.112$	$A_x = 14.5$, $A_y = 14.5$, $A_z = 26.2$	BM86
LiD-{O}	77	$g_x = 2.075$, $g_y = 2.060$, $g_z = 2.005$	$A_x = 15.5$, $A_y = 12.5$, $A_z = 11.7$	BM86
LiH-{A}	13.3			BM86
LiH-{O}	≤ 4.0			BM86

Pd^{2+} $4d^8$. Data tabulation of spin-Hamiltonian parameters ($S = \frac{1}{2}$, $I = \frac{1}{2}$).

Sample	T (°K)	$\tilde{g}$	Ref.
KI	300	2.0	BM73

Rh^0 $4d^9$. Data tabulation of spin-Hamiltonian parameters ($S = \frac{1}{2}$, $I = \frac{1}{2}$).

Sample	T(°K)	$\tilde{g}$	Ref.
MgO	77	2.171	PF74

Pd^+ $4d^9$. Data tabulation of spin-Hamiltonian parameters ($S = \frac{1}{2}$, $I = \frac{1}{2}$).

Sample	T (°K)	Site	$\tilde{g}$	$\tilde{A}^{103}$ (10^{-4} cm^{-1})	Ref.
AgCl	77		$g_\parallel = 2.648$, $g_\perp = 2.160$	$A_\parallel = 16$, $A_\perp = 8$	BM86
MgO	77		2.170		PF74
NH_4Cl (γ-irrad.)	300	I	$g_\parallel = 2.646$, $g_\perp = 2.100$	$A = 21$, $B = 6.0$	BD80
K_2PdCl_4 (γ-irrad.)	77		$g_\parallel = 2.515$, $g_\perp = 2.083$	$\lvert A\rvert = 21.7$, $\lvert B\rvert = 7.8$	BM73
$(NH_4)_2PdCl_4$ (γ-irrad.)	77		$g_\parallel = 2.527$, $g_\perp = 2.086$	$\lvert A\rvert = 20.6$, $\lvert B\rvert = 7.8$	BM73

Ag^{2+} $4d^9$. Data tabulation of spin-Hamiltonian parameters ($S = \frac{1}{2}$, $I = \frac{1}{2}$) (* indicates MHz).

Sample	T (°K)	Site	$\tilde{g}$	$\tilde{A}$ (10^{-4} cm^{-1})	Ref.
AgCl	30		$g_\parallel = 2.165$, $g_\perp = 2.041$	$A_\parallel = 35.4$, $A_\perp = 21.9$	BM83
AgCl (irrad.)	4.2	Tetragonal	$g_\parallel = 2.147$, $g_\perp = 2.040$	$\|A\| = 29$ G, $\|B\| = 26$ G	BM73
$AgBr_{0.001}Cl_{0.999}$ (blue light-irrad.)	4.2–50		$g_x = 2.024$, $g_y = 2.05$, $g_z = 2.133$	$A_x = 24$ G $A_y = 22$ G $A_z = 30$ G	BM83
$AgBr_{0.15}Cl_{0.85}$	60–77		$g_\parallel = 2.078$		BM86
$AgBr_{0.3}Cl_{0.7}$	35–77		$g_\perp = 2.065$		BM86
$CdBr_2$	77		$g_\parallel = 2.072$, $g_\perp = 2.067$	$A = 30.4$	BM86
$CdCl_2$	290		$g_w = 2.095$, $g_u = 2.079$	$\|A\| = 37.5$, $\|B\| \sim 29.5$	BM73
	77		$g_1 = 2.120$, $g_2 = 2.037$, $g_3 = 2.096$		
	77	I	$g_x = g_y = 2.040$, $g_z = 2.174$		BM86
	77	II	$g_x = g_y = 2.116$, $g_z = 2.042$		BM86
	77	III	$g_x = g_y = 2.096$, $g_z = 2.096$		BM86
α-SiO_2			$g_x = 1.992$, $g_y = 1.998$, $g_z = 1.999$	$A_x^{107} = 1098.5$* $A_y^{107} = 1100.5$* $A_z^{107} = 1144.3$* $A_x^{109} = 1266.2$* $A_y^{109} = 1270.8$* $A_z^{109} = 1322.6$*	BM83
$SrCl_2$	78		$g_\parallel = 2.149$, $g_\perp = 2.030$	$A_\parallel^{109} = 41$, $A_\perp^{109} = 31$	BD74
$SrCl_2$ (X-irrad.)	77	Axial	$g_\parallel = 2.150$, $g_\perp = 2.030$	$\|A_\parallel\| = 39$, $\|A_\perp\| = 29$	BD80
$AgClO_4$ (frozen solution in methylcynide) (γ-irrad.)	77		$g = 2.314$	−40 G	BM86
$(AgBr_2Cl_4)^{4-}$	4.2–60		$g_\parallel = 2.110$, $g_\perp = 2.044$	$A_\parallel = 29$, $A_\perp = 19.5$	BM86
$BaO \cdot Al_2O_3 \cdot B_2O_3$ glasses (X-irrad.)					BM86
0%Al_2O_3 · 70%B_2O_3	77		$g_\parallel = 2.327$, $g_\perp = 2.108$		
5%Al_2O_3 · 65%B_2O_3	77		$g_\parallel = 2.308$, $g_\perp = 2.079$		
20%Al_2O_3 · 50%B_2O_3	77		$g_\parallel = 2.273$, $g_\perp = 2.064$		
Naβ-alumina (X-irrad.)	77–300	I	$g_x = 2.023$, $g_y = 2.068$, $g_z = 2.273$		BM83
		II	$g_x = 2.015$, $g_y = 2.058$, $g_z = 2.234$		

Ag^+ $4d^{10}$. Data tabulation of spin-Hamiltonian parameters.

Sample	T (°K)	Site	$\tilde{g}$	D (10^{-4} cm^{-1})	$\tilde{A}^{107,109}$ (GHz)	Ref.
CdS	77	Orthorhombic	$g_0 = 2.018$	$D = 0.52, E/D = 0.17$		BD74
K_2SO_4 (X-irrad.)	77		2.001		$A^{107} = 1.85, A^{109} = 2.13$	BM73
	300		2.001		$A^{107} = 1.79, A^{109} = 2.07$	

Cd^+ $4d^{10}$ $5s^1$. Data tabulation of spin-Hamiltonian parameters (* indicates units of GHz; $S = I = \frac{1}{2}$).

Sample	T (°K)	$\tilde{g}$	$\tilde{A}^{111,113}$ (10^{-4} cm^{-1})	Ref.
K_2SO_4 (X-irrad.)	77	1.999	$A^{111} = 13.501$*, $A^{113} = 14.115$*	BM73
	77	$g_x = 1.996, g_y = 1.998,$ $g_z = 2.060$	$A_x^{111} = 12.618$*, $A_y^{111} = 12.645$*, $A_z^{111} = 12.668$* $A_x^{113} = 13.190$*, $A_y^{113} = 13.220$*, $A_z^{113} = 13.255$*	
$K_2[Cd(CN)_4]$ (γ-irrad.)	77	$g_\parallel = 2.006, g_\perp = 1.989$	$A_\parallel^{113} = 1596$ G, $A_\perp^{113} = 1468$ G	BM86
		$g_\parallel = 1.997, g_\perp = 1.989$	$A_\parallel^{111} = 1525$ G, $A_\perp^{111} = 1403$ G	
	291	$g_\parallel = 1.999, g_\perp = 1.988$	$A_\parallel^{113} = 1653$ G, $A_\perp^{113} = 1528$ G	BM86
		$g_\parallel = 1.999, g_\perp = 1.988$	$A_\parallel^{111} = 1579$ G, $A_\perp^{111} = 1500$ G	
	4.2	$g_x = 1.986, g_y = 1.990,$ $g_z = 1.991$	$A_x^{113} = 1594$ G, $A_y^{113} = 1614$ G, $A_z^{113} = 1706$ G	BM86
		$g_x = 1.987, g_y = 1.990,$ $g_z = 1.991$	$A_x^{111} = 1521$ G, $A_y^{111} = 1546$ G, $A_z^{111} = 1628$ G	
$K_2[Zn(CN)_4]$ (γ-irrad.)	77	$g_x = 1.987, g_y = 1.991,$ $g_z = 1.991$	$A_x^{111} = 1701$ G, $A_y^{111} = 1732$ G, $A_z^{111} = 1816$ G	BM86
		$g_x = 1.987, g_y = 1.990,$ $g_z = 1.991$	$A_x^{113} = 1626$ G, $A_y^{113} = 1658$ G, $A_z^{113} = 1737$ G	
	183	$g_y = 1.997$		
		$g_y = 1.996$		

Sn^{3+} $4d^{10}$ $5s^1$. Data tabulation of spin-Hamiltonian parameters ($S = \frac{1}{2}, I = \frac{1}{2}$).

Sample	T (°K)	$\tilde{g}$	$\tilde{A}^{117,119}$ (10^{-4} cm^{-1})	Ref.
ZnO	14–26	$g_\parallel = 1.988, g_\perp = 1.987$	$\lvert A_\parallel^{117}\rvert = 3181, \lvert A_\perp^{117}\rvert = 3{,}179.2$; $\lvert A_\parallel^{119}\rvert = 3327.1, \lvert A_\perp^{119}\rvert = 3{,}179.2$	BM73

Ag^0 $4d^{10}$ $5s^1$. Data tabulation of spin-Hamiltonian parameters ($S = \frac{1}{2}, I = \frac{1}{2}$).

Sample	T (°K)	$\tilde{g}$	$\tilde{A}^{107}$ (G)	Ref.
KI type glasses (γ-irrad.)	77	2.001	$A^{107} = 1743$ MHz, $A^{109} = 1992$ MHz	BM83
α-SiO_2 (X-irrad.)	20	$g_{11} = 2.000, g_{22} = 1.994, g_{33} = 1.997$	$A_{11}^{107} = -408.7, A_{22}^{107} = 392.6$ G, $A_{33}^{107} = 393.0$	BM83
xKCl·yKBr (X-irrad.) ($x = 99.7, 996.7, 93.7, 87.7\%$; $y = 0.0, 3, 6, 12\%$)	285	1.996	$A^{109} = -656$	BM83
Water glasses (Ag^+-irrad.)	4.2–77	2.002	$A^{109} = -716$	BM83
$AgClO_4$ (frozen solution in methylcynide) (γ-irrad.)	77	$g = 1.997$	$A = -538$	BM86

Ta^{4+} $5d^1$. Data tabulation of spin-Hamiltonian parameters.

Sample	T (°K)	$\tilde{g}$	$\tilde{A}^{183}$ (10^{-4} cm^{-1})	Ref.
TiO_2	1.4, 10	$g_c = 1.945$	$A_{110} < 2.5, A_{\bar{1}10} < 2.5, A_c = 63.9$	PF74

W^{5+} $5d^1$. Data tabulation of spin-Hamiltonian parameters.

Sample	T (°K)	$\tilde{g}$	$\tilde{A}^{183}$ (10^{-4} cm^{-1})	Ref.
TiO_2	63	$g_c = 1.446$	$A_{110} = 92.0, A_{\bar{1}10} = 40.5$	PF74
Cs_2HfCl_6	4.2, 20	$g_\parallel = 1.809, g_\perp = 1.765$	$A^{183} = 124, B^{183} = 65$	BM73
Cs_2ZrCl_6	4.2, 20	$g_\parallel = 1.811, g_\perp = 1.767$	$A^{183} = 124.7, B^{183} = 63.1$	BM73
$Ag_3[W(CN)_8]$ (p)	Room	1.976		LB
$K_3[W(CN)_8] \cdot 0.55H_2O$ (p)	Room	1.98		LB

Re^{6+} $5d^1$. Data tabulation of spin-Hamiltonian parameters.

Sample	T (°K)	$\tilde{g}$	$\tilde{A}^{185,187}$ (10^{-4} cm^{-1})	Ref.
$BaReO_4$ (p)	20	$g_\parallel = 2.00, g_\perp = 1.95$		LB
$CaWO_4$ (n-irrad.)	77	$g_\parallel = 1.849, g_\perp = 1.716$	$A_\parallel^{187} = \pm 30, A_\perp^{187} = \pm 331$	BD80
	4.2	$g_\parallel = 1.855, g_\perp = 1.855$	$A_\parallel^{185} = \pm 42, A_\perp^{185} = \pm 321$ $A_\parallel^{187} = \pm 42, A_\perp^{187} = \pm 324$	BD80
$Ph_4As[RuNCl_4]$	300	$g_\parallel = 1.910, g_\perp = 1.943$	$A_\parallel = 770, A_\perp = 392$	BM83
	110	$g_\parallel = 1.918, g_\perp = 1.947$	$A_\parallel = 770, A_\perp = 391$	BM83

Re^{5+} $5d^2$. Data tabulation of spin-Hamiltonian parameters.

Sample	T (°K)	$\tilde{g}$	D (10^{-4} cm^{-1})	Ref.
K_3ReO_4 (alkali glass)	20	2.08	1200	LB

Re^{4+} $5d^3$. Data tabulation of spin-Hamiltonian parameters.

Sample	T (°K)	$\tilde{g}$	|D| (GHz)	|E| (GHz)	$\tilde{A}^{185,187}$ (10^{-4} cm^{-1})	Ref.
SnO_2	4.2	$g_x = 1.626, g_y = 1.593, g_z = 1.81$	53.2	9.5	$A_x = 730$ G, $A_y = 493$ G, $A_z = 427$ G	BD80
$K_2[PtCl_6]$	4	1.815			398	LB

Ir^{4+} $5d^5$. Data tabulation of spin-Hamiltonian parameters.

Sample	T (°K)	Site	$\tilde{g}$	$\tilde{A}^{191,193}$ (10^{-4} cm^{-1})	Ref.
MgO (γ-irrad.)	77		1.739	$A^{191} = 25.0, A^{193} = 26.7$	ART 72
$K_2[PtBr_6]$	20		$g_\parallel = 1.60, g_\perp = 1.87$		LB
$K_2[PtCl_6]$	20		1.785		LB
$Na_2[PtBr_6] \cdot 6H_2O$	20		$g_x = 2.25, g_y = 2.21, g_z = 0.75$	$A_x = 25.5, A_y = 25.5$	LB
$[IrCl_4]^{4-}$	77	A	$g_\parallel = 1.904, g_\perp = 2.813$		BM86
$[IrCl_6]^{4-}$	77	B	$g_\parallel = 1.867, g_\perp = 2.866$		BM86
$Na_2[PtCl_6] \cdot 6H_2O$	20		$g_x = 2.20, g_y = 2.07, g_z = 1.05$	$A_x = 25.5, A_y = 25.5, A_z = 24$	LB
$(NH_4)_2[IrCl_6]$	20		1.786	25.3	LB
$(NH_4)_2[PtCl_6]$	90–1.5				LB
$Y_3Ca_5O_{12}$	4.2		$g_\parallel = 2.412, g_\perp = 1.442$		BD80

Os^+ $5d^7$. Data tabulation of spin-Hamiltonian parameters.

Sample	T (°K)	Site	$\tilde{g}$	$\tilde{A}^{187,189}$ (10^{-4} cm^{-1})	Ref.
KCl	77–300	A	$g_x = 2.158, g_y = 2.152, g_z = 2.975$	$A_\parallel = 161$	BM83
	77–300	B	$g_x = 2.163, g_y = 2.156, g_z = 1.975$	$A_\parallel = 161, A_\perp = 21.2$	BM83
	77–300	C	$g_x = 2.153, g_y = 2.153, g_z = 1.975$		BM83
	77–300	D	$g_x = 2.158, g_y = 2.158, g_z = 1.975$		BM83
	77–300	E	$g_x = 2.146, g_y = 2.146, g_z = 1.975$		BM83
	77–300	F	$g_x = 2.140, g_y = 2.140, g_z = 1.975$		BM83
	77–300	G	$g_x = 2.152, g_y = 2.152, g_z = 1.975$		BM83

Ir^{2+} $5d^7$. Data tabulation of spin-Hamiltonian parameters.

Sample	T (°K)	Site	$\tilde{g}$	$\tilde{A}^{191,193}$ (10^{-4} cm^{-1})	Ref.
KCl	77	A	$g_x = 2.211, g_y = 2.203, g_z = 1.967$		BD74
	77	B	$g_{iso} \sim 2.006$		

Pt^{3+} $5d^7$. Data tabulation of spin-Hamiltonian parameters.

Sample	T (°K)	$\tilde{g}$	$\tilde{A}^{195}$ (10^{-4} cm^{-1})	Ref.
MgO (X-, γ-irrad.)	4.2–77	$g_\parallel = 1.969, g_\perp = 2.344$	$A_\parallel = 84.1, A_\perp = 26.0$	BM86
$BaTiO_3$	4.2	$g_\parallel = 1.954, g_\perp = 2.452$	$A_\parallel = 0$ cm^{-1}, $A_\perp = \pm 0.0119$ cm^{-1}	BD80
Al_2O_3	79	$g_\perp = 2.200$		PF74
	4.2	$g_\parallel = 2.011, g_\perp = 2.328$		

Pt^{2+} $5d^8$. Data tabulation of spin-Hamiltonian parameters.

Sample	T (°K)	Site	$\tilde{g}$	$\tilde{A}$	Ref.
AgCl	30	$(PtCl_6)^{5-}$	$g_\parallel = 2.815, g_\perp = 2.1$	$A_\parallel = 390.4$	BM83
	30	$(PtCl_6)^{3+} \cdot Ag^+$	$g_\parallel =$ —, $g_\perp = 2.369$	$A_\perp = 288$	
	30	$(PtCl_6)^{3-} \cdot 2Ag^+$	$g_\parallel = 1.949, g_\perp = 2.417$	$A_\parallel = 124.2, A_\perp = 345.0$	
	30	$(PtCl_6)^{3-} \cdot 3Ag^+$	$g_\parallel =$ —, $g_\perp = 2.36$	$A_\perp = 283$	

Pb^{3+} $5d^9$. Data tabulation of spin-Hamiltonian parameters ($S = I = \frac{1}{2}$).

Sample	T (°K)	$\tilde{g}$	$\tilde{A}^{207}$ (10^{-4} cm^{-1})	Ref.
CaO	1.6–300	2.013	$A^{207} = 1.07$	BM73
ZnO	1.6–300	1.999	$A^{207} = 0.80$	BM73
$CaCO_3$ (X-, γ-irrad.)	4.2, 77, 300	$g_\parallel = 2.005, g_\perp = 2.007$	$A_\parallel^{204} = 37.9$ MHz, $A_\perp^{204} = 48.4$ MHz	BM80
$CaWO_4$ (γ-irrad.)	77	2.002	$A^{207} = 1.276$	BM73
$YAl_3(BO_3)_4$	77	$g_\parallel = 2.008, g_\perp = 2.010$	$A_\parallel^{207} = 1.322, A_\perp^{207} = 1.332$	BM86

Tl^{2+} $5d^9$. Data tabulation of spin-Hamiltonian parameters ($S = I = \frac{1}{2}$).

Sample	T (°K)	$\tilde{g}$	$\tilde{A}^{203,205}$ (GHz)	Ref.
CdTe	4.2, 77	-2.035	$\lvert A_{\parallel}^{205}\rvert = 375$ G, $\lvert A_{\perp}^{203}\rvert = 180$ G	BD80
		$a^{203} = 53.3$ GHz, $a^{205} = 53.7$ GHz	$\lvert A_{\parallel}^{207}\rvert = 375$ G, $\lvert A_{\perp}^{207}\rvert = 180$ G	
K_2SO_4 (X-irrad.)	110–300	$g_x = 1.997, g_y = 1.995, g_z = 1.998$	$A_x = 115.35, A_y = 115.42, A_z = 115.72$	BD74
KD_2PO_4 (X-irrad.)	290	$g_{\parallel} = 1.987, g_{\perp} = 1.994$	$A_{\parallel} = 115.4, A_{\perp} = 115.9$	BM86
	77	$g_x = 1.998, g_y = 1.992, g_z = 1.987$	$A_x = 116.1, A_y = 115.5, A_z = 115.0$	
KH_2PO_4 (X-irrad.)	290	$g_{\parallel} = 1.989, g_{\perp} = 1.994$	$A_{\parallel} = 115.7, A_{\perp} = 115.7$	BM86
	77	$g_x = 1.997, g_y = 1.992, g_z = 1.988$	$A_x = 116.7, A_y = 116.2, A_z = 115.7$	

Pb^{+} $5d^{10}$ $6s^1$. Data tabulation of spin-Hamiltonian parameters ($S = I = \frac{1}{2}$).

Sample	T (°K)	$\tilde{g}$	$\tilde{A}^{207}$ (G)	Ref.
$Pb(C_2H_3O_2)_2 \cdot 3H_2O$ (X-irrad.)	77	$g_x = 1.999, g_y = 2.003, g_z = 2.004$	$A_x^{207} = 2157, A_y^{207} = 3491, A_z^{207} = 3752$	BD74

Ce^{4+} $4f^0$. Data tabulation of spin-Hamiltonian parameters.

Sample	T (°K)	Site	$\tilde{g}$	$\tilde{A}^{141}$ (10^{-4} cm^{-1})	Ref.
CeO_2	4.2		$g_x = 2.031,\ g_y = 2.007,\ g_z = 2.033$		BD74
	77		$g_x = 2.018,\ g_y = 2.005,\ g_z = 2.032$		
	50		$g_x = 2.027,\ g_y = 2.005,\ g_z = 2.042$	$A_x = 3.1,\ A_y = 2.3,\ A_z = 2.7$	

La^{2+} $4f^1$. Data tabulation of spin-Hamiltonian parameters.

Sample	T (°K)	Site	$\tilde{g}$	$\tilde{A}^{191,193}$ (10^{-4} cm^{-1})	Ref.
$SrCl_2$ (X-irrad.)	1.1–4.2		$g_1 = 1.881,\ qg_2 = -0.068$	$A_1^{139} = -119.5,\ qA_2^{139} = -18.8,\ A_1^{138} = -110$	BD74

Ce^{3+} $4f^1$. Data tabulation of spin-Hamiltonian parameters.

Sample	T (°K)	Site	$\tilde{g}$	D (10^{-4} cm^{-1})	Ref.
BaO (pc)	<77 K		$\|g\| = 0.934$		REY72
BaO	4.2–100		0.934		BD74
CaO (pc)	<77 K		$\|g\| = 0.796$		REY72
CaO	4.2–100		0.796		BD74
SrO (pc)	<77 K		$\|g\| = 0.895$		REY72
SrO	4.2–100		0.895		BD74
BaF_2	1.8		$g_\parallel = 2.593$, $g_\perp = 1.548$		BD74
CaF_2	4.2	Trigonal	$g_\parallel = 3.3$, $g_\perp = 0.37$		BM73
	1.2–2.18		$g_\parallel = 3.07$, $g_\perp = 1.39$		BM86
	1.8		$g_\parallel = 3.038$, $g_\perp = 1.396$		BD74
	4.2		$g_1 = 1.9$, $g_2 = -1.3$		BD80
SrF_2	NR	Tetragonal	$g_\parallel = 2.868$, $g_\perp = 1.466$		BM86
		Rhombic	$g_1 = 2.722$, $g_2 = 1.739$, $g_3 = 1.294$		BM86
	1.8		$g_\parallel = 2.922$, $g_\perp = 1.465$		BD74
TiO_2	4.2		$g_c = 3.86$	143.2 $b_4^0 = 20.5$ $b_4^2 = -2.4$	PF74
$CeCl_3$	NR		$g_\parallel = 4.037$, $g_\perp = 0.17$		PF74
LaF_3	20		$g_x = 0.32$, $g_y = 0.90$, $g_z = 2.61$		SJ66
PrF_3	1.4–4.2		$g_x = 0.39$, $g_y = 0.946$, $g_z = 2.69$		BM73
$YCl_3 \cdot 6H_2O$			$g_x = 2.67$, $g_y = 2.305$, $g_z = 0.782$		SJ67
$SrTiO_3$	4.2		$g_\parallel = 3.005$, $g_\perp = 1.118$		PF74
$YLiF_3$	1.5–40		$g_\parallel = 2.737$, $g_\perp = 1.475$		BM73
$SrWO_4$	1.5–4.2		$g_\parallel = 2.87$, $g_\perp = 1.45$		BD74
$LaNbO_4$	1.5–45	Monoclinic	$g_x = 0.53$, $g_y = 3.19$, $g_z = 1.52$		BM73
$La_2(SO_4)_3 \cdot 9H_2O$			$g_\parallel = 3.81$, $g_\perp = 0.35$		ROG71
$La_2(SO_4)_3 \cdot 9H_2O$	1.4–40	Upper line	$g_\parallel = 3.95$, $g_\perp = 0.31$		BM73
		Lower line	$g_\parallel = 3.665$, $g_\perp = 0.38$		
$Ca_5(PO_4)_3F$	4.2		$g_\parallel = 3.467$, $g_\perp < 0.6$		BD74
Cs_2NaYCl_6	1.4–4.2		$\|g\| = 1.264$		BM83
Ce ethyl sulfate			$g_\parallel = 0.955$, $g_\perp = 2.185$		PF74
			$g_\parallel = 3.725$, $g_\perp = 0.2$		
$Na_3[Ce(C_7H_3NO_4)_3] \cdot 15H_2O$	4.2	1	$g_x = 1.013$, $g_y = 0.727$, $g_z = 1.773$		BD74
		2	$g_x = 1.026$, $g_y = 0.717$, $g_z = 1.765$		
		3	$g_x = 1.065$, $g_y = 0.683$, $g_z = 1.777$		
		4	$g_x = 1.080$, $g_y = 0.669$, $g_z = 1.767$		

Pr^{3+} $4f^2$. Data tabulation of spin-Hamiltonian parameters.

Sample	T (°K)	$\tilde{g}$	$\tilde{A}^{141}$ (10^{-4} cm^{-1})	Δ (10^{-4} cm^{-1})	Ref.
CaF_2	4.2	1.936	$A = 836$		BM86
$PrCl_3$	NR	$g_\parallel = 1.035$, $g_\perp = 0.1$	$A^{141} = 502.0$	0.02	PF74
Pr-ethyl sulfate	NR	$g_\parallel = 1.525$			PF74

Nd^{3+} $4f^3$. Data tabulation of spin-Hamiltonian parameters (* indicates units of MHz).

Sample	T (°K)	Site	$\tilde{g}$	$\tilde{A}^{143}$ (10^{-4} cm^{-1})	Ref.
CaO	20			$B_4 = -195$, $B_6 = -131$	BM73
CaF_2	4.2		$g_1 = 2.8$, if $g_2 = 0.68$ or $g_1 = 2.1$, if $g_2 = -0.68$		BD80
$CeIr_2$	1.4–35		2.67		BM86
$CeIr_2$ (0.6% Nd^{3+})	1.5		2.57	131 G	BM86
$CeRh_2$	1.4–35		2.60		BM86
$CeRh_2$ (x% Nd^{3+})	1.5				BM86
($x = 0.2$)			2.60	131 G	
($x = 0.4$)			2.60	131 G	
($x = 0.6$)			2.59	131 G	
($x = 1.1$)			2.61	131 G	
$CeRu_2$ (type II superconducting)	1.4–35		2.70	127 G	BM86
$LaIr_2$	1.4–35		2.633	131 G	BM86
$LaIr_2$ (x% Nd^{3+})	1.5				BM86
($x = 0.01$)			2.633	131 G	
($x = 0.05$)			2.633	131 G	
($x = 0.1$)			2.637	131 G	
($x = 1.0$)			2.638	131 G	
($x = 2.0$)			2.643	131 G	
$LaPt_2$	1.5		2.630	131 G	BM86
$LaPt_2$	1.4–35		2.630		BM86
$LaRh_2$	1.4–35		2.640	131 G	BM86

$LaRh_2$ (x% Nd^{3+})	1.5			BM86
($x = 0.01$)		2.639	131 G	
($x = 0.1$)		2.636	131 G	
($x = 0.3$)		2.643	131 G	
($x = 0.5$)		2.636	131 G	
($x = 1.0$)		2.646	131 G	
($x = 2.0$)		2.639	131 G	
$LaRu_2$ (type II superconducting)	1.4–35	2.85		BM86
$LuAl_2$	1.5–20	2.4		BM86
$LuIr_2$ (p)	1.4–4.1	2.62	$A^{143} = 205$ G, $A^{145} = 127$ G	BM83
$LuRh_2$ (p)	1.4–4.1	2.62	$A^{143} = 205$ G, $A^{145} = 127$ G	BM83
$ScAl_2$	1.5–20	2.5		BM86
$ThRu_2$ (type II superconducting)	1.4–35	2.684	127 G	BM86
YAl_2	1.5–20	2.74		BM86
YIr_2	1.4–35	2.640	131 G	BM86
YIr_2 (x% Nd^{3+})	1.5			BM86
($x = 0.01$)		2.630	131 G	
($x = 0.05$)		2.630	131 G	
($x = 0.1$)		2.645	131 G	
($x = 0.3$)		2.640	131 G	
($x = 1.0$)		2.646	131 G	
YPt_2	1.4–35	2.62		BM86
YPt_2 (x% Nd^{3+})	1.5			BM86
($x = 0.1$)		2.025	131 G	
($x = 0.5$)		2.019	131 G	
($x = 1.0$)		2.022	131 G	
YRh_2	1.4–35	2.640	132 G	BM86
YRh_2 (x% Nd^{3+})	1.5			BM86
($x = 0.02$)		2.640	131 G	
($x = 0.05$)		2.640	131 G	
($x = 0.1$)		2.640	131 G	
($x = 1.0$)		2.644	131 G	
$YbAl_2$	1.5–20	2.63		BM86
$Nd_{16.7}Ag_{83.3}$ (amorphous alloys)	>100	2.3		BM86

Nd^{3+} $4f^3$ *(continued)*

Sample	T (°K)	Site	$\tilde{g}$	$\tilde{A}^{143}$ (10^{-4} cm^{-1})	Ref.
$LaBr_3$	4.2		$g_\parallel = 3.853$, $g_\perp = 1.903$	$A^{143} = 403$, $B^{143} = 186$	JW66
				$A^{145} = 251$, $B^{145} = 114$	
LaF_3	4.2		$g_x = 1.356$, $g_y = 1.092$, $g_z = 3.11$		SJ66
LaF_3	4.2			$A = -1.5$ MHz	BM86
$NdCl_3$	NR		$g_\parallel = 3.996$, $g_\perp = 1.763$	$A^{143} = 425$, $B^{143} = 167$	PF74
				$A^{145} = 264$, $B^{145} = 104$	
PrF_3	1.4–4.2		$g_x = 1.500$, $g_y = 1.094$, $g_z = 2.933$		BM73
$YCl_3(Nd^{143})$	1.5		$g_{11} = 1.173$, $g_{22} = 1.652$, $g_{33} = 2.636$	$A_{11}^{143} = 420.1^*$, $A_{22}^{143} = 571.4^*$, $A_{33}^{143} = 816.9^*$	BM86
$YCl_3(Nd^{145})$	1.5		$g_{11} = 1.163$, $g_{22} = 1.654$, $g_{33} = 2.633$	$A_{11}^{145} = 260.3^*$, $A_{22}^{145} = 355.8^*$, $A_{33}^{145} = 507.8^*$	
$YCl_3 \cdot 6H_2O$			$g_x = 0.26$, $g_y = 0.33$, $g_z = 3.50$	$A_z^{143} = 356$, $A_z^{145} = 222$	SJ67
EuGaG	4.2		$g_x = 3.820$, $g_y = 1.965$, $g_z = 1.190$		BD80
$SrTiO_3$	4.2		$g_\parallel = 2.609$, $g_\perp = 2.472$		PF74
$CaWO_4$	4.2		$g_\parallel = 2.035$, $g_\perp = 2.537$		BM86
$CaWO_4$	1.5–4.2			$A = 638.7$, $B = 777.9$	BD74
$SrWO_4$	1.5–4.2		$g_\parallel = 1.53$, $g_\perp = 2.56$	$A = 442$, $B = 798$	BD74
$LiYF_4$	4.2		$g_\parallel = 1.987$, $g_\perp = 2.554$	$A^{143} = 197$, $B^{143} = 263$	BM73
				$A^{145} = 124$, $B^{145} = 164$	
YPO_4	1.7–4.2		$g_\parallel = 1.222$, $g_\perp = 3.080$		BM73
XO, Al_2O_3 (X = Ca)	4.2	I	$g_x = 1.072$, $g_y = 0.489$, $g_z = 5.468$		BD74
		IIA	$g_x = 0.496$, $g_y = 0.484$, $g_z = 4.420$		
		IIB	$g_x = 0.708$, $g_y = 0.449$, $g_z = 4.413$		
		IIIA	$g_x = 1.382$, $g_y = 0.70$, $g_z = 5.209$		
		IIIB	$g_x = 1.390$, $g_y = 0.654$, $g_z = 5.153$		
		IVA	$g_y = 0.3$, $g_z = 4.863$		
		IVB	$g_y = 0.3$, $g_z = 4.890$		
$La_2(SO_4)_3 \cdot 9H_2O$	4.2		$g_\parallel = 3.963$, $g_\perp = 1.743$	$A^{143} = 429$, $B^{143} = 167$	JW66
				$A^{145} = 267$, $B^{145} = 103$	
Nd ethyl sulfate	NR		$g_\parallel = 3.535$, $g_\perp = 2.072$	$Bg_\parallel/Ag_\perp = 0.892$	PF74
BaY_2F_8	1.5–20		$g_x = 1.566$, $g_y = 1.854$, $g_z = 3.277$		BM86
$Y_3Al_5O_{12}$	4.2		$g_x = 1.163$, $g_y = 1.733$, $g_z = 3.958$		BD74
$Ca_5(PO_4)_3F$	4.2		$g_\parallel = 6.024$, $g_\perp = 0.03$		BD74
LuAlG (garnet)	20		$g_x = 1.789$, $g_y = 1.237$, $g_z = 3.834$		PF74
LuGaG (garnet)	20		$g_x = 2.083$, $g_y = 1.323$, $g_z = 3.550$		PF74
YAlG (garnet)	20		$g_x = 1.733$, $g_y = 1.179$, $g_z = 3.915$		PF74
YGaG (garnet)	20		$g_x = 2.027$, $g_y = 1.251$, $g_z = 3.667$		PF74

Sm^{3+} $4f^5$. Data tabulation of spin-Hamiltonian parameters.

Sample	T (°K)	$\tilde{g}$	$\tilde{A}^{147,149}$, $B^{147,149}$ (10^{-4} cm^{-1})	$Bg_{\parallel}/Ag_{\perp}$	Ref.
CaF_2	4.2	$g_1 = 0.43$, $g_2 = 0.28$			BD80
LaF_3		$g_x = 0.23$, $g_y = 0.558$, $g_z = 0.720$			SJ66
$SmCl_3$ (average values)	NR	$g_{\parallel} = 0.584$, $g_{\perp} = 0.613$	$A^{147} = 60.7$, $B^{147} = 245$ $A^{149} = 49.9$, $B^{149} = 202$		PF74
Sm ethyl sulfate	NR	$g_{\parallel} = 0.596$, $g_{\perp} = 0.604$		4.2	PF74
SmB_6	4.2	1.94–1.95			BM86, BD80
$YLiF_4$	1.5–40	$g_{\parallel} = 0.410$, $g_{\perp} = 0.644$			BM83

Eu^{2+} $4f^7$. Data tabulation of spin-Hamiltonian parameters (* indicates unit of GHz).

Sample	T (°K)	$\tilde{g}$	D ($= b_2^0$) (10^{-4} cm^{-1})	E ($= b_2^2$) (10^{-4} cm^{-1})	$\frac{1}{3}F(b_4^0)$	$\frac{1}{4}d(b_6^0)$	$\tilde{A}^{151}$, $\tilde{A}^{153}$ (10^{-4} cm^{-1})	Ref.
Ca metal	5–30	1.993						BD74
CaO	290	1.991			24.0	−1.6	$A^{151} = 29.63$, $A^{153} = 13.05$	PF74
	77	1.992			25.1	−2.1	$A^{151} = 30.09$	
	4.2	1.992			25.7	−15	$A^{151} = 30.16$, $A^{153} = 13.46$	
KBr	300	1.994	320.9	−88.6	$b_4^0 = -10.3$ $b_4^2 = -8.3$ $b_4^4 = 51.4$	$b_6^0 = 0$		BM83
	77	1.994	333.0	−93.8	$b_4^0 = -11.2$ $b_4^2 = -15.3$ $b_4^4 = 56.3$			BM83
	1.2, 300	$g_{\parallel} = 1.996$ $g_{\perp} = 1.993$	319.3	81.9	$b_4^0 = -11.1$ $b_4^2 = -8.8$ $b_4^4 = 49.7$	$b_6^0 = 0.5$	$A^{151} = -30.66$, $A^{153} = -13.25$	BD80
KCl	300	1.994	346.6	−91.4	$b_4^0 = -5.0$ $b_4^2 = -9.4$ $b_4^4 = 28.0$	$b_6^0 = 0$		BM83
	77	1.995	355.8	−95.5	$b_4^0 = -4.8$ $b_4^2 = -7.0$ $b_4^4 = 27.9$			BM83

Eu^{2+} $4f^7$ *(continued)*

Sample	T (°K)	$\tilde{g}$	$D\ (=b_2^0)$ (10^{-4} cm^{-1})	$E\ (=b_2^2)$ (10^{-4} cm^{-1})	$\frac{1}{3}F(b_4^0)$	$\frac{1}{4}d(b_6^0)$	$\tilde{A}^{151}$, $\tilde{A}^{153}$ (10^{-4} cm^{-1})	Ref.
	553	1.991					−31.0	BM86
	290	1.992	343.5	−94.5	$b_4^0=-5.0$ $b_4^2=-4.7$ $b_4^4=29.5$	$b_6^0=0.4$	$A^{151}=-31.3$, $A^{153}=-13.9$	BD80
KI	300	$g_\parallel=1.993$ $g_\perp=1.993$	336.8	89.3	$b_4^0=-15.2$ $b_4^2=-2.8$ $b_4^4=69.7$	$b_6^0=0.4$		BD80
	94	$g_\parallel=1.993$ $g_\perp=1.994$	348.0	86.7	$b_4^0=-16.4$ $b_4^2=-4.8$ $b_4^4=74.3$	$b_6^0=0.4$		BD80
MgO (γ-irrad.)	77	1.989			$240B_4=-291.8$	$5040B_6=14.7$	$\lvert A^{151}\rvert=-31.63$	BM73
	4.2	1.989			−292.6	14.7	$\lvert A^{153}\rvert=-13.95$	
NaBr	300	1.994	387.5	−70.5	$b_4^0=-15.0$ $b_4^2=9.6$ $b_4^4=56.3$	$b_6^0=0.5$		BM83
	77	1.995	405.9	−73.7	$b_4^0=-22.6$ $b_4^2=-14.1$ $b_4^4=61.9$			BM83
	1.2, 300	$g_\parallel=1.992$ $g_\perp=1.993$	384.7	18.3	$b_4^0=-15.1$ $b_4^2=-55.6$ $b_4^4=72.6$	$b_6^0=0.4$	$A^{151}=-30.80$, $A^{153}=-13.10$	BD80
NaCl	553	1.995					−13.0	BM86
	300	1.995	460.4	−77.6	$b_4^0=-8.3$ $b_4^2=-13.8$ $b_4^4=39.0$	$b_6^0=0.2$		BM83
	77	1.995	486.2	−84.4	$b_4^0=-9.2$ $b_4^2=-18.8$ $b_4^4=58.6$			BM83
NaI	300	$g_\parallel=1.993$ $g_\perp=1.994$	352.8	19.2	$b_4^0=-19.2$ $b_4^2=-6.0$ $b_4^4=85.4$	$b_6^0=1.3$		BD80
	94	$g_\parallel=1.996$ $g_\perp=1.992$	373.3	19.2	$b_4^0=-20.8$ $b_4^2=-6.3$ $b_4^4=92.5$	$b_6^0=1.7$		BD80

RbBr	300	1.994	282.2	−102.4	$b_4^0 = -9.4$ $b_4^2 = -17.1$ $b_4^4 = 55.5$	$b_6^0 = 0$		BM83
	300	$g_\parallel = 1.993$ $g_\perp = 1.995$	283.4	117.9	$b_4^0 = -9.6$ $b_4^2 = -11.2$ $b_4^4 = 44.8$	$b_6^0 = 0.4$		BD80
	77	$g_\parallel = 1.993$ $g_\perp = 1.992$	292.8	135.9	$b_4^0 = -10.5$ $b_4^2 = -13.8$ $b_4^4 = 47.5$	$b_6^0 = 0.5$		BD80
	77	1.996	286.4	−103.2	$b_4^0 = -12.2$ $b_4^2 = -16.9$ $b_4^4 = 65.0$			BM83
RbCl	553	1.991					−31.0	BM86
	300	1.995	294.6	−139.8	$b_4^0 = -4.0$ $b_4^2 = -3.4$ $b_4^4 = 25.1$	$b_6^0 = 0.5$		BM83
	295	1.993	293.5	140.3	$b_4^0 = -3.9,$ $b_4^2 = 2.7,$ $b_4^4 = 21.7$	$b_6^0 = 0.5$	$A^{153} = -13.5$	BM86
	77	1.995	300.9	−145.0	$b_4^0 = -4.2$ $b_4^2 = -4.7$ $b_4^4 = 37.2$			BM83
RbI	300	$g_\parallel = 1.993$ $g_\perp = 1.991$	314.3	115.4	$b_4^0 = -13.7$ $b_4^2 = 3.2$ $b_4^4 = 64.4$	$b_6^0 = 0.4$		BD80
	94	$g_\parallel = 2.006$ $g_\perp = 2.008$	327.8	111.8	$b_4^0 = -14.7$ $b_4^2 = -17.8$ $b_4^4 = 65.9$	$b_6^0 = 0.3$		BD80
SnTe	1.5–300	1.99–1.992						BD80
SrO	1.6–77	1.991					$\|A\|^{151} = 29.9$, $\|A\|^{153} = 13.2$	PF74
SmS	4.2, 77	1.99						BD74
	77	2.226			$b_4 = 17$ G		$A^{151} = 29$ G	BM73
	4.2	2.241			$b_4 = 17$ G		$A^{151} = 29$ G	
SmSe	4.2, 77	1.99						BD74
	77	2.124			$b_4 = 11$ G		$A^{151} = 30.8$ G, $A^{153} = 13.6$ G	BM73
	4.2	2.132			$b_4 = 11$ G		$A^{151} = 30.8$ G, $A^{153} = 13.6$ G	
SmTe	4.2, 77	1.99						BD74
	77	2.019			$b_4 = 6$ G		$A^{151} = 30.7$ G, $A^{153} = 13.8$ G	BM73
	4.2	2.016			$b_4 = 6$ G		$A^{151} = 30.7$ G, $A^{153} = 13.8$ G	

Eu^{2+} $4f^7$ *(continued)*

Sample	T (°K)	$\tilde{g}$	D $(= b_2^0)$ $(10^{-4}$ cm$^{-1})$	E $(= b_2^2)$ $(10^{-4}$ cm$^{-1})$	$\frac{1}{3}F(b_4^0)$	$\frac{1}{4}d(b_6^0)$	$\tilde{A}^{151}$, $\tilde{A}^{153}$ $(10^{-4}$ cm$^{-1})$	Ref.
SnTe	1.5–20	1.991			$A^{151} = 2.90$ G			BD74
BaFBr	290	1.992	245.9		$b_4^0 = 9.0$, $b_4^4 = 28.4$	$b_6^0 = 0.4$		BM83
BaFCl	300	1.992	34.1		$b_4^0 = 8.5$, $b_4^4 = 0.0$	$b_6^0 = 0.2$		BM83
BaFI	291	1.991	$b_2^0 = 541.8$		$b_4^0 = 12.0$, $b_4^4 = 43.1$	$b_6^0 = 0.5$		BM83
CaFCl	293	1.988	−933.6		$b_4^0 = 9.3$, $b_4^4 = 0.0$	$b_6^0 = 0.4$		BM83
$Pb_{1-x}Eu_xSe$ $(x = 1.3\%)$	4.2	$g = 1.98$			$b_4 = 0.27$*	$b_6 = -0.0026$*		IC95A, MC95
$Pb_{1-x}Eu_xS$ $(x = 0.8\%)$	4.2	$g = 1.975$			$b_4 = 0.449$*	$b_6 = -0.001$*		IC96
SrFCl	292	1.991	−448.3		$b_4^0 = 9.3$, $b_4^4 = 1.1$	$b_6^0 = 0.4$		BM83
$BaCl_2$	300		24	−450				BD80
	77		33	−467				
BaF_2	NR						$A^{151} = -0.101$*, $A^{153} = -0.004$*	BM73
CaF_2	300	1.995			$b_4 = 55.5$	$b_6 = -0.3$	$A^{151} = 34.3$, $A^{153} = 15.1$	BM86
CdF_2	NR						$A^{151} = -0.102$*, $A^{153} = -0.046$*	BM73
	20, 300	1.992			$b_4 = -53.7$	$b_6 = 0.2$		BM86
$PbBr_2$		$g_x = 1.998$ $g_y = 1.989$ $g_z = 1.994$	$b_2^0 = 177$	$b_2^2 = -527$	$b_4^0 = 5.6$, $b_4^2 = 2$, $b_4^4 = 26$			BD74
$PbCl_2$	Room	1.989	$b_2^0 = 99$	$b_2^2 = -437$				ART 72
SrF_2	NR						$A^{151} = -0.102$*, $A^{153} = -0.045$*	BM73
TiO_2	4.2	$g_c = 15.1$					$A_c \sim 2.7$	PF74
LaF_3	295	1.992	$b_2^0 = 0.870$*	$b_2^2 = 0.62$*	$b_4^0 = 0.023$*, $b_4^2 = 0.04$*, $b_4^4 = 1.12$*			MF84
$SrTiO_3$	300	1.990			105.9	1.1		PF74
	2	1.990	−10.0		106.6	6.7		
$Pb_3(PO_4)_2$ (α-phase, γ-irrad.)	298	$g_1 = 2.003$, $g_2 = 2.003$, $g_3 = 1.989$	$b_2^0 = 862$	$b_2^2 = -68$	$b_4^0 \approx 1$, $b_4^2 = 16$, $b_4^4 = 65$	$b_6^0 = 6$	$A^{151} = 33$, $A^{153} = 14$	BM86
EuB_6	> 30	1.995						BM86
Eu ethyl sulfate	NR	$g_\parallel = 1.991$, $g_\perp = 1.991$						PCF74

Gd^{3+} $4f^7$. Data tabulation of spin-Hamiltonian parameters (units are GHz unless otherwise indicated, * denotes 10^{-4} cm^{-1}, ** denotes cm^{-1}).

Sample	T (°K)	Site	$\bar{g}$	b_2^0 (GHz)	b_2^2 (GHz)	b_4^0 (GHz)	b_4^2 (GHz)	b_4^4 (GHz)	b_6^0 (GHz)	Ref.
Acetates										
$R(CH_3COO)_3 \cdot 4H_2O$										
R = Dy	293		$g_\parallel = 1.990, g_\perp = 1.989$	−0.850	0.818	−0.017				MB83
	140		$g_\parallel = 2.004, g_\perp = 1.999$	−0.907	0.893	−0.022				MB83
R = Er	293		$g_\parallel = 1.993, g_\perp = 2.002$	−0.874	0.817	−0.006				MB83
R = Nd	293		$g_\parallel = 1.998, g_\perp = 1.999$	−0.866	0.857	−0.009				MB83
	95		$g_\parallel = 2.006, g_\perp = 1.970$	−0.891	1.004	−0.008				MB83
R = Sm	293		$g_\parallel = 1.998, g_\perp = 1.991$	−0.864	0.846	−0.009				MB83
	112		$g_\parallel = 2.002, g_\perp = 1.991$	−0.897	1.000	−0.010				MB83
	9.6		$g_\parallel = 1.967, g_\perp = 1.993$	−0.956	0.955	−0.010				MB83
R = Y	293		$g_\parallel = 2.007, g_\perp = 1.999$	−0.932	0.825	−0.006				MB83
	60		$g_\parallel = 2.000, g_\perp = 2.005$	−0.988	0.936	−0.005				MB83
	4.8		$g_\parallel = 2.016, g_\perp = 2.003$	−1.010	0.850	−0.008				MB83
R = Yb	293		$g_\parallel = 1.981, g_\perp = 1.983$	−0.902	0.797	−0.011				MB83
	103		$g_\parallel = 1.997, g_\perp = 1.989$	−0.972	0.843	−0.010				MB83
	8.6		$g_\parallel = 2.029, g_\perp = 1.932$	−0.936	0.674	−0.030				MB83
Arsenate										
$TbAsO_4$	1.5		1.977	$B_2^0 = -1.22$	$B_2^2 = -30$	$B_4^0 = -10$		$B_4^4 = 2.5$		BD80
$TmAsO_4$	1.5		1.980	$B_2^0 = -1.17$	$B_2^2 = -52$	$B_4^0 = -11$		$B_4^4 = 5.4$		BD80
Bromates										
$Dy(BrO_3)_3 \cdot 9H_2O$	295		$g_\parallel = 2.003, g_\perp = 2.003$	−0.416	0.023	0.004	−0.191	−0.085	−0.007	ML92B
	77		$g_\parallel = 2.002, g_\perp = 2.001$	−0.445	0.024	−0.003	−0.177	−0.086	0.0074	
$Eu(BrO_3)_3 \cdot 9H_2O$	295		$g_\parallel = 2.003, g_\perp = 2.003$	−0.493	−0.042	−0.016	0.086	0.019	0.001	ML92B
	77		$g_\parallel = 1.999, g_\perp = 1.999$	−0.528	0.054	−0.025	0.077	0.025	−0.010	
$Pr(BrO_3)_3 \cdot 9H_2O$	295		$g_\parallel = 1.989, g_\perp = 1.985$	−1.921	1.147	0.036	0.017	−0.046	0.002	BMM90
	230		$g_\parallel = 1.996, g_\perp = 1.995$	−1.978	1.040	0.035	−0.001	−0.051	0.005	
	4		$g_\parallel = 1.991, g_\perp = 2.010$	−2.069	0.912	0.037	0.037	−0.056	0.005	
$Sm(BrO_3)_3 \cdot 9H_2O$	295		$g_\parallel = 1.998, g_\perp = 1.998$	−1.950	1.279	0.049	−0.064	−0.360	−0.011	MBF90
	115		$g_\parallel = 1.995, g_\perp = 1.997$	−2.033	0.849	0.037	0.021	0.002	0.002	
	30		$g_\parallel = 1.968, g_\perp = 1.989$	−2.024	0.830	0.044	−0.005	−0.083	−0.010	
Halides										
BaFCl	294		$g = 1.991$	84.3		2.4		58.7	0.3	BM83

Gd^{3+} $4f^7$ *(continued)*

Sample	T (°K)	Site	$\bar{g}$	b_2^0 (GHz)	b_2^2 (GHz)	b_4^0 (GHz)	b_4^2 (GHz)	b_4^4 (GHz)	b_6^0 (GHz)	Ref.
CaFCl	294		$g = 1.991$	−48.2		−4.6		45.5	0.2	BM83
SrFCl	294		$g = 1.992$	30.4		−3.6		49.9	0.2	BM83
$BaCl_2$	300		1.994	90*	−440*					BD80
	77		1.992	84*	−461*					
BaF_2	295		$g_\parallel = 1.993, g_\perp = 1.993$	−1704.4		20.1			−0.1	BD80
BaF_2	77–295	1	1.991	−898.7*		−13.6*	$b_4^3 = 222$*		−0.1*	BD80
	77–295	2	1.991	−1700.4		19.8*	$b_4^3 = -475$*		−0.1*	BD80
CaF_2	77–295	1	1.991	−409.3*		−13.9*	$b_4^3 = 1230$*		−0.1*	BD80
	77–295	2	1.991	−1668.6*		26.2*	$b_4^3 = 704$*		−0.3*	BD80
CaF_2	295		$g_\parallel = 1.992, g_\perp = 1.992$	44.2	49.5	90		−164.2	1.4	BD80
CaF_2	77–295	1	1.991	−614.7		−13.2*	$b_4^3 = 216$*		−0.1*	BD80
	77–295	2	1.991	−1815.0*		−22.4*	$b_4^3 = 575$*		−0.2*	BD80
CaF_2 (Gd^{3+}–H pairs)	290	Tetragonal	NR	−1841*		−23.8*		−114*	−0.7*	BM73
CaF_2 (Gd^{3+}–D pairs)	290	Tetragonal	NR	−1861*		−23.7*		−113*	−0.7*	BM73
CaF_2 (Gd^{3+}–F pairs)	290	Tetragonal		−1485*		−23.1*		−147*	−0.7*	BM73
CdF_2	NR			116 MHz	42 MHz					ART72
CdF_2 (M-doped)										BM83
M = Ag	300			17.0 G	8.77 G	0.0004 G	−0.004 G	−0.003 G		
M = Cs	300			73.5 G	−1.3 G	−0.0007 G	−0.0077 G	−0.0018 G		
M = K	300			27.4 G	−6.7 G	−0.0004 G	−0.0037 G	−0.0022 G		
M = Li	300			5.9 G	−15.4 G	−0.001 G	−0.0002 G	−0.0048 G		
M = Na	300			1.1	−10.0 G	0.0007 G	−0.0012 G	0.0032 G		
M = Rb	300			47.3 G	−4.0 G	−0.0005 G	0.0042 G	−0.0017 G		
$PbCl_2$	room		1.989	99**	−437*					ART72
$SrCl_2$ (Gd^{3+}–M)										BD80
M = Ag	77, 295			$B_2^0 = -29.4$ G	$B_2^2 = -77.8$ G					
M = Ba	77, 295			$B_2^0 = 4.9$ G	$B_2^2 = -5.8$ G					
M = Ca	77, 295			$B_2^0 = 2.2$ G	$B_2^2 = -2.2$ G					

M = Cs	77, 295			$B_2^0 = 14.6$ G	$B_2^2 = -49.6$ G					
M = Cu	77, 295			$B_2^0 = -36.7$ G	$B_2^2 = -100.8$ G					
M = K	77, 295			$B_2^0 = -12.8$ G	$B_2^2 = -50.6$ G					
M = Li	77, 295			$B_2^0 = -29.4$ G	$B_2^2 = -86.1$ G					
M = Na	77, 295			$B_2^0 = -24.9$ G	$B_2^2 = -59.2$ G					
M = Rb	77, 295			$B_2^0 = -1.5$ G	$B_2^2 = -51.1$ G					
M = Tl	77, 295			$B_2^0 = -0.4$ G	$B_2^2 = -56.3$ G					
PbF_2	300		1.990			$b_4 = -40.0$*			$b_6 = -0.1$	BM86
	20					$b_4 = -43.3$*				
SrF_2	293	Cubic	$g = 1.992$			41.0*			0.6*	BD74
		Trigonal	$g = 1.992$	137.6*		−27.55*			1.1*	BD74
		Tetragonal	$g = 1.992$	1123*		19.7*			0.6*	BD74
	295	TET	$g_\parallel = 1.992, g_\perp = 1.992$	−1126.2		20.4		−140.6	0.6	BD80
	295	T4	$g_\parallel = 1.992, g_\perp = 1.992$	−135.6		27.8			−0.2	
	295	T2	$g_\parallel = 1.993, g_\perp = 1.992$	−1815.1	1	22.8			−0.1	
CaF_3	NR		1.990	239*	−5*	5.6*	27*	−43*	−0.14*	SJ66
CeF_3	300		$g = 1.982$	0.735	−0.084	0.016	0.085	0.169	—	MMK81
CeF_3	4.2		$g_\parallel = 2.054$							MB87
	160		$g_\parallel = 1.952$	0.712	−0.40	0.008	−0.35			
	200		$g_\parallel = 1.950$	0.707	−0.57	0.001	−0.28	0.01		
	293		$g_\parallel = 1.950$	0.700	−0.40	0.010	−0.28	0.07		
	373		$g_\parallel = 2.000$	0.690	−0.10	0.010	−0.28	0.07		
$GdCl_3$			$g_\parallel = 1.991, g_\perp = 1.991$					$A^{155} = 3.8$	$A^{157} = 5.0$	PF74
$LaCl_3$	77		1.992	1.8*		0.006*			0.000*	BM86
$LaCl_3$	293.5	Hexagonal	1.992	7.40*		2.11*			0.21*	BM73
LaF_3	300		$g = 2.001$	0.693	−0.085	0.016	0.074	0.125	—	MMK80
										MF84
	64		$g = 1.984$	0.705	−0.001	0.015	0.067	0.119	—	MMK81
	4.5		$g = 1.976$	0.696	−0.001	0.017	0.064	0.115	—	MMK81
NdF_3	300		$g = 2.008$	0.795	−0.149	0.018	0.105	0.128	—	MMK81
PrF_3	300		$g = 1.977$	0.773	−0.081	0.016	0.083	0.117	—	MMK81
	60		$g = 2.001$	0.812	0.000	0.018	0.076	0.068	—	MMK81
	3.1		$g = 1.990$	0.823	−0.040	0.015	0.080	0.120	—	MMK81

Gd^{3+} $4f^7$ *(continued)*

Sample	T (°K)	Site	$\tilde{g}$	b_2^0 (GHz)	b_2^2 (GHz)	b_4^0 (GHz)	b_4^2 (GHz)	b_4^4 (GHz)	b_6^0 (GHz)	Ref.
$RCl_3 \cdot 6H_2O$										
R = Dy	295		$g_x = 1.984, g_y = 1.934, g_z = 2.006$	1.915	−1.268	−0.029	−0.007	0.018	—	MR88
R = Eu	295		$g_x = 1.990, g_y = 1.975, g_z = 2.001$	1.867	−1.143	0.030	−0.008	−0.008	—	MR88
R = Er	295		$g_x = 1.981, g_y = 1.976, g_z = 1.969$	1.876	−1.292	−0.031	0.039	−0.001	—	MR88
R = Ho	295		$g_x = 1.980, g_y = 1.974, g_z = 1.987$	1.886	−1.254	−0.033	0.026	−0.003	—	MR88
R = Nd	295		$g_x = 1.984, g_y = 1.977, g_z = 1.990$	1.861	−1.062	−0.033	0.020	0.006	—	MR88
R = Sm	295		$g_x = 1.987, g_y = 1.986, g_z = 1.991$	1.867	−1.107	−0.032	0.020	0.004	—	MR88
R = Tb	295		$g_x = 1.981, g_y = 1.985, g_z = 1.991$	1.881	−1.190	−0.035	0.021	0.007	—	MR88
R = Tm	295		$g_x = 1.977, g_y = 1.988, g_z = 1.984$	1.872	−1.294	−0.035	0.006	0.003	—	MR88
R = Y	295		$g_x = 1.993, g_y = 1.996, g_z = 1.992$	1.894	−1.254	−0.034	0.028	−0.007	—	MR88
R = Yb	295		$g_x = 2.001, g_y = 1.914, g_z = 1.999$	1.909	−1.328	−0.026	−0.025	0.038	—	MR88
$CsSrCl_3$	$T > 3$–5		1.990			11.6 G			0.4 G	BM83
	$T < 385$		1.990	8 G		11.6 G			0.4 G	BM83
$KCdF_3$	487	Cubic	1.992			$\lvert b_4 \rvert = 1.6$*			$\lvert b_6 \rvert = 0.7$	BM86
$KMgF_3$	297		1.992			$b_4 = -11.15$*			$b_6 = 1.20$*	BM73
	77		1.992			$b_4 = -12.94$*			$b_6 = 1.33$*	
	4.2		1.992			$b_4 = -13.52$*			$b_6 = 1.31$*	
$KZnF_3$	487	Cubic	1.992			$b_4 = -6.97$*			$b_6 = 1.01$	BM86
	300	Cubic	1.992			$b_4 = -8.59$*			$b_6 = 1.13$	BM86
	77	Cubic	1.992			$b_4 = -10.64$*			$b_6 = 1.24$	BM86
	4.2	Cubic	1.992			$b_4 = -11.16$*			$b_6 = 1.27$	BM86
$RbCdF_3$	296	Cubic	1.992			$b_4 = \pm 4.4$*			$b_6 = \mp 0.8$	BM86
	296	Tetra-gonal	1.992	$b_2' = \pm 291$*		$b_4 = \pm 8.3$*			$b_6 = \mp 0.9$	BM86

$LiYF_4$	289		$g_{\parallel} = 1.990, g_{\perp} = 1.991$	−2.486		−0.058		0.305	−0.001	MM88
	140		$g_{\parallel} = 1.991, g_{\perp} = 1.990$	−2.571		−0.059		0.313	0.000	
	4.2		$g_{\parallel} = 1.990, g_{\perp} = 1.990$	−2.600		−0.060		0.315	0.000	
$LiY_{0.9}Yb_{0.1}F_4$	289		$g_{\parallel} = 1.988, g_{\perp} = 1.992$	−2.483		−0.056		0.311	0.001	MM88
	79		$g_{\parallel} = 1.987, g_{\perp} = 1.993$	−2.586		−0.059		0.324	−0.001	
	4.2		$g_{\parallel} = 1.966, g_{\perp} = 1.976$	−2.564		−0.060		0.325	−0.004	
$LiY_{0.8}Yb_{0.2}F_4$	296		$g_{\parallel} = 1.981, g_{\perp} = 1.990$	−2.465		−0.060		0.315	−0.002	MM88
	79			−2.558		−0.059		0.334	−0.000	
			$g_{\parallel} = 1.973, g_{\perp} = 1.988$							
$LiY_{0.7}Yb_{0.3}F_4$	289		$g_{\parallel} = 1.988, g_{\perp} = 1.991$	−2.481		−0.058		0.316	0.001	MM88
	100		$g_{\parallel} = 1.970, g_{\perp} = 1.986$	−2.546		−0.061		0.336	−0.001	
$LiY_{0.6}Yb_{0.4}F_4$	290		$g_{\parallel} = 1.978, g_{\perp} = 1.987$	−2.462		−0.060		0.325	0.002	MM88
$LiY_{0.4}Yb_{0.6}F_4$	290		$g_{\parallel} = 1.985, g_{\perp} = 1.996$	−2.478		−0.060		0.351	−0.001	MM88
$LiY_{0.2}Yb_{0.8}F_4$	286		$g_{\parallel} = 1.976, g_{\perp} = 1.992$	−2.494		−0.065		0.333	−0.000	MM88
$LiY_{0.6}Yb_{0.4}F_4$	290		$g_{\parallel} = 1.972, g_{\perp} = 1.986$	−2.481		−0.068		0.321	−0.003	MM88
Cs_2NaRCl_6										BM83
R = La	296		1.992			$B_4 = -0.25$			$B_6 = 0.38$	
R = Pr	296		1.993			$B_4 = -0.25$			$B_6 = 0.38$	
R = Tb	296		1.993			$B_4 = -0.26$			$B_6 = 0.23$	
R = Y	296		1.993			$B_4 = -0.26$			$B_6 = 0.41$	
	77		1.994			$B_4 = -0.29$			$B_6 = 0.23$	
	4.2		1.995			$B_4 = -0.31$			$B_6 = 0.57$	
R = Yb	296		1.993			$B_4 = -0.27$			$B_6 = 0.40$	
Hydroxides										
$La(OH)_3$	77		1.992	−274.0*		−2.3*			0.5*	BM73
$Y(OH)_3$	300		1.992	−136.2*		−2.0*			0.5*	BM73
	77		1.992	130.4*		−2.2*			0.6*	
Garnets ($A_3B_5O_{12}$)										
LuAlG (garnet)	300		1.989	571.5*	115.4*	−50.2*		40.4*		PF74
LuGaG (garnet)	300		1.99	275.0*	228.0*	−44.9*		39.2*		PF74
	4.2		1.99	279.0*	238.0*	−47.2*		43.3*		PF74
YAlG (garnet)	300		1.990	777.7*	−85.1*	−46.9*		23.2*	3.77*	PF74
	4.2		1.990	776.4*	−96.9*	−48.6*		23.5*	4.62*	PF74
$Y_3Al_2(Al_4)_3 \cdot (YAlG)$	77	a	$g_{\parallel} = 2.003, g_{\perp} = 2.003$	−714.0 MHz		−7.8 MHz		$\lvert b_4^3 \rvert = 0.084$ MHz		BM83
		d	$g_{\parallel} = 2.004, g_{\perp} = 2.004$	−655.3 MHz		0.123 MHz		$\lvert b_4^4 \rvert = 0.064$ MHz		
YGaG (garnet)	300		1.991	440.7*	216.1*	−43.2*	0.3*	36.1*	0.3*	PF74

Gd^{3+} $4f^7$ (*continued*)

Sample	T (°K)	Site	$\bar{g}$	b_2^0 (GHz)	b_2^2 (GHz)	b_4^0 (GHz)	b_4^2 (GHz)	b_4^4 (GHz)	b_6^0 (GHz)	Ref.
YGaG (garnet)	300		1.992	441.3*	−269.7*	−42.2*		21.1*	0.6*	PF74
	4.2		1.992	448.0*	−283.8*	−45.2*		22.1*	4.15*	PF74
Nitrates										
$R(NO_3)_3 \cdot 6H_2O$										
(average values)										
R = La, Ce	295		$g_x = 1.998, g_z = 1.996$	−0.683	0.569	−0.001	0.294	0.000	−0.001	MM78
R = La	77		$g_x = 2.002, g_z = 1.994$	−1.077	1.047	0.008	−0.007	−0.005	0.000	MM78
	4.2		$g_x = 2.005, g_z = 2.010$	−1.129	1.077	0.009	0.018	−0.001	0.004	MM78
R = Pr, Nd, Sm, Dy, Er, Yb, Y	295		$g_x = 1.994, g_z = 1.994$	−1.381	1.682	−0.004	0.002	−0.046	−0.001	MM78
R = Pr, Sm	77		$g_x = 1.989, g_z = 2.008$	−1.863	1.068	−0.005	0.070	0.038	0.000	MM78
	3.5		$g_x = 1.995, g_z = 2.024$	−1.962	1.058	−0.006	−0.016	−0.028	−0.002	MM78
$Nd_2Mg_3(NO_3)_{12} \cdot 24H_2O$	298		$g_\parallel = 1.992, g_\perp = 1.992$	116.6*		0.276*			0.400*	BM86
	77		$g_\parallel = 1.991, g_\perp = 1.992$	128.4*		0.471*			0.840*	
$Nd_2Zn_3(NO_3)_{12} \cdot 24H_2O$	298		$g_\parallel = 1.993, g_\perp = 1.993$	130.9*		0.705*			0.850*	BM83
	77		$g_\parallel = 1.994, g_\perp = 1.993$	120.6*		0.460*			0.400*	
$Pr_2Mg_3(NO_3)_{12} \cdot 24H_2O$	300		$g_\parallel = 1.992, g_\perp = 1.995$	113.7*		0.29*			0.44*	BM83
	77		$g_\parallel = 1.992, g_\perp = 1.995$	123.2*		0.48*			0.79*	BM83
$Pr_2Zn_3(NO_3)_{12} \cdot 24H_2O$	298		$g_\parallel = 1.993, g_\perp = 1.994$	114.6*		0.592*			0.293*	BM83
	77		$g_\parallel = 1.992, g_\perp = 1.994$	123.7*		0.750*			0.600*	BM83
$Sm_2Zn_3(NO_3)_{12} \cdot 24H_2O$	300		$g_\parallel = 1.992, g_\perp = 1.992$	126.6*		0.58*			0.48*	BM83
	77		$g_\parallel = 1.992, g_\perp = 1.992$	137.6*		0.65*			0.74*	BM83
Oxides										
CaO	290, 4.2		1.992			−12.2*			1.2*	PF74
SrO	290		1.991			5.9*				PF74
	70		1.991			6.5*				
	4.2		1.989			5.8*				
MgO	295	Cubic	1.992		$240B_4 = -134.5^*$		$A^{155} = 3.85^*, A^{157} = 5.06^*$		$5040B_6 = 8.5^*$	BM73
	77		1.992		$240B_4 = -139.8^*$		$A^{155} = 3.85^*, A^{157} = 5.06^*$		$5040B_6 = 8.8^*$	
	4.2				$240B_4 = -140.0^*$				$5040B_6 = 8.8^*$	
ThO_2	293		$g = 1.991$	$c = -657.6$ MHz, $d = -4.15$ MHz						BD74
TiO_2	295		$g_c = 1.993$	120.9	22.8	20.2		−182.0	−1.0	PF74
	77		$g_c = 1.994$	107.8		21.5				

	1.8		$g_c = 1.999$							
Al_2O_3	290		1.991							PF74
La_2O_3	4.2, 293		1.992	160.5 G		0.74 G	= 0.07 G		0.01 G	BD80
$BaTiO_3$	425		1.995			≅6*			≅0*	PF74
	300		1.995	−293.6*		4.0*	−2.0*	1.6*		
$GdCrO_3$	180		2.01							BD80
$LaAlO_3$	689		1.991	67.0						PF74
	575		1.991	150.6		5.60		0.9		
	415		1.991	307.0		5.99		1.0		
	293		1.991	371.2		6.17		1.0		
	80		1.991	479.2		6.42		0.9		
	20		1.991	490.7		6.46		0.9		
$SrTiO_3$	300		1.992			−5.7*			0.5*	PF74
	77		1.992	−233.6*		−4.8*		−2.5*	−0.25*	
	4.2		1.992	−362.5*		−3.24*		−4.2*	1.4*	
$EuAsO_4$	6–550					0.001**		0.019**	0.000**	BM86
$EuVO_4$						−0.001**		0.019**	0.000**	BM86
$HfSiO_4$	NR		$g = 1.99$	694.2*	−517.2*	−3.2*		−5.4*	1.0*	BD74
$HoVO_4$			$g_s = 2.025$	−40.4**		−0.00**			0.000**	BM86
$ThSiO_4$	77		$g = 1.992$	−766.6*	−482.4*	−9.0*	−0.5*	−32.5*	0.5*	BD74
	4.2		$g = 1.992$	−767.9*	−481.8*	−9.1*	−0.2*	−34.5*	0.5*	BD74
$YNbO_4$		I	1.99	400 G	120 G					ART72
		II	1.94	400 G	180 G					
$ZrSiO_4$	NR		$g = 1.99$	662.3*	−576.5*	−4.2*		−7.9*	0.7*	BD74
La_2MoO_6	293		$g_x = 1.998, g_y = 1.991, g_z = 1.991$	1.107	−0.528	−0.049	−0.765	−0.048	−0.001	BM86
14%CaO · 86%ZrO_2	293		$g = 1.989$	7.1 G	−1.4 G	0.02 G	0.02 G	0.09 G	−0.005 G	BD74
Sulfates										
$3CdSO_4 \cdot 8H_2O$	298	I	$g_x = 1.994, g_y = 1.993, g_z = 1.992$	313.2*	−27.2*	0.048*	0.009*	0.037*	0.0003*	BM83
		II	$g_x = 1.993, g_y = 1.993, g_z = 1.992$	316.6*	−26.5*	0.067*	0.026*	0.030*	0.0001*	
$La_2(SO_4)_3 \cdot 9H_2O$	300		$g = 2.001$	0.693	−0.085	0.016	0.074	0.125		MMK80
	295		$g_x = 1.998, g_z = 1.993$	−1.802	—	−0.009	—	—	0.000	MM80
	77		$g_x = 2.010, g_z = 1.997$	−1.935	—	−0.010	—	—	0.002	MM80
	3.9		$g_x = 2.002, g_z = 1.990$	−1.927	—	−0.011	—	—	−0.002	MM80

Gd^{3+} $4f^7$ (*continued*)

Sample	T (°K)	Site	$\bar{g}$	b_2^0 (GHz)	b_2^2 (GHz)	b_4^0 (GHz)	b_4^2 (GHz)	b_4^4 (GHz)	b_6^0 (GHz)	Ref.
$R_2(SO_4)_3 \cdot 8H_2O$										
R = Pr, Nd, Sm, Er, Y,										
Yb (average values)	295		$g_x = 1.994, g_z = 1.991$	−1.843	1.113	0.037	0.005	−0.032	0.001	MM79
R = Pr, Sm, Er, Y										
(average values)	77		$g_x = 1.997, g_z = 1.994$	−2.032	0.917	0.038	−0.003	−0.030	−0.002	MM79
	3.8		$g_x = 1.993, g_z = 2.006$	−2.056	0.934	0.039	0.018	−0.032	0.001	MM79
R = Ce	295		$g_x = 1.998, g_z = 1.984$	−1.349	0.821	−0.014	−0.040	−0.055	−0.001	MM80
	77		$g_x = 1.968, g_z = 2.015$	−1.386	0.865	−0.022	0.000	0.045	−0.002	MM80
R = Pr	300		$g_\parallel = 1.945, g_\perp = 1.960$	652.80 G	−481.42 G					BD80
R = Nd	300		$g_\parallel = 1.943, g_\perp = 1.966$	640.42 G	−481.32 G					BD80
$NH_4Ce(SO_4)_2 \cdot 4H_2O$	295		$g_\parallel = 1.991, g_\perp = 1.990$	0.370	−0.004	0.008	0.002	−0.022	0.001	ML92A
$NH_4Ce(SO_4)_2 \cdot 4H_2O$	295		$g_\parallel = 1.991, g_\perp = 1.991$	0.380	−0.040	0.001	0.000	−0.015	0.002	MLM90
$NH_4Eu(SO_4)_2 \cdot 4H_2O$	295		$g_\parallel = 1.996, g_\perp = 1.997$	0.331	−0.003	0.020	−0.042	−0.011	0.003	ML91
$NH_4La(SO_4)_2 \cdot 4H_2O$	295		$g_\parallel = 1.977, g_\perp = 1.977$	0.387	−0.005	−0.009	−0.007	−0.032	0.0002	ML90
$NH_4Nd(SO_4)_2 \cdot 4H_2O$	295		$g_\parallel = 1.994, g_\perp = 1.993$	0.359	0.022	0.008	0.034	−0.043	0.004	MLM90
$NH_4Pr(SO_4)_2 \cdot 4H_2O$	295		$g_\parallel = 1.995, g_\perp = 1.995$	0.347	−0.0003	−0.009	−0.0002	−0.028	0.0006	MS91A
$NH_4Sm(SO_4)_2 \cdot 4H_2O$	295		$g_\parallel = 1.992, g_\perp = 1.992$	0.270	−0.022	0.003	−0.002	−0.012	0.012	MLM90
$NH_4Sm(SO_4)_2 \cdot 4H_2O$	295		$g_\parallel = 1.997, g_\perp = 1.996$	0.340	−0.006	−0.007	−0.003	−0.032	0.010	ML92A
Sulfides										
CdS	296		$g = 1.989$	0.185**						BD74
ThOS	298		$g_\parallel = 1.992, g_\perp = 1.991$	−183.3*		1.740*		28.30*	−0.591*	BM86
La_2O_2S	293		$g_\parallel = 1.992, g_\perp = 1.992$	462.32*		14.80*	$b_4^3 = 10$*		−0.2*	BD74
(Ethyl) Sulphates										
$R(C_2H_5SO_4)_3 \cdot 9H_2O$										BM86, BM83, GT72, ART72
R = Ce	297		$g_\parallel = 1.992, g_\perp = 1.992$	187.378*		−3.858*			0.461*	
R = Dy	297		$g_\parallel = 1.993, g_\perp = 1.993$	157.245*		−3.813*			0.477*	
R = Eu	297		$g_\parallel = 1.992, g_\perp = 1.992$	168.041*		−3.842*			0.472*	
R = Er	297		$g_\parallel = 1.993, g_\perp = 1.993$	151.624*		−3.760*			0.482*	
R = Gd			$g_\parallel = 1.991, g_\perp = 1.991$					$A^{155} = 3.8^*, A^{157} = 5.0^*$		PF74

R = Ho	297	$g_{\parallel} = 1.993, g_{\perp} = 1.993$	154.743*		−3.783*		0.479*	
R = La	297	$g_{\parallel} = 1.992, g_{\perp} = 1.992$	193.829*		−3.846*		0.457*	
	77		0.019**		3.2*		0.4	
R = Lu	297	$g_{\parallel} = 1.992, g_{\perp} = 1.992$	142.755*		−3.651*		0.487*	
R = Nd	297	$g_{\parallel} = 1.992, g_{\perp} = 1.992$	178.969*		−3.866*		0.466*	
R = Pr	297	$g_{\parallel} = 1.992, g_{\perp} = 1.992$	182.838*		−3.865*		0.463*	
R = Sm	297	$g_{\parallel} = 1.992, g_{\perp} = 1.992$	171.194*		−3.852*		0.470*	
R = Tb	297	$g_{\parallel} = 1.993, g_{\perp} = 1.993$	160.392*		−3.825*		0.475*	
R = Tm	297	$g_{\parallel} = 1.993, g_{\perp} = 1.992$	147.511*		−3.721*		0.484*	
R = Yb	297	$g_{\parallel} = 1.992, g_{\perp} = 1.992$	144.623*		−3.680*		0.486*	
Scheelites								
$LiYF_4$	295	$g_{\parallel} = 1.989, g_{\perp} = 1.987$	−2.484		−0.067	0.278	−0.008	MK85
	85	$g_{\parallel} = 1.981, g_{\perp} = 1.983$	−2.514		−0.067	0.424	−0.010	MK85
	5	$g_{\parallel} = 1.980, g_{\perp} = 1.984$	−2.518		−0.067	0.437	−0.015	MK85
$LiYbF_4$	295	$g_{\parallel} = 1.986, g_{\perp} = 1.987$	−2.489		−0.051	0.414	−0.010	MK85
Vanadate								
$TmVo_4$	1.5	1.970	$B_2^0 = -1.77$	$B_2^2 = -2.2$	$B_4^0 = -6.7$	$B_4^4 = 6.0$		BD80
Other compounds								
GdAu	77–300	$g_s = 2.028$						BD74
LaP	2–300	$g = 1.993$						BD74
PrBi	1.3–4.2	2.74						BD80
PrN	1.7–4.2	1.65						BD80
PrSb	1.3–4.2	2.40			$b_4 = 3.25$ G			BM86, BD80
PrTe (p)	1.3–4.2	2.48						BD80
TmBi (p)	1.3–4.2	1.04						BD80
TmN	1.7–4.2	2.45						BD80
TmSb	1.3–4.2	1.56						BD80
YBi	2–4.2	2.005			$b_4 = 47$ G			BD80
$GdCo_2$	380–426	$g = 1.992$						BD74
$CeIr_2$ (x% Gd^{3+})								BM86
($x = 0.043$%)	1.5	2.005						
($x = 0.1$%)	1.5	2.006						
($x = 0.3$%)	1.5	1.996						
($x = 1.0$%)	1.5	1.998						

Gd^{3+} $4f^7$ *(continued)*

Sample	T (°K)	Site	$\bar{g}$	b_2^0 (GHz)	b_2^2 (GHz)	b_4^0 (GHz)	b_4^2 (GHz)	b_4^4 (GHz)	b_6^0 (GHz)	Ref.
$CeRh_2$ (x% Gd^{3+})										BM86
($x = 0.04\%, 0.1\%$)	1.5		1.998							
($x = 0.3\%$)	1.5		1.993							
($x = 1.0\%$)	1.5		1.996							
YIr_2 (x% Gd^{3+})										BM86
($x = 0.02\%$)	1.5		1.993							
($x = 0.04\%$)	1.5		1.998							
($x = 0.05\%$)	1.5		1.995							
($x = 0.08\%$)	1.5		1.995							
($x = 0.35\%$)	1.5		1.994							
YRh_2 (x% Gd^{3+})										BM86
($x = 0.01\%$)	1.5		1.995							
($x = 0.05\%$)	1.5		1.999							
($x = 1.0\%$)	1.5		1.993							
$LaIr_2$ (x% Gd^{3+})										BM86
($x = 0.0025\%, 0.005\%, 0.05\%$)	1.5		1.979							
($x = 0.5\%$)	1.5		1.983							
$LaRh_2$ (x% Gd^{3+})										BM86
($x = 0.01\%, 0.3\%$)	1.5		1.994							
($x = 0.02\%, 0.1\%$)	1.5		1.993							
$LaPt_2$ (Gd = 1.0%)	1.5		2.040							BM86
$LaNi_5$	1.8–15		$g_\parallel = g_\perp = 1.85$	250 G						BM86
$LaNi_5$-hydride (p)	4.2		2.05							BD80
$LaNi_5$-unhydrided	4.2		1.86							BD80
Bi_2Te_3	300		2.003	−111*		−4.75*		$A^{155} = 21.9$*	1.76*	IS95
	≤10		2.004	−122*		−4.52*			1.93*	
$Bi_{2(1-x)}Gd_{2x}Te_3$ ($x = 1.0\%$)	10		2.004	−122*		−4.52*			1.93*	IC95B
La_3In	4.0		$g_{eff} = 2.005$							BD74
$KSc(MoO_4)_2$	293		$g_\parallel = 1.992, g_\perp = 1.992$	−35.65*		8.88*	$b_4^3 = 79.5$*		−1.42*	BD80
$RbSc(MoO_4)_2$	293		$g_\parallel = 1.992, g_\perp = 1.992$	−47.54*		7.51*	$b_4^3 = 74.0$*		−1.38*	BD80
$Pb_3(PO_4)_2$	473	β-phase	$g_z = 2.025, g_\perp = 1.974$	990*		10.0*	$b_4^3 = -0.3$*		−1.7*	BM86
	298	α-phase	$g_1 = 2.006, g_2 = 1.989, g_3 = 1.982$	930*	−77*	8.3*	0	30*	−0.7*	BM86

$Pb_3(PO_4)_2$	298	α-phase	$g_x = 1.982, g_y = 2.006,$ $g_z = 1.989$	$\|D\| = 298.0$ G	$\|E\| = 82.0$ G	8.9 G	0 G	32 G	0.8 G	BD80
	473	β-phase	$g_x = 1.973, g_y = 1.973,$ $g_z = 2.025$	$\|D\| = 318.0$ G	$\|E\| = 0$ G	11.0 G			1.8 G	BD80
$Pb_5Ge_3O_{11}$	295	Trigo- nal	1.991	0.814		−0.12	$b_4^3 = \pm 0.25$		0.004	BD80
		Tri- clinic		0.82		−0.14			0.004	BD80
$Gd_xY_{1-2}As$ (*p*)	15–300		1.995							BD80
$Gd_xY_{1-x}P$ (*p*)	15–300		1.995							BD80
$Gd_xY_{1-2}Sb$ (*p*)	15–300		1.995							BD80

Tb^{4+} $4f^7$. Data tabulation of spin-Hamiltonian parameters (* indicates units of GHz).

Sample	T (°K)	Site	$\tilde{g}$	b_ℓ^m	Ref.
Pr_2O_3	293		$g_\parallel = 1.999$, $g_\perp = 2.026$	$b_2^0 = 0.145$ cm^{-1}, $b_4^0 = 0.002$ cm^{-1}, $b_6^0 = 0.0004$ cm^{-1}	BM83
	150		$g_\parallel = 1.994$, $g_\perp = 2.026$	$b_2^0 = 0.153$ cm^{-1}, $b_4^0 = 0.003$ cm^{-1}, $b_6^0 = 0.0001$ cm^{-1}	BM83
$BaTbO_3$	293		1.998	$\|b_4\| = 250$ MHz, $\|b_6\| = 42$ MHz	BM83
$CaWO_4$	4.2		$g_\parallel = 2.009$	$b_2^0 = 1.912$*, $b_4^0 = -0.422$*, $b_4^4 = -1.827$*, $b_6^0 = 0.006$*	BM83

Tb^{3+} $4f^8$. Data tabulation of spin-Hamiltonian parameters (* indicates units of GHz).

Sample	T (°K)	Site	$\tilde{g}$	b_ℓ^m	Δ (10^{-4} cm^{-1})	$\tilde{A}^{159}$ (10^{-4} cm^{-1})	Ref.
CaF_2 (γ-irrad.)	NR	I	$g_\parallel = 1.989$	$b_2^0 = 1.14$*, $b_4^0 = 0.69$*, $b_6^0 = 0.98$*		0.11*	BM73
		II	$g_\parallel = 1.989$	$b_2^0 = 1.12$*, $b_4^0 = 0.65$*, $b_6^0 = 0.83$*		0.11*	BM73
PbF_2	4.2		0.35		$\Delta_1 = 1.45$ cm^{-1} $\Delta_2 = 4.34$ cm^{-1}		BM83
$TbCl_3$	NR		$g_\parallel = 17.78$, $g_\perp < 0.1$		0.201	2120.0	PF74
YPO_4	1.6		10.60			1220.0	BD80
$CaWO_4$	4.2	I	17.777		8.131 GHz	$A = 6.284$ GHz	BM83
		II	17.666		26.546 GHz	$A = 6.417$ GHz	BM83
$La(C_2H_5SO_4)_3 \cdot 9H_2O$	1.6–10		$g_\parallel = 17.69$		0.458 cm^{-1}	$A = 0.209$ cm^{-1}	BM83
Tb ethyl sulfate	NR		$g_\parallel = 17.72$, $g_\perp < 0.3$				PF74

Dy^{3+} $4f^9$. Data tabulation of spin-Hamiltonian parameters.

Sample	T (°K)	Site	$\tilde{g}$	$\tilde{A}^{161}$, $\tilde{A}^{163}$ (10^{-4} cm^{-1})	Ref.
Ag (film on quartz)	1.5–4.2		7.67		BM83
Al (film on quartz)	1.5–4.2		7.59		BM83
Ir (splatted)	1.4–4.2		7.49	$A^{163} = 82$ G	BD74
Rhodium	1.4–4.2		7.5	$A = 85$ G	BM73
Zircon	4.2		$g_\parallel = 1.132$, $g_\perp = 9.974$	$A^{161} = 30.7$, $B^{161} = 282$	BM73
				$A^{163} = 40$, $B^{163} = 397$	
Th-Dy alloys (p)	1.4–4.2		7.61	$A^{161} = 57$ G, $A^{173} = 80$ G	BD74
BaO	4.2		6.633		REY74
CaO	20		$g_\parallel = 3.09$, $g_\perp \sim 15$		PF74
	4.2		6.583	$A_z^{161} = 182$, $A_z^{163} = 260$	REY74
MgO	4.2		6.539	$A_z^{161} = 184$, $A_z^{163} = 257$	REY74
SrO	4.2		6.572	$A_z^{161} = 184$, $A_z^{163} = 261$	REY74
CaF_2	1.8		$g_\parallel = 36.4$, $\|g_\perp\| < 10$		BM86
	4.2		7.45	$\Delta = E(\Gamma_7)–E(\Gamma_8) = 9.5$ cm^{-1}	BD80
CdF_2	4.2	Cubic (Γ_7)	7.425		CC76
		Cubic (Γ_8)	2.451		
		Orthorhombic	$g_1 = 2.103$, $g_2 = 8.661$, $g_3 = 10.185$		
$LuAl_2$	1.5–20		7.733		BM86
$ScAl_2$	1.5–20		7.715		BM86
$ScAl_2$ (p)	4.2–20		7.49		BD80
$YbAl_2$	1.5–20		7.670		BM86
$LaCl_3$	2.1		$g_\parallel = 6.15$, $g_\perp = 3.327$		BM83
LaF_3	4.2		$g_x = 5.52$, $g_y = 0.96$, $g_z = 13.8$		SJ66
$LaBe_{13}$	1.8–4.2		7.42	$\Delta_1 \geq 6$ K	BM83
$ThBe_{13}$	1.8–4.2		7.46		BM83
$YCl_3 \cdot 6H_2O$			$g_x = 2.00$, $g_y = 1.52$, $g_z = 16.52$	$A_z^{161} = 471$, $A_z^{163} = 657$	SJ67
$KMgF_3$	1.5, 4.2		$g = 6.544$	$A_z^{161} = 194$, $A_z^{163} = 274.2$	BM73
$KZnF_3$	1.7–36	Cubic	6.505		BD80
		Trigonal	$g_\parallel = 3.531$, $g_\perp = 7.828$		
$LiTmF_4$	1.6–7		$g_\parallel = 1.207$, $g_\perp = 9.372$		BM86
	1.6–7		$g_\parallel = 3.7$, $g_\perp = 4.5$		BM86
$LiYF_4$	4.2		$g_\parallel = 1.112$, $g_\perp = 9.219$	$A^{161} = 31.6$, $B^{161} = 263$	BM73, BM86
				$A^{163} = 41.5$, $B^{163} = 366$	

$LiYF_4$	1.6–7	$g_\parallel = 1.112$, $g_\perp = 9.219$		BM86
$YAsO_4$	1.7, 4.2	$g_\parallel = 1.845$, $g_\perp = 9.60$ (lowest Kramers doublet)		BD74
	1.7, 4.2	$g_\parallel = 1.250$, $g_\perp = 9.60$ (excited Kramers doublet)		
$YLiF_4$	1.5–40	$g_\parallel = 4.65$, $g_\perp = 4.77$	$\Delta_1 = 19$ K, $\Delta_2 = 57$ K	BM83
YVO_4	1.5–4.2	$g_\parallel = 1.92$, $g_\perp = 9.3$		BM73
LuAlG (garnet)	20	$g_x = 2.29$, $g_y = 0.91$, $g_z = 16.6$		PF74
LuGaG (garnet)	20	$g_x = 13.45$, $g_y = 0.57$, $g_z = 3.41$		PF74
YAlG (garnet)	20	$g_x = 0.73$, $g_y = 0.40$, $g_z = 18.2$		PF74
YGaG (garnet)	20	$g_x = 11.07$, $g_y = 1.07$, $g_z = 7.85$		PF74

Ho^{3+} $4f^{10}$. Data tabulation of spin-Hamiltonian parameters.

Sample	T (°K)	$\tilde{g}$	$\tilde{A}^{165}$ (10^{-4} cm^{-1})	δ (GHz)	Ref.
CaF_2	4.2	$g_\parallel = 14.8$	$A^{165} = 9.8$ GHz	50.8	BM73
$HoCl_3$	NR	$g_\parallel = 16.01$, $g_\perp \approx 0.0$	$A^{165} = 3510.0$		PF74
$YAsO_4$	1.6	15.30	$A = 3314.0$	$\Delta_0 \leq 0.17$ cm^{-1}	BD80
Ho ethyl sulfate	NR	$g_\parallel = 15.36$			PF74

Ho^{2+} $4f^{11}$. Data tabulation of spin-Hamiltonian parameters.

Sample	T (°K)	$\tilde{g}$	$\tilde{A}$ (10^{-4} cm^{-1})	Ref.
$SrCl_2$	4.2	$g = -5.91$	$A = -0.133$	BD80

Er^{3+} $4f^{11}$. Data tabulation of spin-Hamiltonian parameters.

Sample	T (°K)	Site	$\tilde{g}$	$Bg_{\parallel}/Ag_{\perp}$	$\tilde{A}^{167}$ (10^{-4} cm^{-1})	Ref.
Ag (film on quartz)	1.5–4.2		6.82			BM83
Ag (film on PMMA)	1.5–4.2		6.85			BM83
Al (film on quartz)	1.5–4.2		6.80			BM83
Al (film on PMMA)	1.5–4.2		6.78			BM83
Ce (metal powder)	1.4–4.2		6.77			BM83
Ir (splatted, powder)	1.4–4.2		6.73			BD74
Rhodium (metal)	1.4–4.2		6.80		$A = 75.5$ G	BM73
Zircon	4.2		$g_{\parallel} = 3.703$, $g_{\perp} = 6.971$		$A^{167} = 130.0$, $B^{167} = 246.7$	BM73
Cu:Er (metal)	1.4–4.2	Cubic	6.86		$A/B = 2.5$	BM73
Pd:Er (alloy)	4.2		6.5			BD74
Pt:Er (alloy)	4.2		6.5			BD74
Pt:Er (metal)	1.4–4.2	Cubic	6.50		$A/B = 3.5$	BM73
Rh:Er (metal)	1.4–4.2	Cubic	6.78		$A/B = 2.2$	BM73
Th:Er (alloy)	1.4–4.2		6.84		$A^{167} = 75$ G	BD74
Th:Er (metal)	1.4–4.2	Cubic	6.84		$A/B = 2.0$	BM73
CaO	20, 4.2		$g_{\parallel} = 4.73$, $g_{\perp} = 7.86$			PF74, SJ67
MgO	20		$g_x = 4.62$, $g_y = 3.86$, $g_z = 3.6$			PF74
LaSb ($\Gamma_8^{(1)}$ ground state)	1.6–4.2		−5.669			BD80
PrSb ($\Gamma_8^{(1)}$ ground state)	1.6–4.2		−6.576			BD80
TmSb ($\Gamma_8^{(1)}$ ground state)	1.6–4.2		−5.669			BD80
YSb ($\Gamma_8^{(1)}$ ground state)	1.6–4.2		−5.669			BD80
$ScH_{1.91}$	2	Cubic	6.77			BM86
ScH_x (p)	2	I	$g_{\parallel} = 9.0$, $g_{\perp} = 5.42$			BM86
		II	$g_{\parallel} = 11.0$, $g_{\perp} = 4.05$			
CaF_2	5	H^-, GS	$g_{\parallel} = 7.465$			BM86
		H^-, ES	$g_{\parallel} = 1.683$, $g_{\perp} = 9.09$			
		D^-, GS	$g_{\parallel} = 7.330$			
		H^-, ES	$g_{\parallel} = 1.680$, $g_{\perp} = 9.03$			
	5	$C_{4v}(F^-)$GS:	$g_{\parallel} = 7.81$, $g_{\perp} = 6.259$			BM86
		ES:	$g_{\parallel} = 1.735$, $g_{\perp} = 9.19$		$B_2^0 = 632.8$, $B_4^0 = 1{,}028.0$, $B_6^0 = 451.8$, $B_4^2 = -1{,}794.4$	

		$C_{4v}(H^-)$GS:	$g_\parallel = 7.48$, $g_\perp = 6.23$		
		ES:	$g_\parallel = 1.61$, $g_\perp = 9.20$	$B_2^0 = 796.8$, $B_4^0 = 1{,}025.6$, $B_6^0 = 579.7$, $B_4^2 = -1{,}794.4$	
CdF_2 (M^+ compensated)					BD74
$M^+ = Li^+$	4.2		$g_x = 5.412$, $g_y = 5.397$, $g_z = 9.071$		
$M^+ = Na^+$	4.2		$g_x = 5.898$, $g_y = 5.882$, $g_z = 8.274$		
$M^+ = Ag^+$	4.2		$g_x = 5.315$, $g_y = 6.473$, $g_z = 8.402$		
$M^+ = K^+$	4.2		$g_x = 4.999$, $g_y = 6.564$, $g_z = 8.333$		
	4.2		$g_x = 6.759$, $g_y = 6.759$, $g_z = 6.759$		
$CePt_2$	1.4–50		$g = 6.78$		BM86
$LuAl_2$	1.5–20		6.846		BM86
$ScAl_2$	1.5–20		6.840		BM86
$ScAl_2$ (p)	4.2–20		6.73		BD80
TiO_2	4.2		$A_g \sim 15.1$	$A_C \sim 2.7$	PF74
$YbAl_2$	1.5–20		6.880		BM86
$ErCl_3$	NR		$g_\parallel = 1.989$, $g_\perp = 8.757$	$A^{167} = 66.4$, $B^{167} = 304$	PF74, SJ67
LaF_3 (garnet)	20		$g_x = 2.99$, $g_y = 4.91$, $g_z = 10.89$		SJ66
PrF_3	1.4, 4.2		$g_x = 2.801$, $g_y = 4.48$, $g_z = 11.36$		BM73
YCl_3	1.5–2.1		$g_\parallel = 3.350$, $g_\perp = 2.857$		BM73
$YCl_3{\cdot}6H_2O$			$g_z = 13.74$		PF74
$LaBe_{13}$	1.7–16		6.74		BM83
$LuBe_{13}$	1.7–16		6.74		BM83
$ThBe_{13}$	1.7–16		6.76		BM83
$KMgF_3$ (cubic)	1.5		$g_{110} = 4.029$, $g_{111} = 3.995$, $g_{110} = 3.974$	$A^{161} = 30$, $B^{161} = 263$	BM73
	1.5	Axial I(A)	$g_\parallel = 4.216$, $g_\perp = 7.886$	$A^{163} = 40$, $B^{163} = 366$	
		Axial (B)	$g_\parallel = 4.394$, $g_\perp = 7.08$		
		Axial II:	$g_\parallel = 2.682$, $g_\perp = 8.362$		
$KZnF_3$	1.7–3.6	Cubic	3.925–4.065		BD80
		Trigonal	$g_\parallel = 4.390$, $g_\perp = 7.813$		BD80

Er^{3+} $4f^{11}$. Data tabulation of spin-Hamiltonian parameters.

Sample	T (°K)	Site	$\tilde{g}$	$Bg_{\parallel}/Ag_{\perp}$	$\tilde{A}^{167}$ (10^{-4} cm^{-1})	Ref.
Er ethyl sulfate	NR		$g_{\parallel} = 1.47$, $g_{\perp} = 8.85$	1.0		PF74
$CaWO_4$ (Na compensated)	NR	A	$g_1 = 13.58$, $g_2 = 2.91$, $g_3 = 0.62$			BD74
		B	$g_1 = 9.98$, $g_2 = 6.66$, $g_3 = 2.32$			
		C	$g_1 = 13.3$, $g_2 = 2.68$, $g_3 = 1.98$			
$HfSiO_4$	4.2	Tetragonal	$g_{\parallel} = 4.316$, $g_{\perp} = 6.682$		$A = 151.3$, $B = 232.6$	BD74
$LiTmF_4$	1.2–20		$g_{\parallel} = 2.960$, $g_{\perp} = 8.074$			BM86
$LiTmF_4$	1.6–7		$g_{\parallel} = 2.960$, $g_{\perp} = 8.074$			BM86
$LiYF_4$	1.6–7		$g_{\parallel} = 3.137$, $g_{\perp} = 8.105$			BM86
$LiYF_4$	4.2		$g_{\parallel} = 3.137$, $g_{\perp} = 8.105$		$A = 108.6$, $B = 280$	BM73
$SrWO_4$	1.5–4.2		$g_{\parallel} = 0.83$, $g_{\perp} = 8.47$			BD74
$ThSiO_4$	4.2	Tetragonal	$g_{\parallel} = 4.802$, $g_{\perp} = 6.309$		$A = 167.0$, $B = 219.4$	BD74
	4.2	Orthorhombic	$g_x = 0.62$, $g_y = 14.85$, $g_z = 2.795$			
$ZrSiO_4$	4.2	Tetragonal	$g_{\parallel} = 3.718$, $g_{\perp} = 6.997$		$A = 130.4$, $B = 243.8$	BD74
	4.2	Orthorhombic	$g_x = 0.31$, $g_y = 15.08$, $g_z = 2.145$			
$YAsO_4$	1.6, 4.2		$g_{\parallel} = 6.639$, $g_{\perp} = 5.177$		$A = 223$, $B = 182$	BM73
$YLiF_4$	8		$g_{\parallel} = 8.18$, $g_{\perp} = 4.43$			BM83
YPO_4	4.2		$g_{\parallel} = 6.42$, $g_{\perp} = 4.81$		$A = 214$, $B = 163.4$	BM73
LuAlG (garnet)	20		$g_x = 6.93$, $g_y = 4.12$, $g_z = 8.43$			PF74
LuGaG (garnet)	20		$g_x = 3.183$, $g_y = 3.183$, $g_z = 12.62$			PF74
YAlG (garnet)	20		$g_x = 7.75$, $g_y = 3.71$, $g_z = 7.35$			PF74
YGaG (garnet)	20		$g_x = 4.69$, $g_y = 4.03$, $g_z = 10.73$			PF74
Th_4H_{15} (p)	1.6–30		$g_{\parallel} = 3.40$, $g_{\perp} = 8.20$			BM86
$Ca_5(PO_4)_3F$	4.2		$g_{\parallel} = 16.48$, $g_{\perp} \leq 1.0$			BD74

Tm^{3+} $4f^{12}$. Data tabulation of spin-Hamiltonian parameters.

Sample	T (°K)	$\tilde{g}$	$\tilde{A}^{159}$ (10^{-4} cm^{-1})	Ref.
Thulium ethyl sulfate	4.2, 6.2, 7.2		$A/g = 110.7$	BM73
$YAsO_4$	1.6	8.03	$A = 910.0$	BD80
YVO_4	1.6	10.22	$A = 115.0$	BD80
$Tm(C_2H_5SO_4)_3 \cdot 9H_2O$	4.2	$g_{\parallel} = 13.69$	$A = 1.533$	BD74
$Y_2(C_2H_5SO_4)_3 \cdot 9H_2O$	4.2	$g_{\parallel} = 13.69$	$A = 1.533$	BD74

Tm^{2+} $4f^{13}$. Data tabulation of spin-Hamiltonian parameters.

Sample	T (°K)	$\tilde{g}$	$\tilde{A}^{169}$ (10^{-4} cm^{-1})	Ref.
$KMgF_3$	1.5, 4.2	$g = 2.591$	$A_{1/2} = 283.0$	BM73

$\mathbf{Yb^{3+}}$ $\mathbf{4f^{13}}$. Data tabulation of spin-Hamiltonian parameters.

Sample	T (°K)	Site	$\tilde{g}$	$\tilde{A}^{171,173}$ (10^{-4} cm^{-1})	Ref.
BaO	4.2		2.593	$A_z^{171} = 692$, $A_z^{173} = 192$	REY74
CaO	20		2.585	$A_{171} = 698$	PF74
	4.2		2.586	$A_z^{171} = 698$	REY74
MgO	4.2		2.566	$A_z^{171} = 681$, $A_z^{173} = 188$	REY74
SrO	4.2		2.585	$A_z^{171} = 685$, $A_z^{173} = 190$	REY74
ZnO	NR	Substitutional	$g_\parallel = 1.311$, $g_\perp = 4.421$		ART72
		Interstitial	$g_\parallel = 4.812$, $g_\perp = 2.390$		ART72
$Pb_{1-x}Yb_xTe$ ($x = 1.0\%$)	4.2		$g = 2.52$	$A^{171} = 59$ mT, $A^{173} = 18.5$ mT	IC96
$Pb_{1-x}Yb_xSe$ ($x = 0.3\%$)	4.2		$g = 2.585$	$A^{171} = 138$ mT, $A^{173} = 39$ mT	IC96
$Pb_{1-x}Yb_xS$ ($x = 1.0\%$)	4.2		$g = 2.545$	$A^{171} = 60$ mT, $A^{173} = 19$ mT	IC96
CaF_2	NR	Tetragonal		$A_\parallel = 12.37$ MHz, $A_\perp = 10.97$ MHz	BD80
CdF_2	4.2	Center I	$g_\parallel = 3.667$, $g_\perp = 0.28$		BM73
		Center II	$g_\parallel = 3.219$, $g_\perp = 0.40$		
		Center III	$g_\parallel = 3.164$, $g_\perp = 0.40$		
CdF_2	1.5–78		$g = 3.436$	$A_z^{171} = 878$, $A_z^{173} = 240$	BM73
CeO_2	4.2, 12, 20.4	Cubic	3.429		BM86
		Trigonal	$g_\parallel = 4.152$, $g_\perp = 3.060$		
$LaRh_2$	1.4–50		4.45	$A = 570$ G	BM86
α-PbF_2	4.2		3.42	$A^{171} = 905$, $A^{173} = 246$	BD80
PbF_2	1.7–20		3.429	$A^{171} = 902.8$, $A^{173} = -243.2$	BM83
SrF_2	1.7–4.2		$g_\parallel = 2.813$, $g_\perp = 3.746$	$A^{171} = 557$ G, $A^{173} = 153$ G	BD80
				$B^{171} = 557$ G, $B^{173} = 155$ G	
	4.2		$g_\parallel = 1.349$, $g_\perp = 4.420$		BM83
ThO_2	4.2, 12, 20.4	Cubic	3.424		BM86
		Trigonal	$g_\parallel = 4.495$, $g_\perp = 2.872$		
LaF_3	20		$g_x = 3.76$, $g_y = 5.20$, $g_z = 1.21$		BR61, SJ66
PrF_3	1.4–4.2		$g_x = 3.47$, $g_y = 5.427$, $g_z = 1.205$	$A_y^{171} = 1,460$	BM73
$YCl_3{\cdot}6H_2O$			$g_x < 0.5$, $g_y < 0.5$, $g_z = 5.57$		SJ67
$KMgF_3$	1.5	Cubic	$g = 2.584$	$A_{1/2}^{171} = 685$, $A_{5/2}^{173} = 189$	BM73
		Axial I	$g_\parallel = 1.844$, $g_\perp = 2.896$		
		Axial II	$g_\parallel = 1.078$, $g_\perp = 4.377$		
$KZnF_3$	1.7–36	Cubic	$g = 2.591$		BD80
		Trigonal	$g_\parallel = 1.822$, $g_\perp = 2.903$		

$SrTiO_3$	77		$g_\perp = 2.67$		PF74
	65		$g_\parallel = 2.25$, $g_\perp = 2.70$		
	50		$g_\parallel = 2.170$, $g_\perp = 2.720$		
	2		$g_\parallel = 2.10$, $g_\perp = 2.785$	$A^{171} = 530.0$, $B^{171} = 720.0$ $B^{175} = 11.0$	
$HfSiO_4$	4.2	Axial	$g_\parallel = 6.998$, $g_\perp = 0.4$		BD74
	4.2	Orthorhombic	$g_x = 3.208$, $g_y = 2.503$, $g_z = 3.071$		
$LaNbO_4$	1.5–45		$g_x = 5.35$, $g_y = 1.90$, $g_z = 1.86$		BM73
$LiYF_4$	4.2		$g_\parallel = 1.331$, $g_\perp = 3.917$	$A^{171} = 338$, $A_z^{173} = 1028$ $B_z^{171} = 93$, $A_z^{173} = 284$	BM73
$ThSiO_4$	4.2	Axial	$g_\parallel = 0.91$, $g_\perp = 3.519$		BD74
	4.2	Orthorhombic	$g_x = 4.573$, $g_y = 1.09$, $g_z = 0.815$		
$YAsO_4$	1.6–4.2		$g_\parallel = 0.53$, $g_\perp = 3.58$		BM73
(Y, Yb) AsO_4	1–4.2		$g_\parallel = 0.57$, $g_\perp = 3.58$		BD74
YPO_4	1.7–4.2		$g_\parallel = 1.522$, $g_\perp = 3.125$	$A_z^{171} = 398$, $B_z^{173} = 110$ $A_z^{171} = 832$, $B_z^{173} = 251$	BM73
$ZrSiO_4$	4.2	Orthorhombic	$g_x = 3.292$, $g_y = 2.530$, $g_z = 2.782$		BD74
Yttrium ethylsulfate	1.2–4.2		$g_\parallel = 3.328$		BM73
EuGaG	4.2		$g_x = 2.68$, $g_y = 3.66$, $g_z = 3.81$		BD80
LuAlG (garnet)	20		$g_x = 3.842$, $g_y = 3.738$, $g_z = 2.594$		PF74
LuGaG (garnet)	20		$g_x = 3.653$, $g_y = 3.559$, $g_z = 2.994$		PF74
YAlG (garnet)	20		$g_x = 3.87$, $g_y = 3.78$, $g_z = 2.47$		PF74
YGaG (garnet)	20		$g_x = 3.73$, $g_y = 3.60$, $g_z = 2.85$		PF74
Cs_2HfCl_6	NR		$g = 2.583$	$A^{171} = 572.1$ G, $A^{173} = 157.3$ G	BD74
Cs_2NaYCl_6	1.8		$g = (-)2.586$	$A = -2.065$ GHz, $E(\Gamma_8)-E(\Gamma_6) = 225\ cm^{-1}$	BD74

U^{5+} $5f^1$. Data tabulation of spin-Hamiltonian parameters.

Sample	T (°K)	Site	$\tilde{g}$	Ref.
LiF	77	Tetragonal I	$\lvert g_{\parallel}\rvert = 0.253$, $\lvert g_{\perp}\rvert = 0.472$	BM86
		Tetragonal II	$\lvert g_{\parallel}\rvert < 0.1$, $\lvert g_{\perp}\rvert = 0.587$	
		Rhombic I	$\lvert g_x\rvert = 0.517$, $\lvert g_y\rvert = 0.387$, $\lvert g_z\rvert = 0.20$	
		Rhombic II	$\lvert g_x\rvert = 0.562$, $\lvert g_y\rvert = 0.563$, $\lvert g_z\rvert = 0.337$	
		Rhombic III	$\lvert g_x\rvert = 1.110$, $\lvert g_y\rvert = 0.537$, $\lvert g_z\rvert = 0.20$	
		Rhombic IV	$\lvert g_x\rvert = 0.937$, $\lvert g_y\rvert = 0.15$, $\lvert g_z\rvert = 0.543$	
CaF_2 (γ-irrad.)	77	Rh 1	$\lvert g_x\rvert = 0.22$, $\lvert g_y\rvert = 0.821$, $\lvert g_z\rvert = 1.388$	BM83
		Rh 2	$\lvert g_x\rvert = 0.20$, $\lvert g_y\rvert = 1.610$, $\lvert g_z\rvert = 0.880$	
CaF_2	77	I	$g_x = 0.22$, $g_y = 0.821$, $g_z = 1.388$	BM83
		II	$g_x = 0.20$, $g_y = 1.614$, $g_z = 0.880$	
CaF_2	77		$\lvert g_x\rvert = 0.540$, $\lvert g_y\rvert = 2.775$, $\lvert g_z\rvert = 0.471$	BM86
$BaUO_{3.50}$	295		2.08	BM83
$CaUO_{3.8}$	295		2.04	BM83
$MgUO_{3.8}$	295		2.10	BM83
$SrUO_{3.67}$	295		2.14	BM83
BaU_3O_{11}	295		2.08	BM83
$MgU_3O_{8.90}$	295		2.10	BM83
$SrU_4O_{12.80}$	295		2.31	BM83
CaU_4O_{1200}	295		2.61	BM83

U^{4+} $5f^2$. Data tabulation of spin-Hamiltonian parameters.

Sample	T (°K)	Site	$\tilde{g}$	$\tilde{A}^{235}$ (10^{-4} cm^{-1})	Ref.
CaF_2	4.2	I	$g_{\parallel} = 2.02$	$A = 54$	BM86
		II	$g_{\parallel} = 560$	$A = 117.3$	

U^{3+} $5f^3$. Data tabulation of spin-Hamiltonian parameters. ($S = \frac{1}{2}$)

Sample	T (°K)	$\tilde{g}$	Ref.
BaF_2	1.6–10	$g_\parallel = 3.334$, $g_\perp = 2.115$	BM83
CaF_2 (uv irrad.)	77	$g_\parallel = 3.579$, $g_\perp = 1.88$	BM86
CaF_2	4.2	$g_\parallel = 1.793$, $g_\perp = -3.152$	BM83

Np^{4+} $5f^3$. Data tabulation of spin-Hamiltonian parameters.

Sample	T (°K)	$\tilde{g}$	Ref.
Cs_2ZrCl_6	1.5–1.65	$g = 0.40$	BM83

Am^{4+} $5f^5$. Data tabulation of spin-Hamiltonian parameters. (The ground state is Γ_7.)

Sample	T (°K)	$\tilde{g}$	$\tilde{A}^{241,243}$ (10^{-4} cm^{-1})	Ref.
ThO_2	1.5–4.2	1.286	$A^{241} = 45.7$, $A^{243} = 45.3$	BM73

Pu^{3+} $5f^5$. Data tabulation of spin-Hamiltonian parameters. (The ground state is Γ_7.)

Sample	T (°K)	$\tilde{g}$	$\tilde{A}^{239}$ (10^{-4} cm^{-1})	Ref.
$SrCl_2$	1.5–4.2	1.120	127.9	BM73
ThO_2	1.5–4.2	1.312	65.4	BM73

Am^{2+} $5f^7$. Data tabulation of spin-Hamiltonian parameters. Here c is a fourth-order crystal-field parameter. The results are on an excited state (Γ_7) resonance.

Sample	T (°K)	g	c (cm^{-1})	$\tilde{A}^{241}$ ($I = \frac{5}{2}$)	$\tilde{A}^{243}$ ($I = \frac{5}{2}$)	Ref.
$SrCl_2$		1.928	−1.92	83.9 G	8.32 G	ART 72
SrF_2	4.2	$g_J = 1.925$				BD74

Cm^{3+} $5f^7$. Data tabulation of spin-Hamiltonian parameters. Here c is a fourth-order crystal field parameter.

Sample	T (°K)	$\tilde{g}$	c (cm^{-1})	$\tilde{A}^{245}$ ($I = \frac{7}{2}$) (G)	μ^{245} ($I = \frac{5}{2}$)	Ref.
$SrCl_2$		1.926	−1.83	19.0	$\pm 0.5\,\mu_N$	ART 72
$SrCl_2$	4.2, 25	$g_J = 1.928$				BD74
SrF_2		$g_J = 1.926$				BD74
ThO_2		$g_J = 1.924$				BD74

Cf^{3+} $5f^9$. Data tabulation of spin-Hamiltonian parameters ($I = \frac{9}{2}$).

Sample	T (°K)	$\tilde{g}$	$\tilde{A}^{249}$ (10^{-4} cm^{-1})	Ref.
$Cs_2NaLuCl_6$	4.2	$g = 6.273$	75.2	BD80

(*continued from page 153*)
symmetry, $\|$ and $\perp$ subscripts indicate the values along and perpendicular to the symmetry axis. The following abbreviations have been used: G = Gauss, GHz = gigahertz, HT = high temperature, kG = kiloGauss, LT = low temperature, LHT = liquid-helium temperature, LNT = liquid-nitrogen temperature, RT = room temperature, MHz = megahertz, NR = not reported, P = polycrystalline. Finally, although the listed data do not represent a complete collection, they still include a reasonably good volume of published data, especially for the Co^{2+}, Fe^{3+}, Gd^{3+}, and Ni^{2+} ions for which no comprehensive listing of data appears in the literature.

REFERENCES TO TRANSITION ION DATA TABULATION

ART72. J. O. Artman, Magn. Reson. Rev. **1**, 169 (1972). (Review article.)

BA94. V. Bindilatti, A. N. Anisimov, N. F. Oliveira, Jr., Y. Shapira, M. Goiran, F. Yang, S. Isber, M. Averous, and M. Demicniuk, Phys. Rev. **B50**, 16464 (1994).

BD74. H. A. Buckmaster and D. B. Delay, Magn. Reson. Rev. **3**, 127 (1974). (Review article.)

BD80. H. A. Buckmaster and D. B. Delay, Magn. Reson. Rev. **5**, 121 (1980). (Review article.)

BM73. H. A. Buckmaster, Magn. Reson. Rev. **2**, 273 (1973). (Review article.)

BM83. H. A. Buckmaster, Magn. Reson. Rev. **8**, 283 (1983). (Review article.)

BM86. H. A. Buckmaster, Magn. Reson. Rev. **11**, 81 (1986). (Review article.)

BMM90. G. Bacquet, S. K. Misra, L. E. Misiak, and F. Fabre, Solid State Commun. **75**, 369 (1990).

BR61. J. M. Baker and R. S. Rubins, Proc. Phys. Soc. (London) **78**, 1133 (1961).

CC76. T.-T. Chang and M. I. Cohen, J. Chem. Phys. **64**, 5255 (1976).

CJ88. P. Chand, V. K. Jain, and G. C. Upreti, Magn. Reson. Rev. **14**, 49 (1988). (Review article.)

CM71. J. G. Clarke and J. A. Mackinnon, Canad. J. Phys. **49**, 1539 (1979).

CU85. P. Chand, G. C. Upreti, S. K. Misra, and M. Bartkowski, J. Chem. Phys. **82**, 5307 (1985).

GT72. L. E. Gerkin and D. L. Thorsell, J. Thorsell, J. Chem. Phys. **57**, 2665 (1972).

IA95. S. Isber, M. Averous, Y. Shapira, V. Bindilatti, A. N. Anisimov, N. F. Oliveira, Jr., V. M. Orera, and M. Demianiuk, Phys. Rev. **B51**, 15211 (1995).

IC95A. S. Isber, S. Charar, V. Mathet, C. Fau, M. Averous, and Z. Galacki, Phys. Rev. **B52**, 1678 (1995).

IC95B. S. Isber, S. Charar, V. Mathet, C. Fau, and M. Averous, Phys. Rev. **B51**, 15578 (1995).

IC96. S. Isber, S. Charar, X. Gratens, C. Fau, M. Averous, S. K. Misra, and Z. Golacki, Phys. Rev. **B54**, 7634 (1996).

IS95. S. Isber, S. Charar, V. Mathet, C. Fau, and M. Averous, Phys. Rev. **B51**, 15578 (1995).

JW66. D. R. Johnston and E. Y. Wong, J. Chem. Phys. **45**, 1073 (1966).

Koh79. R. P. Kohin, Magn. Reson. Rev. **5**, 75 (1979). (Review article.)

KM88. S. Z. Korczak and S. K. Misra, Acta Phys. Polon. **A73**, 45 (1988).

LB. Landolt-Börnstein, Zahlenwerte und Funktionen aus Naturwissenchaften und Technik, Neue Serie, Band 12, Ergänzungshand 4 (1973, 1974) zu Band 2, Teilband b: Paramagnetische Elektronen resonanz, E. König und G. König (Springer-Verlag, Berlin, 1984).

MB83. S. K. Misra, M. Bartkowski, and P. Mikolajczak, J. Chem. Phys. **78**, 5369 (1983).

MB87. S. K. Misra, G. Bacquet, and F. Fabre, Solid State Commun. **62**, 729 (1987).

MBF90. S. K. Misra, G. Bacquet, and J. Frandon, J. Phys.: Conden. Matter **2**, 5603 (1990).

MW87. S. K. Misra, C. Wang, S. Han, and S. Z. Korczack, Phys. Rev. **B36**, 3542 (1987).

MC95. S. K. Misra, Y. Chang, V. Petkov, S. Isber, S. Charar, C. Fau, M. Averous, and Z. Galacki, J. Phys.: Conden. Matter 7, 7897 (1995).

MCK95. S. K. Misra, Y. Chang, and J. Kotlinski, J. Phys.: Conden. Matter 7, 3327 (1995).

MF84. S. K. Misra, F. Fabre, J. Frandon, and P. Mikolajczak, Solid State Commun. **50**, 541 (1984)

MGS77. S. K. Misra, P. Giguère, and G. R. Sharp, J. Chem. Phys. **66**, 1758 (1977).

MK85. S. K. Misra, M. Kahrizi, P. Mikolajczak, and L. Misiak, Phys. Rev. **B32**, 4738 (1985).
MK87. S. K. Misra and S. Z. Korczak, J. Phys. C: Solid State Phys. **20**, 4485 (1987).
MK91. S. K. Misra and S. Z. Korczak, Physica **B168**, 143 (1991).
ML90. S. K. Misra and X. Li, J. Phys.: Conden. Matter **2**, 9613 (1990).
ML91. S. K. Misra and X. Li, J. Phys.: Conden. Matter **3**, 9741 (1991).
ML92A. S. K. Misra and X. Li, J. Phys. Rev. **B45**, 2927 (1992).
ML92B. S. K. Misra and X. Li, J. Phys.: Conden. Matter **4**, 3659 (1992).
ML92C. S. K. Misra and X. Li, J. Phys. Rev. **B45**, 12883 (1992).
MLM90. S. K. Misra, X. Li, E. Misiak, and C. Wang, Physica **B167**, 209 (1990).
MSO90. S. K. Misra, J. Sun, and U. Orhun, Phys. Status. Sol. (b) **162**, 585 (1990).
MLW91. S. K. Misra, X. Li and C. Wang, J. Phys.: Conden. Matter **3**, 8479 (1991).
MM78. S. K. Misra and P. Mikolajczack, J. Chem. Phys. **69**, 3095 (1978).
MM79. S. K. Misra and P. Mikolajczack, J. Chem. Phys. Solids **40**, 477 (1979).
MM80. S. K. Misra and P. Mikolajczack, J. Chem. Phys. Solids **41**, 617 (1980).
MM88. L. E. Misiak and S. K. Misra, Phys. Rev. **B38**, 8673 (1988).
MM93. S. K. Misra and L. E. Misiak, Phys. Rev. **B48**, 13579 (1993).
MMC94. S. K. Misra, L. E. Misiak, and J. Capobianco, J. Phys.: Conden. Matter **6**, 3955 (1994).
MMK81. S. K. Misra, P. Mikolajczack, and S. Z. Korczak, J. Chem. Phys. **74**, 922–926 (1981).
MMCB94. S. K. Misra, L. E. Misiak, J. Capobianco, and M. Bettinelli, J. Magn. Reson. **109**, 216 (1994).
MKK92. S. K. Misra, M. Kahrizi, and S. Z. Korczak, Physica **B182**, 186 (1992).
MMC94. S. K. Misra, L. E. Misiak, and P. Chand, Physica **B202**, 31 (1994).
MR88. S. K. Misra and C. Rudowicz, Phys. Statist. Sol. **147**, 680 (1988).
MS74. J. A. Mackinnon and G. Stevenson, Canad. J. Phys. **52**, 91 (1974).
MS75. J. A. Mackinnon and G. Shannon, Canad. J. Phys. **53**, 841 (1975).
MS76. S. K. Misra and G. R. Sharp, J. Chem. Phys. **65**, 3056 (1976).
MS90. S. K. Misra and J. Sun, Phys. Rev. **B42**, 8601 (1990).
MS90A. S. K. Misra and J. Sun, Physica **B162**, 331 (1990).
MS91. S. K. Misra and J. Sun, Phys. Rev. **B44**, 11862 (1991).
MS91A. S. K. Misra, and J. Sun, Phys. Status. Sol. (b), **162**, 265 (1991).
MSL91. S. K. Misra, J. Sun, and X. Li, Physica **B168**, 170 (1991).
MSrev91. S. K. Misra and J. Sun, Magn. Reson. Rev. **16**, 57 (1991). (Review article.)
MW89A. S. K. Misra and C. Wang, Phys. Rev. **B39** 8832 (1989).
MW89B. S. K. Misra and C. Wang, Phys. Status. Sol. (b) **154**, 259 (1989).
MW89C. S. K. Misra and C. Wang, Physica **B159**, 321 (1989).
MW89D. S. K. Misra and C. Wang, Physica **B160**, 129 (1989).
MW90. S. K. Misra and C. Wang, Phys. Rev. **B41**, 1 (1990).
MWrev90. S. K. Misra and C. Wang, Magn. Reson. Rev. **14**, 157 (1990). (Review article.)
PF74. *CRC Handbook of Spectroscopy*, Chapter on Electron Spin Resonance by C. P. Poole, Jr., and H. A. Farach, pp. 217–314 (1974). CRC, Boca Raton, FL.
RDH89. R. S. Rubins, J. E. Drumheller, and L. S. Hutton, J. Chem. Phys. **91**, 3614 (1989).
REY72. R. W. Reynolds, L. A. Boatner, Y. Chen, and M. M. Abraham, Phys. Rev. Lett. **29**, 18 (1972).
REY74. R. W. Reynolds, L. A. Boatner, Y. Chen, and M. M. Abraham, J. Chem. Phys. **60**, 1593 (1974).
ROG71. R. N. Rogers, P. I. Biagi, and A. T. Long, J. Chem. Phys. **55**, 5417 (1971).
SJ66. M. B. Schulz and C. D. Jeffries, Phys. Rev. **149**, 270 (1966).
SJ67. M. B. Schulz and C. D. Jeffries, Phys. Rev. **159**, 277 (1967).
VJ90. V. K. Jain, Magn. Reson. Rev. **14**, 261 (1990). (Review article.)
WM68. G. A. Woonton and J. A. Mackinnon, Canad. J. Phys. **46**, 59 (1968).
YCRC93. T. H. Yeom, Y. M. Chang, C. Rudowicz, and S. H. Choh, Solid State Commun. **87**, 245 (1993).
YRC94. T. H. Yeom, C. Rudowicz, and S. H. Choh, Ferroelectrics **156**, 249 (1994).

CHAPTER IX

EVALUATION OF SPIN-HAMILTONIAN PARAMETERS FROM EPR DATA OF SINGLE CRYSTALS

Sushil K. Misra

Physics Department, Concordia University, Montreal, Quebec H3G 1M8, Canada

A. INTRODUCTION

A number of papers dealing with the evaluation of spin-Hamiltonian parameters (SHPs) from EPR data of single crystals have been published over the years. Conventionally, perturbation expressions giving theoretical dependence of resonant EPR line positions on SHPs have been used for this purpose. However, the disadvantage here is that since the EPR line positions are those obtained only for the special orientations of the external magnetic field, e.g., parallel to the X, Y, or Z magnetic axes, it is not possible to estimate the individual values of all SHPs. This is because the line positions for a chosen orientation of the external magnetic field depend only upon certain combinations of some SHPs. In addition, only the absolute values of some of the parameters can be determined. Furthermore, one is restricted by the approximations of the perturbation technique.

As far as computer evaluation of SHPs using exact numerical diagonalization of the spin-Hamiltonian (SH) matrix is concerned, it is certainly possible to estimate accurately the individual values of all SHPs provided that data corresponding to several orientations of the external magnetic field are simultaneously fitted in a rigorous least-squares fitting (LSF) manner. Since the number of simultaneously fitted data points is large, errors of estimated SHPs are reduced considerably.

It is the objective of this chapter to provide a unified description of the perturbation, brute-force, and least-squares fitting techniques, which can be used for evaluation of SHPs from single-crystal EPR data.

B. PERTURBATION APPROACH

A number of publications have dealt with the perturbation calculation of eigenvalues of SH for electron-nuclear spin-coupled systems.[1–36] (References 1–36 are here listed in chronological order, except for refs. 15, 29–34.) The earliest articles were confined to zero- and first-order considerations of fine-structure (FS) and hyperfine-structure (HFS) energies[1–4] and some reviews and surveys are available which provide general background information.[8,27–29] Second-order perturbation theory has been applied to spin systems in which the $\tilde{\mathbf{g}}^2$ and hyperfine tensors have coincident axes,[6,12] as well as to those systems with noncoincident axes.[11,13–15,17–19,21] Orthorhombic,[12] monoclinic,[23] and triclinic[23] symmetries, as well as nonuniaxial anisotropy[32] and low symmetries, in general,[20,22,27,29] have been treated. Extension to third order in perturbation has been discussed,[10,16,24,33–36] but not widely applied. Special approaches, based on the perturbation treatment involving circularly polarized radiation with reference to the asymmetric nature of the g-matrix,[5] the nature of the A-matrix,[2] the application of seventh-order degenerate state perturbation theory,[9] the half-field spectra,[16] the intensities of allowed and forbidden transitions,[17] the electron-nuclear double resonance (ENDOR),[7,25] the Zeeman terms cubic in the spin (HS^3),[32] and methods to evaluate SHPs[20,26,29–31] have been described. References are available for the LSF[29–31] and brute-force[26] methods treated in this chapter.

The eigenvalues calculated by perturbation can, in principle, be used to evaluate SHPs from EPR or ENDOR data. The resonance condition requires that for an observed transition the difference in the pair of eigenvalues, participating in resonance, be equal to $h\nu$, where ν is the frequency of microwave radiation and h is Planck's constant. The "allowed" ($\Delta M = \pm 1, \Delta m = 0$) and sometimes forbidden ($\Delta M \neq \pm 1$, $\Delta m = 0$; $\Delta M = \pm 1, \Delta m \neq 0$; or $\Delta M \neq \pm 1, \Delta m \neq 0$) transitions. (Here M and m denote, respectively, the electronic and nuclear magnetic quantum numbers.) The resonant magnetic field values corresponding to a transition can be obtained by equating the energy eigenvalue difference $|E_{M,m} - E_{M',m'}|$ to $h\nu$, the quantum of energy absorbed in the transition $M, m \leftrightarrow M', m'$ resposible for the resonance. Many papers contain the expressions for the eigenvalues or the resonant-field values. Chambers et al.[37] gave expressions for the resonant-field values appropriate for a spin Hamiltonian corresponding to rhombic or lower symmetry and containing the fine-structure terms up to fourth order, including an isotropic hyperfine interaction term. Tennant[21] and McGavin et al.[38] gave eigenvalues for any orientation of the external magnetic field for a fine-structure interaction, including the terms up to sixth order in the crystal field; particular attention was paid to inclusion of crossterms (simultaneous contribution of two crystal-field potentials to eigenvalues, i.e., terms depending on products of two SHPs corresponding to different crystal field potentials). They claimed excellent agreement with data, even for cases where the off-diagonal terms in the spin Hamiltonian were large. Upreti[39] also gave expressions for fine structure, as well as for hyperfine structure, eigenvalues for an arbitrary orientation of the external magnetic field corresponding to orthorhombic symmetry: terms only up to second order in fine-structure crystal-field potential were taken into account, but hyperfine, nuclear Zeeman and nuclear quadrupole interactions were considered in detail. The final eigenvalue expressions, from which the resonant-field values for an EPR transition can be obtained, for an arbitrary orientation of the external magnetic field with respect to the crystal-field axes, are described below.

The following spin Hamiltonian covers most of the cases encountered in practice

$$\mathscr{H} = \mathscr{H}_{\text{f}} + \mathscr{H}_{\text{hf}},$$

where

$$\mathcal{H}_{\mathrm{f}} = \mu_{\mathrm{B}}\mathbf{S}\cdot\mathbf{g}\cdot\mathbf{H} + B_2^0O_2^0 + B_2^2O_2^2 + B_4^0O_4^0 + B_4^2O_4^2 + B_4^3O_4^3 + C_4^3R_4^3 \quad . \\ + B_4^4O_4^4 + B_6^0O_6^0 + B_6^2O_6^2 + B_6^4O_6^4 + B_6^6O_6^6 + B_6^3O_6^3 + C_6^3R_6^3 \tag{1a}$$

and

$$\mathcal{H}_{\mathrm{hf}} = \mathbf{S}\cdot\mathbf{A}\cdot\mathbf{I} - g_{\mathrm{N}}\mu_{\mathrm{N}}H\cdot I + Q'\{I_z^2 - \tfrac{1}{3}I(I+1)\} + Q''(I_x^2 - I_y^2). \tag{1b}$$

In (1) $\mathcal{H}_{\mathrm{f}}$ represents the fine-structure part, while $\mathcal{H}_{\mathrm{hf}}$ represents the hyperfine-structure part of the spin Hamiltonian. **S** and **I** are the electronic and nuclear spins, respectively; μ_{B} and μ_{N} are the Bohr and nuclear magnetons, respectively; g_{N} is the nuclear g factor; **A** is the hyperfine-interaction tensor, and Q', Q'' represent the nuclear-quadrupole-interaction constants. In the expression for $\mathcal{H}_{\mathrm{f}}$, the O_l^m are the operator equivalents as defined by Abragam and Bleaney[8] and R_4^3 is an imaginary spin operator,[21] often neglected due to its small contribution.

After transforming the crystal field and hyperfine terms of (1) to Zeeman diagonal coordinates, the perturbation technique can be employed to arrive at the eigenvalues $E_{M,m}$. In particular, to second order in perturbation,

$$E_{M,m} = H'M + E_{\mathrm{f}}(M) + E_{\mathrm{hf}}(M,m), \tag{2}$$

$$\begin{aligned} E_{\mathrm{f}}(M) = {} & \tfrac{1}{2}\{(3a_{33}{}^2 - 1)B_2^0 + C_2^{2'}B_2^2\}f_1(S,M) + \tfrac{1}{8}(35a_{33}{}^4 - 30a_{33}{}^2 + 3) \\ & \times B_4^0 f_2(S,M) + \left(\frac{\sqrt{231}}{4}\{(231a_{33}{}^6 - 315a_{33}{}^4 + 105a_{33}{}^2 - 5\right)B_6^0 \\ & + (33a_{33}{}^4 - 18a_{33}{}^2 + 1)C_2^{2'}B_6^2 + (11a_{33} - 1)C_4^{4'}B_6^4 + C_6^{6'}B_6^6\})g_{60} \\ & + \tfrac{1}{8}\{(7a_{33}{}^2 - 1)C_2^{2'}B_4^2 + a_{13}a_{33}C_3^{3'}B_4^3 + a_{23}a_{33}S_3^{3'}C_4^3 + C_4^{4'}B_4^4\} \\ & \times f_2(S,M) + \frac{XX^*}{4H'}f_3(S,M) + \frac{YY^*}{8H'}f_4(S,M) + \frac{AA^*}{56H'}f_5(S,M) \\ & + \frac{BB^*}{56H'}f_6(S,M) + \frac{CC^*}{24H'}f_7(S,M) + \frac{DD^*}{64H'}f_8(S,M) \\ & + \frac{(AX^* + A^*X)}{4\sqrt{14}H'}f_9(S,M) + \frac{(BY^* + B^*Y)}{8\sqrt{7}H'}f_{10}(S,M) \\ & + FF^*(g_{6-2}{}^2/e_4 + g_{62}{}^2/e_3) + (FY^* + F^*Y)(g_{2-2}g_{6-2}/e_4 + g_{22}g_{62}/e_3) \\ & + (BF^* + B^*F)(g_{4-2}g_{6-2}/e_4 + g_{42}g_{62}/e_3) + HH^*(g_{6-4}{}^2/e_8 + g_{64}{}^2/e_7) \\ & + (DH^* + D^*H)(g_{4-4}g_{6-4}/e_8 + g_{44}g_{64}/e_7) + KK^*(g_{6-6}{}^2/e_{10} + g_{66}{}^2/e_9). \end{aligned} \tag{3}$$

Here, the magnetic flux density, H, is at an angle $\cos^{-1}(l,m,n)$ to the directions of the principal values, g_x, g_y, g_z, of the g-matrix, $H' = g\mu_{\mathrm{B}}H, a_{13} = lg_x/g$, $a_{23} = mg_y/g$,

TABLE 1. Values of $f_n(S, M)$ and $g_{l\pm m}$ required in (3).

Function	Value
$f_1(S, M)$	$3M^2 - S(S+1)$
$f_2(S, M)$	$35M^4 + 25M^2 - 30M^2S(S+1) - 6S(S+1) + 3S^2(S+1)^2$
$f_3(S, M)$	$S_{0-1}{}^2(2M-1)^2 - S_{01}{}^2(2M+1)$
$f_4(S, M)$	$S_{0-1}{}^2S_{-1-2}{}^2 - S_{01}{}^2S_{12}{}^2$
$f_5(S, M)$	$S_{0-1}{}^2\{f_5^-(S, M)\}^2 - S_{01}{}^2\{f_5^+(S, M)\}^2$
$f_5^{\mp}(S, M)$	$14M^3 \mp 21M^2 + 19M - 6MS(S+1) \pm 3S(S+1) \mp 6$
$f_6(S, M)$	$S_{0-1}{}^2S_{-1-2}{}^2\{f_6^-(S, M))\}^2 - S_{01}{}^2S_{12}{}^2\{f_6^+(S, M)\}^2$
$f_6^{\mp}(S, M)$	$7M^2 \mp 14M - S(S+1) + 9$
$f_7(S, M)$	$S_{0-1}{}^2S_{-1-2}{}^2S_{-2-3}{}^2(2M-3)^2 - S_{01}{}^2S_{12}{}^2S_{23}{}^2(2M+3)^2$
$f_8(S, M)$	$S_{0-1}{}^2S_{-1-2}{}^2S_{-2-3}{}^2S_{-3-4}{}^2 - S_{01}{}^2S_{12}{}^2S_{23}{}^2S_{34}{}^2$
$f_9(S, M)$	$S_{0-1}{}^2(2M-1)f_5^-(S, M) - S_{01}{}^2(2M+1)f_5^+(S, M)$
$f_{10}(S, M)$	$S_{0-1}{}^2S_{-1-2}{}^2f_6^-(S, M) - S_{01}{}^2S_{12}{}^2f_6^+(S, M)$

$g_{2\pm 2} = \frac{1}{2}S_{0\pm 1}S_{\pm 1\pm 2}$,

$g_{4\pm 2} = \frac{1}{2}\sqrt{\frac{1}{7}}S_{0\pm 1}S_{0\pm 2}\{7M^2 \pm 14M + [9 - S(S+1)]\}$,

$g_{4\pm 4} = \frac{1}{4}S_{0\pm 1}S_{\pm 1\pm 2}S_{\pm 2\pm 3}S_{\pm 3\pm 4}$,

$g_{6\pm 2} = \frac{1}{8}\sqrt{\frac{5}{11}}S_{0\pm 1}S_{\pm 1\pm 2}\{33M^4 \pm 132M^3 + 3[91 - 6S(S+1)]M^2$
$\pm 6[47 - 6S(S+1)]M + [120 - 26S(S+1) + S^2(S+1)^2]\}$,

$g_{6\pm 4} = \frac{1}{4}\sqrt{\frac{3}{22}}S_{0\pm 1}S_{\pm 1\pm 2}S_{\pm 2\pm 3}S_{\pm 3\pm 4}\{11M^2 \pm 44M + [50 - S(S+1)]\}$,

$g_{6\pm 6} = \frac{1}{8}S_{0\pm 1}S_{\pm 1\pm 2}S_{\pm 2\pm 3}S_{\pm 3\pm 4}S_{\pm 4\pm 5}S_{\pm 5\pm 6}$

S_{ab} $\{S(S+1) - (M+a)(M+b)\}^{1/2}$.

$a_{33} = ng_z/g$; $C_n^{n'} + iS_n^{n'} = (a_{13} - ia_{23})^n$, and $g = (l^2g_x{}^2 + m^2g_y{}^2 + n^2g_z{}^2 + 2lmg_xg_y + 2lng_xg_z + 2mng_yg_z)^{1/2}$.

The $f_n(S, M)$ $(n = 1, 10)$ as well as the g_{im} are given in Table 1. In a plane containing a principal g-matrix direction the coefficients $X, Y, A, \ldots$ may be written explicitly in terms of a_{13}, a_{23}, a_{33} and the crystal-field coefficients $B_2^0, B_2^2, B_4^0, \ldots$. In general directions this can only be done for the modulus square coefficients $XX^*, YY^*, AA^*, \ldots$. The expressions for X, Y, A, B, C and D in the (g_x, g_y) and (g_x, g_z) planes are given in Table 2 and should cover most cases of practical interest. The explicit forms for the coefficients Y, B, D, F, H, K, that are required for inclusion of the parameters B_6^m are given in Table 3 (only for the magnetic field orientation in the directions g_x, g_y, and g_z.) Finally, the e-denominators are as follows: $e_3 = -e_4 = 2H'$, $e_7 = -e_8 = 4H'$, and $e_9 = -e_{10} = 6H'$.

The expressions for the resonant fields for the "allowed" transitions for $S = \frac{7}{2}$ (e.g., Gd^{3+}, Eu^{2+}) can be expressed to fourth order in perturbation for the magnetic field orientation along the z-axis as follows:[40]

$$g\mu_B H = g\mu_B H^{(2)} + g\mu_B H^{(4)}, \tag{4}$$

TABLE 2. Values of the coefficients X, Y, A, B, C, D, in the (a) $g_x g_y$ and (b) $g_x g_z$ planes. $a_{13}, a_{23}, a_{33}, C_n^{n'}$, and $S_n^{n'}$ are defined in (3).

Coefficient		Value
(a)	X	$-iS_2^2 B_2^2$
	Y	$\frac{3}{2}B_2^0 + \frac{1}{2}C_2^{2'}B_2^2$
	A	$-\frac{i\sqrt{7}}{2\sqrt{2}}\{S_2^{2'}B_4^2 - 2S_4^{4'}B_4^4\} - \frac{\sqrt{7}}{4\sqrt{2}}\{C_3^{3'}B_4^3 + S_3^{3'}C_4^3\}$
	B	$\frac{\sqrt{7}}{2}\{-5B_4^0 + C_2^{2'}B_4^2 + C_4^{4'}B_4^4\} - \frac{i\sqrt{7}}{4}\{S_3^{3'}B_4^3 - C_3^{3'}C_4^3\}$
	C	$\frac{\sqrt{i}}{2\sqrt{2}}\{7S_2^{2'}B_4^2 + 2S_4^{4'}B_4^4\} - \frac{3}{4\sqrt{2}}\{C_3^{3'}B_4^3 - S_3^{3'}C_4^3\}$
	D	$\frac{1}{4}\{35B_4^0 + 7C_2^{2'}B_4^2 + C_4^{4'}B_4^4\} - \frac{i}{4}\{S_3^{3'}B_4^3 - C_3^{3'}C_4^3\}$
(b)	X	$a_{13}a_{33}(-3B_2^0 + B_2^2)$
	Y	$\frac{3}{2}{a_{13}}^2 B_2^0 + \frac{1}{2}(1 + {a_{33}}^2)B_2^2$
	A	$\frac{\sqrt{7}}{4\sqrt{2}}a_{13}\{-20a_{33}(7{a_{33}}^2 - 3)B_4^0 + 4A_{33}(7{a_{33}}^2 - 4)B_4^2 + a_{13}(4{a_{33}}^2 - 1)B_4^3$ $+ 4{a_{13}}^2 a_{33}B_4^4\} - \frac{i3\sqrt{7}}{4\sqrt{2}}{a_{13}}^2 a_{33}C_4^3,$
	B	$\frac{\sqrt{7}}{\sqrt{2}}\{5a_{13}^2(7{a_{33}}^2 - 1)B_4^0 + (7{a_{33}}^4 - 6{a_{33}}^2 + 2)B_4^2 + a_{13}{a_{33}}^3 B_4^3$ $+ {a_{13}}^2(1 + {a_{33}}^2)B_4^4\} - \frac{i\sqrt{7}}{4}a_{13}(3{a_{33}}^2 - 1)C_4^3,$
	C	$\frac{1}{4\sqrt{2}}\{-140{a_{13}}^3 a_{33}B_4^0 - 28a_{13}{a_{33}}^3 B_4^2 + (4{a_{33}}^4 + 3{a_{33}}^2 - 3)B_4^3$ $+ 4a_{13}a_{33}({a_{33}}^2 + 3)B_4^4\} - \frac{i}{4\sqrt{2}}a_{33}(9{a_{33}}^2 - 5)C_4^3,$
	D	$\frac{1}{4}(35{a_{13}}^4 B_4^0 + 7{a_{13}}^2(1 + {a_{33}}^2)B_4^2 - A_{13}a_{33}({a_{33}}^2 + 3)B_4^3$ $+ ({a_{33}}^4 + 6{a_{33}}^2 + 1)B_4^4\} + \frac{i}{4}a_{13}(3{a_{33}}^2 + 1)C_4^3.$

where to second-order perturbation, $g\mu_B H^{(2)}$ of (4) are

$$\begin{aligned}
\pm\tfrac{7}{2} \leftrightarrow \pm\tfrac{5}{2}: g\mu_B H^{(2)} &= h\nu \mp (6b_2^0 + 20b_4^0 + 6b_6^0) + E[45/(1 \pm 3y) \\
&\quad - 21/(1 \pm 5y)], \\
\pm\tfrac{5}{2} \leftrightarrow \pm\tfrac{3}{2}: g\mu_B H^{(2)} &= h\nu \mp (4b_2^0 - 10b_4^0 - 14b_6^0) + E[60/(1 \pm y) \\
&\quad - 21/(1 \pm 5y) - 45/(1 \pm 3y)], \\
\pm\tfrac{3}{2} \leftrightarrow \pm\tfrac{1}{2}: g\mu_B H^{(2)} &= h\nu \mp (2b_2^0 - 12b_4^0 + 14b_6^0) + E[21/(1 \pm 5y) \\
&\quad - 45/(1 \pm 3y) - 120y/(1 - y^2)], \\
+\tfrac{1}{2} \leftrightarrow -\tfrac{1}{2}: g\mu_B H^{(2)} &= h\nu + E[90/(1 - 9y^2) - 120/(1 - y^2)],
\end{aligned} \tag{5}$$

TABLE 3. Values of the coefficients of (3) in principal directions of the g-tensor.

Coefficient	g_x direction (upper sign) g_y direction (lower sign)	g_z direction
diag(6)	$\frac{\sqrt{231}}{4}(-5B_6^0 \pm B_6^2 - B_6^4 \pm B_6^6)$	$4\sqrt{231}B_6^0$
Y	$\frac{3}{2}B_2^0 \pm \frac{1}{2}B_2^2$	B_2^2
B	$\frac{\sqrt{7}}{2}(-5B_4^0 \pm B_4^2 + B_4^4)$	$\sqrt{7}B_4^2$
D	$\frac{1}{4}(35B_4^0 \pm 7B_4^2 + B_4^4)$	$2B_4^4$
F	$\frac{\sqrt{11}}{8\sqrt{5}}(105B_6^0 \mp 17B_6^2 + 5B_6^4 \pm 15B_6^6)$	$\frac{4\sqrt{11}}{\sqrt{5}}B_6^2$
H	$\frac{\sqrt{22}}{8\sqrt{3}}(-63B_6^0 \pm 3B_6^2 + 13B_6^4 \pm 3B_6^6)$	$\frac{2\sqrt{22}}{\sqrt{3}}B_6^4$
K	$\frac{1}{8}(231B_6^0 \pm 33B_6^2 + 11B_6^4 \pm B_6^6)$	$4B_6^6$

where $y = b_2^0/g\mu_B H^{(2)}$, $E = (b_2^2)^2/18g\mu_B H^{(2)}$. The parameters b_{2r}^s are related to the parameters B_{2r}^s of (1) in the following way: $b_2^s = 3B_2^s$ $(s = 0, 2)$; $b_4^s = 60B_4^s$ $(s = 0, 2, 4)$; $b_6^s = 1260B_6^s$ $(s = 0, 2, 4, 6)$. To fourth order in perturbation, $g\mu_B H^{(4)}$ of (4) are

$$\begin{aligned}
\pm\tfrac{7}{2} \leftrightarrow \pm\tfrac{5}{2} : g\mu_B H^{(4)} &= g\mu_B H^{(2)} + x[150/(1 \pm 3y)^2(1 \pm y) - 225/(1 \pm 3y)^3 \\
&\quad - 70/(1 \pm 5y)^2(1 \pm 3y) + 49/(1 \pm 5y)^3], \\
\pm\tfrac{5}{2} \leftrightarrow \pm\tfrac{3}{2} : g\mu_B H^{(4)} &= g\mu_B H^{(2)} + x[150/(1 \pm y)^2(1 \mp y) - (7/(1 \pm 5y)^2 \\
&\quad + 20/(1 \pm y)^2)(-7/(1 \pm 5y) + 20/(1 \pm y)) \\
&\quad - 150/(1 \pm 3y)^2(1 \pm y) + 225/(1 \pm 3y)^3], \\
\pm\tfrac{3}{2} \leftrightarrow \pm\tfrac{1}{2} : g\mu_B H^{(4)} &= g\mu_B H^{(2)} + x[70/(1 \mp y)^2(1 \mp 3y) \\
&\quad - 150/(1 \pm y)^2(1 \mp y) - (15/(1 \pm 3y)^2 \\
&\quad + 20/(1 \mp y)^2)(-15/(1 \pm 3y) + 20/(1 \mp y)) \\
&\quad + (7/(1 \pm 5y)^2 + 20/(1 \pm y)^2)(-7/(1 \pm 5y) \\
&\quad + 20/(1 \pm y))], \\
+\tfrac{1}{2} \leftrightarrow -\tfrac{1}{2} : g\mu_B H^{(4)} &= g\mu_B H^{(2)} - x[70/(1 + y)^2(1 + 3y) \\
&\quad + 70/(1 - y)^2(1 - 3y) + (20/(1 + y^2) \\
&\quad + 15/(1 - 3y)^2)(-20/(1 + y) + 15/(1 - 3y)) \\
&\quad + (20/(1 - y)^2 + 15/(1 + 3y)^2)(-20/(1 - y) \\
&\quad + 15/(1 + 3y))].
\end{aligned} \tag{6}$$

In (6) $x = E^2/18g\mu_B H^{(4)}$, $y = b_2^0/g\mu_B H^{(4)}$ $(E = (b_2^2)^2/18g\mu_B H^{(4)})$. (The x and y in (5) for $g\mu_B H^{(2)}$ here should also be replaced by these expressions.)

In order to express the resonant fields for the external magnetic field orientation along the x-axis one needs to transform the SHP in (5) and (6) as follows:[41]

$$
\begin{aligned}
&b_2^0 \to (-b_2^0 + b_2^2)/2,\\
&b_2^2 \to -(3b_2^0 + b_2^2)/2,\\
&b_4^0 \to (3b_4^0 - b_4^2 + b_4^4)/8,\\
&b_4^2 \to (5b_4^0 - b_4^2 - b_4^4)/2,\\
&b_4^4 \to (35b_4^0 + 7b_4^2 + b_4^4)/8,\\
&b_6^0 \to (-5b_6^0 + b_6^2 - b_6^4 + b_6^6)/16,\\
&b_6^2 \to (-105b_6^0 + 17b_6^2 - 5b_6^4 - 15b_6^6)/32,\\
&b_6^4 \to (-63b_6^0 + 3b_6^2 + 13b_6^4 + 3b_6^6)/16,\\
&b_6^6 \to (-231b_6^0 - 33b_6^2 - 11b_6^4 - b_6^6)/32.
\end{aligned}
$$

As for $S = \frac{5}{2}$ (e.g., Mn^{2+}, Fe^{3+}) the expressions for the energy differences W $(= h\nu)$ to fourth order in perturbation containing only the second order SHP are available for the various allowed transitions.[37] They are

$$
\begin{aligned}
W(M: \pm\tfrac{5}{2} \leftrightarrow \pm\tfrac{3}{2}) &= H' \mp 12B_2^0 + 4(B_2^2)^2/H' \pm 18B_2^0(B_2^2)^2/H'^2\\
&\quad + 325(B_2^0)^2(B_2^2)^2/H'^3 + 28(B_2^2)^4/H'^3,\\
W(M: \pm\tfrac{3}{2} \leftrightarrow \pm\tfrac{1}{2}) &= H' \mp 6B_2^0 - 5(B_2^2)^2/H' \pm 99B_2^0(B_2^2)^2/H'^2\\
&\quad + 405(B_2^0)^2(B_2^2)^2/H'^3 - \tfrac{5}{4}(B_2^2)^4/H'^3,\\
W(M: \tfrac{1}{2} \leftrightarrow -\tfrac{1}{2}) &= H' - 8(B_2^2)^2/H' + 648(B_2^0)^2(B_2^2)^2/H'^3 + 56(B_2^2)^4/H'^3. \qquad (7)
\end{aligned}
$$

In (7) $H' = \mu_B gH$. One should substitute for B_2^0 and B_2^2, when $\vec{H}\|\vec{x}$ or $\vec{H}\|\vec{y}$, respectively, as follows:

$$\tfrac{3}{2}(B_2^2 - B_2^0) \quad \text{and} \quad -\tfrac{1}{2}(3B_2^0 + B_2^2) \qquad (\vec{H}\|\vec{x})$$

or

$$\tfrac{3}{2}(B_2^0 + B_2^2) \quad \text{and} \quad \tfrac{1}{2}(3B_2^0 - B_2^2). \qquad (\vec{H}\|\vec{y})$$

The case of low site symmetry requires the evaluation of various noncoincident tensors appearing in the spin Hamiltonian. This is discussed in detail in Section E.5, with a full review of various techniques published in the literature.

As far as the hyperfine energy is concerned, to second order in perturbation,[39] we have

$$
\begin{aligned}
E_{\mathrm{hf}}(M,m) = {} & AMm + \frac{B^2}{2g\mu_{\mathrm{B}}H}[m\{M^2 - S(S+1)\} + M\{I(I+1) - m^2\}] \\
& - g_{\mathrm{N}}\mu_{\mathrm{N}}Hm + P[m^2 - \tfrac{1}{3}I(I+1)] \\
& + \frac{|\lambda|^2}{(2g\mu_{\mathrm{B}}H)^2}\frac{2Am}{M}[\{M^2 - S(S+1)\}^2 - M^2] \\
& + \frac{|\rho|^2}{(g\mu_{\mathrm{B}}H)^2}2AMm[2M^2 + 1 - 2S(S+1)] \\
& + \frac{A^2\sigma}{(2g\mu_{\mathrm{B}}H)^2}[\{S(S+1) - M(M+1)\}\{I(I+1) - m(m-1)\}(2m+1) \\
& - \{S(S+1) - M(M-1)\}\{I(I+1) - m(m+1)\}(2M-1)] \\
& + \frac{A^3}{(2g\mu_{\mathrm{B}}H)^2}[\{S(S+1) - M(M+1)\}\{I(I+1) - m(m-1)\} \\
& \times (m - M - 1) \\
& + \{S(S+1) - M(M-1)\}\{I(I+1) - m(m+1)\}(M - m - 1)]. \qquad (8)
\end{aligned}
$$

In (8), the terms beyond the second have been computed by neglecting the small anisotropy in the hyperfine tensor (i.e., $A \approx B$) and in the g tensor (i.e., $g \approx g_{\parallel} \approx g_{\perp}$).

In the expression for E_{hf} in (8):[39]

$$
\begin{aligned}
\sigma &= \tfrac{3}{2}[B_2^0(3\cos^2\theta - 1) + B_2^2\sin^2\theta\cos 2\phi], \\
\rho &= \tfrac{1}{4}[3B_2^0\sin^2\theta + B_2^2(\cos^2\theta\cos 2\phi - \sin^2\phi + 2iB_2^2\cos\theta\sin 2\phi], \\
\lambda &= \sin\theta[(3B_2^0 - B_2^2\cos 2\phi)\cos\theta + iB_2^2\sin 2\phi], \\
P &= \tfrac{1}{2}[Q'(3\cos^2\theta - 1) + 3Q''\sin^2\theta\cos 2\phi].
\end{aligned}
$$

In the above expression θ is the angle made by the external magnetic field $\vec{H}$ with the principal (z) axis of the O_2^0 spin operator, and the projection of $\vec{H}$ in the xy-plane makes an angle ϕ with the x-axis.

C. BRUTE-FORCE METHODS

In these methods, although exact numerical diagonalization of the spin-Hamiltonian matrix on a digital computer is performed, no well-defined criteria are used to arrive at the final set of SHP corresponding to the absolute minimum chi-square, starting from an initial set of SHP. (Hence the title "brute-force.") Thus, these methods require long computer times, although exact eigenvalues are computed at each stage. The methods described in Sections C.1 and C.2 below were applied specifically to cases of fine-structure splittings only. However, there is no reason why they cannot be applied to cases exhibiting both fine and hyperfine structures.

1. Variation of One Parameter at a Time[42]

In this procedure the value of one parameter is varied at a time over a selected range in several steps starting with a chosen value while the values of other parameters are kept fixed, and resonant fields measured for only one orientation of the external magnetic field are considered. All the parameters are thus covered in descending order of their magnitudes. The sum of the squares of the differences (SD) of the calculated eigenvalues of the levels participating in resonance for the various transitions, calculated by diagonalizing the SH, from that of the quantum of energy of the microwave radiation is calculated for each new choice of the parameter being varied along with the fixed values of the remaining parameters. The particular value of the parameter (the optimum value) being varied that yields the minimum value of SD is then retained for the variation of other parameters. After having determined the optimum values of all the parameters in this manner, the next iteration is carried out by judiciously narrowing the searching ranges of the parameters about their optimum values. The iteration process is repeated until the SD is reduced to a value consistent with experimental uncertainties.

This method has the following disadvantages:

(i) It cannot be successfully applied to cases characterized by large off-diagonal elements in the SH matrix.
(ii) Altering the value of one parameter to obtain the minimum value of the SD changes the values of the previously determined values of the parameters that corresponded to the minimum SD so that they do not correspond any more to the minimum SD. Thus, many more iterations are required before the fit becomes self-consistent.
(iii) Since the resonant-field values used to determine SHPs here are those measured for one orientation of the external field, a good fit along one orientation of the field does not yield the SHPs which are consistent with the spectrum in an orthogonal direction of the field, especially when the off-diagonal elements in the SH matrix are large.

2. Variation of Parameters in Subgroups[43]

This brute-force method does not suffer from the disadvantages of the method described in Section A above, where only one parameter is varied at a time. It is capable of handling cases in which there are large off-diagonal elements in the SH matrix, as well as providing the orthogonal-axis consistency. In addition, economy in computer time is achieved here, as in practice one does not have to make as many diagonalizations of the SH matrix as are required when only one parameter is varied at a time. Briefly, in this method, those parameters, whose values are of the same order of magnitude, are varied simultaneously.[43] Thus the SHPs are divided into subgroups according to orders of magnitude of their values; the various subgroups are treated in descending order of magnitude of their values. Such a grouping is obtained by including in the same subgroup all B_l^m with the same l value. This is necessary, since one is simultaneously considering resonant magnetic fields, corresponding to two (or more) directions; under transformation of coordinate axes the O_l^m (l fixed) transform amongst themselves.

The basic steps of this procedure are as follows:

(i) *Choice of the Initial Values of Parameters.* The large parameters may be estimated by a perturbation approach. The initial values of the smaller parameters may be chosen to be zero.

(ii) *Choice of Resonant-Field Values.* One must use at least as many resonant magnetic field values corresponding to two, or more orientations as there are parameters; usually these are the fields corresponding to the "allowed" transitions; if there are not enough of these, field values corresponding to the "forbidden" transitions may be included. Thus the number of data points used for fitting is increased thereby reducing parameter errors, in addition to achieving the orthogonal-axis consistency.

(iii) *Computation Procedure.* The various computational steps for the estimation of SHPs are summarized as follows:

(a) Choose the initial values of the parameters and their ranges. Usually the same range may be chosen for all B_l^m with the same l value.

(b) Arrange the subgroups in decreasing order of magnitude of values. The parameters are then varied about their initial values according to

$$B_l^m = B_l^m \text{ (initial)} + n_l \Delta B_l^m,$$

where $n_l = -m_l, -(m_l - 1), \ldots, 0, \ldots, (m_l - 1), m_l$; m_l is an appropriately chosen integer depending upon the computer time available, and ΔB_l^m is the increment in the value of B_l^m at each step.

(c) In each iteration the various parameters are considered in subgroups having values of the same order of magnitude, with all possible combinations of the values of its parameters, as described in (b). Of these, the one that yields the lowest chi-square (SD; defined in Section C.1 above) is retained to serve as the initial set of values for that subgroup in the next iteration, as well as for the variation of parameters in other subgroups in the same iteration. The particular set of values of all the parameters in any iteration which yields the lowest chi-square (SD_{min}) is used as the set of initial values for the next iteration. The SD_{min} is now compared with an initially chosen small value V_{min}, consistent with experimental uncertainties. If $SD_{min} \leq V_{min}$, the computation is terminated; otherwise one goes to the next iteration as described in step (d) below.

(d) In the next iteration, the searching ranges of the parameters are reduced (which is equivalent to reducing ΔB_l^m by appropriate factors f_l; $f_l \leq m_l$ in order to ensure adequate overlap of the values of parameters between successive iterations), and one continues as in step (c) above.

The above method was, indeed, successfully applied in ref. 43 to the evaluation of SHPs for Gd^{3+}-doped $SmCl_3 \cdot 6\,H_2O$, characterized by large off-diagonal elements in the SH matrix and orthogonal-axis consistency was, in fact, achieved. As mentioned earlier, this case could not be satisfactorily treated by the technique described in Section C.1 by varying one parameter at a time. The least-squares fitting procedure, although much quicker, yielded the same values of the parameters. This is discussed below.

D. LEAST-SQUARES FITTING (LSF) METHOD

1. Introduction

In the least-squares fitting method one does indeed use a mathematical criterion to arrive at the final set of parameters in an iterative manner, starting from an initial set of parameters.[19,44] However, here one requires the values of the first and second derivatives of chi-square, which in turn depend on the derivatives of the eigenvalues with respect to SHPs which are difficult to evaluate rigorously, as analytic expressions for the eigenvalues are not available, especially for arbitrary orientations of the external magnetic field, and for cases characterized by the presence of large off-diagonal elements in the SH matrix. However, this difficulty can be circumvented by the use of Feynman's theorem,[45] using numerically evaluated eigenvalues and eigenvectors of the SH matrix on a digital computer as shown below. This economises the computer time substantially, as the vector whose components are SHPs, is transformed in one step in the hyperspace of SHPs in each iteration.

The various details of the LSF method as outlined in ref. 46 will be given in Section D.2 below as applied specifically to the fitting of EPR data, followed by a discussion showing the equivalence of the various methods. However, before going to Section D.2, a brief review of the general LSF procedure is presented here, basically to illustrate the underlying technique, as well as to introduce the notation used subsequently.

The column vector $\vec{a}^m$ (components are $a_1^m, a_2^m, \ldots, a_N^m$) describing the set of N parameters corresponding to the absolute minimum value of chi-square (S', defined by (12) below), can be obtained from the column vector $\vec{a}^I$ describing the initially chosen values of the parameters as follows:[44,47]

$$\vec{a}^m = \vec{a}^I - (D'')^{-1} D', \tag{9}$$

where D' is the (column) vector whose elements are the first derivatives of S' with repect to the parameters, evaluated at $\vec{a}^I$, and D'' is the matrix whose elements are the second derivatives of S' with respect to the parameters evaluated at $\vec{a}^m$. Thus

$$D'_k = (\partial S'/\partial a_k)_{\vec{a}^I}, \tag{10}$$

$$D''_{k_1 k_2} = (\partial^2 S'/\partial a_{k_1}\, \partial a_{k_2})_{\vec{a}^m}. \tag{11}$$

The chi-square value (S'), corresponding to a set of parameters described by the (column) vector $\vec{a}$ is defined as

$$S' = \sum_i (f_i(x_i, \vec{a}) - F_i)^2/\sigma_i^2 \tag{12}$$

(the index i covers all data points). In (12) each F_i represents the experimental value for a given experimental variable x_i, σ_i is the corresponding weight factor (related to standard deviation), and $f_i(x_i, \vec{a})$ is the value calculated with the parameters $\vec{a}$, corresponding to the experimental F_i.

Since one does not know $\vec{a}^m$ to begin with, the elements of the matrix D'' are evaluated with respect to $\vec{a}^I$ (referred to as $D''(\vec{a}^I)$); then one obtains a new set of parameters described by the vector $\vec{a}^f$, where

$$\vec{a}^f = \vec{a}^I - (D''(\vec{a}^I))^{-1}\vec{D}' \tag{13}$$

in place of (9). Thus the difference vector $\Delta\vec{a} = \vec{a}^I - \vec{a}^f$, which will be referred to later, can be expressed as

$$\Delta\vec{a} = D''(\vec{a}^I)^{-1}D', \tag{14}$$

which gives, upon multiplying both sides by $D''(\vec{a}^I)$,

$$D''(\vec{a}^I)\Delta\vec{a} = D'. \tag{15}$$

In the fitting, (13) is used iteratively until a sufficiently small S', consistent with experimental uncertainty, is obtained.

2. Details of the LSF Method

The single-ion spin-Hamiltonian $\mathscr{H}$ can be expressed, in the notation of Abragam and Bleaney,[8] as

$$\mathscr{H} = \mu_B \mathbf{H} \cdot \mathbf{g} \cdot \mathbf{S} + \sum_{l,m} B_l^m O_l^m \quad (l \text{ even}, |m| \le l). \tag{16}$$

In the analysis of EPR data one is interested in evaluating the values of the parameters a_j [g_{ij} and B_l^m of (16)] from the resonant magnetic field values obtained for one, or more, orientations of the external magnetic field.

The chi-square function S' for EPR data is

$$S' = \sum_i [(|\Delta E_i| - h\nu_i)^2/\sigma_i^2], \tag{17}$$

where $\Delta E_i = E_{i'} - E_{i''}$ ($E_{i'}$ and $E_{i''}$ are the energies of the levels between which the microwave energy $h\nu_i$ is absorbed at resonance, and ν_i is the corresponding klystron frequency which is usually the same for all i. The $E_{i'}$ and $E_{i''}$ are obtained by substituting the (numerical) values of the parameters and the resonant-field value H_i into the spin Hamiltonian (16) and diagonalizing it on the computer. In (17) σ_i is the weight factor for the ith line position.

In the use of the LSF method, one needs the values of the first and second derivatives of S' with respect to the parameters. This enables one to construct D' and

D'' as discussed in Section D.1. From (17) these are

$$\frac{\partial S'}{\partial a_j} = 2\sum_i \frac{\Delta E_i}{|\Delta E_i|}\frac{(|\Delta E_i| - h\nu_i)}{\sigma_i^2}\left(\frac{\partial E_i'}{\partial a_j} - \frac{\partial E_i''}{\partial a_j}\right), \tag{18}$$

$$\frac{\partial^2 S'}{\partial a_j \partial a_l} = 2\sum_i \frac{1}{\sigma_i^2}\left[\left(\frac{\partial E_{i'}}{\partial a_j} - \frac{\partial E_{i''}}{\partial a_j}\right)\left(\frac{\partial E_{i'}}{\partial a_l} - \frac{\partial E_{i''}}{\partial a_l}\right)\right.$$
$$\left. + \frac{\Delta E_i}{|\Delta E_i|}(|\Delta E_i| - h\nu_i)\left(\frac{\partial^2 E_{i'}}{\partial a_j\,\partial a_l} - \frac{\partial^2 E_{i''}}{\partial a_j\,\partial a_l}\right)\right]. \tag{19}$$

From (18) and (19) it is seen that the problem reduces to the evaluation of the first and second derivatives of $E_{i'}$ and $E_{i''}$ with respect to the parameters. These can be evaluated, numerically, from Feynman's theorem,[45] according to which

$$\frac{\partial E_i}{\partial a_j} = \left\langle \varphi_i \left| \frac{\partial \mathscr{H}}{\partial a_j} \right| \varphi_i \right\rangle, \tag{20}$$

where the $|\varphi_i\rangle$ is the eigenvector of the SH matrix $\mathscr{H}$ having the eigenvalue E_i:

$$\mathscr{H}|\varphi_i\rangle = E_i|\varphi_i\rangle. \tag{21}$$

Exploitation of (20) leads to[46]

$$\frac{\partial^2 E_i}{\partial a_j \partial a_l} = \sum_n{}' \left\langle \varphi_i \left| \frac{\partial \mathscr{H}}{\partial a_l} \right| \varphi_n \right\rangle \left\langle \varphi_n \left| \frac{\partial \mathscr{H}}{\partial a_j} \right| \varphi_i \right\rangle \Big/ (E_i - E_n) + \text{c.c.} \tag{22}$$

In (22) the sum over n goes over all the eigenvalues of $\mathscr{H}$, the prime over the summation sign indicates that $n \neq i$, and c.c. stands for complex conjugate.

Finally, the derivatives can be expressed in a form amenable to computer evaluation:

$$\frac{\partial E_i}{\partial a_j} = \mathrm{Tr}\left[\frac{\partial \mathscr{H}}{\partial a_j}(|\varphi_i\rangle \otimes \langle \varphi_i|)\right] \tag{23}$$

and

$$\frac{\partial^2 E_i}{\partial a_j \partial a_l} = \sum_n{}' \mathrm{Tr}\left[\frac{\partial \mathscr{H}}{\partial a_l}(|\varphi_n\rangle \otimes \langle \varphi_n|)\frac{\partial \mathscr{H}}{\partial a_j}(|\varphi_i\rangle \otimes \langle \varphi_i|)\right] \Big/ (E_i - E_n) + \text{c.c.} \tag{24}$$

In (23) and (24), Tr stands for the trace of a matrix, $|\varphi_i\rangle \otimes \langle \varphi_i|$ is the $(2S+1) \times (2S+1)$ matrix obtained by taking the outer product of the column (eigen) vector $|\varphi_i\rangle$ with itself, whose elements are

$$(|\varphi_i\rangle \otimes \langle \varphi_i|)_{pq} = |\varphi_i\rangle_p |\varphi_i\rangle_q^*, \tag{25}$$

and $\partial\mathscr{H}/\partial q_j$ are the spin operators either proportional to components of $\vec{S}$ or O_l^m as found from (16).

The following steps may, then, be used for the evaluation of SHPs from EPR data by the LSF technique:

(i) Make an initial estimate of the SHP (components of $\vec{a}^I$) by the second-order perturbation theory. Choose the values of SHPs that cannot be so determined to be zero.

(ii) For orthogonal-axis consistency consider simultaneously the resonant field values corresponding to two, or more, orientations of the external magnetic field. If an insufficient number of "allowed" ($|\Delta M| = 1$) lines is available, consider "forbidden" ($|\Delta M| > 1$) lines also. The greater the number of data points used for fitting the smaller is the parameter error. For each resonant-field value H_i, calculate the elements of the SH matrix using the $\vec{a}^I$, determined in step (i) above. Compute the eigenvalues E_i and the eigenvectors $|\varphi_i\rangle$ using an appropriate computer subroutine.

(iii) Using the E_i and $|\varphi_i\rangle$, evaluate $S', \partial S'/\partial a_j$, and $\partial^2 S'/\partial a_j\, \partial a_l$ from (17), (18), (19), (23), (24), and (25).

(iv) Calculate the matrices D' and D'', defined in (10) and (11) above.

(v) Compute the eigenvalues and eigenvectors of D''. To avoid nonconvergence, if any eigenvalue (say mth) is negative or zero, then replace it as follows:

$$\begin{aligned}(D_d'^{-1})_{mm} &= 1/|(D_d)_{mm}| \qquad \text{for} \quad (D_d)_{mm} < 0,\\ &= 0 \qquad\qquad\qquad\; \text{for} \quad (D_d)_{mm} = 0,\end{aligned}$$

where D_d is the diagonal matrix obtained by the diagonalization of D'', i.e., $D_d = UD''(\vec{a}^1)U^+$, where U is an appropriate unitary matrix, so that $D''(\vec{a}^I)^{-1} = U^+ D_d^{-1} U$ to be used in (11). Normally, if $\vec{a}^I$ is sufficiently close to $\vec{a}^m$ all the eigenvalues of $D''(\vec{a}^I)$ are positive definite and the LSF procedure quickly leads to convergence.

Calculate a new set of parameters constituting the vector $\vec{a}^f$ from (13) using the initial set of values of $\vec{a}^I$ of step (i), and with the resulting $\vec{a}^f$ to calculate a new S'. If this S' is less than the old S', go to step (vi); otherwise, use the strategies described in Section D.4 below.

(vi) If the last $S' \leq S'_{\min}$, where $S'_{\min}$ is a small value consistent with experimental uncertainties, terminate the computation; $\vec{a}^f$ as obtained in step (v) above then describes the values of parameters giving the best overall fit to the data. Otherwise, go back to step (ii) using the $\vec{a}^f$ obtained in step (v) above in the place of $\vec{a}^I$.

The LSF method was successfully applied[46] to the fitting of EPR data for Gd^{3+}-doped $SmCl_3 \cdot 6\,H_2O$, characterized by large off-diagonal elements in the SH matrix. It yielded the same set of parameters as that obtained by the use of the brute-force method varying parameters in subgroups (Section D.2), although using the LSF method required only $\frac{1}{68}$ of the time on a CYBER/CDC computer, of that taken by the brute-force method. Of course, the brute-force method, varying only one parameter at a time,[42] failed for this case.

3. Calculation of Parameter Errors

The procedure for the determination of parameter errors in the fitting of EPR data, based on standard statistical analysis,[48] has been well described by Misra and Subramanian.[49] According to statistical analysis the parameter errors can be expressed as[49]

$$\delta a_j = \sqrt{\varepsilon_{jj}}, \tag{26}$$

where ε_{jj} is the jth diagonal element of the error matrix ε which is defined as follows:

$$\varepsilon = A^{-1}, \quad A_{ij} = \tfrac{1}{2}(\partial^2 S'/\partial a_i \partial a_j)_{\vec{a}}, \tag{27}$$

where S' is the chi-square defined by (12) above, the required second derivatives can be computed for EPR by the use of (19), (23), and (24), when all σ_i are equal the parameter errors can be estimated from the following expression:[48]

$$(\delta a_j)^2 = [S'(1)/(N' - N)]\varepsilon_{jj}(1), \tag{28}$$

where $\varepsilon(1)$ is the error matrix obtained when all $\sigma_i = 1, N'$ is the number of lines simultaneously fitted, and N is the number of parameters used. This equation also expresses the fact that for N' large, the parameter errors are inversely proportional to the square root of N'. That means that the larger the number of lines fitted, the smaller is the parameter error.

In general, σ_i's have to be appropriately estimated for EPR data for use in (27). Here σ_i represents the uncertainty in the determination of the quantity $(|\Delta E_i| - h\nu_i)$ appearing in (17). This may arise due to uncertainty in the determination of the center of the first-derivative line, as well as that in the determination of the orientation θ_i of the magnetic field with respect to an axis in a chosen plane. Finally,[49]

$$\sigma_i = \left[\left(\frac{\partial \Delta E_i}{\partial H}\right)^2 \delta H_i^2 + \left(\frac{\partial \Delta E_i}{\partial \theta}\right)^2 \delta \theta_i^2\right]^{1/2}, \tag{29}$$

where

$$\frac{\partial \Delta E_i}{\partial H} = \mathrm{Tr}\left[\left(\frac{\partial \mathscr{H}}{\partial H}\right)(|\varphi_{i'}\rangle \otimes \langle \varphi_{i'}| - |\varphi_{i''}\rangle \otimes \langle \varphi_{i''}|)\right]$$

and

$$\frac{\partial \Delta E_i}{\partial \theta} = \mathrm{Tr}\left[\left(\frac{\partial \mathscr{H}}{\partial \theta}\right)(|\varphi_{i'}\rangle \otimes \langle \varphi_{i'}| - |\varphi_{i''}\rangle \otimes \langle \varphi_{i''}|)\right].$$

In (25) i' and i'' designate the levels participating in resonance and $\mathscr{H}|\varphi_i\rangle = E_i|\varphi_i\rangle$.

4. General Strategies to Achieve Convergence

If the initial set of parameters $\vec{a}^I$ is close enough to that corresponding to the absolute minimum of chi-square, $\vec{a}^m$, the LSF iteration procedure quickly leads to convergence to $\vec{a}^m$. If this is not the case, the following strategies may be used to achieve convergence. If these do not work, one should carefully reestimate the initial values of the parameters $\vec{a}^I$.

a. Use of Interpolated Fields

This technique has been proposed for the fitting of EPR data. Basically, in this technique, one computes using the set of initial parameters, theoretical resonant fields corresponding to the actual resonant fields to be fitted. These may be calculated either from perturbation expressions, or by a brute-force technique. In the latter, numerical diagonalization of the SH matrix is performed for successively increasing magnetic field values until the calculated energy difference for a chosen field value ΔE_i becomes equal to $h\nu_i$; this field value is the calculated field value correspoding to the given transition. The differences between the calculated and experimental field values are then divided into parts by a judiciously chosen integer n_I (>1). The fitting to the parameters is then made sucessively with respect to these interpolated fields, starting with the field lying closest to the calculated fields until such time one finally reaches the observed fields. The parameters $\vec{a}^I$ are thus modified in a step-wise manner toward $\vec{a}^m$, using the sets of interpolated fields. The disadvantage of this technique is the long computer time taken in calculating the theoretical fields by the brute-force method. Furthermore, Jain and Upreti,[50] who exploited the LSF technique as proposed by Uhrin,[51] found that in the case of nonconvergence, the technique of interpolated fields ($n_I > 1$) did not work.

b. Use of Interpolated Frequencies

Recognizing the difficulty in computing theoretical fields, it appears easier to compute theoretical frequencies to be compared with the observed klystron frequencies, as these theoretical frequencies are proportional to the differences between the energies corresponding to the levels participating in resonance. One can then approach toward $\vec{a}^m$ starting with $\vec{a}^I$ in a step-wise fashion using the interpolating frequencies, in the same way as one proceeded with the use of interpolated fields described in D.4.a above. It should be remarked that this technique also does not always bring convergence.

c. Use of Binary Chop[46]

In the case when the value of S'^f (the chi-square obtained with $\vec{a}^f$ in a certain iteration) is greater that S'^I (the chi-square value obtained with the starting parameters for that interation) one can improve the situation by replacing $\Delta\vec{a} \equiv D''(a^I)^{-1} \times D'$ to be used in (13) for computing $\vec{a}^f$, by $\Delta\vec{a}/2$, and then comparing the resulting S'^f with S'^I. If this $S'^f < S'^I$, one proceeds with this $\vec{a}^f$ in the usual manner, otherwise one bisects $\Delta\vec{a}$ again, and so on, until the S'^f obtained using the new $\vec{a}^f$ is less than S'^I.

E. OTHER APPLICATIONS OF THE LSF METHOD

The LSF method described in Section D was developed specifically for a spin Hamiltonian with fine-structure terms only, with the assumption that the principal axes of the g tensor are coincident with those of the crystal-field b_2^m tensor (i.e., one only had to consider g_z, g_x, and g_y). The LSF technique can, however, be extended to cover the cases when there are hyperfine terms present in the spin Hamiltonian, as well as noncoincident anisotropic g and A tensors. On the other hand, one could compute rigorously the intensities of EPR lines for an arbitrary orientation of the external magnetic field; since the intensities depend on SHPs, the line intensities, along with the line positions can be fitted in an LSF manner to evaluate SHPs. The line positions can also be estimated from the known value of the line position at a close orientation of the external magnetic field by the LSF technique to compute the angular variation of EPR spectra.[55] (For more details see Section E.3.)

1. Electron-Nuclear Spin-Coupled Systems[57]

When the SH contains both the fine structure (FS) and hyperfine-structure (HFS) terms there are certain difficulties encountered in the application of the LSF method.

(i) The HFS terms cause further splitting of the FS energy levels. Thus, the various energy levels for these electron-nuclear spin-coupled (ENSC) systems are not at all sufficiently spaced unlike those for systems with only electron spins. Thus there may be many pairs of energy levels whose differences are reasonably close to $h\nu_i$. Then, it may be extremely difficult to identify the energy levels participating in resonance for a given EPR line on the basis of energy difference, especially when the initial set of SHPs is far from that corresponding to the absolute minimum of chi-square. Thus in order to achieve convergence of the LSF procedure, it is essential that the energy levels participating in resonance are correctly specified.

(ii) Since here one is dealing with rather a large SH matrix (of dimension $(2S+1)(2I+1) \times (2S+1)(2I+1)$; e.g., 36×36 for Mn^{2+}, $S = I = \frac{5}{2}$) it is important to choose good initial values of SHPs to save computer time in arriving at convergence quickly as discussed below in Section E.1.a.

a. *Estimation of Initial Values of FS SHPs*

These can be obtained by using the line positions that correspond to the presence of FS terms only. These lines are fitted by the LSF technique to evaluate the FS SHPs. This is done by taking the line positions $H(M,m)$ for multiplets corresponding to HFS terms (same M, all various m) and averaging them to obtain the resulting FS line position had the HFS interaction been turned to zero. To this end, one can take advantage of the perturbation expressions as given in Section B above. Specifically, the contribution H to the resonant-field value for the transition $M, m \leftrightarrow M-1, m$ by the HFS terms $AS_zI_z + B(S_xI_x + S_yI_y)$ for $\vec{H} \| \vec{z}$ is given by[30,52]

$$H_{\rm hf}(M,m) = -Am - \frac{B^2}{2H_0}[I(I+1) - m^2 + m(2M-1)], \tag{30}$$

where $H_0 = (h\nu/g\mu_{\rm B})$ [when $\vec{H} \| \vec{x}$ replace A by B and B^2 by $(A^2+B^2)/2$ in (30)]. The resulting FS line position $H_{\rm f}(M)$ ($H(M,m) = H_{\rm f}(M) + H_{\rm hf}(M,m)$) from the HFS multiplet can now be obtained from (30). For example, for Mn^{2+}:

$$H_{\rm f}(M) = \tfrac{3}{4}[H(M,\tfrac{5}{2}) + H(M,-\tfrac{5}{2})] - \tfrac{1}{8}\{[H(M,\tfrac{3}{2}) + H(M,-\tfrac{3}{2})] + [H(M,\tfrac{1}{2}) + H(M,-\tfrac{1}{2})]\}.$$

b. Estimation of HFS Parameters

The initial values of the large HFS parameters A and B should be carefully chosen and the initial values of the remaining parameters should be set equal to zero. As seen from (30) the value of A is very closely the average of all the separations of the HFS transitions for $\vec{H}\|\vec{z}$ in the $2S$ FS groups, as the average of m for $m = -I, -(I-1), \ldots, (I-1), I$ is zero. B can be similarly estimated for $\vec{H}\|\vec{x}$.

c. Identification of Energy Levels Participating in Resonance

After fitting $H_f(M)$ to evaluate the FS parameters the electronic magnetic quantum numbers for the transitions corresponding to each $H_f(M)$ become known. It is now easy to decide which particular hyperfine energy levels participate in resonance when the HFS interaction is turned on, i.e., for a resonant EPR line position as actually observed. Depending upon the sign of A and B the $(2I+1)$ hyperfine energy levels corresponding to a chosen value of M for $\vec{H}\|\vec{z}$ (or nearly $\|\vec{z}$) occur in order of increasing energy as follows:[21]

$$m = -I, -(I-1), \ldots, (I-1), I \qquad \text{if} \quad MA > 0,$$

whereas they will be

$$m = I, (I-1), \ldots, -(I-1), -I \qquad \text{if} \quad MA < 0,$$

for $\vec{H}\|\vec{x}, A$ is to be replaced by B.

d. Construction of the SH Matrix for ENSC Systems

The SH $\mathscr{H}(S,I)$ for ENSC systems consists of terms of the type $\mathscr{H}'(S,I) = \mathscr{H}'(S)\mathscr{H}'(I), \mathscr{H}''(S)$, and $\mathscr{H}''(I)$. The SH matrix elements corresponding to these terms can then be expressed in the direct-product space as

$$\langle M'm'|\mathscr{H}'(S,I)|M'',m''\rangle = \langle M'|\mathscr{H}'(S)|M''\rangle\langle m'|\mathscr{H}'(I)|m''\rangle,$$

$$\langle M'm'|\mathscr{H}''(S)|M'',m''\rangle = \langle M'|\mathscr{H}''(S)|M''\rangle\langle m'|\,\mathbb{I}_I\,|m''\rangle,$$

and

$$\langle M'm'|\mathscr{H}''(I)|M'',m''\rangle = \langle M'|\,\mathbb{I}_S\,|M''\rangle\langle m'|\mathscr{H}''(I)|m''\rangle,$$

where $\mathbb{I}_S$ and $\mathbb{I}_I$ represent the unit operators in the electronic and nuclear spin spaces, respectively.

e. Absolute Signs of Parameters

The LSF procedure gives the correct relative signs of all the FS, as well as of all the HFS, parameters. Thus knowing the absolute sign of one FS parameter, i.e., B_2^0 and one HFS parameter, i.e., A, the absolute signs of all parameters can be deduced. The absolute sign of B_2^0 can be determined from the relative intensities of lines at liquid helium temperature (LHT).[8] If the intensity of the high-field lines for $\vec{H}\|\vec{z}$ increases (decreases) relative to the low-field lines as the temperature is reduced the absolute sign of B_2^0 is positive (negative). The absolute sign of A can be determined either from hyperfine interaction data, or from the overall splitting of the hyperfine sets.[52] If the overall splitting of the hyperfine sets for $\vec{H}\|\vec{z}$ decreases with increasing field then B_2^0 and A have the same sign, otherwise they have opposite signs. For $\vec{H}\|\vec{x}$, similar considerations apply for the sign of B.

The LSF technique as described in this section has been successfully applied, e.g., to Mn^{2+}-doped $ZnSiF_6 \cdot 6\,H_2O$.[53]

2. Fitting of ENDOR Data

The considerations of Section E.1 above for the determination of SHPs from EPR data for ENSC systems may be applied to the determination of SHPs from electron nuclear double resonance (ENDOR) data. In ENDOR, one saturates an FS transition $M' \leftrightarrow M''$, and observes the ENDOR frequencies corresponding to absorption between the HF levels corresponding to the FS levels M' and M''. Thus the data to be fitted are the ENDOR frequencies which depend on the differences between appropriate energy levels for the ENSC system for a given orientation of the external magnetic field with respect to the crystal-field axes. Hence, in the expression for chi-square, ΔE_i are the energy differences between the levels participating in ENDOR, while ν_i are the observed ENDOR frequencies. With this cognizance, the LSF technique as developed above for the evaluation of SHPs from EPR data for ENSC systems can be applied similarly to the analysis of ENDOR data. For example, it was successfully applied to the case of Fe^{3+}-doped guanidinium aluminum sulfate hexahydrate.[54]

3. Angular Variation of EPR Spectra

Here, for a given set of SHPs, one computes the resonant fields for different orientations of the external magnetic field in a given plane. For $\vec{H}\|\vec{z}$, one can calculate these fields from perturbation expressions provided that the off-diagonal elements in the SH matrix are not large. However, for arbitrary orientations of the external magnetic field it is difficult to calculate perturbation expressions. Another method is the application of the brute-force technique. However, this requires exorbitantly long computer time, especially when S and I are large.

The LSF technique not only provides accurate values of the calculated fields, it is also computer efficient.[55] The known resonant-field values $H_r(\theta)$ for a given orientation (θ) of the magnetic field are used as initial values for computing the corresponding fields $H_r(\theta+\delta\theta)$ at an angle $\delta\theta$ from it, using the LSF criterion. Specifically.

$$H_r(\theta+\delta\theta) = \text{iterative limit of } \left[H' - \left(\frac{\partial^2 S'}{\partial H^2}\right)_{H'}^{-1}\left(\frac{\partial S'}{\partial H}\right)_{H'}\right]. \tag{31}$$

In (31) one starts with $H' = H_r(\theta)$; S' is the chi-square defined by (17). Thus starting from $H_r(\theta)$ one can determine $H_r(\theta+\delta\theta)$. One can use these $H_r(\theta+\delta\theta)$ to compute

$H_r(\theta + 2\delta\theta)$. In this way, starting from an initial set of $H_r(\theta)$ the LSF technique can be used to calculate fields for various magnetic field orientations in a given plane, by appropriately choosing a small $\delta\theta$.

a. Computation of the Initial Resonant Fields $H_r(\theta)$

These can be either computed by the use of perturbation expressions (Section B) for $\vec{H}\|\vec{z}$, or $\vec{H}\|\vec{x}$, provided that the off-diagonal elements in the SH matrix are not large, or by the brute-force technique mentioned above, especially when the off-diagonal elements in the SH matrix are large.

b. Computation of the First and Second Derivatives of S' with Respect to H Required in (31)

Applying the procedure outlined in Section D.2, these can be expressed as follows:

$$\frac{\partial S'}{\partial H} = 2\frac{\Delta E_i}{|\Delta E_i|}(|\Delta E_i| - h\nu)\left(\frac{\partial E_i'}{\partial H} - \frac{\partial E_i''}{\partial H}\right),$$

$$\frac{\partial^2 S'}{\partial H^2} = 2\left(\frac{\partial E_i'}{\partial H} - \frac{\partial E_i''}{\partial H}\right)^2 + 2(|\Delta E_i| - h\nu)\frac{\Delta E_i}{|\Delta E_i|}\left(\frac{\partial^2 E_i'}{\partial H^2} - \frac{\partial^2 E_i''}{\partial H^2}\right), \tag{32}$$

where $\Delta E_i = E_i' - E_i''$; E_i' and E_i'' are the energies of the levels participating in resonance. The derivatives of E_i', E_i'' required in (32) can be expressed, using the procedure of Section D.2, as follows:

$$\frac{\partial E_i}{\partial H} = \mathrm{Tr}\left[\left(\frac{\partial \mathscr{H}}{\partial H}\right)(|\varphi_i\rangle \otimes \langle\varphi_i|)\right],$$

$$\frac{\partial^2 E_i}{\partial H^2} = \sum_n{}' \mathrm{Tr}\left[\frac{\partial \mathscr{H}}{\partial H}(|\varphi_n\rangle \otimes \langle\varphi_n|)\frac{\partial \mathscr{H}}{\partial H}(|\varphi_i\rangle \otimes \langle\varphi_i|)\right]\Big/(E_i - E_n). \tag{33}$$

In (33) $|\varphi_i\rangle$ is the eigenvector of $\mathscr{H}$ with the eigenvalue $E_i : \mathscr{H}|\varphi_i\rangle = E_i|\varphi_i\rangle$; index n runs over all levels except for i (indicated by a prime over the summation sign).

c. Problems Encountered in the Application of the LSF Method and Their Solutions

(i) When two or more lines converge to the same resonant-field value at a particular orientation θ (case of "degeneracy"), then for a subsequent orientation $\theta + \delta\theta$ of the magnetic field, only one of the lines can be computed by the use of (31), unless one checks the various eigenvalues of the SH matrix extremely carefully. This arises because there is now only one initial value available to be used in (31). In order to overcome this problem one judiciously changes the interval $\delta\theta$ to a different (smaller) value at the preceding orientation $\theta - \delta\theta$ so the situation of degeneracy is circumvented. Another way is to use the brute-force method at the subsequent orientation $\theta + \delta\theta$ to find the two or more correponding resonant fields.

(ii) There are some lines which do not continue up to $\theta = 0°$ (e.g., Z-axis) or $\theta = 90°$ (e.g., X-axis), or both, e.g., the "loop" or "lobe" lines. The existence of these lines can be checked by the application of the brute-force method for a judiciously chosen θ (say θ_0), at which these lines exist. The field values thus obtained may be used for studying their angular variation by the use of (31) by decreasing and increasing the value of θ in steps of $\delta\theta$, starting from θ_0 at which the resonant field values are known.

(iii) When a resonant-field value, computed by the use of (31), becomes negative even for a sufficiently small $\delta\theta$, that indicates that this line terminates at that orientation (θ), e.g., the "lobe" lines do not continue beyond a certain θ.

(iv) If it is found that a computed resonant-field value changes by a relatively large magnitude for a particular choice of $\delta\theta$, greater accuracy may be obtained by decreasing $\delta\theta$ so that the change in the field value is not "too" large.

(v) At low magnetic fields sometimes the eigenvalues, arranged in increasing or decreasing values, corresponding to a given transition do not occur in the same order for $\theta + \delta\theta$ as for θ. This is not surprising, as the computer output gives the eigenvalues in increasing, or decreasing values. The correct pairs of eigenvalues (i' and i'') can be checked by examining the expectation value $\langle \varphi_{i'} | \mathscr{H}' | \varphi_{i''} \rangle$, where $\mathscr{H}'$ represents the interaction of the microwave magnetic field with the electronic magnetic moment. The value of this matrix element for incorrect pairs of eigenvalues will be negligible.

The above procedure was successfully applied to the calculation of the angular variation of EPR spectra for the purely electronic systems of Gd^{3+}-doped $NdCl_3 \cdot 6H_2O$ AT X- and Q-band frequencies ($S = \frac{7}{2}$), X-band ESR spectrum of Fe^{3+}-doped guinidinium aluminum sulfate hexahydrate ($S = \frac{5}{2}$), and the ENSC system of Mn^{2+}-doped $Ni(CH_3COO)_2 \cdot 4H_2O$ ($S = I = \frac{5}{2}$).

4. Calculation and Fitting of Line Intensities

By computing EPR line intensities and comparing them with those observed, one has an extra check on the computed SHPs. On the other hand, line intensities themselves can be used in the LSF technique to evaluate SHPs, especially in those situations where an insufficient number of lines is available for the evaluation of SHPs. Then using both the intensities and line positions one doubles the number of data points used for the evaluation of parameters, thus decreasing the parameter errors. Such a situation occurs, e.g., at liquid-helium temperatures, when one obtains data only for one or two orientations of the external magnetic field. Such a procedure has already been proposed.[56]

a. Intensity Operator[56]

By definition the intensity operator Γ_{mn} is such that its trace with the density matrix ρ gives the intensity I_{mn} of the line corresponding to the transition $m \leftrightarrow n$.

$$I_{mn} = \mathrm{Tr}(\rho\Gamma_{mn}). \tag{34}$$

In (34), $\rho = \sum_j p_j P_j$ is the density matrix, where $P_j = |\varphi_j\rangle \otimes \langle\varphi_j|$ is the projection operator and

$$p_j = \exp(-E_j/k_B T) \Big/ \sum_i \exp(-E_i/k_B T)$$

is the normalized population of the jth energy level with the energy E_j, k_B and T being the Boltzmann constant and absolute temperature, respectively. At high temperatures $\rho = \sum_j P_j/(2S+1)$, as the populations of all the levels are equal for the presence of FS terms only. When both FS and HFS terms are present $\rho = \sum_j P_j/[(2S+1)(2I+1)]$.

The intensity operator is[56]

$$\Gamma_{mn} = K(\mathscr{H}_{\rm rf} P_n \mathscr{H}_{\rm rf} P_m - \mathscr{H}_{\rm rf} P_m \mathscr{H}_{\rm rf} P_n), \tag{35}$$

where the constant K is the same for all the lines and depends only upon the amplitude of the rf magnetic field and the characteristics of the electronic detection system of the spectrometer, and $\mathscr{H}_{\rm rf}$ denotes the interaction Hamiltonian of the time-dependent rf field inducing transitions with the magnetic moment of the spins. In general,

$$\mathscr{H}_{\rm rf} = [g_\parallel \mu_B (H_{\rm rf})_z S_z + g_\perp \mu_B (H_{\rm rf})_x S_x] \cos \omega t, \tag{36}$$

where $g_\parallel$ and $g_\perp$ are the g factors parallel and perpendicular to the z-axis, $H_{\rm rf}$ is the amplitude of the rf magnetic field, and ω is the angular frequency of the rf field.

b. Fitting of Line Intensities[56]

The LSF technique as described in Section D.2 above can be applied to the inclusion of both the line intensities and line positions provided that the chi-square, S', is appropriately described, and one knows how to compute its first and second derivative for the evaluation of $\Delta\vec{a}$ (14). As explicitly evaluated

$$S' = S'_{\rm pos} + S'_{\rm int},$$

where $S'_{\rm pos}$ is the chi-square corresponding to the EPR line positions; the expressions for its derivatives as required in the LSF method have already been described in Section D.2. As for $S'_{\rm int}$, the contribution of the line intensities, the following are noted:[56]

$$S'_{\rm int} = \sum_{mn} (I_i^{cn} - I_i^{mn})^2 / \sigma_i''^2. \tag{37}$$

In (37), I_i^{cn}, I_i^{mn}, and σ_i'' are, repectively, the normalized calculated intensity (normalized for each external-field orientation separately), the normalized measured intensity (normalized the same way as I_{cn}), and the weight factor for the ith line intensity. The required derivatives are

$$\frac{\partial S'_{\rm int}}{\partial a_k} = \sum_i \frac{2(I_i^{cn} - I_i^{mn})}{\sigma_i''^2} \frac{\partial I_i^{cn}}{\partial a_k} \tag{38}$$

and

$$\frac{\partial^2 S'_{\rm int}}{\partial a_k\,\partial a_l} = \sum_i \left(\frac{2}{\sigma_i''^2}\right)\left[(I_i^{cn} - I_i^{mn})\frac{\partial^2 I_i^{cn}}{\partial a_k\,\partial a_l} + \frac{\partial I_i^{cn}}{\partial a_k}\frac{\partial I_i^{cn}}{\partial a_l}\right]. \tag{39}$$

For use in (38) and (39), the following expressions are required.

c. *Normalized Intensity and its Derivatives*

The calculated normalized intensity I_i^{cn} and the measured normalized intensity I_i^{mn} for a given orientation of the external magnetic field are

$$I_i^{cn} = I_i^c \Big/ \sum_j I_j^c, \qquad I_i^{mn} = I_i^m \Big/ \sum_j I_j^m. \tag{40}$$

In (40), j covers all lines considered for each different orientation of the external magnetic field. From (34) and (35)

$$\frac{\partial I_i^{cn}}{\partial a_k} = \frac{\left(\sum_j I_j^c\right)\frac{\partial I_i^c}{\partial a_k} - I_i^c \sum_j \frac{\partial I_j^c}{\partial a_k}}{\left(\sum_j I_j^c\right)^2}$$

and

$$\begin{aligned}\frac{\partial^2 I_i^{cn}}{\partial a_k\,\partial a_l} = &\left(\sum_j I_j^c\right)\left[\left(\sum_j I_j^c\right)\left(\frac{\partial^2 I_i^c}{\partial a_k\,\partial a_l}\right) + \left(\sum_j \frac{\partial I_j^c}{\partial a_l}\right)\frac{\partial I_i^c}{\partial a_k} - \left(\sum_j \frac{\partial I_j^c}{\partial a_k}\right)\frac{\partial I_i^c}{\partial a_l}\right.\\ &\left. - I_i^c\left(\sum_j \frac{\partial^2 I_j^c}{\partial a_k\,\partial a_l}\right)\right] - 2\left\{\left(\sum_j I_j^c\right)\frac{\partial I_i^c}{\partial a_k} - I_i^c \sum_j \frac{\partial I_j^c}{\partial a_k}\right\}\\ &\times \left(\sum_j \frac{\partial I_j^c}{\partial a_l}\right)\Big/\left(\sum_j I_j^c\right)^3.\end{aligned} \tag{41}$$

For use in (41), as obtained from (35), we have

$$\frac{\partial I_i^c}{\partial a_k}\left[\left(\frac{\partial p_{i'}}{\partial a_k} - \frac{\partial p_{i''}}{\partial a_k}\right)\mathrm{Tr}(\mathscr{H}_{\rm rf} P_{i''} \mathscr{H}_{\rm rf} P_{i'} + (p_{i'} - p_{i''})\frac{\partial f_{ij}}{\partial a_k}\right], \tag{42}$$

where

$$\frac{\partial f_{ij}}{\partial a_k} = \mathrm{Tr}\left(\mathscr{H}_{\rm rf}\frac{\partial P_{i''}}{\partial a_k}\mathscr{H}_{\rm rf} P_{i'} + \mathscr{H}_{\rm rf} P_{i''}\mathscr{H}_{\rm rf}\frac{\partial P_{i'}}{\partial a_k} + \frac{\partial \mathscr{H}_{\rm rf}}{\partial a_k} P_{i''}\mathscr{H}_{\rm rf} P_{i'} + \mathscr{H}_{\rm rf} P_{i''}\frac{\partial \mathscr{H}_{\rm rf}}{\partial a_k} P_{i'}\right). \tag{43}$$

For use in the above equations the derivatives of the projection operators P_j and the normalized thermal populations p_j are required; these are given later by (45)–(48).

From (42) we obtain for the second derivatives

$$\frac{\partial^2 I_i^c}{\partial a_k\,\partial a_l} = \left(\frac{\partial^2 p_{i'}}{\partial a_k\,\partial a_l} - \frac{\partial^2 p_{i''}}{\partial a_k\,\partial a_l}\right)\mathrm{Tr}(\mathscr{H}_{\mathrm{rf}} P_{i''} \mathscr{H}_{\mathrm{rf}} P_{i'}) + \left(\frac{\partial p_{i'}}{\partial a_k} - \frac{\partial p_{i''}}{\partial a_k}\right)$$
$$\times\,\mathrm{Tr}\left(\frac{\partial \mathscr{H}_{\mathrm{rf}}}{\partial a_l} P_{i''}\mathscr{H}_{\mathrm{rf}} P_{i'} + \mathscr{H}_{\mathrm{rf}}\frac{\partial P_{i''}}{\partial a_l}\mathscr{H}_{\mathrm{rf}} P_{i'} + \mathscr{H}_{\mathrm{rf}} P_{i''}\frac{\partial \mathscr{H}_{\mathrm{rf}}}{\partial a_l} P_{i'}\right.$$
$$\left. + \mathscr{H}_{\mathrm{rf}} P_{i''}\mathscr{H}_{\mathrm{rf}}\frac{\partial P_{i'}}{\partial a_l}\right) + \left(\frac{\partial p_{i'}}{\partial a_l} - \frac{\partial p_{i''}}{\partial a_l}\right)\frac{\partial f_{ij}}{\partial a_k} + (p_{i'} - p_{i''})\frac{\partial^2 f_{ij}}{\partial a_k\,\partial a_l}, \tag{44}$$

where

$$\frac{\partial^2 f_{ij}}{\partial a_k\,\partial a_l} = \mathrm{Tr}\left(\frac{\partial \mathscr{H}_{\mathrm{rf}}}{\partial a_l}\frac{\partial P_{i''}}{\partial a_k}\mathscr{H}_{\mathrm{rf}} P_{i'} + \mathscr{H}_{\mathrm{rf}}\frac{\partial^2 P_{i''}}{\partial a_k\,\partial a_l}\mathscr{H}_{\mathrm{rf}} P_{i'}\right.$$
$$+ \mathscr{H}_{\mathrm{rf}}\frac{\partial P_{i''}}{\partial a_k}\frac{\partial \mathscr{H}_{\mathrm{rf}}}{\partial a_l} P_{i'} + \mathscr{H}_{\mathrm{rf}}\frac{\partial P_{i''}}{\partial a_k}\mathscr{H}_{\mathrm{rf}}\frac{\partial P_{i'}}{\partial a_l} + \frac{\partial \mathscr{H}_{\mathrm{rf}}}{\partial a_l} P_{i''}\mathscr{H}_{\mathrm{rf}}\frac{\partial P_{i'}}{\partial a_k}$$
$$+ \mathscr{H}_{\mathrm{rf}}\frac{\partial P_{i''}}{\partial a_l}\mathscr{H}_{\mathrm{rf}}\frac{\partial P_{i'}}{\partial a_k} + \mathscr{H}_{\mathrm{rf}} P_{i''}\frac{\partial \mathscr{H}_{\mathrm{rf}}}{\partial a_l}\frac{\partial P_{i''}}{\partial a_k} + \mathscr{H}_{\mathrm{rf}} P_{i''}\mathscr{H}_{\mathrm{rf}}\frac{\partial^2 P_{i'}}{\partial a_k\,\partial a_l}$$
$$+ \frac{\partial \mathscr{H}_{\mathrm{rf}}}{\partial a_k}\frac{\partial P_{i''}}{\partial a_l}\mathscr{H}_{\mathrm{rf}} P_{i'} + \frac{\partial \mathscr{H}_{\mathrm{rf}}}{\partial a_k} P_{i''}\frac{\partial \mathscr{H}_{\mathrm{rf}}}{\partial a_l} P_{i'} + \frac{\partial \mathscr{H}_{\mathrm{rf}}}{\partial a_k} P_{i''}\mathscr{H}_{\mathrm{rf}}\frac{\partial P_{i'}}{\partial a_l}$$
$$\left. + \frac{\partial \mathscr{H}_{\mathrm{rf}}}{\partial a_l} P_{i''}\frac{\partial \mathscr{H}_{\mathrm{rf}}}{\partial a_k} P_{i'} + \mathscr{H}_{\mathrm{rf}}\frac{\partial P_{i''}}{\partial a_l}\frac{\partial \mathscr{H}_{\mathrm{rf}}}{\partial a_k} P_{i'} + \mathscr{H}_{\mathrm{rf}} P_{i''}\frac{\partial \mathscr{H}_{\mathrm{rf}}}{\partial a_k}\frac{\partial P_{i'}}{\partial a_l}\right).$$

For use in (42)–(44) the derivatives of the projection operators are

$$\frac{\partial P_i}{\partial a_k} = -(\mathscr{H} - E_i \mathrm{II})^{-1}\left(\frac{\partial \mathscr{H}}{\partial a_k} - \frac{\partial E_i}{\partial a_k}\mathrm{II}\right)P_i \tag{45}$$

and

$$\frac{\partial^2 P_i}{\partial a_k\,\partial a_l} = (\mathscr{H} - E_i \mathrm{II})^{-1}\left[\frac{\partial^2 E_i}{\partial a_k\,\partial a_l}P_i - \left(\frac{\partial \mathscr{H}}{2\partial a_l} - \frac{\partial E_i}{\partial a_l}\mathrm{II}\right)\frac{\partial P_i}{\partial a_k}\right]. \tag{46}$$

In (48) and (46) II is the unit matrix. The derivatives of eigenvalues as required in (44) and (45) can be computed using (23) and (24).

Finally, the derivatives of normalized populations as required in (43) and (44) are

$$\frac{\partial p_i}{\partial a_k} = \frac{p_i}{kT}\left\{\left(\sum_j e^{-E_j/k_{\mathrm{B}}T}\frac{\partial E_j}{\partial a_k}\right)\bigg/ z - \frac{\partial E_i}{\partial a_k}\right\}, \qquad \text{where} \quad z = \sum_j e^{-E_j/k_{\mathrm{B}}T}, \tag{47}$$

and

$$\frac{\partial^2 p_i}{\partial a_k a_l} = \frac{\partial p_i/\partial a_l}{k_B T}\left\{\left(\sum_j e^{-E_j/k_B T}\frac{\partial E_j}{\partial a_k}\right)\Big/ z - \frac{\partial E_i}{\partial a_k}\right\}$$

$$+\frac{p_i}{k_B T}\left\{-\sum_j\left(\frac{e^{-E_j/k_B T}}{kT}\frac{\partial E_j}{\partial a_l}\frac{\partial E_j}{\partial a_k} + e^{-E_j/k_B T}\frac{\partial^2 E_j}{\partial a_k \partial a_l}\right)\Big/ z\right.$$

$$\left.+\left(\sum_j e^{-E_j/k_B T}\frac{\partial E_j}{\partial a_k}\right)\left(\sum_j \frac{e^{-E_j/k_B T}}{k_B T}\frac{\partial E_j}{\partial a_l}\right)\Big/ z^2 - \frac{\partial^2 E_i}{\partial a_k \partial a_l}\right\}. \tag{48}$$

d. Limits of Applicability of the Method

The line-intensity expression and its application to LSF as described is only applicable to S-state ions and EPR induced by the rf magnetic field. It is not applicable to the cases of Jahn–Teller ions, as in these cases the line is broadened by strain; the peak position comes from nonzero strain sites and the transition probabilities in different regions of the line are also sometimes significantly strain dependent.

The intensity operator described above has been successfully applied to the computation of the line intensities for Gd^{3+}-doped $SmCl_3 \cdot 6\,H_2O$.[56]

5. Evaluation of Noncoincident Anisotropic $\tilde{\mathbf{g}}^2$, $\tilde{\mathbf{A}}^2$, $\tilde{\mathbf{D}}$, $\tilde{\mathbf{P}}$, and $\tilde{\mathbf{g}}_n{}^2$ Tensors

The objective of this section is to discuss, in detail, a rigorous evaluation of SHPs from EPR/ENDOR data for the low symmetry situation, characterized by noncoincident tensors in the SH. The following discussion will be concerned specifically with EPR data. However, the details apply equally well to ENDOR data where, instead of measuring transition magnetic fields (resonant line positions) at a fixed microwave frequency, one measures resonant frequencies between the pair of energy levels participating in resonance, keeping the external magnetic field constant.

a. Spin Hamiltonian

For the present considerations the following spin Hamiltonian, representing the interaction of one electron spin with several (N) nuclear spins, will be used:[19]

$$\mathscr{H} = \mathscr{H}_{\rm ze} + \mathscr{H}_{\rm zf} + \mathscr{H}_{\rm hf} + \mathscr{H}_{\rm nq} + \mathscr{H}_{\rm zn}. \tag{49}$$

In (49), $\mathscr{H}_{\rm ze}, \mathscr{H}_{\rm zf}, \mathscr{H}_{\rm hf}, \mathscr{H}_{\rm nq}$, and $\mathscr{H}_{\rm zn}$ are, respectively, the electron Zeeman term, the zero-field term, the hyperfine term, the nuclear quadropole term, and the nuclear Zeeman term. They are expressed as follows:

$$\mathscr{H}_{\rm ze} = \mu_B \mathbf{S}^T \cdot \tilde{\mathbf{g}} \cdot \mathbf{H}, \tag{50a}$$

$$\mathscr{H}_{\rm zf} = \mathbf{S}^T \cdot \tilde{\mathbf{D}} \cdot \mathbf{S}, \tag{50b}$$

$$\mathscr{H}_{\mathrm{hf}} = \sum_{r=1}^{N} \mathbf{I}_r^{\mathrm{T}} \cdot \tilde{\mathbf{A}}_r \cdot \mathbf{S}, \tag{50c}$$

$$\mathscr{H}_{\mathrm{nq}} = \sum_{r=1}^{N} \mathbf{I}_r^{\mathrm{T}} \cdot \tilde{\mathbf{P}}_r \cdot \mathbf{I}_r, \tag{50d}$$

and

$$\mathscr{H}_{\mathrm{zn}} = -\sum_{r=1}^{N} \mu_{\mathrm{N}} \mathbf{I}_r^{\mathrm{T}} \cdot \tilde{\mathbf{g}}_{\mathrm{n}r} \cdot \mathbf{H}. \tag{50e}$$

In (50a)–(50e), the summation runs over the N nuclei, T denotes transpose, $\tilde{\mathbf{g}}, \tilde{\mathbf{D}}, \tilde{\mathbf{A}}, \tilde{\mathbf{P}}_r$, and $\tilde{\mathbf{g}}_{\mathrm{n}r}$ are either matrices or tensors, and bold sanserif denotes a tensor or matrix.

It should be noted that, as discussed by Abragam and Bleaney,[8] $\tilde{\mathbf{g}}, \tilde{\mathbf{g}}_{\mathrm{n}r}, \tilde{\mathbf{A}}_r$ are not really tensors; thus they should be referred to as matrices, while $\tilde{\mathbf{g}}^2 = \tilde{\mathbf{g}}^{\mathrm{T}} \cdot \tilde{\mathbf{g}}, \tilde{\mathbf{A}}^2 = \tilde{\mathbf{A}}^{\mathrm{T}} \cdot \tilde{\mathbf{A}}$, and $\tilde{\mathbf{g}}_{\mathrm{n}r}^2 = \tilde{\mathbf{g}}_{\mathrm{n}r}^{\mathrm{T}} \cdot \tilde{\mathbf{g}}_{\mathrm{n}r}$ are, indeed, symmetric tensors.[30] In what follows the terms tensor and matrix for these quantities will be used interchangeably. As for the tensors $\tilde{\mathbf{D}}$ and $\tilde{\mathbf{P}}_r$, they are symmetric[5] and traceless, i.e., $D_{xx} + D_{yy} + D_{zz} = P_{rxx} + P_{ryy} + P_{rzz} = 0$. Thus $\tilde{\mathbf{D}}$ and $\tilde{\mathbf{P}}_r$ are each represented by five independent elements.

In order to compute the eigenvalues of (49) using perturbation theory, the electron and nuclear spins are chosen to be quantized along the directions $\tilde{\mathbf{g}} \cdot \boldsymbol{\eta}/g$ and $\tilde{\mathbf{A}}_r \cdot \tilde{\mathbf{g}} \cdot \boldsymbol{\eta}/gK_r$, respectively; here $g = (\tilde{\boldsymbol{\eta}}^{\mathrm{T}} \cdot \tilde{\mathbf{g}}^{\mathrm{T}} \cdot \tilde{\mathbf{g}} \cdot \tilde{\boldsymbol{\eta}})^{1/2}, K_r = (\boldsymbol{\eta}^{\mathrm{T}} \cdot \tilde{\mathbf{g}}^{\mathrm{T}} \cdot \tilde{\mathbf{A}}_r^{\mathrm{T}} \cdot \tilde{\mathbf{A}}_r \cdot \tilde{\mathbf{g}} \cdot \boldsymbol{\eta})^{1/2}/\mathbf{g}$, and $\boldsymbol{\eta}$ is the unit vector along $\mathbf{H}$. For these quantization axes, the various terms of SH, (49), can be expressed as[18,19]

$$\mathscr{H}_{\mathrm{ze}} = \mu_{\mathrm{B}} g \mathbf{H} S_u, \tag{51a}$$

$$\begin{aligned}\mathscr{H}_{\mathrm{zf}} = (\mathbf{u}^{\mathrm{T}} \cdot \tilde{\mathbf{D}} \cdot \mathbf{u}) {S_u}^2 + \tfrac{1}{4} D_0 (S_+ S_- + S_- S_+) + \tfrac{1}{2} D_1^* (S_u S_+ + S_+ S_u) \\ + \tfrac{1}{2} D_1 (S_u S_- + S_- S_u) + \tfrac{1}{4} D_2^* {S_+}^2 + \tfrac{1}{4} D_2 {S_-}^2,\end{aligned} \tag{51b}$$

$$\begin{aligned}\mathscr{H}_{\mathrm{hf}} + \mathscr{H}_{\mathrm{zn}} = \sum_r \{ K_r(S_u) I_{kr} + \tfrac{1}{2} A_{1r}^* S_+ I_{kr} + \tfrac{1}{2} A_{1r} S_- I_{kr} \\ + \tfrac{1}{4} A_{+r} S_+ I_{-r} + \tfrac{1}{4} A_{+r}^* S_- I_{+r} + \tfrac{1}{4} A_{-r}^* S_+ I_{+r} + \tfrac{1}{4} A_{-r} S_- I_{-r} \},\end{aligned} \tag{51c}$$

$$\begin{aligned}\mathscr{H}_{\mathrm{nq}} = \sum_r \{ (\mathbf{k}_r^{\mathrm{T}} \cdot \tilde{\mathbf{P}}_r \cdot \mathbf{k}_r) {I_{kr}}^2 + \tfrac{1}{4} P_{0r} (I_{+r} I_{-r} + I_{-r} I_{+r}) \\ + \tfrac{1}{2} P_{1r}^* (I_{kr} I_{+r} + I_{+r} I_{kr}) + \tfrac{1}{2} P_{1r} (I_{kr} I_{-r} + I_{-r} I_{kr}) + \tfrac{1}{4} P_{2r}^* {I_{+r}}^2 + \tfrac{1}{4} P_{2r} {I_{-r}}^2 \}.\end{aligned} \tag{51d}$$

In (51a)–(51d), $\mathbf{u} = \tilde{\mathbf{g}} \cdot \boldsymbol{\eta}/g, \mathbf{i}, \mathbf{j}$ are mutually perpendicular unit vectors orthogonal to $\boldsymbol{\eta}$, $S_{\pm} = (S_x \pm i S_y), D_0 = -(\mathbf{u}^{\mathrm{T}} \cdot \tilde{\mathbf{D}} \cdot \mathbf{u})$, $D_1 = (\mathbf{u}^{\mathrm{T}} \cdot \tilde{\mathbf{D}} \cdot \mathbf{i}) + i(\mathbf{u}^{\mathrm{T}} \cdot \tilde{\mathbf{D}} \cdot \mathbf{j})$; $D_2 = (\mathbf{i}^T \cdot \tilde{\mathbf{D}} \cdot \mathbf{i}) - (\mathbf{j}^T \cdot \tilde{\mathbf{D}} \cdot \mathbf{j}) + 2i(\mathbf{i}^{\mathrm{T}} \cdot \tilde{\mathbf{D}} \cdot \mathbf{j})$, $\mathbf{K}_r(S_u) = (\mathbf{A}_r \cdot \tilde{\mathbf{g}}/g) S_u - \mathbf{g}_{\mathrm{n}r} \boldsymbol{\mu}_{\mathrm{N}} \mathbf{H}\mathbf{E}$ ($\mathbf{E}$ is the unit tensor); $\mathbf{k}_r(S_u) = \mathbf{K}_r(S_u) \cdot \boldsymbol{\eta}/(\boldsymbol{\eta}^T \cdot \mathbf{K}_r^{\mathrm{T}}(S_u) \cdot \mathbf{K}_r(S_u) \cdot \boldsymbol{\eta})^{1/2}$; $A_{1r} = (\mathbf{k}_r^{\mathrm{T}} \cdot \tilde{\mathbf{A}}_r \cdot \mathbf{i}) + i(\mathbf{k}_r^{\mathrm{T}} \cdot \tilde{\mathbf{A}}_r \cdot \mathbf{j})$; $A_{\pm r} = [(\xi_r^{\mathrm{T}} \cdot \tilde{\mathbf{A}}_r \cdot \mathbf{i}) \pm \mathbf{i}(\boldsymbol{\eta}_r^{\mathrm{T}} \cdot \tilde{\mathbf{A}}_r \cdot \mathbf{j})] + i[(\boldsymbol{\eta}_r^{\mathrm{T}} \cdot \tilde{\mathbf{A}}_r \cdot \mathbf{i}) \mp (\xi_r^{\mathrm{T}} \cdot \tilde{\mathbf{A}}_r \cdot \mathbf{j})]$ (the mutually perpendicular unit vectors ξ_r and $\boldsymbol{\eta}_r$ are orthogonal to $\mathbf{k}_r(S_u)$ and $I_{\pm r} = (I_{\xi r} \pm \mathbf{i} I_{\eta r})$; $P_{0r} = -(\mathbf{k}_r^{\mathrm{T}} \cdot \tilde{\mathbf{P}}_r \cdot \mathbf{k}_r)$; $p_{1r} = (\mathbf{k}_r^{\mathrm{T}} \cdot \tilde{\mathbf{P}}_r \cdot \xi_r) + i(\mathbf{k}_r^{\mathrm{T}} \cdot \tilde{\mathbf{P}}_r \cdot \boldsymbol{\eta}_r)$; and $P_{2r} = (\xi_r^{\mathrm{T}} \cdot \tilde{\mathbf{P}}_r \cdot \xi_r) - (\boldsymbol{\eta}_r^{\mathrm{T}} \cdot \tilde{\mathbf{P}}_r \cdot \boldsymbol{\eta}_r) + 2i(\xi_r^{\mathrm{T}} \cdot \tilde{\mathbf{P}}_r \cdot \boldsymbol{\eta}_r)$. In the above definitions the multiplying factor $i = \sqrt{-1}$, subscripts x and y denote components along $\mathbf{i}$ and $\mathbf{j}$, respectively, and * denotes complex conjugate.

With the expressions given by (51a)–(51d), it is easy to calculate the eigenvalues using a perturbation approximation; the terms $\mathscr{H}_{ze} + \mathscr{H}_{hf} + \mathscr{H}_{zn}$ are used as zero-order terms and the direct-product representation in the space of electron-nuclear spin-coupled systems is used.[18,19]

b. Eigenvalues

(i) *Perturbation Approach*

The eigenvalues of the SH, given by (49), with the various terms expressed by (51a)–(51d), have been calculated by several researchers using perturbation approximations. To second order of perturbation, taking into account nuclear Zeeman and certain cross terms, they can be expressed as follows:

$$E(M;\{m\}) = E^{(0)}(M;\{m\}) + E^{(1)}(M;\{m\}) + E^{(2)}(M;\{m\}). \tag{52}$$

In (52) the superscripts 0, 1, and 2 represent, respectively, the zero-, first-, and second-order perturbation contributions, and $\{m\}$ stands for $m_1, m_2, \ldots, m_N$. Here M stands for the electronic magnetic quantum number and $m_1, m_2, \ldots$ stand for the nuclear magnetic quantum numbers for nuclei $1, 2, \ldots$, respectively. Explicitly:[18,19]

$$E^{(0)}(M;\{m\}) = \mu_B gHM - \sum_r G_r \mu_N m_r, \tag{53a}$$

$$\begin{aligned} E^{(1)}(M;\{m\}) &= \tfrac{1}{2}\mathrm{Tr}(\tilde{\mathbf{D}})[S(S+1) - M^2] + \tfrac{1}{2}d_1[3M^2 - S(S+1)] \\ &\quad + \sum_r \{K_r M m_r + \tfrac{1}{2}p_{1r}[3m_r{}^2 - I_r(I_r+1)]\}, \end{aligned} \tag{53b}$$

$$\begin{aligned} E^{(2)}(M;\{m\}) &= (d_2 - d_1{}^2)[8M^2 + 1 - 4S(S+1)]M/(2\mu_B gH) \\ &\quad + [\mathrm{Tr}(\tilde{\mathbf{D}}^2) - 2d_2 + d_1{}^2 - 2d_{-1}\,\mathrm{Det}(\tilde{\mathbf{D}})] \\ &\quad \times [2S(S+1) - 2M^2 - 1]M/(8\mu_B gH) \\ &\quad + \sum_r \{[\{\mathrm{Tr}(\tilde{\mathbf{A}}_r^{\mathrm{T}} \cdot \tilde{\mathbf{A}}_r) - k_r{}^2\}M\{I_r(I_r+1) - m_r{}^2\}/2 \\ &\quad - \mathrm{Det}(A_r)\{S(S+1) - M^2\}m_r/K_r + (k_r{}^2 - K_r{}^2)Mm_r^2 \\ &\quad + 2(e_r - d_1)K_r\{3M^2 - S(S+1)\}m_r]/(2\mu_B gH) \\ &\quad + \mathrm{Tr}(P_r)[I_r(I_r+1) - m_r{}^2]/2\} \\ &\quad + \sum_{i>j}(L_{ij} - K_i K_j)Mm_i m_j/(\mu_B gH) \\ &\quad + M^{-1}\sum_r \{\mu_N{}^2 H^2(g_r{}^2 - G_r{}^2)m_r/(2K_r) \\ &\quad + (p_{2r} - p_{1r}{}^2)[8m_r{}^2 - 4I_r(I_r+1) + 1]m_r/(2K_r) \\ &\quad + [\mathrm{Tr}(\tilde{\mathbf{P}}_r{}^2) - 2p_{2r} + p_{1r}{}^2 - 2p_{-1r}\,\mathrm{Det}(\tilde{\mathbf{P}}_r)] \\ &\quad \times [2I_r(I_r+1) - 2m_r{}^2 - 1]m_r/(8K_r) \\ &\quad - \mu_N H(q_r - G_r p_{1r})[3m_r{}^2 - I_r(I_r+1)]/K_r\}, \end{aligned} \tag{53c}$$

where

$$\boldsymbol{\eta} = \mathbf{H}/|\mathbf{H}|,$$

$$g^2 = \boldsymbol{\eta}^{\mathrm{T}} \cdot \tilde{\mathbf{g}}^{\mathrm{T}} \cdot \tilde{\mathbf{g}} \cdot \boldsymbol{\eta},$$

$$g^2 d_n = \boldsymbol{\eta}^{\mathrm{T}} \cdot \tilde{\mathbf{g}}^{\mathrm{T}} \cdot \tilde{\mathbf{D}}^n \cdot \tilde{\mathbf{g}} \cdot \boldsymbol{\eta} \quad (n = \text{integer}),$$

$$g^2 K_r{}^2 = \boldsymbol{\eta}^{\mathrm{T}} \cdot \tilde{\mathbf{g}}^{\mathrm{T}} \cdot \tilde{\mathbf{A}}_r^{\mathrm{T}} \cdot \tilde{\mathbf{A}}_r \cdot \tilde{\mathbf{g}} \cdot \boldsymbol{\eta},$$

$$g^2 K_r{}^2 k_r{}^2 = \boldsymbol{\eta}^{\mathrm{T}} \cdot \tilde{\mathbf{g}}^{\mathrm{T}} \cdot \tilde{\mathbf{A}}_r^{\mathrm{T}} \cdot \tilde{\mathbf{A}}_r \cdot \tilde{\mathbf{A}}_r^{\mathrm{T}} \cdot \tilde{\mathbf{A}}_r \cdot \tilde{\mathbf{g}} \cdot \boldsymbol{\eta},$$

$$g^2 K_r{}^2 e_r = \boldsymbol{\eta}^{\mathrm{T}} \cdot \tilde{\mathbf{g}}^{\mathrm{T}} \cdot (\tilde{\mathbf{D}} \cdot \tilde{\mathbf{A}}_r^{\mathrm{T}} \cdot \tilde{\mathbf{A}}_r + \tilde{\mathbf{A}}_r^{\mathrm{T}} \cdot \tilde{\mathbf{A}}_r \cdot \tilde{\mathbf{D}}) \cdot \tilde{\mathbf{g}} \cdot \boldsymbol{\eta}/2,$$

$$g_r{}^2 = \boldsymbol{\eta}^{\mathrm{T}} \cdot \tilde{\mathbf{g}}_{\mathrm{n}r}^{\mathrm{T}} \cdot \tilde{\mathbf{g}}_{\mathrm{n}r} \cdot \boldsymbol{\eta},$$

$$g K_r G_r = \boldsymbol{\eta}^{\mathrm{T}} \cdot (\tilde{\mathbf{g}}^{\mathrm{T}} \cdot \tilde{\mathbf{A}}_r^{\mathrm{T}} \cdot \tilde{\mathbf{g}}_{\mathrm{n}r} + \tilde{\mathbf{g}}_{\mathrm{n}r}^{\mathrm{T}} \cdot \tilde{\mathbf{A}}_r \cdot \tilde{\mathbf{g}}) \cdot \boldsymbol{\eta}/2,$$

$$g^2 K_r{}^2 p_{nr} = \boldsymbol{\eta}^{\mathrm{T}} \cdot \tilde{\mathbf{g}}^{\mathrm{T}} \cdot \tilde{\mathbf{A}}_r^{\mathrm{T}} \cdot \tilde{\mathbf{P}}_r^n \cdot \tilde{\mathbf{A}}_r \cdot \tilde{\mathbf{g}} \cdot \boldsymbol{\eta} \quad (n = \text{integer}),$$

$$g K_r q_r = \boldsymbol{\eta}^{\mathrm{T}} \cdot (\tilde{\mathbf{g}}^{\mathrm{T}} \cdot \tilde{\mathbf{A}}_r^{\mathrm{T}} \cdot \tilde{\mathbf{P}}_r \cdot \tilde{\mathbf{g}}_{\mathrm{n}r} + \tilde{\mathbf{g}}_{\mathrm{n}r}^{\mathrm{T}} \cdot \tilde{\mathbf{P}}_r \cdot \tilde{\mathbf{A}}_r \cdot \tilde{\mathbf{g}}) \cdot \boldsymbol{\eta}/2,$$

$$g^2 K_i K_j L_{ij} = \boldsymbol{\eta}^{\mathrm{T}} \cdot \tilde{\mathbf{g}}^{\mathrm{T}} \cdot (\tilde{\mathbf{A}}_i^{\mathrm{T}} \cdot \tilde{\mathbf{A}}_i \cdot \tilde{\mathbf{A}}_j^{\mathrm{T}} \cdot \tilde{\mathbf{A}}_j + \tilde{\mathbf{A}}_j^{\mathrm{T}} \cdot \tilde{\mathbf{A}}_j \cdot \tilde{\mathbf{A}}_i^{\mathrm{T}} \cdot \tilde{\mathbf{A}}_i) \cdot \tilde{\mathbf{g}} \cdot \boldsymbol{\eta}/2. \tag{54}$$

In (53b) and (53c) Det stands for determinant, while Tr stands for trace of a matrix. The Zeeman term (53a) has been assumed to be dominant. The term in (53c), with the factor M^{-1} has been derived under the assumption that the first-order hyperfine energy, $K_r M m_r$, for each nucleus is larger than the corresponding nuclear Zeeman term, $\mu_{\mathrm{N}} \mathbf{I}_r^{\mathrm{T}} \cdot \mathbf{g}_{\mathrm{n}r} \cdot \mathbf{H}$, or quadrupole interaction $\mathbf{I}_r^{\mathrm{T}} \cdot \mathbf{P}_r \cdot \mathbf{I}_r$.

It is easily seen from (53a)–(53c) that from EPR data only the tensors $\tilde{\mathbf{g}}^{\mathrm{T}} \cdot \tilde{\mathbf{g}}, \tilde{\mathbf{A}}_r^{\mathrm{T}} \cdot \tilde{\mathbf{A}}_r, \tilde{\mathbf{g}}_{\mathrm{n}r}^{\mathrm{T}} \cdot \tilde{\mathbf{g}}_{\mathrm{n}r}$ (rather than $\tilde{\mathbf{g}}, \tilde{\mathbf{A}}_r$, and $\tilde{\mathbf{g}}_{\mathrm{n}r}$), $\tilde{\mathbf{D}}$, and $\tilde{\mathbf{P}}_r$ can be evaluated. For, these determine the single-crystal EPR line positions, calculated from the resonance condition $\Delta E = E_1 - E_2 = h\nu$, where E_1, E_2 are the eigenvalues of the pair of levels participating in resonance.

Using the eigenvalues (53a)–(53c) the resonance condition for "allowed" transitions $(M, m_i, m_j \leftrightarrow M-1, m_i, m_j)$ can be expressed as

$$\begin{aligned}
\Delta E^{(\mathrm{a})} = h\nu = {}& \mu_{\mathrm{B}} g H - \mathrm{Tr}(\tilde{\mathbf{D}})(2M-1)/2 + 3d_1(2M-1)/2 \\
&+ (d_2 - d_1{}^2)[1 - 4S(S+1) + 8(3M^2 - 3M + 1)]/(2\mu_{\mathrm{B}} g H) \\
&+ [\mathrm{Tr}(\tilde{\mathbf{D}}^2) - 2d_2 + d_1{}^2 - 2d_{-1}\,\mathrm{Det}(\tilde{\mathbf{D}})][2S(S+1) \\
&- 1 - 2(3M^2 - 3M + 1)]/(8\mu_{\mathrm{B}} g H) \\
&+ \sum_r \{K_r m_r + [\{\mathrm{Tr}(\tilde{\mathbf{A}}_r^{\mathrm{T}} \cdot \tilde{\mathbf{A}}_r) - k_r{}^2\}\{I_r(I_r+1) - m_r{}^2\}/2 \\
&+ \mathrm{Det}(\tilde{\mathbf{A}}_r)(2M-1)m_r/K_r + (k_r{}^2 - K_r{}^2)m_r{}^2 \\
&+ 2(e_r - d_1)K_r\{3(2M-1)\}m_r]/(2\mu_{\mathrm{B}} g H) + \sum_{i>j}(L_{ij} - K_i K_j)m_i m_j/(\mu_{\mathrm{B}} g H)
\end{aligned}$$

$$- \sum_r \{\mu_n^2 H^2(g_r{}^2 - G_r{}^2)m_r/(2K_r) + (p_{2r} - p_{1r}{}^2)[8m_r{}^2 - 4I_r(I_r+1)+1]m_r/(2K_i)$$

$$+ [\mathrm{Tr}(\tilde{\mathbf{P}}_r{}^2) - 2p_{2r} + p_{1r}{}^2 - 2p_{-1r}\,\mathrm{Det}(\tilde{\mathbf{P}}_r)][2I_r(I_r+1) - 2m_r{}^2 - 1]m_r/(8K_r)$$

$$- \mu_{\mathrm{N}} H(q_r - G_r p_{1r})[3m_r{}^2 - I_r(I_r+1)]/K_r\}/(M(M-1)). \qquad (55)$$

The resonance condition for "forbidden" transitions $(M_i, m_i, m_j \leftrightarrow (M-1), m_i \pm n_i, m_j \pm n_j; n_i, n_j > 0)$ can be likewise expressed as follows:

$$\Delta E^{(f)} = h\nu = \Delta E^{(a)} + \sum_i \{\mp K_i(M-1)n_i$$

$$+ (1/(2g\mu_{\mathrm{B}}H))[(n_i/2)\{\mathrm{Tr}(\tilde{\mathbf{A}}_i^{\mathrm{T}} \cdot \tilde{\mathbf{A}}_i) - k_i{}^2\}(M-1)(n_i \pm 2m_i)$$

$$+ ((\mathrm{Det}(A_i)n_i/K_i)\{\pm S(S+1) \mp (M-1)^2\}$$

$$- n_i(k_i{}^2 - K_i^2)(M-1)(n_i \pm 2m_i)$$

$$- 2n_i(e_i - d_1)K_i\{\pm 3(M-1)^2 \mp S(S+1)\}]$$

$$\pm G_i\mu_{\mathrm{N}} H n_i + (n_i/2)\,\mathrm{Tr}(\tilde{\mathbf{P}}_i)[n_i \pm 2m_i] - p_{1i}(3n_i/2)[n_i \pm 2m_i]\}$$

$$- (1/(g\mu_{\mathrm{B}}H)) \sum_{i>j} (L_{ij} - K_iK_j)(M-1)(\pm m_j n_i \pm m_i n_j + n_i n_j)$$

$$+ (1/M(M-1)) \sum_i \{\mp(\mu_{\mathrm{N}}^2 H^2 M n_i/(2K_i))(g_i^2 - G_i^2)$$

$$\pm ((p_{2i} - p_{1i}{}^2)/(2K_i))[\{4I_i(I_i+1) - 1\}Mn_i$$

$$+ 8Mn_i(\pm 3m_i^2 - 3m_i n_i \mp n_i{}^2)]$$

$$+ \{[\mathrm{Tr}(\tilde{\mathbf{P}}_i{}^2) - 2p_{2i} + p_{1i}{}^2 - 2p_{-1i}\,\mathrm{Det}(\tilde{\mathbf{P}}_i)]/(8K_i)\}[\mp\{2I_i(I_i+1) - 1\}Mn_i$$

$$+ 2Mn_i(\mp 3m_i{}^2 + 3m_i n_i \pm n_i{}^2)]$$

$$+ (3\mu_{\mathrm{N}} HMn_i/K_i)(q_i - G_i p_{1i})(n_i \pm 2m_i)\}. \qquad (56)$$

Complexities Associated with the Use of Second-Order-Perturbed Eigenvalues in the Application of LSF Procedure

In this section the intricacies of the energy differences $\Delta E^{(a)}$ and $\Delta E^{(f)}$ for the "allowed" and "forbidden" transitions are discussed with the use of special coordinate systems. "Forbidden" transitions can be treated in terms of a special coordinate system for the presence of one nucleus only. (The specific details of the application of the LSF procedure for this coordinate system are given later.)

"Allowed" transitions. An examination of the eigenvalue difference $\Delta E^{(a)}$ for "allowed" transitions, (55), reveals that $\Delta E^{(a)}$ depends upon the quantities g^2, d_n, $k_i{}^2$, $K_i{}^2$, and e_i, defined by (54). It follows then that these are the tensors $\tilde{\mathbf{g}}^2 = \tilde{\mathbf{g}}^{\mathrm{T}} \cdot \tilde{\mathbf{g}}$

and $\tilde{\mathbf{A}}_i^2 = \tilde{\mathbf{A}}_i^{\mathrm{T}} \cdot \tilde{\mathbf{A}}_i$ (rather than $\tilde{\mathbf{g}}, \tilde{\mathbf{A}}_i$) which determine the single-crystal "allowed" EPR line positions. Thus, these tensors can be evaluated from the "allowed" line positions. Further, the calculation of K_i^2 and k_i^2 involves the tensors $\tilde{\mathbf{g}}^{\mathrm{T}} \cdot \tilde{\mathbf{A}}_i^{\mathrm{T}} \cdot \tilde{\mathbf{A}}_i \cdot \tilde{\mathbf{g}}$ $(= \tilde{\mathbf{g}}^{\mathrm{T}} \cdot \tilde{\mathbf{A}}_i^2 \cdot \tilde{\mathbf{g}})$ and $\tilde{\mathbf{g}}^{\mathrm{T}} \cdot \tilde{\mathbf{A}}_i^{\mathrm{T}} \cdot \tilde{\mathbf{A}}_i \cdot \tilde{\mathbf{A}}_i^{\mathrm{T}} \cdot \tilde{\mathbf{A}}_i \cdot \tilde{\mathbf{g}}$ $(= \tilde{\mathbf{g}}^{\mathrm{T}} \cdot \tilde{\mathbf{A}}_i^2 \cdot \tilde{\mathbf{A}}_i^2 \cdot \tilde{\mathbf{g}} = \tilde{\mathbf{g}}^{\mathrm{T}} \cdot \tilde{\mathbf{A}}_i^4 \cdot \tilde{\mathbf{g}})$, respectively. In order to calculate them, it may appear at first sight that they depend upon the elements of the $\tilde{\mathbf{g}}$ matrix. That this is not true can be easily seen by the use of the coordinate system in which the $\tilde{\mathbf{g}}^2$ tensor is diagonal (i.e., the principal-axes coordinate system of $\tilde{\mathbf{g}}^2$); more details can be found in ref. 4. In this coordinate system the $\tilde{\mathbf{g}}$ matrix is also diagonal; its diagonal elements are the square roots of the corresponding elements of the diagonal tensor $\tilde{\mathbf{g}}^2$.[30] The calculation of the quantities $g^2, d_n, k_i{}^2, K_i{}^2$, and e_i requires the use of the direction cosines of $\boldsymbol{\eta}$ (the unit vector along $\mathbf{H}$), the three diagonal elements of the (diagonal) matrix $\tilde{\mathbf{g}}^2$, the six independent elements of the symmetric tensor $\tilde{\mathbf{A}}^2$, and the five independent elements of the symmetric tensor $\tilde{\mathbf{D}}$ (since $\mathrm{Tr}\,\tilde{\mathbf{D}} = D_{x'x'} + D_{y'y'} + D_{z'z'} = 0$).[5] Then, using the "allowed" EPR line positions, one can evaluate the three nonzero elements of the diagonal matrix obtained by the diagonalization of $\tilde{\mathbf{g}}^2$ and the three Euler angles (θ, ϕ, ψ), required to rotate the laboratory axes to coincide with the coordinate system in which the $\tilde{\mathbf{g}}^2$ tensor is diagonal [the angles (θ, ϕ, ψ) also determine the direction cosines of $\boldsymbol{\eta}$ in the coordinate system in which the $\tilde{\mathbf{g}}^2$ tensor is diagonal, discussed later in (ii) below] the six independent elements of the $\tilde{\mathbf{A}}_i^2$, and the five independent elements of the $\tilde{\mathbf{D}}$ tensor.

"*Forbidden*" *transitions*. These transitions correspond to the transition energy $\Delta E^{(f)}$ $(\equiv E(M, \{m_i\} - E(M-1, \{m_i \pm n_i\})$, where { } covers the indices $i = 1, 2, \ldots, N)$, which can be easily calculated from (53) and (54), and is expressed by (56). An examination of $\Delta E^{(f)}$, (56), reveals that $\Delta E^{(f)}$ is determined by the symmetric tensors $\tilde{\mathbf{g}}_{ni}{}^2$ (each characterized by six independent elements) and $\tilde{\mathbf{P}}_i$ (five independent elements since $\mathrm{Tr}(\tilde{\mathbf{P}}_i) = 0$),[5] along with $\tilde{\mathbf{g}}^2$ and $\tilde{\mathbf{A}}_i{}^2$ tensors. $\tilde{\mathbf{g}}_{ni}{}^2$ and $\tilde{\mathbf{P}}_i$ can only be determined from the "forbidden" line positions as the "allowed" line positions do not depend upon them; on the other hand, $\tilde{\mathbf{g}}^2$ and $\tilde{\mathbf{A}}_i{}^2$ can, indeed, be determined from the "allowed" EPR line positions as described previously. It is seen that the various quantities G_i, p_{ni}, and q_i, describing $\Delta E^{(f)}$, given by (54), can be easily calculated if one were to use the coordinate system in which the $\tilde{\mathbf{g}}_{ni}{}^2$ tensor is diagonal, provided that there is only one nuclear spin present ($L_{ij} = 0$ for this case, and the summation over i disappears). Otherwise, one has to use an arbitrary coordinate system, which is common to all spins. In order to calculate the quantities G_i, p_{ni}, g_{ni}, and L_{ij} (54) one requires the tensor $\tilde{\mathbf{A}}_i \cdot \tilde{\mathbf{g}}$ (and its transpose), $\tilde{\mathbf{P}}_i, \tilde{\mathbf{A}}_j{}^2$, and $\mathbf{g}_{ni}{}^2$. For the presence of only one nuclear spin (say ith nuclear spin), in the coordinate system in which $\tilde{\mathbf{g}}_{ni}{}^2$ is diagonal, $\tilde{\mathbf{A}}_i \cdot \tilde{\mathbf{g}}$ can be expressed as $U_i^{\dagger}(\tilde{\mathbf{A}}_i \cdot \tilde{\mathbf{g}})_{\mathrm{d}} U_i$, where U_i is the 3×3 matrix of direction cosines relating the coordinate system in which the tensor $(\tilde{\mathbf{A}}_i \cdot \tilde{\mathbf{g}})^2$ $\{= \tilde{\mathbf{g}}^{\mathrm{T}} \cdot \tilde{\mathbf{A}}_i^{\mathrm{T}} \cdot \tilde{\mathbf{A}}_i \cdot \tilde{\mathbf{g}} = \tilde{\mathbf{g}}^{\mathrm{T}} \cdot \tilde{\mathbf{A}}_i{}^2 \cdot \tilde{\mathbf{g}}\}$ is diagonal. U_i depends on the Euler angles $(\Theta_i', \Phi_i', \Psi_i')$ required to rotate the coordinate system in which $\tilde{\mathbf{g}}^{\mathrm{T}} \cdot \tilde{\mathbf{A}}_i{}^2 \cdot \tilde{\mathbf{g}}$ is diagonal to coincide with the coordinate system in which g_{ni}^2 is diagonal, while $(\mathbf{A}_i \cdot \tilde{\mathbf{g}})_{\mathrm{d}}$ is the diagonal matrix whose elements are the square roots of the matrix obtained by the diagonalization of $\tilde{\mathbf{g}}^{\mathrm{T}} \cdot \tilde{\mathbf{A}}_i{}^2 \cdot \tilde{\mathbf{g}}$. As for $\tilde{\mathbf{P}}_i$, its five independent elements are to be expressed in the coordinate system in which $\tilde{\mathbf{g}}_{ni}{}^2$ is diagonal. More details are given later.

(ii) *Exact Matrix Diagonalization*

The eigenvalues and eigenvectors of the SH matrix given by (49), (50a)–(50e) can be rigorously computed numerically by the use of a suitable diagonalizing subroutine (e.g., JACOBI or EIGERS) on a digital computer, provided that, as seen above, the six independent elements of each of the $\tilde{\mathbf{g}}^2, \tilde{\mathbf{A}}_i{}^2$, and $\tilde{\mathbf{g}}_{nr}{}^2$ tensors, as well as the five independent elements of each of the traceless symmetric matrices $\tilde{\mathbf{D}}$ and $\tilde{\mathbf{P}}_r$ are known. For, as seen in context with the perturbation calculation of eigenvalues, these particular tensors, indeed determine the eigenvalues. That is to say that $\tilde{\mathbf{g}}, \tilde{\mathbf{A}}_r$, and $\tilde{\mathbf{g}}_{nr}$ do not, in fact, determine eigenvalues. This is equivalent to requiring that the tensors $\tilde{\mathbf{g}}, \tilde{\mathbf{A}}_r$, and $\tilde{\mathbf{g}}_{nr}$ in (50a), (50c), and (50e) should be replaced by the corresponding symmetric tensors, defined as

$$\mathbf{O}^{(s)} = \mathbf{L}^{-1}(\mathbf{O}_{\mathrm{d}}^2)^{1/2}\mathbf{L}, \tag{57}$$

where $(\mathbf{O}_{\mathrm{d}}^2)^{1/2}$ is the diagonal matrix whose elements are the square roots of the diagonal matrix $\mathbf{O}_{\mathrm{d}}^2$ $(= \mathbf{L}\mathbf{O}^2\mathbf{L}^{-1})$ obtained by the diagonalization of $\mathbf{O}^2 = \mathbf{O}^{\mathrm{T}} \cdot \mathbf{O}$ $(\mathbf{O} = \mathbf{g}, \mathbf{A}_r, \mathbf{g}_{nr})$, and $\mathbf{L}$ is the matrix that diagonalizes $\mathbf{O}^2$; thus the column vectors of $\mathbf{L}$ are the direction cosines of the principal directions of $\mathbf{O}^2$. Then each $\mathbf{O}^{(s)}$ is a symmetric matrix, thus characterized by six independent elements. Alternatively, each $\mathbf{O}^{(s)}$ can be described in terms of the three diagonal elements of $(\mathbf{O}_{\mathrm{d}}^2)^{1/2}$, and three Euler angles (θ, ϕ, ψ) corresponding to the three successive rotations required to bring the laboratory frame in coincidence with the principal axes of the tensor $\mathbf{O}^2$.[58]

Explicitly,

$$\mathbf{L}^{-1} = \begin{pmatrix} \cos\psi\cos\phi - \cos\theta\sin\phi\sin\psi & -\sin\psi\cos\phi - \cos\theta\sin\phi\cos\psi & \sin\theta\sin\phi \\ \cos\psi\sin\phi + \cos\theta\cos\phi\sin\psi & -\sin\psi\sin\phi + \cos\theta\cos\phi\cos\psi & -\sin\theta\cos\phi \\ \sin\theta\sin\psi & \sin\theta\cos\psi & \cos\theta \end{pmatrix}. \tag{58}$$

Finally, in order to find the eigenvalues by numerical diagonalization of the SH matrix, the following SH is appropriate

$$\mathscr{H} = \mu_{\mathrm{B}} S^{\mathrm{T}} \cdot \tilde{\mathbf{g}}^{(\mathrm{s})} \cdot \mathbf{H} + \mathbf{S}^{\mathrm{T}} \cdot \tilde{\mathbf{D}} \cdot \mathbf{S} + \sum_{i=1}^{N} (\mathbf{I}_i^{\mathrm{T}} \cdot \tilde{\mathbf{A}}_i^{(\mathrm{s})} \cdot \mathbf{S} - \mu_{\mathrm{N}} \mathbf{I}_i^{\mathrm{T}} \cdot \tilde{\mathbf{g}}_{\mathrm{n}i}^{(\mathrm{s})} \cdot \mathbf{H} + \mathbf{I}_i^{\mathrm{T}} \cdot \tilde{\mathbf{P}}_i \cdot \mathbf{I}_i). \tag{59}$$

It should be noted that in (59) each of $\mathbf{g}^{(\mathrm{s})}, \tilde{\mathbf{A}}_i^{(\mathrm{s})}$, and $\tilde{\mathbf{g}}_{\mathrm{n}i}^{(\mathrm{s})}$ is characterized by six parameters in terms of three eigenvalues of each of $\tilde{\mathbf{g}}^2, \tilde{\mathbf{A}}^2$, and $\tilde{\mathbf{g}}_{\mathrm{n}i}^2$ and three corresponding Euler angles, while each of $\tilde{\mathbf{D}}, \tilde{\mathbf{P}}_i$ is characterized by five elements, instead of six, due to the fact that trace $(\tilde{\mathbf{D}}) = \mathrm{trace}(\tilde{\mathbf{P}}_i = 0)$. These independent parameters can be evaluated by the application of the LSF technique, described below in Section E.5.c.

c. Evaluation of SHPs by the LSF Technique

The application of the LSF technique will now be illustrated for the case of electron spin $S = \frac{1}{2}$ (i.e., $\mathbf{S}^{\mathrm{T}} \cdot \mathbf{D} \cdot \mathbf{S} = 0$) coupled to one nuclear spin $I \geq \frac{1}{2}$.

(i) *First-Order Perturbation*

This procedure is well described by Misra[30] for the presence of one nuclear spin. Here, the eigenvalues, calculated to first order in perturbation, neglecting the nuclear Zeeman interaction for $S = \frac{1}{2}, I \geq \frac{1}{2}$, are from (53b)

$$E^{(I)}(M, m) = \mu_{\mathrm{B}} g H M + K M m, \tag{60}$$

where the electronic magnetic quantum number $M = \pm\frac{1}{2}$ and the nuclear quantum number $m = -I, -(I-1), \ldots, (I-1), I$. For "allowed" transitions ($\Delta M = \pm 1; \Delta m = 0$, i.e., $M = \frac{1}{2} \leftrightarrow -\frac{1}{2}, m \leftrightarrow m$) the energy differences are

$$\Delta E^{(I)}(m) = \mu_{\mathrm{B}} g H + K m. \tag{61}$$

Now, using (54), the parameter g can be expressed as

$$g = \{(\tilde{\mathbf{g}}^2)_{xx} l^2 + 2(\tilde{\mathbf{g}}^2)_{xy} lm + (\tilde{\mathbf{g}}^2)_{yy} m^2 + 2(\tilde{\mathbf{g}}^2)_{yz} mn + (\tilde{\mathbf{g}}^2)_{zz} n^2 + 2(\tilde{\mathbf{g}}^2)_{xz} ln\}^{1/2}. \tag{62}$$

In (62) l, m, n are the direction cosines of $\boldsymbol{\eta}$ $(= \mathbf{H}/|\mathbf{H}|)$ with respect to the laboratory frame (x, y, z). On the other hand, using (54), one can express K in the principal-axes system (x', y', z') of the tensor $\tilde{\mathbf{g}}^2$ as

$$K = \{(l' g_{x'})^2 (\tilde{\mathbf{A}}^2)_{x'x'} + 2(l' g_{x'})(m' g_{y'})(\tilde{\mathbf{A}}^2)_{x'y'} + (m' g_{y'})^2 (\tilde{\mathbf{A}}^2)_{y'y'} + 2(l' g_{x'})(n' g_{z'})(\tilde{\mathbf{A}}^2)_{x'z'} + (n' g_{z'})^2 (\tilde{\mathbf{A}}^2)_{z'z'} + 2(m' g_{y'})(n' g_{z'})(\tilde{\mathbf{A}}^2)_{y'z'}\}^{1/2}. \tag{63}$$

In (63) $g_{x'}, g_{y'}, g_{z'}$ are the square roots of the principal values of the $\tilde{\mathbf{g}}^2$ tensor, while l', m', n' are the direction cosines of $\boldsymbol{\eta}$ in the principal axes system of $\tilde{\mathbf{g}}^2$.

From (60) it is seen that there are $(2I + 1)$ "allowed" hyperfine lines for any orientation of $\mathbf{H}$, whose center lies at the magnetic field value $H_0 = h\nu/\mu_{\mathrm{B}} g$, as found from the resonance condition. Thus, in the first-order perturbation approximation, the centers of hyperfine sets only depend on $\tilde{\mathbf{g}}^2$, they do not depend on $\tilde{\mathbf{A}}^2$. Examining (62) for g, it is seen that at least six data points in three mutually perpendicular planes are required to determine the elements of the symmetric tensor $\tilde{\mathbf{g}}^2$. In practice, one fits as many centers of hyperfine sets for the magnetic field orientation in three mutually perpendicular planes as possible, but certainly greater than, or equal to, six in order to evaluate the elements of the $\tilde{\mathbf{g}}^2$ tensor. The $\tilde{\mathbf{g}}^2$ tensor, so estimated, can then be diagonalized to obtain its principal values and its eigenvectors; the square roots of the principal values are the elements $g_{x'}, g_{y'}, g_{z'}$, while the eigenvectors represent the direction cosines of the principal axes of the $\tilde{\mathbf{g}}^2$ tensor with respect to the laboratory frame (x, y, z). Knowing these direction cosines, one can now calculate l', m', n', the direction cosines of $\boldsymbol{\eta}$. The values of the set $(g_{x'}, g_{y'}, g_{z'}; l', m', n')$ so obtained can now be used in (63) as initial values to evaluate the elements of $\tilde{\mathbf{A}}^2$ tensor in an LSF fitting of hyperfine lines observed for several orientations of the external magnetic field in three mutually perpendicular planes, as described above. The matrix of $\tilde{\mathbf{A}}^2$ so obtained can then be diagonalized to obtain the principal values of $\tilde{\mathbf{A}}^2$, the resulting eigenvectors are the direction cosines of the principal axes of $\tilde{\mathbf{A}}^2$ with respect to the principal axes of $\tilde{\mathbf{g}}^2$ tensor. For more details, see refs. 30 and 31.

(ii) *Second-Order Perturbation*

For the case $S = \frac{1}{2}$ (i.e., $\mathbf{S}^{\mathrm{T}} \cdot \tilde{\mathbf{D}} \cdot \mathbf{S} = 0$), neglecting the nuclear Zeeman interaction, considering the case of one electron spin coupled to one nuclear spin, the energy differences for "allowed" transitions $(\Delta M = \pm 1, \Delta m = 0)$, calculated from (53a)–(53c), are

$$\Delta E^{(\mathrm{II})}(m) = \mu_{\mathrm{B}} g H + Km + [\{\mathrm{Tr}(\tilde{\mathbf{A}}^2) - k^2\} \times \{I(I+1) - m^2\}/2 + (k^2 - K^2)m^2]/(2\mu_{\mathrm{B}} g H). \tag{64}$$

The quantities g and K have been explicitly expressed by (62) and (63). In order to express k^2 explicitly, as given by (54), it appears that, in general, one needs to know the nine elements of the $\tilde{\mathbf{g}}$ matrix, if it is asymmetric. However, in any case, only six parameters are, indeed, required. These are the three eigenvalues of the symmetric $\tilde{\mathbf{g}}^2$ tensor and the three Euler angles defining the direction cosines of the three principal directions of the $\tilde{\mathbf{g}}^2$-tensor with respect to the laboratory frame. These direction cosines are the column vectors of the matrix $\mathbf{L}^{-1}$, expressed in terms of the Euler angles (θ, ϕ, ψ), given by (58).

Finally, k^2 can be expressed as

$$\begin{aligned} k^2 = \{&(l' g_{x'})^2 (\tilde{\mathbf{A}}^4)_{x'x'} + 2(l' g_{x'})(m' g_{y'})(\tilde{\mathbf{A}}^4)_{x'y'} \\ &+ (m' g_{y'})^2 (\tilde{\mathbf{A}}^4)_{y'y'} + 2(l' g_{x'})(n' g_{z'})(\tilde{\mathbf{A}}^4)_{x'z'} + (n' g_{z'})^2 (\tilde{\mathbf{A}}^4)_{z'z'} \\ &+ 2(m' g_{y'})(n' g_{z'})(\tilde{\mathbf{A}}^4)_{z'z'}\}/(g^2 K^2). \end{aligned} \tag{65}$$

In writing (65), the fact that $\tilde{\mathbf{A}}^4$ is a symmetric tensor has been taken into account. In (65), (l', m', n') are the direction cosines of $\mathbf{B}$ in the principal-axes system (x', y', z') of $\tilde{\mathbf{g}}^2$. These can be expressed as

$$\begin{pmatrix} l' \\ m' \\ n' \end{pmatrix} = \begin{pmatrix} l_{x''} & m_{x''} & n_{x''} \\ l_{y''} & m_{y''} & n_{y''} \\ l_{z''} & m_{z''} & n_{z''} \end{pmatrix} \begin{pmatrix} l \\ m \\ n \end{pmatrix}, \tag{66}$$

where (l, m, n) are the direction cosines of $\mathbf{B}$ in the laboratory frame (x, y, z) and $(l_{\alpha''}, m_{\alpha''}, n_{\alpha''}; \alpha = x, y, z)$ are the direction cosines of the principal directions of the $\tilde{\mathbf{g}}^2$ tensor with respect to the laboratory frame. The $\tilde{\mathbf{A}}^4$ matrix, appearing in (65) is defined as $\tilde{\mathbf{A}}^4 = \tilde{\mathbf{A}}^2 \cdot \tilde{\mathbf{A}}^2 = \tilde{\mathbf{A}}^{\mathrm{T}} \cdot \tilde{\mathbf{A}} \cdot \tilde{\mathbf{A}}^{\mathrm{T}} \cdot \tilde{\mathbf{A}}$. Thus:

$$(\tilde{\mathbf{A}}^4)_{\alpha\beta} = \sum_{\gamma} \tilde{\mathbf{A}}^2_{\alpha\gamma} \tilde{\mathbf{A}}^2_{\gamma\beta}, \qquad \alpha, \beta, \gamma = x, y, z. \tag{67}$$

(It is seen from (67) that $\tilde{\mathbf{A}}^4$ is a symmetric tensor, using the fact that $\tilde{\mathbf{A}}^2$ is a symmetric tensor.)

Equations (64)–(66) reveal that the energy differences $\Delta E^{(\mathrm{II})}(m)$ depend on 12 parameters. These are the three eigenvalues of the matrix $\tilde{\mathbf{g}}^2$, the three Euler angles required to rotate the laboratory axes to coincide with the principal axes of the $\mathbf{g}^2$ tensor, and the six independent elements of the symmetric tensor $\tilde{\mathbf{A}}^2$. The derivatives of the eigenvalues required in the LSF procedure can then be appropriately calculated using the dependence of $\Delta E^{(\mathrm{II})}(m)$ on these 12 parameters. Further, it is seen that because of the dependence on m^2 of the third term on the right-hand side of (64), the centers of the hyperfine sets (hyperfine resonant line positions) for any orientation of **H** depend both on $\tilde{\mathbf{g}}^2$ and $\tilde{\mathbf{A}}^2$, unlike the case of first-order perturbed eigenvalues where the dependence of the centers of the hyperfine sets is only on $\tilde{\mathbf{g}}^2$. Thus, for the application of the LSF procedure, using second-order perturbed eigenvalues, one has to fit simultaneously the 12 parameters $g_{x'}, g_{y'}, g_{z'}, \theta, \phi, \psi; \tilde{\mathbf{A}}_{\alpha\beta}{}^2$ $(\alpha, \beta = x, y, z)$. The initial values of the first six parameters may be chosen to be those obtained by the LSF method using eigenvalue expressions, calculated using the first-order perturbation approximation, and fitting the centers of hyperfine sets, as described in Section E.1. The initial values of the Euler angles may be obtained from the eigenvectors (direction cosines) of the principal values of $\tilde{\mathbf{g}}^2$ matrix, by solving (57).

In order to use the LSF procedure, one should identify the various quantities that determine the energy differences between the pair of levels participating in resonance. These are, in turn, functions of the particular SHPs for the "allowed" and "forbidden" transitions, as seen in Sections (i) and (ii) above.

It is found helpful in some cases to use different special coordinate axes for the fitting of "allowed" and "forbidden" transitions; this is described below. An arbitrary coordinate system can also be used; however, the application of the LSF procedure becomes much more complicated in this case; the details are given later.

Use of Special Coordinate Axes

"Allowed" line positions. From (55), giving the energy difference $\Delta E^{(\mathrm{a})}$ between the levels participating in resonance for the "allowed" transitions, one first needs to express explicitly the quantities $\boldsymbol{\eta}, \tilde{\mathbf{g}}^2, K_i{}^2$, and $k_i{}^2$, on which $\Delta E^{(\mathrm{a})}$ depends. These quantities, in turn, depend on SHPs. They are described, in the principal-axes system of the tensor $\tilde{\mathbf{g}}^2$, as follows. $\boldsymbol{\eta}$ is the unit vector along the Zeeman field **H**. Its direction cosines (l, m, n) in the laboratory axes are thus known. In order to express (l', m', n'), its direction cosines in the principal-axes system of $\tilde{\mathbf{g}}^2$, the transformation matrix U, which is a function of the Euler angles (θ, ϕ, ψ) required to rotate the laboratory axes to coincide with the $\tilde{\mathbf{g}}^2$ principal axes, are required:

$$\boldsymbol{\eta} = \begin{pmatrix} l' \\ m' \\ n' \end{pmatrix} = U \begin{pmatrix} l \\ m \\ n \end{pmatrix}. \tag{68}$$

In (68) the transpose of the matrix U is related to the matrix $\mathbf{L}^{-1}$ given by (58) as follows:

$$U = (\mathbf{L}^{-1})^{\dagger}. \tag{69}$$

Here † denotes adjoint. Then, using (54) which defines $\tilde{\mathbf{g}}, K_i{}^2, k_i{}^2, d_n$, and e_i, one can express

$$g^2 = \sum_{\alpha'} a_{\alpha'}{}^2, \tag{70a}$$

$$K_i{}^2 g^2 = \sum_{\alpha',\beta'} (2 - \delta_{\alpha'\beta'}) A_{\alpha'\beta'}{}^2 a_{\alpha'} a_{\beta'}, \tag{70b}$$

$$k_i{}^2 K_i{}^2 g^2 = \sum_{\alpha',\beta'} (2 - \delta_{\alpha'\beta'}) A_{\alpha'\beta'}{}^4 a_{\alpha'} a_{\beta'}, \tag{70c}$$

where

$$A_{\alpha'\beta'}{}^4 = (\mathbf{A}^2 \cdot \mathbf{A}^2)_{\alpha'\beta'} = \sum_{\gamma'} A_{\alpha'\gamma'}{}^2 A_{\gamma'\beta'}{}^2, \tag{70d}$$

$$e_i K_i{}^2 g^2 = \frac{1}{2} \sum_{\alpha',\beta'} a_{\alpha'} a_{\beta'} \sum_{\gamma'} (D_{\alpha'\gamma'} A_{i\alpha'\beta'} + A_{i\alpha'\gamma'} D_{\gamma'\beta'}), \tag{70e}$$

and

$$d_n g^2 = \sum_{\alpha'\beta'} a_{\alpha'} a_{\beta'} \sum_{\gamma'} (\mathbf{D}^n)_{\alpha'\beta'}, \qquad n = -1, 1, 2, \tag{70f}$$

where

$$(\mathbf{D}^2)_{\alpha'\beta'} = \sum_{\gamma'} D_{\alpha'\gamma'} D_{\gamma'\beta'}. \tag{70g}$$

In (70a)–(70g) the summations on α', β', γ' are over x', y', z', labeling the principal axes of the $\tilde{\mathbf{g}}^2$ tensor. The tensors $\tilde{\mathbf{A}}^2$ and $\tilde{\mathbf{A}}^4$ ($\equiv \tilde{\mathbf{A}}^2 \cdot \tilde{\mathbf{A}}^2$) and, $\tilde{\mathbf{D}}$ are symmetric, i.e., $A_{\alpha'\beta'}{}^2 = A_{\beta'\alpha'}{}^2$, $A_{\alpha'\beta'}{}^4 = A_{\beta'\alpha'}{}^4$, $D_{\alpha'\beta'} = D_{\beta'\alpha'}$. In order to calculate $(\tilde{\mathbf{D}}^{-1})_{\alpha'\beta'}$, required in the calculation of d^{-1} (70f), one first notes that $\tilde{\mathbf{D}}$ is a 3×3 matrix. Now, the inverse R of a matrix F can be expressed as[58]

$$R = F^{-1} = (\mathrm{Det}(F))^{-1} \begin{pmatrix} F_{yy}F_{zz} - F_{yz}F_{zy} & F_{xz}F_{zy} - F_{xy}F_{zz} & F_{xy}F_{yz} - F_{xz}F_{yy} \\ F_{yz}F_{zx} - F_{yx}F_{zz} & F_{xx}F_{zz} - F_{xz}F_{zx} & F_{xz}F_{yx} - F_{xx}F_{yz} \\ F_{yx}F_{zy} - F_{yy}F_{zx} & F_{xy}F_{zx} - F_{xx}F_{zy} & F_{xx}F_{yy} - F_{xy}F_{yx} \end{pmatrix}, \tag{71}$$

where the matrix multiplying $(\mathrm{Det}\, F)^{-1}$ on the right-hand side of (71) is defined as $\mathrm{Adj}(F)$, the adjoint matrix to the matrix F. Since, d_{-1} depends on $\tilde{\mathbf{D}}^{-1}$, the inverse of the matrix $\tilde{\mathbf{D}}$, it would, therefore, according to (71), be undefined if the determinant of $\tilde{\mathbf{D}}$ vanishes. If this occurs, one can use the alternative expression $d_{-1}\,\mathrm{Det}(\tilde{\mathbf{D}}) = \mathrm{Tr}(\mathrm{Adj}(\tilde{\mathbf{D}})) - d_1\,\mathrm{Tr}(\tilde{\mathbf{D}}) + d_2$.[18,19]

Finally, the $a_{\alpha'}$ required in (70a)–(70c), (70e), and (70f) are given as

$$a_{x'} = l' g_{x'}, \qquad a_{y'} = m' g_{y'}, \qquad a_{z'} = n' g_{z'}. \tag{72}$$

In (72), $g_{x'}$, $g_{y'}$, $g_{z'}$ are the square roots of the principal values of $\tilde{\mathbf{g}}^2$, while (l', m', n') are the direction cosines of $\mathbf{H}$ (the components of $\boldsymbol{\eta}$) in the principal-axes system of $\tilde{\mathbf{g}}^2$; these are calculated by the use of (68) and (69).

The 17 parameters (for the presence of one nuclear spin) which can be evaluated from the LSF fitting of the "allowed" EPR line positions (for $S \geq 1$), observed in three mutually perpendicular planes are thus the three diagonal elements $\tilde{\mathbf{g}}_{\alpha'}{}^2$ $(\alpha' = x', y', z')$, the three Euler angles (θ, ϕ, ψ), the components of $\tilde{\mathbf{A}}_{\alpha'\beta'}{}^2$ (six independent elements), and the components $D_{\alpha'\beta'}$ (five independent elements). To this end, one needs to use the expression for $\Delta E^{(a)}$, (55), along with the definitions of the various quantities therein, as given by (70a)–(70g), (71), and (72)

"*Forbidden*" *line positions*. The energy difference $\Delta E^{(f)}$, (56), between the pair of levels participating in resonance for "forbidden" transitions requires quantities, over and above those described previously in context with the "allowed" line positions. These are given as follows, for the presence of only one nuclear spin (denoted by i), using the coordinate axes $(x_{i'}, y_{i'}, z_{i'})$ in which the $\tilde{\mathbf{g}}_{ni}{}^2$ tensor is diagonal:

$$g_{ni}{}^2 = \sum_{\alpha_i'} b_{\alpha_i'}{}^2, \tag{73a}$$

$$G_i K_i g = \frac{1}{2} \sum_{\alpha_i' \beta_i'} \eta_{\alpha_i'} \eta_{\beta_i'} \{ (\tilde{\mathbf{A}}_i \cdot \tilde{\mathbf{g}})_{\beta'\alpha'} (\tilde{\mathbf{g}}_{ni})_{\beta'} + (\tilde{\mathbf{A}}_{i'} \cdot \tilde{\mathbf{g}})_{\alpha'\beta'} (\tilde{\mathbf{g}}_{ni})_{\alpha'} \}, \tag{73b}$$

$$p_{ni} K_i{}^2 g^2 = \sum_{a_i' \beta_i'} \eta_{\alpha_i'} \eta_{\beta_i'} \sum_{\gamma_i' \delta_i'} (\tilde{\mathbf{A}}_i \cdot \tilde{\mathbf{g}}_{ni})_{\gamma_i' \alpha_i'} (\tilde{\mathbf{A}}_i \cdot \tilde{\mathbf{g}}_{ni})_{\delta_i' \beta_i'} (\tilde{\mathbf{P}}_i^n)_{\gamma_i' \delta_i'}, \tag{73c}$$

$$g_i K_i g = \frac{1}{2} \sum_{\alpha_i' \beta_i'} \eta_{\alpha_i'} \eta_{\beta_i'} \{ (\tilde{\mathbf{A}}_i \cdot \tilde{\mathbf{g}})_{\gamma_i' \alpha_i'} (\tilde{\mathbf{P}}_i)_{\gamma_i' \beta_i'} (\tilde{\mathbf{g}}_{ni})_{\beta_i'} + (\tilde{\mathbf{A}}_i \cdot \tilde{\mathbf{g}})_{\gamma_i' \beta_i'} (\tilde{\mathbf{P}}_i)_{\alpha_i' \gamma_i'} (\tilde{\mathbf{g}}_{ni})_{\alpha_i'} \}. \tag{73d}$$

In (73a)–(73d), $\eta_{\alpha_i'}$ are the components of $\boldsymbol{\eta}$, the unit vector along the Zeeman field $\mathbf{H}$; these are, thus, the direction cosines of the external Zeeman field in the principal-axes system of $\tilde{\mathbf{g}}_{ni}^2$ tensor. $\boldsymbol{\eta}$ can be expressed in terms of the direction cosines (l, m, n) of $\mathbf{H}$ in the laboratory axes as follows:

$$\boldsymbol{\eta}_i = \begin{pmatrix} l_{i'} \\ m_{i'} \\ n_{i'} \end{pmatrix} = U_i \begin{pmatrix} l \\ m \\ n \end{pmatrix}. \tag{74}$$

In (74), the matrix U_i can be described in the same way as the matrix U (69), if one replaces θ, ϕ, ψ by Θ_i, Φ_i, Ψ_i, respectively; here $(\Theta_i, \Phi_i, \Psi_i)$ are the Euler angles required to rotate the laboratory axes to coincide with the $\tilde{\mathbf{g}}_{ni}{}^2$ principal axes.

In (73a), $b_{\alpha_i'}$ are given as $b_{x_i'} = l_i' g_{nx_i'}$, $b_{y_i'} = m_i' g_{ny_i'}$, $b_{z_i'} = n_i' g_{nz_i'}$, where $g_{nx_i'}$, $g_{ny_i'}$, $g_{nz_i'}$ are the square roots of the principal values of $\tilde{\mathbf{g}}_{ni}{}^2$.

The $\tilde{\mathbf{A}}_i \cdot \tilde{\mathbf{g}}$ (and thus $\tilde{\mathbf{g}}^{\mathrm{T}} \cdot \tilde{\mathbf{A}}_i^{\mathrm{T}}$), required in (73b)–(73d), to be expressed in the $\tilde{\mathbf{g}}_{ni}{}^2$ principal-axes system, are obtained from $\tilde{\mathbf{g}}^{\mathrm{T}} \cdot \tilde{\mathbf{A}}_i^2 \cdot \tilde{\mathbf{g}}$ as evaluated from the "allowed" EPR line positions in the $\tilde{\mathbf{g}}^2$ principal-axes system, as follows:

$$\tilde{\mathbf{A}}_i \cdot \tilde{\mathbf{g}} = U_i U U' (\tilde{\mathbf{g}}^{\mathrm{T}} \cdot \tilde{\mathbf{A}}_i^2 \cdot \tilde{\mathbf{g}})_{\mathrm{d}}^{1/2} U'^{\dagger} U^{\dagger} U_i^{\dagger}. \tag{75}$$

In (75), $(\tilde{\mathbf{g}}^{\mathrm{T}} \cdot \tilde{\mathbf{A}}_i^2 \cdot \tilde{\mathbf{g}})_{\mathrm{d}}{}^{1/2}$ is the diagonal matrix, whose elements are the square roots of the matrix obtained from the diagonalization of the unitary matrix $\tilde{\mathbf{g}}^{\mathrm{T}} \cdot \tilde{\mathbf{A}}_i^2 \cdot \tilde{\mathbf{g}}$, as determined in the $\tilde{\mathbf{g}}^2$ principal-axes system, using the "allowed" EPR line positions:

$$(\tilde{\mathbf{g}}^{\mathrm{T}} \cdot \tilde{\mathbf{A}}_i^2 \cdot \tilde{\mathbf{g}})_{\mathrm{d}} = U'^{\dagger} (\tilde{\mathbf{g}}^{\mathrm{T}} \cdot \tilde{\mathbf{A}}_i^2 \cdot \tilde{\mathbf{g}}_i) U'. \tag{76}$$

The matrices U', $U_{i'}$, and U used in (75) are defined by (76), (74), and (69), respectively.

Use of Arbitrary Coordinate Axes

The use of special coordinate systems, described above, simplified the expressions of the various quantities required to determine $\Delta E^{(\mathrm{a})}$ and $\Delta E^{(\mathrm{f})}$. However, it is possible to use an entirely arbitrary coordinate system although its use is more complicated than those of the special coordinate systems, described in the preceding section. Further, for the presence of more than one nucleus, no special coordinate system is found to be particularly useful. For the case of an arbitrary coordinate system, the following observations, as made from an examination of (54)–(56), applicable to the case of the presence of more than one nuclear spin, are found to be helpful. The quantities that can be determined from the "allowed" EPR line positions in an arbitrary coordinate system are the components of the symmetric tensors $\tilde{\mathbf{g}}^{\mathrm{T}} \cdot \tilde{\mathbf{g}}$, $\tilde{\mathbf{g}}^{\mathrm{T}} \cdot \tilde{\mathbf{A}}_i^2 \cdot \tilde{\mathbf{g}}$ $(= \tilde{\mathbf{g}}^{\mathrm{T}} \cdot \tilde{\mathbf{A}}_i^{\mathrm{T}} \cdot \tilde{\mathbf{A}}_i \cdot \tilde{\mathbf{g}})$, and $\tilde{\mathbf{g}}^{\mathrm{T}} \cdot \tilde{\mathbf{D}} \cdot \tilde{\mathbf{g}}$. On the other hand, the "forbidden" line positions enable the determination of the components of $\tilde{\mathbf{g}}_{ni}{}^2$ and $\tilde{\mathbf{P}}_i$, provided that one uses

$$\tilde{\mathbf{A}}_i \cdot \tilde{\mathbf{g}} = U'' (\tilde{\mathbf{g}}^{\mathrm{T}} \cdot \tilde{\mathbf{A}}^2 \cdot \tilde{\mathbf{g}}^2)_{\mathrm{d}}{}^{1/2} U''^{\dagger}, \tag{77a}$$

and

$$\tilde{\mathbf{g}}_{ni} = U_i (\tilde{\mathbf{g}}_{ni}{}^2)_{\mathrm{d}}{}^{1/2} U_i^{\dagger}. \tag{77b}$$

In (77a), $(\tilde{\mathbf{g}}^{\mathrm{T}} \cdot \tilde{\mathbf{A}}^2 \cdot \tilde{\mathbf{g}})_{\mathrm{d}}^{1/2}$ is the diagonal matrix whose elements are the square roots of the elements of the matrix obtained by the diagonalization of $\tilde{\mathbf{g}}^{\mathrm{T}} \cdot \tilde{\mathbf{A}}^2 \cdot \tilde{\mathbf{g}}$, already determined from the "allowed" EPR line positions, i.e.,

$$(\tilde{\mathbf{g}}^{\mathrm{T}} \cdot \tilde{\mathbf{A}}^2 \cdot \tilde{\mathbf{g}})_{\mathrm{d}} = U''^{\dagger} (\tilde{\mathbf{g}}^{\mathrm{T}} \cdot \tilde{\mathbf{A}}^2 \cdot \tilde{\mathbf{g}}) U'', \tag{78}$$

where U'' is the unitary matrix effecting the diagonalization of $\tilde{\mathbf{g}}^{\mathrm{T}} \cdot \tilde{\mathbf{A}}^2 \cdot \tilde{\mathbf{g}}$. Furthermore, in (77b), $(\tilde{\mathbf{g}}_{ni}{}^2)_{\mathrm{d}}{}^{1/2}$ is the diagonal matrix, whose elements are the square roots of the elements of the matrix obtained by the diagonalization of $\tilde{\mathbf{g}}_{ni}{}^2$, i.e.,

$$(\tilde{\mathbf{g}}_{ni}{}^2)_{\mathrm{d}} = U_i^{\dagger} (\tilde{\mathbf{g}}_{ni}{}^2) U_i, \tag{79}$$

where U_i is the unitary matrix that effects the diagonalization of $\tilde{\mathbf{g}}_{ni}{}^2$.

Explicitly, for use in (54)–(56), in terms of the above-mentioned quantities $\tilde{\mathbf{g}}^2, \tilde{\mathbf{g}}^{\mathrm{T}} \cdot \tilde{\mathbf{A}}^2 \cdot \tilde{\mathbf{g}}$, $\tilde{\mathbf{g}}^{\mathrm{T}} \cdot \tilde{\mathbf{D}} \cdot \tilde{\mathbf{g}}, \tilde{\mathbf{g}}_{ni}{}^2, \tilde{\mathbf{P}}_i$, determinable from EPR data, using $(\tilde{\mathbf{g}}^2)^{-1} = (\tilde{\mathbf{g}}^{\mathrm{T}} \cdot \tilde{\mathbf{g}})^{-1} = \tilde{\mathbf{g}}^{-1} \cdot (\tilde{\mathbf{g}}^{\mathrm{T}})^{-1}$, one can express

$$g^2 = \boldsymbol{\eta}^{\mathrm{T}} \cdot (\tilde{\mathbf{g}}^2) \cdot \boldsymbol{\eta}, \tag{80a}$$

$$d_n g^2 = \boldsymbol{\eta}^{\mathrm{T}} \cdot (\tilde{\mathbf{g}}^{\mathrm{T}} \cdot \tilde{\mathbf{D}} \cdot \tilde{\mathbf{g}}) \cdot (\tilde{\mathbf{g}}^2)^{-1} (\tilde{\mathbf{g}}^{\mathrm{T}} \cdot \tilde{\mathbf{D}} \cdot \tilde{\mathbf{g}}) \cdot (\tilde{\mathbf{g}}^2)^{-1} \cdots (\tilde{\mathbf{g}}^{\mathrm{T}} \cdot \tilde{\mathbf{D}} \cdot \tilde{\mathbf{g}}) \cdot \boldsymbol{\eta}, \tag{80b}$$

[$(\tilde{\mathbf{g}}^{\mathrm{T}} \cdot \tilde{\mathbf{D}} \cdot \tilde{\mathbf{g}})$ appears n times with $(\tilde{\mathbf{g}}^2)^{-1}$ in between],

$$K_i{}^2 g^2 = \boldsymbol{\eta}^{\mathrm{T}} \cdot (\tilde{\mathbf{g}} \cdot \tilde{\mathbf{A}}_i{}^2 \cdot \tilde{\mathbf{g}}) \cdot \boldsymbol{\eta}, \tag{80c}$$

$$k_i{}^2 K_i{}^2 g^2 = \boldsymbol{\eta}^{\mathrm{T}} \cdot (\tilde{\mathbf{g}}^{\mathrm{T}} \cdot \tilde{\mathbf{A}}_i{}^2 \cdot \tilde{\mathbf{g}})(\tilde{\mathbf{g}}^2)^{-1} (\tilde{\mathbf{g}}^{\mathrm{T}} \cdot \tilde{\mathbf{A}}_i{}^2 \cdot \tilde{\mathbf{g}}) \cdot \boldsymbol{\eta}, \tag{80d}$$

$$\begin{aligned} e_i K_i{}^2 g^2 = \tfrac{1}{2} \boldsymbol{\eta}^{\mathrm{T}} \cdot [&(\tilde{\mathbf{g}}^{\mathrm{T}} \cdot \tilde{\mathbf{D}} \cdot \tilde{\mathbf{g}})(\tilde{\mathbf{g}}^2)^{-1} (\tilde{\mathbf{g}}^{\mathrm{T}} \cdot \tilde{\mathbf{A}}^2 \cdot \tilde{\mathbf{g}}) \\ &+ (\tilde{\mathbf{g}}^{\mathrm{T}} \cdot \tilde{\mathbf{A}}^2 \cdot \tilde{\mathbf{g}})(\tilde{\mathbf{g}}^2)^{-1} (\tilde{\mathbf{g}}^{\mathrm{T}} \cdot \tilde{\mathbf{D}} \cdot \tilde{\mathbf{g}})] \cdot \boldsymbol{\eta}, \end{aligned} \tag{80e}$$

$$g_{ni}{}^2 = \boldsymbol{\eta}^{\mathrm{T}} \cdot (\tilde{\mathbf{g}}_{ni}{}^2) \cdot \boldsymbol{\eta}, \tag{80f}$$

$$G_i K_i g = \tfrac{1}{2} \boldsymbol{\eta}^{\mathrm{T}} \cdot [(\tilde{\mathbf{A}}_i \cdot \tilde{\mathbf{g}})^{\mathrm{T}} \cdot (\tilde{\mathbf{g}}_{ni}) + (\tilde{\mathbf{g}}_{ni}^{\mathrm{T}}) \cdot (\tilde{\mathbf{A}}_i \cdot \tilde{\mathbf{g}})] \cdot \boldsymbol{\eta}, \tag{80g}$$

$$p_{ni} K_i{}^2 g^2 = \boldsymbol{\eta}^{\mathrm{T}} \cdot (\tilde{\mathbf{A}}_i \cdot \tilde{\mathbf{g}})^{\mathrm{T}} \cdot (\tilde{\mathbf{P}}_i)^n \cdot (\tilde{\mathbf{A}}_i \cdot \tilde{\mathbf{g}}) \cdot \boldsymbol{\eta}, \tag{80h}$$

$$q_i K_i g = \tfrac{1}{2} \boldsymbol{\eta}^{\mathrm{T}} \cdot [(\tilde{\mathbf{A}}_i \cdot \tilde{\mathbf{g}})^{\mathrm{T}} \cdot (\tilde{\mathbf{P}}_i) \cdot (\tilde{\mathbf{g}}_{ni}) + (\tilde{\mathbf{g}}_{ni})^{\mathrm{T}} \cdot (\tilde{\mathbf{P}}_i) \cdot (\tilde{\mathbf{A}}_i \cdot \tilde{\mathbf{g}})] \cdot \boldsymbol{\eta}, \tag{80i}$$

$$\begin{aligned} L_{ij} K_i K_j g^2 = \tfrac{1}{2} \boldsymbol{\eta}^{\mathrm{T}} \cdot [&(\tilde{\mathbf{g}}^{\mathrm{T}} \cdot \tilde{\mathbf{A}}_i{}^2 \cdot \tilde{\mathbf{g}}) \cdot (\tilde{\mathbf{g}}^2)^{-1} \cdot (\tilde{\mathbf{g}}^{\mathrm{T}} \cdot \tilde{\mathbf{A}}_j{}^2 \cdot \tilde{\mathbf{g}}) \\ &+ (\tilde{\mathbf{g}}^{\mathrm{T}} \cdot \tilde{\mathbf{A}}_j^2 \cdot \tilde{\mathbf{g}}) \cdot (\tilde{\mathbf{g}}^2)^{-1} \cdot (\tilde{\mathbf{g}}^{\mathrm{T}} \cdot \tilde{\mathbf{A}}_i^2 \cdot \tilde{\mathbf{g}})] \cdot \boldsymbol{\eta}. \end{aligned} \tag{80j}$$

p_{-1i}, as given by (80h) requires the inverse matrix $\tilde{\mathbf{P}}_i{}^{-1}$. This can be calculated using (71). However, if $\mathrm{Det}(\tilde{\mathbf{P}}_i) = 0$, one can use the alternative expression $\mathrm{Tr}(\mathrm{Adj}(\tilde{\mathbf{P}}_i) - P_1 \, \mathrm{Tr}(\tilde{\mathbf{P}}_i) + P_2)$ in place of $p_{-1i} \, \mathrm{Det}(\tilde{\mathbf{P}}_i)$.[18,19]

The quantities determinable from EPR data are enclosed within parentheses in (80a)–(80i).

As far as the evaluation of the required traces and determinants is concerned, using the property that $\mathrm{Tr}(EF) = \mathrm{Tr}(FE)$ and $\mathrm{Det}(EF) = \mathrm{Det}(E) \cdot \mathrm{Det}(F)$, the following is noted:

$$\mathrm{Tr}(\tilde{\mathbf{A}}^2) = \mathrm{Tr}[(\tilde{\mathbf{g}}^{\mathrm{T}} \cdot \tilde{\mathbf{A}}^2 \cdot \tilde{\mathbf{g}})(\tilde{\mathbf{g}}^{\mathrm{T}} \cdot \tilde{\mathbf{g}})^{-1}] = \mathrm{Tr}[(\tilde{\mathbf{g}}^{\mathrm{T}} \cdot \tilde{\mathbf{A}}^2 \cdot \tilde{\mathbf{g}}) \cdot (\tilde{\mathbf{g}}^2)^{-1}], \tag{81a}$$

$$\mathrm{Det}(\tilde{\mathbf{A}}) = \{\mathrm{Det}(\tilde{\mathbf{g}}^{\mathrm{T}} \cdot \tilde{\mathbf{A}}^2 \cdot \tilde{\mathbf{g}}) / \mathrm{Det}(\tilde{\mathbf{g}}^{\mathrm{T}} \cdot \tilde{\mathbf{g}})\}^{1/2}. \tag{81b}$$

Simultaneous LSF Fitting of Both the "Allowed" and "Forbidden" Line Positions

It should be noted that both the "allowed" and "forbidden" line positions can be simultaneously fitted, in the LSF procedure, to evaluate the various tensor components. As for the choice of the coordinate system, one could either use any one of the two special coordinate systems described above (principal-axes systems of $\tilde{\mathbf{g}}^2$, or those of $\tilde{\mathbf{g}}_{ni}{}^2$), or an entirely arbitrary coordinate system. Although, the simultaneous fitting of both the "allowed" and "forbidden" line-position groups is a bit more complicated than the LSF fitting of only one of these alone, one can determine the components of all the tensors at once by a simultaneous fitting of both these groups of lines; as well, the parameter errors for this case are also reduced, since the numbers of lines fitted has been increased.[49]

The relevant steps to be followed for the application of the LSF procedure, as well as an illustrative example, are well described in ref. 59.

d. Numerical Evaluation of the Derivatives Required in the LSF Procedure

Application of the LSF technique, using matrix diagonalization, to the case of noncoincident tensors poses certain problems since the eigenvalues, which determine the resonant line positions, do not depend on the elements of the $\tilde{\mathbf{g}}$, $\tilde{\mathbf{A}}$ tensors, which are not symmetric in general; rather, they depend on the elements of the symmetric matrices $\tilde{\mathbf{g}}^2$, $\tilde{\mathbf{A}}^2$, $\tilde{\mathbf{g}}_n{}^2$ which do not appear in the SH given by (49), (50a)–(50c). Thus, these are the elements of the $\mathbf{g}^2$, $\mathbf{A}^2$ matrices which can be determined from the data, rather than the elements of the matrices $\tilde{\mathbf{g}}$, $\tilde{\mathbf{A}}$. For, one can only determine those quantities from a physical measurement on which the measured property depends. On the other hand, if one were to use the symmetric matrices $\mathbf{O}^{(s)}$ $\{= \mathbf{g}^{(s)}, \mathbf{A}^{(s)}, \mathbf{g}_n^{(s)}\}$, as defined by (57), the six elements of each of which consist of three principal eigenvalues of the $\mathbf{O}^2 \equiv \mathbf{O}^{\mathrm{T}} \cdot \mathbf{O}$ matrices, and the three corresponding Euler angles, required to rotate the laboratory axes (x, y, z) to coincide with the principal axes of the $\mathbf{O}^2$ tensor, then one could use the SH given by (59) for the purpose of matrix diagonalization. In this case, one needs six independent elements defining each of the matrices $\mathbf{g}^{(s)}$, $\mathbf{A}^{(s)}$, and $\mathbf{g}_n^{(s)}$ and five independent elements of each of the symmetric matrices $\mathbf{P}^{(s)}$ and $\mathbf{Q}^{(s)}$, since their traces are zero, i.e., $P_{xx} + P_{yy} + P_{zz} = Q_{xx} + Q_{yy} + Q_{zz} = 0$. Consequently, the SH given by (59) can be used, in an LSF manner, to evaluate SHPs from resonant line positions obtained for the magnetic field orientation in three mutually perpendicular planes.

For the applicability of the LSF procedure the required derivatives can be calculated using Feynman's theorem[45]

$$\begin{aligned} \partial E_i/\partial a_j &= \langle i|\partial \mathscr{H}/\partial a_j|i\rangle \\ &= \mathrm{Tr}[(\partial \mathscr{H}/\partial a_j)(|i\rangle \otimes \langle i|)]. \end{aligned} \tag{82}$$

In (82), $|i\rangle$ is the eigenvector of $\mathscr{H}$ corresponding to the eigenvalue E_i, i.e., $\mathscr{H}|i\rangle = E_i|i\rangle$; and $|i\rangle \otimes \langle i|$ is the outer product of the eigenvector $|i\rangle$ with itself; its elements are thus given as

$$(|i\rangle \otimes \langle i|)_{kl} = |i\rangle_k |i\rangle_l^*. \tag{83}$$

In (83), $|i\rangle_k$ denotes the kth element of column vector $|i\rangle$. As for the operator $\partial\mathscr{H}/\partial a_j$, which is the derivative of $\mathscr{H}$ with respect to the parameter a_j, it can be evaluated for different terms of the SH, which depend on different parameters, as follows. $\mathscr{H}_{\mathrm{zf}}$ can be expressed as

$$\mathcal{H}_{\text{zf}} = \mathbf{S}^{\text{T}} \cdot \tilde{\mathbf{D}} \cdot \mathbf{S} = \sum_{\alpha,\beta=x,y,z} D_{\alpha\beta} S_\alpha S_\beta. \tag{84}$$

Then,

$$\partial(\mathbf{S}^{\text{T}} \cdot \tilde{\mathbf{D}} \cdot \mathbf{S})/\partial D_{\alpha\beta} = S_\alpha S_\beta + S_\beta S_\alpha \qquad (\alpha \neq \beta), \tag{85a}$$

and

$$\partial(\mathbf{S}^{\text{T}} \cdot \tilde{\mathbf{D}} \cdot \mathbf{S})/\partial D_{\alpha\alpha} = S_\alpha{}^2 - S_z{}^2 \qquad (\alpha = x, y). \tag{85b}$$

In writing (85a) it has been taken into account that $\tilde{\mathbf{D}}$ is symmetric, i.e., $D_{\alpha\beta} = D_{\beta\alpha}$. In (85b) the fact that $D_{zz} = -D_{xx} - D_{yy}$ has been taken into account. In a similar manner, noting that $P_{zz} = -P_{xx} - P_{yy}$, the derivatives of $\mathcal{H}_{\text{nq}}$ for the presence of a single nucleus are

$$\partial(\mathbf{I}^{\text{T}} \cdot \tilde{\mathbf{P}} \cdot \mathbf{I})/\partial P_{\alpha\beta} = I_\alpha I_\beta + I_\beta I_\alpha \qquad (\alpha \neq \beta), \tag{86a}$$

and

$$\partial(\mathbf{I}^{\text{T}} \cdot \tilde{\mathbf{P}} \cdot \mathbf{I})/\partial P_{\alpha\alpha} = I_\alpha{}^2 - I_z{}^2 \qquad (\alpha = x, y). \tag{86b}$$

As for the derivatives of the electronic Zeeman term $\mathcal{H}_{\text{ze}} = \mu_{\text{B}} \mathbf{S}^{\text{T}} \cdot \mathbf{g}^{(\text{s})}$. I with respect to the elements of $g^{(\text{s})}$, equation (57) defining $g^{(\text{s})}$, is taken into account. Then, each element of the matrix $g^{(\text{s})}$ is a function of six parameters $\{a_j\} \equiv (g_1^{\text{d}}, g_2^{\text{d}}, g_3^{\text{d}}; \theta, \phi, \psi)$, where g_i^{d} $(i = 1, 2, 3)$ are the square roots of the principal values of the $\mathbf{g}^2$ matrix, and (θ, ϕ, ψ) are the Euler angles needed to rotate the laboratory axes to coincide with the principal axes of the $\mathbf{g}^2$ tensor. Explicitly,

$$\mathcal{H}_{\text{ze}} = \sum_{\alpha,\beta=x,y,z} \tfrac{1}{2}\mu_{\text{B}}(H_\alpha S_\beta + H_\beta S_\alpha) g_{\alpha\beta}^{(\text{s})}. \tag{87}$$

In writing (87), the fact that $\mathbf{g}^{(\text{s})}$ is a symmetric matrix has been used. The derivatives of $\mathcal{H}_{\text{ze}}$ can now be expressed as

$$\partial\mathcal{H}_{\text{ze}}/\partial a_j = \sum_{\alpha,\beta} \tfrac{1}{2}\mu_{\text{B}}(H_\alpha S_\beta + H_\beta S_\alpha)\partial g_{\alpha\beta}^{(\text{s})}/\partial a_j. \tag{88}$$

Finally, using the derivatives $\partial g_{\alpha\beta}^{(\text{s})}/da_j$ from the known dependence of $g_{\alpha\beta}^{(\text{s})}$ on a_j as given previously, the right-hand side of (88) can be calculated.

Similarly, for the hyperfine term $\mathcal{H}_{\text{hf}} = \mathbf{S}^{\text{T}} \cdot \mathbf{A}^{(\text{s})} \cdot \mathbf{I}$,

$$\partial\mathcal{H}_{\text{hf}}/\partial a_k = \sum_{\alpha,\beta} \tfrac{1}{2}(S_\alpha I_\beta + S_\beta I_\alpha)\partial A_{\alpha\beta}^{(\text{s})}/\partial a_k, \tag{89}$$

where a_k stands for any of the six parameters $\{a_k\} \equiv (A_1{}^{\text{d}}, A_2{}^{\text{d}}, A_3{}^{\text{d}}; \alpha, \beta, \gamma)$; with

$A_i^{\rm d}, i = 1, 2, 3$, being the square roots of the eigenvalues of the matrix $\tilde{\mathbf{A}}^2$, and (α, β, γ) being the Euler angles required to rotate the laboratory axes to coincide with the principal axes of the $\tilde{\mathbf{A}}^2$ tensor.

Finally, the derivatives of the nuclear Zeeman term $\mathscr{H}_{\rm zn} = -\mu_{\rm N}\mathbf{I}^{\rm T} \cdot \tilde{\mathbf{g}}_{\rm n}^{(\rm s)} \cdot \mathbf{H}$, can be evaluated with respect to the six elements $(g_{{\rm n}i}^{\rm d}, i = 1, 2, 3; \xi, \eta, \zeta)$ where $g_{{\rm n}i}^{\rm d}, i = 1, 2, 3$, are the square roots of the eigenvalues of the symmetric tensor $\tilde{\mathbf{g}}_{\rm n}^2$, and (ξ, η, ζ) are the Euler angles required to rotate the laboratory axes to coincide with the principal axes of the $\tilde{\mathbf{g}}_{\rm n}^2$ tensor, in the same manner as that utilized to evaluate the derivatives of $\mathscr{H}_{\rm ze}$.

e. General Remarks

In the evaluation of SHPs from EPR line positions for the case of noncoincident tensors, appropriate to low-symmetry, one can either use the perturbation approach, or exact numerical diagonalization of the SH matrix. However, in either case, one can only estimate those parameters from the data on which the eigenvalue differences actually depend. For the case of perturbation, for "allowed" transitions, these are the $\tilde{\mathbf{g}}^2$ and $\tilde{\mathbf{A}}^2$ tensors for $S = \frac{1}{2}$ and $I \geq \frac{1}{2}$. The principal values of $\tilde{\mathbf{g}}^2$ and $\tilde{\mathbf{A}}^2$ tensors and their direction cosines can be obtained by numerical diagonalizations of the corresponding matrices. The eigenvalues of a matrix represent its principle values, while the normalized eigenvectors represent the corresponding direction cosines. The square roots of the eigenvalues of $\tilde{\mathbf{g}}^2$ and $\tilde{\mathbf{A}}^2$ matrices are the principal values of g and A matrices, respectively.

For the case of $S > \frac{1}{2}$, $I \geq \frac{1}{2}$, one can estimate the components of the tensors $\tilde{\mathbf{g}}^2$, $\tilde{\mathbf{g}}^{\rm T} \cdot \tilde{\mathbf{A}}^2 \cdot \tilde{\mathbf{g}}$, and $\tilde{\mathbf{g}}^{\rm T} \cdot \tilde{\mathbf{D}} \cdot \tilde{\mathbf{g}}$ from "allowed" line positions if one takes into account the corresponding differences of eigenvalues, given by (53a) and (53c) and the definitions of the quantities $\tilde{\mathbf{g}}^2$, d_n, K^2, k^2, given by (54). For the case of only one nucleus, the following relations are found to be useful:

$$\text{(i)}\ \tilde{\mathbf{g}}^{\rm T} \cdot \tilde{\mathbf{A}}^2 \cdot \tilde{\mathbf{A}}^2 \cdot \tilde{\mathbf{g}} = \tilde{\mathbf{g}}^{\rm T} \cdot \tilde{\mathbf{A}}^2 \cdot \tilde{\mathbf{g}} \cdot (\tilde{\mathbf{g}}^{\rm T} \cdot \tilde{\mathbf{g}})^{-1} \cdot \tilde{\mathbf{g}}^{\rm T} \cdot \tilde{\mathbf{A}}^2 \cdot \tilde{\mathbf{g}} = (\tilde{\mathbf{g}}^{\rm T} \cdot \tilde{\mathbf{A}}^2 \cdot \tilde{\mathbf{g}}) \cdot (\tilde{\mathbf{g}}^2)^{-1} \cdot (\tilde{\mathbf{g}}^{\rm T} \cdot \tilde{\mathbf{A}}^2 \cdot \tilde{\mathbf{g}});$$

$$\text{(ii)}\ \tilde{\mathbf{g}}^{\rm T} \cdot \tilde{\mathbf{D}}^2 \cdot \tilde{\mathbf{g}} = \tilde{\mathbf{g}}^{\rm T} \cdot \tilde{\mathbf{D}} \cdot \tilde{\mathbf{g}} \cdot (\tilde{\mathbf{g}}^{\rm T} \cdot \tilde{\mathbf{g}})^{-1} \cdot \tilde{\mathbf{g}}^{\rm T} \cdot \tilde{\mathbf{D}} \cdot \tilde{\mathbf{g}} = (\tilde{\mathbf{g}}^{\rm T} \cdot \tilde{\mathbf{D}} \cdot \tilde{\mathbf{g}}) \cdot (\tilde{\mathbf{g}}^2)^{-1} \cdot (\tilde{\mathbf{g}}^{\rm T} \cdot \tilde{\mathbf{D}} \cdot \tilde{\mathbf{g}});$$

$$\text{(iii)}\ \mathrm{Tr}(\tilde{\mathbf{A}}^2) = \mathrm{Tr}\{(\tilde{\mathbf{g}}^{\rm T} \cdot \tilde{\mathbf{A}}^2 \cdot \tilde{\mathbf{g}}) \cdot (\tilde{\mathbf{g}}^2)^{-1}\};$$

$$\text{(iv)}\ \mathrm{Det}(\tilde{\mathbf{A}}) = \{\mathrm{Det}(\tilde{\mathbf{A}})^2\}^{1/2} = \{\mathrm{Det}(\tilde{\mathbf{g}}^{\rm T} \cdot \tilde{\mathbf{A}}^2 \cdot \tilde{\mathbf{g}})/\mathrm{Det}(\tilde{\mathbf{g}}^2)\}^{1/2};\ \text{and}$$

$$\text{(v)}\ \tilde{\mathbf{A}}^2 = \tilde{\mathbf{V}}^{-1} \cdot (\tilde{\mathbf{g}}_{\rm d})^{-1} \cdot \tilde{\mathbf{V}} \cdot \tilde{\mathbf{V}}^{-1} \cdot (\tilde{\mathbf{g}}^{\rm T} \cdot \tilde{\mathbf{A}}^2 \cdot \tilde{\mathbf{g}}) \cdot \tilde{\mathbf{V}} \cdot \tilde{\mathbf{V}}^{-1} (\tilde{\mathbf{g}}_{\rm d})^{-1} \cdot \tilde{\mathbf{V}}.$$

In expressing (iii) the facts that $(\tilde{\mathbf{g}}^2)^{-1} = (\tilde{\mathbf{g}}^{\rm T} \cdot \tilde{\mathbf{g}})^{-1}$ and $\mathrm{Tr}(CD) = \mathrm{Tr}(DC)$ have been used. For (iv) the fact that $\mathrm{Det}(CD) = \mathrm{Det}(C) \times \mathrm{Det}(D)$ has been used. In (iv) the matrix $\tilde{\mathbf{V}}$ diagonalizes the $\tilde{\mathbf{g}}^2$ matrix, i.e., $\tilde{\mathbf{g}}_{\rm d}^2 = \tilde{\mathbf{V}} \tilde{\mathbf{g}}^2 \tilde{\mathbf{V}}^{-1}$. The above relations are useful when the laboratory axes are employed for the description of the various matrices.

From these relations it is seen that for the second-order perturbed eigenvalues, (53c), the eigenvalue differences for the "allowed" transitions, neglecting the nuclear Zeeman term, depend upon the matrices $\tilde{\mathbf{g}}^2$ and $\tilde{\mathbf{g}}^T \cdot \tilde{\mathbf{A}}^2 \cdot \tilde{\mathbf{g}}$. The quantities $g_n{}^2$, G, p_n, and q (54) cannot be determined from "allowed" line positions, since the corresponding eigenvalue differences do not depend upon them. However, the line positions for the hyperfine "forbidden" transition do depend on them. They can, then, be evaluated from the "forbidden" line positions, using the values of $\tilde{\mathbf{g}}^2$, $\tilde{\mathbf{A}}^2$, and $\tilde{\mathbf{D}}$ tensors already determined from the fitting of "allowed" line positions, from the second-order eigenvalues, given by (53c). The following relation is useful for the presence of more than one nucleus:

$$\tilde{\mathbf{g}}^T \cdot (\tilde{\mathbf{A}}_i^T \cdot \tilde{\mathbf{A}}_i \cdot \tilde{\mathbf{A}}_j^T \cdot \tilde{\mathbf{A}}_j + \tilde{\mathbf{A}}_j^T \cdot \tilde{\mathbf{A}}_j \cdot \tilde{\mathbf{A}}_i^T \cdot \tilde{\mathbf{A}}_i) \cdot \tilde{\mathbf{g}} = (\tilde{\mathbf{g}}^T \cdot \tilde{\mathbf{A}}_i{}^2 \cdot \tilde{\mathbf{g}}) \cdot (\tilde{\mathbf{g}}^2)^{-1} \cdot (\tilde{\mathbf{g}}^T \cdot \tilde{\mathbf{A}}_j{}^2 \cdot \tilde{\mathbf{g}}) + (\tilde{\mathbf{g}}^T \cdot \tilde{\mathbf{A}}_j{}^2 \cdot \tilde{\mathbf{g}}) \cdot (\tilde{\mathbf{g}}^2)^{-1} \cdot (\tilde{\mathbf{g}}^T \cdot \tilde{\mathbf{A}}_i{}^2 \cdot \tilde{\mathbf{g}}).$$

Finally, it should be noted that, as discussed in Sections E.5.b and E.5.c, the EPR data cannot be used to determine the individual elements of $\tilde{\mathbf{g}}$ and $\tilde{\mathbf{A}}$ matrices; only the eigenvalues and eigenvectors (direction cosines of the principal values) of $\tilde{\mathbf{g}}^2$ and $\tilde{\mathbf{A}}^2$ matrices can be determined from EPR data.

Although the present description has been confined to the second-order electron spin operator $\mathbf{S}^T \cdot \tilde{\mathbf{D}} \cdot \mathbf{S}$, higher-order electron spin operators $O_l^m, l = 4, 6, \ldots (-l \leq m \leq l)$ for $S \geq 2^8$ can be easily incorporated in the LSF procedure. Consideration of higher-order O_l^m in the LSF procedure has been well described in a general way previously.[46] See also McGavin et al.,[26] who considered an SH with several non-coincident tensors of higher than second order, i.e., fourth order, in **S** and **I**:

$$\mathscr{H} = \mu_B \mathbf{H} \cdot \tilde{\mathbf{g}} \cdot \mathbf{S} + \mathbf{S} \cdot \tilde{\mathbf{D}} \cdot \mathbf{S} + \mathbf{I} \cdot \tilde{\mathbf{A}} \cdot \mathbf{S} + \mathbf{I} \cdot \tilde{\mathbf{P}} \cdot \mathbf{I} - \mu_N \mathbf{H} \cdot \tilde{\mathbf{g}}^{(\mathbf{I})} \cdot \mathbf{I} + \sum_m B_4^m O_4^m(\mathbf{S}) + \sum_m C_4^m O_4^m(\mathbf{I}), \tag{90}$$

and were successful in applying the LSF technique for successive refinement of SHPs when

$$\mu_B \mathbf{H} \cdot \tilde{\mathbf{g}} \cdot \mathbf{S} \gg \mathbf{I} \cdot \tilde{\mathbf{A}} \cdot \mathbf{S} \sim \mathbf{S} \cdot \tilde{\mathbf{D}} \cdot \mathbf{S} \gg \mathbf{I} \cdot \tilde{\mathbf{P}} \cdot \mathbf{I} \sim \mu_N \mathbf{H} \cdot \tilde{\mathbf{g}}^{(\mathbf{I})} \cdot \mathbf{I}. \tag{91}$$

Basically, in their approach, perturbation expressions are used for the eigenvalues, and their dependence on the tensors $\tilde{\mathbf{g}}$, $\tilde{\mathbf{D}}$, $\tilde{\mathbf{A}}$, $\tilde{\mathbf{P}}$, and $\tilde{\mathbf{g}}_n$ is exploited. EPR line positions for a minimum of six orientations (two in each of three orthogonal planes) of the external magnetic field are required to determine the six independent elements of each of the tensors $\tilde{\mathbf{g}}^2$, $\tilde{\mathbf{K}}^2$, $\tilde{\mathbf{d}}_n$, $\tilde{\mathbf{p}}_{nr}$, $\tilde{\mathbf{G}}_r$, as given by (54). First, only the tensor $\tilde{\mathbf{g}}^2$ is determined neglecting other terms in the SH. It is then diagonalized

$$(\tilde{\mathbf{g}}^2)_{\text{diag}} = \mathbf{L}^{-1}(\tilde{\mathbf{g}}^2)\mathbf{L}. \tag{92}$$

The components of this tensor are then calculated in the laboratory frame by the inverse transformation

$$\tilde{\mathbf{g}} = \mathbf{L}(\tilde{\mathbf{g}}^2)_{\text{diag}}{}^{1/2}\mathbf{L}^{-1}. \tag{93}$$

This procedure is then repeated to determine the elements of the tensors $\tilde{\mathbf{A}}^2$, $\tilde{\mathbf{D}}$, $\tilde{\mathbf{P}}$, and $\tilde{\mathbf{g}}_n{}^2$, successively. Finally, it is noted that for the case of the lowest, i.e., triclinic, symmetry, all possible m values consistent with a given ℓ occur in the SH.

F. DISCUSSION

There are several aspects of evaluation of SHPs from EPR data using the LSF techniques in conjunction with numerical diagonalization of the SH matrix which are noteworthy.

(i) When the eigenvalues and eigenvectors of the SH matrix are computed, no problem arises if all of the elements of this matrix are real. For this case well-known subroutines, such as JACOBI, which deal with real symmetric matrices can be used. When the elements of the SH matrix are complex due to the presence of spin operators like S_y, or the presence of $O_l^{-|m|}$ spin operators in the SH, then computer subroutines such as EISPACK can be used to determine the real eigenvalues and complex eigenvectors of the SH matrix. If these subroutines are not available, it is still possible to use a symmetric matrix with real elements, A', whose dimension is twice that of the SH matrix

$$A' = \left(\begin{array}{c|c} \mathrm{R} & \mathrm{I} \\ \hline -\mathrm{I} & \mathrm{R} \end{array}\right),$$

where R and I are the matrices composed of the real and imaginary parts of the SH matrix, respectively, constructed in a one-to-one correspondence with the SH matrix. The (real) eigenvalues of the A' matrix are those of the SH matrix occurring twice. The two real eigenvectors of A' corresponding to an eigenvalue of A' contain the real ($\mathbf{v}_1$) and imaginary ($\mathbf{v}_2$) parts of the complex eigenvectors ($\mathbf{v}_1 + i\mathbf{v}_2$) of the SH matrix:

$$\left(\begin{array}{c} \mathbf{v}_1 \\ \hline -\mathbf{v}_2 \end{array}\right) \quad \text{and} \quad \left(\begin{array}{c} \mathbf{v}_2 \\ \hline \mathbf{v}_1 \end{array}\right).$$

(ii) The SH matrix can be expressed in either the "crystal," or "magnetic," coordinate system. It is recommended that the "crystal" representation be used, because the number of terms in the SH is then minimum. Many more spin-operator terms are present in the SH in "magnetic" coordinates than for "crystal" coordinates, and it is necessary to use complicated tables of trigonometric functions based on "rotational" matrices required to transform the spin operators from the "crystal" coordinates to the "magnetic" coordinates.[60] However, one of the advantages of using "magnetic" coordinates is for perturbation calculations, because the Zeeman term can be taken to be the zero-order term. Another advantage of using "magnetic" coordinates is that the coefficient of the spin-Hamiltonian term determines the intensity of HF "forbidden" transition relative to the "allowed" HF transition. A good discussion of operator equivalents has been provided by Rudowicz,[61,62] as well as for the relations between arbitrary SHP in various axes systems. Many of the details outlined in this section are discussed and illustrated in detail by Misra,[63] in a context with a quick simulation of powder spectrum using the "homotopy" technique in conjunction with matrix diagonalization.

(iii) Finally, it should be noted that the use of the perturbation technique has almost become obsolete, because exact matrix diagonalization is now easy to achieve computationally on PCs. However, perturbation expressions are useful to investigate the role and relative importance of individual energy terms, which depend on the SHPs.[64]

REFERENCES

[1] G. T. Trammell, H. Zeldes, and R. Livingston, Phys. Rev. **110**, 630 (1958).
[2] H. M. McConnell, Proc. Nat. Acad. Sci. (USA) **44**, 766 (1958).
[3] H. Zeldes and R. Livingston, J. Chem. Phys. **35**, 563 (1961).
[4] J. A. Weil and J. H. Anderson, J. Chem. Phys. **35**, 1410 (1961).
[5] F. K. Kneubühl, Physik der Konden. Mat. **1**, 410 (1963).
[6] R. E. D. McClung, Canad. J. Phys. **46**, 2271 (1968).
[7] J. M. Baker, E. R. Davies, and J. P. Hurrell, Proc. Roy. Soc. (London) A **308**, 403 (1968).
[8] A. Abragam and B. Bleaney, *Electron Paramagnetic Resonance of Transition Ions* (Clarendon Press, Oxford, 1970).
[9] C. R. Byfleet, D. P. Chong, J. A. Hebden, and C. A. McDowell, J. Magn. Reson. **2**, 69 (1970).
[10] R. M. Golding, R. H. Newman, A. D. Rae, and W. C. Tennant, J. Chem. Phys. **57**, 1912 (1971).
[11] R. M. Golding and W. C. Tennant, Mol. Phys. **25**, 1163 (1972).
[12] U. Sakaguchi, Y. Arata, and S. Fujiwara, J. Magn. Reson. **9**, 118 (1973).
[13] W. C. Lin, Mol. Phys. **25**, 247 (1973).
[14] A. Rockenbauer and P. Simon, J. Magn. Reson. **11**, 217 (1973).
[15] C. P. Keijzers, G. F. M. Paulussen, and E. de Boer, Mol. Phys. **29**, 1973 (1975).
[16] R. M. Golding and W. C. Tennant, Mol. Phys. **28**, 167 (1974).
[17] A. Rockenbauer and P. Simon, Mol. Phys. **28**, 1113 (1974).
[18] M. Iwasaki, J. Magn. Reson. **16**, 417 (1974).
[19] J. A. Weil, J. Magn. Reson. **18**, 113 (1975).
[20] S. D. Pandey and P. Venkateswarlu, J. Magn. Reson. **17**, 137 (1975).
[21] W. C. Tennant, Mol. Phys. **31**, 1505 (1976).
[22] A. M. Grekhov and A. B. Roitsin, Phys. Stat. Sol. (B) **74**, 323 (1976).
[23] J. R. Pilbrow, J. Magn. Reson. **31**, 5041 (1978).
[24] J. R. Pilbrow, J. Magn. Reson. **31**, 479 (1978).
[25] C. A. Hutchison Jr. And T. E. Orlowski, J. Chem. Phys. **73**, 1 (1980).
[26] D. C. McGavin, R. A. Palmer, W. A. Singers, and W. C. Tennant, J. Magn. Reson. **40**, 69 (1980).
[27] J. R. Pilbrow and M. R. Lowrey, Rep. Prog. Phys. **43**, 433 (1980).
[28] A. B. Roitsin, Phys. Stat. Sol. (B) **104**, 11 (1981).
[29] S. K. Misra, Magn. Reson. Rev. **10**, 285 (1986).
[30] S. K. Misra, Physica **124B**, 53 (1984).
[31] S. K. Misra, J. Bandet, G. Bacquet, and T. E. McEnally, Phys. Stat. Sol. (A) **80**, 581 (1983).
[32] M. J. Mombourquette, W. C. Tennant, and J. A. Weil, J. Chem. Phys. **85**, 68 (1986).
[33] E. Meirovitch and R. Poupko, J. Phys. Chem. **82**, 1920 (1978).
[34] G. D. Markham, B. D. N. Rao, and G. H. Reed, J. Magn. Reson. **33**, 595 (1979).
[35] S. K. Misra, Physica B **203**, 193 (1994).
[36] S. K. Misra, Physica B **240**, 183 (1997).
[37] R. P. Feynman, Phys. Rev. **56**, 340 (1939).
[38] D. G. McGavin, R. A. Palmer, W. A. Singers, and W. C. Tennant, Report No. CD2308, Chemistry Division, Department of Scientific and Industrial Research, New Zealand, 1980.

[39] G. C. Upreti, J. Magn. Reson. **13**, 336 (1974).
[40] G. B. Singh and P. Venkateswarlu, Proc. Indian Acad. Sci. **A65**, 211 (1967).
[41] D. A. Jones, J. M. Baker, and D. F. D. Pope, Proc. Phys. Soc. **74**, 249 (1959).
[42] H. A. Buckmaster, R. Chatterjee, J. C. Dering, D. J. I. Fry, Y. H. Shing, J. D. Skirrow, and B. Venkatesan, J. Magn. Reson. **4**, 113 (1971).
[43] S. K. Misra and G. R. Sharp, J. Magn. Reson. **23**, 191 (1976).
[44] J. B. Scarborough, *Numerical Mathematical Analysis* (Johns Hopkins Press, Baltimore, 1966).
[45] R. P. Feynman, Phys. Rev. **56**, 340 (1939).
[46] S. K. Misra, J. Magn. Reson. **23**, 403 (1976).
[47] J. R. Wolberg, *Prediction Analysis* (Van Nostrand, Princeton, NJ, 1967).
[48] P. B. Bevington, *Data Reduction and Error Analysis for the Physical Sciences* (McGraw Hill, New York, 1969).
[49] S. K. Misra and S. Subramanian, J. Phys. C **15**, 7199 (1982).
[50] A. K. Jain and G. C. Upreti, Mol. Phys. **34**, 273 (1977).
[51] J. Uhrin, Czech. J. Phys. B **27**, 595 (1977).
[52] B. Bleaney and D. J. E. Ingram, Proc. Roy. Soc. (London) A **205**, 336 (1951).
[53] S. K. Misra and M. Jalochowski, Physics B **119**, 295 (1983).
[54] S. K. Misra and D. Van Ormondt, Phys. Rev. B **30**, 6327 (1984).
[55] S. K. Misra and P. Vasilopoulos, J. Phys. C **13**, 1083 (1980).
[56] S. K. Misra, J. Phys. C **12**, 5221 (1979).
[57] S. K. Misra, Physica **121B**, 193 (1983).
[58] H. Goldstein, *Classical Mechanics* (Addison-Wesley, Cambridge, MA, 1951).
[59] S. K. Misra, Physica B **151**, 433 (1988).
[60] H. A. Buckmaster, R. Chatterjee, and Y. H. Shing, Phys. Stat. Sol. (a) **13**, 9 (1972).
[61] C. Rudowicz, J. Phys. C **18**, 1415 (1985).
[62] C. Rudowicz, J. Magn. Reson. **63**, 95 (1985).
[63] S. K. Misra, J. Magn. Reson. **137**, 83 (1999).
[64] S. K. Misra, Arab. J. Sci. Engng. **13**, 255 (1988). This paper also lists the particular features of the various papers published on the perturbation calculation of the eigenvalues of SH for electron-nuclear spin-coupled systems.

CHAPTER X
ESR IMAGING

Gareth R. Eaton and Sandra S. Eaton

Department of Chemistry and Biochemistry, University of Denver, Denver, Colorado 80208, USA

The term "imaging" evokes the pictures of anatomy obtained by NMR imaging (NMRI) of proton density in a body. Analogous images of the spatial distribution of unpaired electron density can be obtained by ESR imaging (sometimes called ESRI). In order to obtain spatial information it is necessary to have a technique to distinguish between resonances that come from different regions of the sample. Most commonly this is achieved with magnetic field gradients, but there also are a variety of techniques for localized detection. ESR imaging is not limited to mapping spatial distributions of paramagnetic centers—other dimensions can be time, spectral lineshape, or relaxation time. More generally, any parameter that can impact the ESR spectrum is a potential dimension in an image. Furthermore, if the spectral lineshape varies through the sample, a spatial dimension in the image or localized detection is needed to avoid artifacts.

ESR imaging has developed more slowly than NMR imaging for several reasons.

(1) NMR imaging is based primarily upon protons whose effective concentrations are often of the order of 50 M or more. ESR spectroscopy, including imaging, is usually performed on samples with mM, or lower, concentrations to minimize spectral broadening due to spin–spin interaction. Radicals of physiological interest are present at concentrations far less than a mM. Thus although the magnetic moment for an unpaired electron is about 658 times larger than for a proton, this only partially compensates for the decreased concentration of species to be examined. The net result is greatly decreased sensitivity.

(2) ESR linewidths for "sharp" signals typically are of the order of Gauss (1 Gauss = 10^{-4} Tesla $\sim 2.8 \times 10^6$ Hz) which is much broader than NMR lines and compounds the problems of sensitivity. In addition, the broader the line, the higher the gradient that is needed to achieve the same spatial resolution. Thus the gradients required for ESR imaging are much larger than for NMR imaging.

(3) ESR is typically performed at about 9 GHz to improve sensitivity. Energy of this frequency has poor penetration into large aqueous objects, so to examine samples larger than a few mm requires a lower microwave frequency, which is turn decreases the Boltzmann factor. Despite these challenges there has been substantial activity and progress in the past 15 years as outlined in the following paragraphs and tables.

TABLE 1. CW technique development and image reconstruction.

Technique	Comments	Ref.
1D spatial	Two specks of DPPH, gradient coils, gradient = 1500 G/cm	42
1D spatial	Ferromagnetic wedges to create gradient	43
1D spatial	Modulated gradients	44–47
1D spatial	Gradient coils, distinguish power saturation behavior	48
1D spatial	Permanent magnet to create gradient	49
Localized modulation coils	Move miniature coil inside resonator	50, 51
Gradient in modulation phase	Technique tested with phantom of DPPH samples, direct, Hadamard and Fourier analyses were examined	52
Localized spectra, 1D spatial	Modulated gradients	53
Dynamic imaging of diffusion	1D gradients, Fourier analysis of diffusion coefficient	28, 54, 55
Lineshape deconvolution	Solid DPPH, Fourier deconvolution	56
Lineshape deconvolution	Convolution difference method, 1 nitroxide line	57
Hyperfine deconvolution	Fourier deconvolution, calculated functions	58, 59
Lineshape deconvolution	Minimum negativity constraint	60
Hyperfine deconvolution	Nitroxide, Fourier deconvolution	6, 61, 62
Deconvolution	Practical considerations	63
Hyperfine deconvolution	Nitroxide, spline-based deconvolution	64
Hyperfine deconvolution	Nitroxide, forward subtraction	65
Overmodulate to mask lineshape changes	Nitroxide	66
Gradients	Pulsed to obtain 1200 G/cm	67
Gradients	Coil inside resonator, 2000 G/cm	68
Gradients	Pulsed to obtain 3300 G/cm	69
Gradients	Coil design, 1D	70, 71
Gradients	Coil design, 1D and 2D	72
Gradients	Two coils for 2D	73
Gradients	2D, generation with multipolar magnet	74–76
Gradients	Generation with ferromagnetic materials	77–79
Gradients	Wires on quartz tube inside resonator, 2D	80–82
Gradients	Three coils for 3D	83
Gradients	Three coils for 3D with Bruker TM cavity	84
Image reconstruction	General approaches, details for 2D	85
2D spatial	Rotate sample, 1 gradient coil	86
2D spatial	Rotate magnet, 1 gradient coil	87
2D spatial	Rotate gradients	88
2D spatial	Two gradient coils	73
2D spatial	Reconstruction from 1D modulated gradient projections	89
2D data acquisition	Select projections to maximize information	90
2D spatial–spatial image reconstruction	Compare convoluted back-projection, iterative reconstruction with least squares, and iterative reconstruction with maximum entropy	91
2D image reconstruction	Interpolation of regularly spaced and adaptively measured projections	92
Wire array in resonator	Localized detection by varying local magnetic field	93, 94
Aperture in resonator	2D spatial, move sample relative to aperture	95–98
2D spectral–spatial	Reconstruction from stepped gradients	99–102
2D spectral–spatial	Reconstruction from incomplete data sets	103

TABLE 1 (*continued*)

Technique	Comments	Ref.
2D spectral–spatial	Compare algorithms for reconstruction from incomplete data sets	104, 105
2D spectral–spatial	Images of portions of metal spectra	106
2D spectral–spatial	Incomplete versus complete sets of projections	107
2D spectral–spatial	Spectral dimension accurately gives multiline signal	108
2D spectral–spatial	Overview	109
2D spectral–spatial	Modulated gradients combined with field scan	110, 111
2D spectral–spatial	Resolution compared to 1D spatial	112
3D spatial	Reconstructed from stepped gradient data	113, 114
3D spectral–spatial-spatial	Reconstructed from stepped gradient data	115, 116
3D spectral–spatial-spatial	Comparison of CW and spin echo detection	117, 118
4D spectral–spatial	Calculations for feasibility	119
4D spectral–spatial	Reconstructed from stepped gradient data	120
Portable spectrometer	Localized modulation or aperture in resonator	121, 122

TABLE 2. Imaging at frequencies other than X-band.

Frequency	Experiment	Ref.
Low	Analysis of hyperfine splitting	123–125
Low	Analysis of hyperfine splitting, gradients at low frequency	126
Field-cycled PEDRI	UV-generated radicals, observe by spin trapping	124, 127
250 MHz	Spectrometer design, gradient coils, nitroxide in mouse, 1D spatial	128
250 MHz	Spectrometer design, 2D spectral spatial of nitroxide in phantom, mouse	129
250 MHz	Spectra of spin-trapped radicals in irradiated mouse, 2D spectral–spatial image of trapped radicals in phantom	125
250 MHz	Resonator design	130
250 MHz	Spectrometer design, gradient coils, automatic adjustment of tuning, coupling, and phase control	131
280 MHz	Spectrometer design, multipolar magnet, 1D spatial of nitroxide	132
280 MHz	1D and 2D spatial images of nitroxide in rat	133
280 MHz	Multipolar magnet, 2D spatial of nitroxide in phantoms	134
280 MHz	Multipolar magnet, 2D spatial of nitroxide in rat	135
300 MHz	Spectrometer design, gradient coils for three dimensions	83
300 MHz	Pulsed spectrometer, FID detection, gradient coils	136
700–800 MHz	3D spatial imaging with gradient coils of nitroxide in rat head	137
700 MHz	2D spatial, Fe bis(dithiocarbamato)nitrosyl complex in rat head, 37 °C, gradient coils	138
L-band	2D spatial, spin-labeled dextran in mouse	139
L-band	2D spatial, impact of dielectric on signal intensity	140
1.0 GHz	1D spatial, impact of dielectric on microwave B_1	141

TABLE 2 (*continued*)

Frequency	Experiment	Ref.
1.1 GHz	Impact of lossy material on EPR lineshape	142
1.1 GHz	Localized detection, surface coil	143
1.2 GHz	1D spatial, distinguish signals from individual lithium phthalocyanin crystals for oximetry	144
1.2 GHz	3D spatial, nitroxide in rat tail	145, 146
1.3 GHz	2D spatial, image nitroxide in dead mouse	147
1.3 GHz	3D spectral–spatial–spatial of nitroxide in perfused rat heart	148, 149
1.3 GHz	3D spatial, nitroxide in phantoms	150
1.3 GHz	3D spatial, image of phantom and of isolated heart, resolution is better for glucose char than for nitroxide	151
1.3 GHz	3D spatial, poly-nitroxyl albumin in ischemic heart	152
1.3 GHz	3D spatial, trapped NO in ischemic rat brain	153
1.3 GHz	3D spatial, gated imaging of nitroxide in perfused beating heart	154
1.3 GHz	4D spectral–spatial of glucose char in isolated rat heart	120
1.4–1.8 GHz	Gradient coils, 1D and 2D spatial, nitroxide in phantom, celery stalks	155–157
1.5 GHz	2D spatial image of alanine dosimeter	158
1.5 GHz	2D spatial gated image of nitroxide in model of beating heart	159
1.6 GHz	2D spatial, surface coil, nitroxide in polyacrylamide rod	160
1.6 GHz	2D spatial, surface coil, nitroxide in polycarbonate or polystyrene rod	161
1.6 GHz	2D spatial, surface coil, image nitroxide in murine tumor	162
1.6 GHz	2D spatial, surface coil, radicals in coal	163
1.9 GHz	Surface coil, 1D, image of nitroxide	164
2.0 GHz	Surface coil, 2D, nitroxide in plant stem	165
140 GHz	Ferromagnetic materials create higher gradients at higher magnetic field	79

Reviews of ESR imaging include a book,[1] the proceedings of a conference,[2] general discussions,[3–11] annotated bibliographies,[12–13] articles that focus on comparison of methods,[14–19] biological applications,[20–24] dermatology,[25,26] diffusion,[27,28] defects in solids,[29] geosciences,[30] materials science,[31,32] magnetic field distributions,[33] microscopy,[34,35] small animals,[36] oximetry,[37] proton electron double resonance,[38] resolution,[39] sensitivity,[40] and a tutorial on imaging with gradients.[41] Literature coverage in this chapter is through early 1996.

Papers that report software for image reconstruction and development of hardware for imaging with CW ESR and magnetic field modulation are summarized in Table 1 in order of increasing numbers of dimensions included in the image. Within topic groupings, papers are listed in chronological order. Table 2 covers imaging at frequencies other than X-band. Table 3 summarizes experiments in which detection schemes, other than CW, with magnetic field modulation were used.

TABLE 3. Imaging with detection techniques other than magnetic field modulated CW.

Detection method	Signal detected	Experiment	Ref.
Adiabatic fast passage	Irradiated SiO_2	1D, ferromagnetic rod to create gradient	77
Adiabatic fast passage	Irradiated Si	Spectra at several gradients, calculate spatial distribution	166
Longitudinal detection	DPPH	1D, 300 MHz	167
Nonresonant coil	DPPH	Spectral–spatial, pulsed gradient	69
ENDOR	Nitroxide	1D, gradient coils, effect of O_2 and viscosity	168
ENDOR	Nitroxide	2D spatial, gradient coils, effect of O_2	169, 170
Spin echo	Irradiated quartz	1D spatial, pulsed gradients on between first and second pulses	171
Spin echo	Coal, irradiated quartz	1D spatial, relaxation time selectivity	172
Spin echo	Coal, irradiated quartz	Flip-angle gradients	173
Spin echo	$(Fluoranthenyl)_2AsF_6$	1D spatial, pulsed gradients on during time 2τ	174, 175
Spin echo	Coal, irradiated quartz, nitroxide	T_1 selectivity	176
Spin echo	Nitroxide	1D spatial, sample frozen at 77 K	177
Spin echo	Coal, irradiated quartz, nitroxide	2D and 3D spectral–spatial, compare spin echo and CW detected images	117, 118
Spin echo	Irradiated quartz	2D spatial, pulsed gradients	178
Spin echo	Coal, irradiated quartz	2D spatial, gradient produced by array of wires in resonator	179
FID	$(Fluoranthenyl)_2AsF_6$	1D spatial, pulsed versus CW gradients	180
FID	Nitroxide	2D spectral–spatial, phase encoding gives better image than frequency encoding, pulsed gradients	181
FID	Lithium phthalocyanin, solvated electrons	1D and 2D spatial, FID-detected, 300 MHz, pulsed gradient	136
ELDOR	Nitroxide	Pulsed, 2D with stepped gradients	178
PEDRI (proton–electron double resonance)	Proton NMR	Impact of nitroxide on intensities in image	182–184
Field cycled PEDRI	Proton NMR	Irradiated electrons at different field than used to observe protons, impact of nitroxide	185, 186
Field cycled PEDRI	Proton NMR	Use nitroxides, examine viscosity	187
Overhauser effect	Proton NMR	Impact of O_2, Fremy's salt	188–191
Optical detection	Photoexcited Paraterphenyl-d_{14}	Radicals in radiation tracks	192

TABLE 4. Applications at X-band.

Application	Technique, dimensions, features	Signal observed	Ref.
	Biological tissues and solutions		
Peroxidase assay	Modulated gradients, 1D	Toluidine blue	45, 193
Irradiated tooth	2D spatial	CO_3^{3-} radicals	194, 195
Distribution in adipose versus muscle tissue	2D spatial, gradient coils	Nitroxide	196
Desiccated mouse melanoma	Aperture in resonator	Melanin radicals	197
Spin-labeled stearic acid in vesicle	Modulated gradients, 1D	Nitroxide	198
Spatial distribution in phospholipid multilayer	Modulated gradients, localized spectra and 1D distribution	Nitroxide	53
Oximetry	2D spatial, gradient coils	Nitroxide	199, 200
Oximetry	2D spectral–spatial, gradient coils	Nitroxide	201, 202
Permeability of cells in sphaeroids	2D spatial, gradient coils	Nitroxide + broadening agent	203, 204
	Diffusion		
Diffusion into zeolite	Modulated gradient, 1D spatial	Nitroxide	45, 205
Diffusion into zeolite and catalysts	2D spatial and 2D spectral–spatial	Nitroxide	206
Diffusion into porous catalyst	Ferromagnetic wedges, 2D spatial	Perchlorotriphenyl-methyl radical	207
Diffusion into silica gel	Spin echo, 1D spatial	Nitroxide at 77 K	177
Diffusion in decalin	Ferromagnetic wedge, 1D spatial	Perchlorotriphenyl-methyl radical	43, 208
Diffusion in toluene	Rotate gradients, 2D spatial	Perchlorotriphenyl-methyl radical	88
Diffusion in viscous solvent	2D spectral–spatial	Nitroxide	209
Diffusion into solution of polymer in toluene	2D spectral–spatial	Deuterated nitroxide	210, 211
Diffusion into solution of polymer in DMF	2D spectral–spatial	Mo^V	212

Diffusion in hydrogels	Gradient coils, 2D spectral–spatial	Nitroxide	213
Diffusion in anisotropic fluids	1D spatial with gradient coils, deconvolute hyperfine	Nitroxide	61
Diffusion in anisotropic fluids	1D spatial with ferromagnetic wedges, Fourier analysis of diffusion coefficient	Nitroxide	214
Diffusion in liquid crystal	1D spatial, Fourier analysis	Nitroxide	54
Diffusion in smectic-A phase	1D spatial, Fourier analysis	Nitroxide	215
Diffusion in model membrane	1D spatial, gradient coils	Nitroxide	216
diffusion in model membrane in liquid crystalline state	1D spatial, gradient coils, Fourier analysis	Nitroxide	217
Microscopic and macroscopic diffusion in model membrane	2D spectral–spatial	Nitroxide	218
Diffusion into maize leaves	Gradient coils, 1D spatial	Nitroxide	219
Diffusion into wheat seed	Gradient coils, 2D spatial, overmodulate to mask linewidth changes through sample	Nitroxide	220–222
Diffusion into wheat seeds	Gradient coils, 2D spatial	Radicals from humic materials	223
Diffusion in adipose tissue	1D spatial, simulate profile	Nitroxide	224
Diffusion into rat lung tissue	1D spatial, simulate profile	Nitroxide	225
Diffusion into rat liver tissue	1D spatial, simulate profile	Nitroxide	226
Diffusion into pig ear skin	1D spatial, gradient coils	Nitroxide	227–230
Diffusion into hairless-mice skin	1D spatial and 2D spectral–spatial, modulated gradients	Nitroxide	165 231–233
	Electrochemistry		
Electrochemistry	2D spectral–spatial	Semiquinone	234, 235
Electrochemistry	2D spectral–spatial	*p*-Phenylene-diamine	108
Electrochemistry in Nafion	2D spectral–spatial	Polypyrrole	236
	Electromagnetic fields		
Magnetic field around superconductor	1D spatial	Nitroxide	237
Gradients from ferromagnetics	1D spatial	Nitroxide, DPPH	238
Perturbation of magnetic field by high susceptibility objects	2D spectral–spatial	Polypyrrole	239
Microwave magnetic field in resonator	Gradient coils, 1D and 2D spatial	O_2 at low pressure, 77 K	240

TABLE 4 (*continued*)

Application	Technique, dimensions, features	Signal observed	Ref.
Impact of dielectrics on microwave magnetic field	1D spatial, calculations	Irradiated SiO_2, DPPH	141
Eddy currents in wires in EPR resonator	2D spatial	Polypyrrole	241
Irradiated substances			
Irradiated alanine crystal	2D spatial, isolated line	Alanine radical	73
Irradiated alanine powder	Localized modulation	Alanine radical	242
Irradiated alanine dosimeter	2D spectral–spatial	Alanine radical	243–245
Irradiated sucrose	2D spectral–spatial	Sucrose radical	246
Irradiated $Ca_3(PO_4)_2$	2D spatial, 1 gradient coil, rotate magnet	Trapped H atoms	87
Irradiated H_2SO_4 ice	Gradient coil, pulsed, 1D spatial	Trapped H atoms	245, 247
Irradiated H_2SO_4 ice	Spectral–spatial	Trapped H atoms	248
Irradiated polypropylene	2D spectral–spatial	Radical from light stabilizer	243, 246
Irradiated SiO_2	2D spatial, 1 gradient coil, rotate magnet	Color centers	87
Irradiated SiO_2	2D spectral–spatial	E′ defects	249–251
Minerals and geological samples			
Aragonite, barite	2D spatial, move relative to aperture	Radiation defects	252
Apatite	2D spatial, move relative to aperture	Hole centers	253
Coral	2D spatial, move relative to aperture	CO_2^- and SO_3^-	254
Diamond	2D spatial, 1 gradient coil, rotate sample	Nitrogen centers	86
Diamond	2D spatial, 1 gradient coil, rotate magnet	Nitrogen centers	87
Synthetic diamond	2D spatial, move relative to aperture	Nitrogen and nickel impurities	255
Fossil crinoid	2D spatial, move relative to aperture	Mn^{2+}	95
Zircon single crystal	2D spatial, move relative to aperture	Gd, hole centers	252, 256

Other materials			
Mesophase carbon	2D spatial	Organic radicals	257
Catalyst	2D spatial, move relative to aperture	V^{4+} and Mo^{5+}	258
Photoannealing of radicals in glass	1D, ferromagnetic wedge to create gradient, time dependence at one point in projection	Trapped electrons	259
Pore structure of ceramics	Gradient coils, 2D spatial	Nitroxide	260
Pore structure of ceramics	Gradient coils, 3D spatial	Nitroxide	261
Thermal annealing of poly-silicon film	Move relative to aperture, 1D	Dangling bond	262
Polymerization of photoresist	Gradient coils, 1D spatial, deconvoluted multiline spectrum	Photo-generated radicals	263
Oxidation of radicals in coal	Surface coil + gradient, 2D spatial	Radicals in coal	163
PTFE oxidation	1D spatial	Peroxy radicals	264
Polymer degradation	Gradient coils, 2D spatial	Nitroxide, phenoxide	265
Polypropylene oxidation	Modulated gradients 1D spatial	Peroxy and alkyl radicals	266
Polypropylene oxidation	Gradient coils, 1D spatial	Peroxy radicals	267
Polyisoprene + NO_2	Gradient coils, 2D spatial	R_2NO radicals	268
Other			
Short-lived radical in flow cell	1D spatial, gradient coil	Ascorbic acid radical	269, 270
Vanadyl complex in frozen liquid crystal	1D spatial, pulsed gradient	Vanadyl bis(acetylacetonate)	67, 271
Solid NO_2 on rod	2D spatial, gradient coils	NO_2	271, 272

Because of the interest in biomedical applications of ESR imaging and the lower lossiness of biological samples at lower microwave frequency there has been, and continues to be, a substantial effort to develop instrumentation at lower frequencies than X-band, and these applications, are summarized in Table 2, listed in order of increasing microwave frequency.

As the instrumentation and software has become more sophisticated, there has been an increasing emphasis on applications. Applications at X-band are summarized in Table 4. These are listed in the following groupings: biological tissues and solutions, diffusion, electrochemistry, electromagnetic fields, irradiated substances, minerals and geological samples, other materials, and other.

ACKNOWLEDGMENTS

The work in the author's laboratory on ESR imaging was supported by NSF grants CHE8722804 and CHE9103262.

REFERENCES

[1] G. R. Eaton, S. S. Eaton, and K. Ohno, editors, *EPR Imaging and in Vivo EPR* (CRC Press, Boca Raton, FL, 1991).
[2] Physica Medica **5**, 2–4 (1989).
[3] O. E. Yakimchenko and Ya. S. Lebedev, Khim. Fiz. **2**, 445 (1983). (Translated as Soviet J. Chem. Phys. **2**, 745 (1985).)
[4] S. S. Eaton and G. R. Eaton, Spectroscopy **1**, 32 (1986).
[5] K. Ohno, Appl. Spectr. Rev. **22**, 1 (1986).
[6] G. R. Eaton and S. S. Eaton, in *Electron Magnetic Resonance in the Solid State*, Canadian Society for Chemistry Symposium Series, volume 1, edited by J. Weil (Canadian Society for Chemistry, Ottawa, 1987), p. 638.
[7] K. Ohno, Magn. Reson. Rev. **11**, 275 (1987).
[8] G. R. Eaton and S. S. Eaton, Bull. Magn. Reson. **10**, 22 (1988).
[9] M. Schara, Magn. Reson. Med. Biol. **1**. 245 (1988).
[10] A. Sotgiu, G. Gualtieri, F. Momo, and P. L. Indovina, Phys. Med. **4**, 149 (1988).
[11] L. J. Berliner, in *Magnetic Resonance Microscopy: Methods and Applications in Materials Science, Agriculture, and Biomedicine*, edited by B. Blümich and W. Kuhn (VCH, Weinheim, 1992), p. 151.
[12] S. S. Eaton and G. R. Eaton, ESR Specialist Periodical Reports **12b**, 176 (1991).
[13] S. S. Eaton and G. R. Eaton, ESR Specialist Periodical Reports **15**, 169 (1996).
[14] U. Ewert, T. Herrling, and W. Schneider, Exp. Technol. Physik (Jena) **36**, 289 (1988).
[15] U. Ewert, T. Herrling, and W. Schneider, 24th Ampere Congress, Poznan 1988 (Elsevier, Amsterdam, 1989), p. 281.
[16] U. Ewert and T. Herrling, *Electron Magnetic Resonance of Disordered Systems*—89 (Pruvetz, Bulgaria, 1990).
[17] L. J. Berliner and H. Fujii, Makromolekulare Chemie—Macromolecular Symposia **34**, 263 (1990).
[18] S. S. Eaton and G. R. Eaton, in *Modern Pulsed and Continuous-Wave Electron Spin Resonance*, edited by L. Kevan and M. K. Bowman (Wiley, New York, 1990), p. 405.
[19] G. R. Eaton and S. S. Eaton, in *Microscopic and Spectroscopic Imaging of the Chemical State*, edited by M. D. Morris (Marcel Dekker, New York, 1993), Ch. 11, p. 395.
[20] L. J. Berliner, Arch. Biol. Med. Exp. **22**, 119 (1989).
[21] H. Swartz, in *Advanced EPR*, edited by A. Hoff (Elsevier, Amsterdam, 1990), p. 753.
[22] S. Colacicchi, M. Ferrari, and A. Sotgiu, Internat. J. Biochem. **24**, 205 (1992).

[23] S. Colacicchi, M. Alecci, G. Gualtieri, V. Quaresima, C. L. Ursini, M. Ferrari, and A. Sotgiu, J. Chem. Soc. Perkin Trans. **2**, 2077 (1993).
[24] M. Ferrari, V. Quaresima, C. L. Ursini, M. Alecci, and A. Sotgiu, Internat. J. Rad. Oncol. Biol. Phys. **29**, 421 (1994).
[25] J. Fuchs, N. Groth, T. Herrling, and L. Packer, Meth. Enzymol. **233**, 140 (1994).
[26] T. Herrling, N. Groth, K.-U. Thiessenhusen, J. Fuchs, and U. Ewert, in *Magnetic Resonance Microscopy: Methods and Applications in Materials Science, Agriculture, and Biomedicine*, edited by B. Blümich and W. Kuhn (VCH, Weinheim, 1992), p. 563.
[27] J. H. Freed, Ann. Rev. Biophys. Biomol. Struct. **23**, 1 (1994).
[28] J. K. Moscicki, Y.-K. Shin, and J. H. Freed, in *EPR Imaging and in vivo EPR*, edited by G. R. Eaton, S. S. Eaton, and K. Ohno (CRC Press, Boca Raton, FL, 1991), Ch. 19.
[29] M. J. R. Hoch, in *EPR Imaging and in vivo EPR*, edited by G. R. Eaton, S. S. Eaton, and K. Ohno (CRC Press, Boca Raton, FL, 1991), Ch. 20.
[30] M. Ikeya, M. Furusawa, H. Ishii, and T. Miki, Appl. Magn. Reson. **1**, 70 (1990).
[31] K. Ohno, in *Electronic Magnetic Resonance of the Solid State*, edited by J. A. Weil, M. K. Bowman, and K. F. Preston (Canadian Society for Chemistry, Ottawa, 1987), p. 629.
[32] M. Ikeya, Ann. Rev. Mater. Sci. **21**, 45 (1991).
[33] Ya. S. Lebedev and O. Ye. Yakimchenko, in *EPR Imaging and in vivo EPR*, edited by G. R. Eaton, S. S. Eaton, and K. Ohno (CRC Press, Boca Raton, FL, 1991), Ch. 23.
[34] M. Ikeya, in *Magnetic Resonance Microscopy: Methods and Applications in Materials Science, Agriculture, and Biomedicine*, edited by B. Blümich and W. Kuhn (VCH, Weinheim, 1992), p. 133.
[35] M. Ikeya, *New Applications of Electron Spin Resonance: Dating, Dosimetry, and Microscopy* (World Scientific, Singapore, 1993), Ch. 14.
[36] G. R. Cherryman, A. D. Stevens, and C. M. Smith, in *The Encyclopedia of Nuclear Magnetic Resonance*, edited by D. M. Grant and R. K. Harris (Wiley, New York, 1995), p. 2486.
[37] H. M. Swartz and J. F. Glockner, in *EPR Imaging and in vivo EPR*, edited by G. R. Eaton, S. S. Eaton, and K. Ohno (CRC Press, Boca Raton, FL, 1991), Ch. 24.
[38] D. J. Lurie, in *The Encyclopedia of Nuclear Magnetic Resonance*, edited by D. M. Grant and R. K. Harris (Wiley, New York, 1995), p. 2481.
[39] M. J. R. Hoch and U. Ewert, in *EPR Imaging and in vivo EPR*, edited by G. R. Eaton, S. S. Eaton, and K. Ohno (CRC Press, Boca Raton, FL, 1991), Ch. 15.
[40] U. Ewert, in *EPR Imaging and in vivo EPR*, edited by G. R. Eaton, S. S. Eaton, and K. Ohno (CRC Press, Boca Raton, FL, 1991), Ch. 16.
[41] G. R. Eaton and S. S. Eaton, Concepts Magn. Reson. **7**, 49 (1995).
[42] W. Karthe and E. Wehrsdorfer, J. Magn. Reson. **33**, 107 (1979).
[43] E. V. Gal'tseva, O. Ye. Yakimchenko, and Ya. S. Lebedev, Chem. Phys. Lett. **99**, 301 (1983).
[44] T. Herrling, N. Klimes, W. Karthe, U. Ewert, and B. Ebert, J. Magn. Reson. **49**, 203 (1982).
[45] B. Ebert, T. Hanke, and N. Klimes, Studia Biophysica **103**, 161 (1984).
[46] U. Ewert and T. Herrling, J. Magn. Reson. **61**, 11 (1985).
[47] T. Herrling, in *EPR Imaging and in vivo EPR*, edited by G. R. Eaton, S. S. Eaton, and K. Ohno (CRC Press, Boca Raton, FL, 1991), Ch. 5.
[48] S. S. Eaton and G. R. Eaton, J. Magn. Reson. **59**, 474 (1984).
[49] A. Nicula, S. Nicula, L. Giurgiu, and I. Ursu, Studia Univ. Babes—Bolyai, Physica **31**, 3 (1986).
[50] T. Miki and M. Ikeya, Japan. J. Appl. Phys. **26**, L1495 (1987).
[51] T. Miki and M. Ikeya, J. Magn. Reson. **80**, 502 (1988).
[52] E. Szczpeniak and J. P. Hornak, J. Magn. Reson. **A104**, 315 (1993).
[53] T. Páli, B. Ebert, and L. I. Horváth, J. Magn. Reson. **96**, 491 (1992).
[54] J. K. Moscicki, Y.-K. Shin, and J. H. Freed, J. Magn. Reson. **84**, 554 (1989).
[55] J. K. Moscicki, Y. K. Shin, and J. H. Freed, Physica Medica **5**, 229 (1989).
[56] M. J. R. Hoch, J. Phys. C **14**, 5659 (1981).
[57] H. Fujii and L. J. Berliner, J. Magn. Reson. **68**, 377 (1986).
[58] K. Ohno, J. Magn. Reson. **50**, 145 (1982).

[59] A. Sotgiu, D. Gazzillo, and F. Momo, J. Phys. C **20**, 6297 (1987).
[60] F. Momo, Physica Medica **5**, 217 (1989).
[61] J. P. Hornak, J. K. Moscicki, D. J. Schneider, and J. H. Freed, J. Chem. Phys. **84**, 3387 (1986).
[62] F. Momo, S. Colacicchi, and A. Sotgiu, Meas. Sci. Technol. **4**, 60 (1993).
[63] U. Ewert and K.-U. Thiessenhusen, in *EPR Imaging and in vivo EPR*, edited by G. R. Eaton, S. S. Eaton, and K. Ohno (CRC Press, Boca Raton, FL, 1991), Ch. 11.
[64] G. Placidi, M. Alecci, and A. Sotgiu, Rev. Sci. Instrum. **65**, 58 (1994).
[65] P. Kuppusamy and J. L. Zweier, Magn. Reson. Med. **35**, 316 (1996).
[66] O. E. Yakimchenko, A. I. Smirnov, and Ya. S. Lebedev, Appl. Magn. Reson. **1**, 1 (1990).
[67] K. Ohno, Japan. J. Appl. Phys. **23**, L224 (1984).
[68] M. Ikeya and T. Miki, Japan. J. Appl. Phys. **26**, L929 (1987).
[69] K. Ohno and T. Murakami, J. Magn. Reson. **79**, 343 (1988).
[70] F. Momo, O. Adriani, G. Gualtieri, and A. Sotgiu, J. Phys. E **21**, 565 (1988).
[71] R. W. Quine, G. Rinard, D. E. Rugg, G. R. Eaton, and S. S. Eaton, Comput. Phys. **6**, 656 (1992).
[72] R. W. Quine, G. R. Eaton, K. Ohno, and S. S. Eaton, in *EPR Imaging and in vivo EPR*, edited by G. R. Eaton, S. S. Eaton, and K. Ohno (CRC Press, Boca Raton, FL, 1991), Ch. 3.
[73] K. Ohno, J. Magn. Reson. **64**, 109 (1985).
[74] A. Sotgiu, J. Appl. Phys. **59**, 689 (1986).
[75] A. Sotgiu, G. Gualtieri, and F. Momo, J. Phys. D **20**, 1574 (1987).
[76] M. Alecci, G. Gualtieri, A. Sotgiu, L. Testa, and V. Varoli, Meas. Sci. Technol. **2**, 32 (1991).
[77] O. E. Yakimchenko, I. N. Kartsivadze, B. V. Ozherelev, and Ya. S. Lebedev, Dokl. Akad. Nauk SSSR **268**, 384 (1983), p. 32, in translation.
[78] A. I. Smirnov, E. N. Degtyarev, O. E. Yakimchenko, and Ya. S. Lebedev, Inst. Exp. Technol. **34**, 202 (1991).
[79] A. I. Smirnov, O. G. Poluectov, and Ya. S. Lebedev, J. Magn, Reson. **97**, 1 (1992).
[80] M. Furusawa and M. Ikeya, Japan. J. Appl. Phys. **30**, L1682 (1991).
[81] M. Furusawa and M. Ikeya, Appl. Radiat. Isotop. **44**, 381 (1993).
[82] M. Yamamoto and M. Ikeya, Japan J. Appl. Phys. **33**, 4887 (1994).
[83] J. A. Brivati, A. D. Stevens, and M. C. R. Symons, J. Magn. Reson. **92**, 480 (1991).
[84] G. Gualtieri, G. Placidi, A. Sotgiu, and S. Del Monaco, Rev. Sci. Instrum. **66**, 3715 (1995).
[85] R. K. Woods, W. B. Hyslop, R. B. Marr, and P. C. Lauterbur, in *EPR Imaging and in vivo EPR*, edited by G. R. Eaton, S. S. Eaton, and K. Ohno (CRC Press, Boca Raton, FL, 1991), Ch. 10.
[86] M. J. R. Hoch and A. R. Day, Solid State Commun. **30**, 211 (1979).
[87] W. Zommerfelds and M. J. R. Hoch, J. Magn. Reson. **67**, 177 (1986).
[88] A. I. Smirnov, E. N. Degtyarev, O. E. Yakimchenko, and Ya. S. Lebedev, Prib. Tekh. Eksp. 100 (1988).
[89] T. Páli, L. Sass, L. I. Horváth, and B. Ebert, J. Magn. Reson. **86**, 338 (1990).
[90] G. Placidi, M. Alecci, and A. Sotgiu, J. Magn. Reson. **B108**, 50 (1995).
[91] C. M. Smith and A. D. Stevens, British J. Radiol. **67**, 1186 (1994).
[92] G. Placidi, M. Alecci, and A. Sotgiu, J. Magn. Reson. **B110**, 75 (1996).
[93] M. Ikeya and H. Ishii, J. Magn. Reson. **88**, 130 (1990).
[94] H. Miyamaru and M. Ikeya, Appl. Radiat. Isotop. **44**, 397 (1993).
[95] M. Furusawa and M. Ikeya, Anal. Sci. (Japan) **4**, 649 (1988).
[96] M. Ikeya, M. Furusawa, and M. Kasuya, Scanning Microscopy **4**, 245 (1990).
[97] M. Furusawa and M. Ikeya, Japan. J. Appl. Phys. **29**, 270 (1990).
[98] A. Hochi, M. Furusawa, and M. Ikeya, Appl. Rad. Isotop. **44**, 401 (1993).
[99] U. Ewert and T. Herrling, Chem. Phys. Lett. **129**, 516 (1986).
[100] M. M. Maltempo, J. Magn. Reson. **69**, 156 (1986).
[101] M. M. Maltempo, S. S. Eaton, and G. R. Eaton, J. Magn. Reson. **72**, 449 (1987).
[102] E. D. A. Stemp, G. R. Eaton, S. S. Eaton, and M. M. Maltempo, J. Phys. Chem. **91**, 6467 (1987).
[103] M. M. Maltempo, S. S. Eaton, and G. R. Eaton, J. Magn. Reson. **77**, 75 (1988).
[104] M. M. Maltempo, S. S. Eaton, and G. R. Eaton, Physica Medica **5**, 223 (1989).

[105] M. M. Maltempo, S. S. Eaton, and G. R. Eaton, in *EPR Imaging and in vivo EPR*, edited by G. R. Eaton, S. S. Eaton, and K. Ohno (CRC Press, Boca Raton, FL, 1991), Ch. 14.
[106] M. M. Maltempo, S. S. Eaton, and G. R. Eaton, J. Magn. Reson. **85**, 303 (1989).
[107] G. R. Eaton, S. S. Eaton, and M. M. Maltempo, Appl. Rad. Isotop. **40**, 1227 (1989).
[108] M. Sueki, S. S. Eaton, and G. R. Eaton, Anal. Sci. (Japan) **7**, 571 (1991).
[109] M. M. Maltempo, S. S. Eaton, and G. R. Eaton, in *EPR Imaging and in vivo EPR*, edited by G. R. Eaton, S. S. Eaton, and K. Ohno (CRC Press, Boca Raton, FL, 1991), Ch. 13.
[110] T. Herrling, K.-U. Thiessenhusen, and U. Ewert, J. Magn. Reson. **100**, 123 (1992).
[111] U. Ewert, in *EPR Imaging and in vivo EPR*, edited by G. R. Eaton, S. S. Eaton, and K. Ohno (CRC Press, Boca Raton, FL, 1991), Ch. 12.
[112] M. Sueki, S. S. Eaton, G. R. Eaton, and M. M. Maltempo, Physica Medica **5**, 93 (1989).
[113] S. Colacicchi, P. L. Indovina, F. Momo, and A. Sotgiu, J. Phys. E **21**, 910 (1988).
[114] R. K. Woods, G. C. Bacic, P. C. Lauterbur, and H. M. Swartz, J. Magn. Reson. **84**, 247 (1989).
[115] S. S. Eaton, M. M. Maltempo, E. D. A. Stemp, and G. R. Eaton, Chem. Phys. Lett. **142**, 576 (1987).
[116] M. Sueki, R. W. Quine, S. S. Eaton, G. R. Eaton, M. M. Maltempo, Physica Medica **5**, 77 (1989).
[117] M. Sueki, G. R. Eaton, and S. S. Eaton, Pure Appl. Chem. **62**, 229 (1990).
[118] M. Sueki, G. R. Eaton, and S. S. Eaton, Appl. Magn. Reson. **1**, 20 (1990).
[119] R. K. Woods, W. B. Hyslop, H. M. Swartz, Physica Medica **5**, 121 (1989).
[120] P. Kuppusamy, M. Chzhan, A. Samouilov, P. Wang, and J. L. Zweier, J. Magn. Reson. **B107**, 116 (1995).
[121] M. Ikeya and M. Furusawa, Appl. Radiat. Isotop. **40**, 845 (1989).
[122] M. Ikeya, K. Megro, H. Miyamaru, and H. Ishii, Appl. Magn. Reson. **2**, 663 (1991).
[123] D. J. Lurie, I. Nicholson, and J. R. Mallard, J. Magn. Reson. **95**, 405 (1991).
[124] D. J. Lurie, I. Nicholson, J. S. McLay, and J. R. Mallard, Appl. Magn. Reson. **3**, 917 (1992).
[125] H. J. Halpern. S. Pou, M. Peric, C. Yu, E. Barth, G. M. Rosen, J. Amer. Chem. Soc. **115**, 218 (1993).
[126] D. G. Gillies, L. H. Sutcliffe, and M. R. Symms, J. Chem. Soc. Farad. Trans. **90**, 2671 (1994).
[127] D. J. Lurie, J. McLay, I. Nicholson, and J. R. Mallard, J. Magn. Reson. **95**, 191 (1991).
[128] H. J. Halpern. D. P. Spencer, J. van Polen, M. K. Bowman, A. C. Nelson, E. M. Dowey, and B. A. Teicher, Rev. Sci. Instrum. **60**, 1040 (1989).
[129] H. J. Halpern and M. K. Bowman, *EPR Imaging and in vivo EPR*, edited by G. R. Eaton, S. S. Eaton, and K. Ohno (CRC Press, Boca Raton, FL, 1991), Ch. 6.
[130] A. D. Stevens, British J. Radiol. **67**, 1243 (1994).
[131] A. D. Stevens and J. A. Brivati, Meas. Sci. Technol. **5**, 793 (1994).
[132] M. Alecci, S. D. Penna, A. Sotgiu, L. Testa, and I. Vannucci, Rev. Sci. Instrum. **63**, 4263 (1992).
[133] M. Alecci, M. Ferrari, V. Quaresima, A. Sotgiu, and C. L. Ursini, Biophys. J. **67**, 1274 (1994).
[134] M. Alecci, S. D. Penna, A. Sotgiu, and L. Testa, Appl. Magn. Reson. **3**, 909 (1992).
[135] V. Quaresima, M. Alecci, M. Ferrari, and A. Sotgiu, Biochem. Biophys. Res. Commun. **183**, 829 (1992).
[136] J. Bourg, M. C. Krishna, J. B. Mitchell, R. G. Tschudin, T. J. Pohida, W. S. Friauf, P. D. Smith, J. Metcalfe, F. Harrington, and S. Subramanian, J. Magn. Reson. **B102**, 112 (1993).
[137] S.-I. Ishida, S. Matusmoto, H. Yokoyama, N. Mori, H. Kumashiro, N. Tsuchihashi, T. Ogata, M. Yamada, M. Ono, T. Kitajima, H. Kamada, and E. Yoshida, Magn. Reson. Imag. **10**, 109 (1992).
[138] T. Yoshimura, S. Fujii, H. Yokoyama, and H. Kamada, Chem. Lett. 309 (1995).
[139] H. Yoshioka, H. Tanizawa, T. Ogata, and S. Kazama, Biol. Pharm. Bull. **18**, 1572 (1995).
[140] T. Ogata, Y. Ishikawa, M. Ono, and L. J. Berliner, J. Magn. Reson. **97**, 616 (1992).
[141] M. Sueki, G. A. Rinard, S. S. Eaton, and G. R. Eaton, J. Magn. Reson. **A118**, 173 (1996).
[142] J. Jiang, K. J. Liu, T. Walczak, and H. M. Swartz, J. Magn. Reson. **B106**, 230 (1995).
[143] G. Bacic, M. J. Nilges, R. L. Magin, T. Walczak, and H. M. Swartz, Magn. Reson. Med. **10**, 266 (1989).
[144] A. I. Smirnov, S.-W. Norby, R. B. Clarkson, T. Walczak, and H. M. Swartz, Magn. Reson. Med. **30**, 213 (1993).
[145] M. Alecci, S. Colacicchi, P. L. Indovina, F. Momo, P. Pavone, and A. Sotgiu, Magn. Reson. Imag. **8**, 59 (1990).

[146] M. Alecci, S. Colacicchi, P. L. Indovina, F. Momo, P. Pavone, and A. Sotgiu, Physica Medica **5**, 99 (1989).
[147] K. Takeshita, H. Utsumi, and A. Hamada, Biochem. Biophys. Res. Commun. **177**, 874 (1991).
[148] P. Kuppusamy, M. Chzhan, K. Vij, M. Shteynbuk, D. J. Lefer, E. Giannella, and J. L. Zweier, Proc. Nat. Acad. Sci. U.S.A. **91**, 3388 (1994).
[149] P. Kuppusamy and J. L. Zweier, Curr. Top. Biophys. **18**, 3 (1994).
[150] P. Kuppusamy, M. Chzhan, and J. L. Zweier, J. Magn. Reson. **B106**, 122 (1995).
[151] P. Kuppusamy, P. Wang, and J. L. Zweier, Magn. Reson. Med. **34**, 99 (1995).
[152] P. Kuppusamy, P. Wang, and J. L. Zweier, Biochemistry **35**, 7051 (1996).
[153] P. Kuppusamy, S. T. Ohnishi, Y. Numagami, T. Ohnishi, and J. L. Zweier, J. Cereb. Blood Flow Metab. **15**, 899 (1995).
[154] P. Kuppusamy, M. Chzhan, P. Wang, and J. L. Zweier, Magn. Reson. Med. **35**, 323 (1996).
[155] H. Fujii and L. J. Berliner, Magn. Reson. Med. **2**, 275 (1985).
[156] H. Nishikawa, H. Fujii, and L. J. Berliner, J. Magn. Reson. **62**, 79 (1985).
[157] L. J. Berliner and H. Fujii, Science **227**, 517 (1985).
[158] S. Colacicchi, S. Onori, E. Petetti, and A. Sotgiu, Appl. Radiat. Isotop. **44**, 391 (1993).
[159] L. Testa, G. Gualtieri, and A. Sotgiu, Phys. Med. Biol. **38**, 295 (1993).
[160] L. J. Berliner and H. Fujii, J. Magn. Reson. **69**, 68 (1986).
[161] L. J. Berliner, X. Wan, and H. Fujii, J. Polymer Sci., Polymer Lett. **24**, 587 (1986).
[162] L. J. Berliner, H. Fujii, X. Wan, and S. J. Lukiewicz, Magn. Reson. Med. **4**, 380 (1987).
[163] M. Liang, X. Wan, and L. J. Berliner, Fuel **68**, 1340 (1989).
[164] L. J. Berliner and H. Fujii, in *Magnetic Resonance in Biology and Medicine*, edited by G. Govil, C. L. Khetrapal, and A. Saran (Tata McGraw-Hill, New Delhi, India, 1985), p. 101.
[165] J. Fuchs, H. J. Freisleben, N. Groth, T. Herrling, G. Zimmer, R. Milbradt, and L. Packer, Free Rad. Res. Commun. **15**, 245 (1991).
[166] Yu. V. Gorelkinskii and A. A. Kim, Izv. Akad. Nauk Kazakh. SSR, Ser. Fiz.-Mat. **24** (1985) (CA 103: 114780e).
[167] I. Nicholson, F. J. L. Robb, and D. J. Lurie, J. Magn. Reson. **B104**, 284 (1994).
[168] E. G. Janzen and Y. Kotake, J. Magn. Reson. **69**, 567 (1986).
[169] Y. Kotake, U. M. Oehler, and E. G. Janzen, J. Chem. Soc. Faraday I **84**, 3275 (1988).
[170] E. G. Janzen, Y. Kotake, and U. M. Oehler, in *EPR Imaging and in vivo EPR*, edited by G. R. Eaton, S. S. Eaton, and K. Ohno (CRC Press, Boca Raton, FL, 1991), Ch. 9.
[171] A. D. Milov, A. Yu. Pusep, S. A. Dzuba, and Yu. D. Tsvetkov, Chem. Phys. Lett. **119**, 421 (1985).
[172] S. S. Eaton and G. R. Eaton, J. Magn. Reson. **67**, 73 (1986).
[173] G. R. Eaton and S. S. Eaton, J. Magn. Reson. **67**, 561 (1986).
[174] G. G. Maresch, A. Grupp, M. Mehring, J. U. v. Schutz, and H. C. Wolf, J. Physique **46**, 461 (1985).
[175] D. J. Sloop, H.-L. Yu, and T.-S. Lin, Chem. Phys. Lett. **124**, 456 (1986).
[176] G. R. Eaton and S. S. Eaton, J. Magn. Reson. **71**, 271 (1987).
[177] A. A. Sukhoroslov, R. I. Samoilova, and A. D. Milov, Appl. Magn. Reson. **2**, 577 (1991).
[178] U. Ewert, R. H. Crepeau, S. Lee, C. R. Dunnam, D. Xu, and J. H. Freed, Chem. Phys. Lett. **184**, 34 (1991).
[179] M. Yamamoto and M. Ikeya, Japan. J. Appl. Phys. **34**, L115 (1995).
[180] G. G. Maresch, M. Mehring, and S. Emid, Physica **138B**, 261 (1986).
[181] U. Ewert, R. H. Crepeau, C. R. Dunham, D. Xu, S. Lee, and J. H. Freed, Chem. Phys. Lett. **184**, 25 (1991).
[182] D. J. Lurie, D. M. Bussell, L. H. Bell, and J. R. Mallard, J. Magn. Reson. **76**, 366 (1988).
[183] D. J. Lurie, I. Nicholson, M. A. Foster, and J. R. Mallard, Philos. Trans. Roy. Soc. London. Ser. A **333**, 453 (1990).
[184] D. J. Lurie, J. M. S. Hutchison, L. H. Bell, I. Nicholson, D. M. Bussell, and J. R. Mallard, Physica Medica **5**, 85 (1989).
[185] D. J. Lurie, J. M. S. Hutchison, L. H. Bell, I. Nicholson, D. M. Bussell, and J. R. Mallard, J. Magn. Reson. **84**, 431 (1989).

[186] D. J. Lurie, I. Nicholson, and J. R. Mallard, J. Magn. Reson. **94**, 197 (1991).
[187] I. Nicholson, D. J. Lurie, and F. J. L. Robb, J. Magn. Reson. **B104**, 250 (1994).
[188] D. Grucker, Magn. Reson. Med. **14**, 140 (1990).
[189] D. Grucker, V. Ringeisen, and J. Chambron, Bull. Magn. Reson. **12**, 65 (1990).
[190] D. Grucker and J. Chambron, Magn. Reson. Im. **11**, 691 (1993).
[191] D. Grucker and J. Chambron, Physica Medica **5**, 329 (1989).
[192] B. M. Odintsov and R. G. Yakhin, JETP Lett. **57**, 143 (1993).
[193] P. Mohr, Th. Hanke, W. Kuhn, and B. Ebert, Biomed. Biochim. Acta **42**, 663 (1983).
[194] M. Ikeya and M. Furusawa, Oral. Rad. **4**, 31 (1988).
[195] H. Ishii and M. Ikeya, Japan. J. Appl. Phys. **29**, 871 (1990).
[196] G. Bacic, T. Walczak, F. Demsar, and H. M. Swartz, Magn. Reson. Med. **8**, 209 (1988).
[197] H. Katsuda, T. Kobayashi, H. Saito, T. Matsunaga, and M. Ikeya, Chem. Pharm. Bull. **38**, 2838 (1990).
[198] T. Páli, B. Ebert, and L. I. Horváth, Biochim. Biophys. Acta **904**, 346 (1987).
[199] F. Demsar, T. Walczak, P. D. Morse II, G. Bacic, Z. Zolnai, and H. M. Swartz, J. Magn. Reson. **76**, 224 (1988).
[200] G. Bacic, F. Demsar, Z. Zolnai, and H. M. Swartz, Magn. Reson. Med. Biol. **1**, 55 (1988).
[201] R. K. Woods, J. W. Dobrucki, J. F. Glockner, P. D. Morse, II, and H. M. Swartz, J. Magn. Reson. **85**, 50 (1989).
[202] H. M. Swartz and J. F. Glockner, Physica Medica **5**, 109 (1989).
[203] J. W. Dobrucki, F. Demsar, T. Walczak, R. K. Woods, G. Bacic, and H. M. Swartz, British J. Cancer **61**, 221 (1990).
[204] J. W. Dobrucki, R. M. Sutherland, and H. M. Swartz, Magn. Reson. Med. **19**, 42 (1991).
[205] K. Ulbricht, T. Herrling, U. Ewert, and B. Ebert, Colloids and Surfaces **11**, 19 (1984).
[206] K. Ulbricht, U. Ewert, T. Herrling, K. U. Thiesssenhusen, G. Aebli, J. Völter, and W. Schneider, in *EPR Imaging and in vivo EPR*, edited by G. R. Eaton, S. S. Eaton, and K. Ohno (CRC Press, Boca Raton, FL, 1991), Ch. 22.
[207] O. E. Yakimchenko, E. N. Degtyarev, V. N. Parmon, and Ya. S. Lebedev, J. Phys. Chem. **99**, 2048 (1995).
[208] O. E. Yakimchenko, E. V. Galtseva, and Ya. S. Lebedev, Khim. Fiz. **4**, 1563 (1985) (CA 104: 24517r).
[209] E. D. A. Stemp, G. R. Eaton, S. S. Eaton, and M. M. Maltempo, J. Chem. Soc. Chem. Commun. 61 (1988).
[210] S.-C. Kweon, Z. Gao, P. Eagle, S. Schlick, J. Labsky, and J. Pilar, Polymer Mater. Sci. Engng. **71**, 169 (1994).
[211] Z. Gao, J. Pilar and S. Schlick, J. Phys. Chem. **100**, 8430 (1996).
[212] Z. Kruczala, Z. Gao, and S. Schlick, J. Phys. Chem. **100**, 11427 (1996).
[213] S. Schlick, J. Pilar, S.-C. Kweon, J. Vacik, and J. Labsky, Macromol. **28**, 5780 (1995).
[214] D. A. Cleary, Y.-K. Shin, D. J. Schneider, and J. H. Freed, J. Magn. Reson. **79**, 474 (1988).
[215] J. K. Moscicki, Y.-K. Shin, and J. H. Freed, J. Chem. Phys. **99**, 634 (1993).
[216] Y.-K. Shin and J. H. Freed, Biophys. J. **55**, 537 (1989).
[217] Y.-K. Shin, J. K. Moscicki, and J. H. Freed, Biophys. J. **57**, 445 (1990).
[218] Y.-K. Shin, U. Ewert, D. E. Budil, and J. H. Freed, Biophys. J. **59**, 950 (1991).
[219] F. Demsar, M. Schara, and M. Kveder, Farm. Vestn. **37**, 131 (1986).
[220] A. I. Smirnov, O. E. Yakimchenko, S. I. Aksenov, E. A. Golovina, G. I. Likhtenshtein, and Ya. S. Lebedev, Fiz. Rast. **35**, 663 (1988) (translation in Soviet Plant Physiol. **35**, 516 (1988)).
[221] E. A. Golivina, A. I. Smirnov, O. E. Yakimchenko, and S. I. Aksenov, Soviet Plant. Physiol. **38**, 90 (1991).
[222] O. V. Trofanchuk, R. G. Yakhin, A. V. Anisimov, K. L. Aminov, D. R. Aleev, and B. M. Odintsov, Russian J. Phys. Chem. **67**, 1345 (1993).
[223] A. I. Smirnov, O. E. Yakimchenko, H. A. Golovina, S. Kh. Bekova, and Ya. S. Lebedev, J. Magn. Reson. **91**, 386 (1991).
[224] F. Demsar, P. Cevc, and M. Schara, J. Magn. Reson. **69**, 258 (1986).
[225] M. Kveder and M. Schara, Farm. Vestn. **39**, 217 (1988).

[226] F. Demsar, H. M. Swartz, and M. Schara, Magn. Reson. Med. Biol. **1**, 17 (1988).
[227] V. Gabrijelcic, M. Šentjurc, and J. Kristl, Internat. J. Pharm. **62**, 75 (1990).
[228] J. Kristl, S. Pecar, J. Korbar-Smid, F. Demsar, and M. Schara, Drug. Devel. Indust. Pharm. **15**, 1423 (1989).
[229] V. Gabrijelcic, M. Šentjurc, and M. Schara, Internat. J. Pharm. **102**, 151 (1994).
[230] F. Demsar and J. Kristl, Physica Medica **5**, 321 (1989).
[231] J. Fuchs, H. J. Freisleben, N. Groth, T. Herrling, G. Zimmer, R. Milbradt, and L. Packer, Free Rad. Res. Commun. **15**, 245 (1991).
[232] J. Fuchs, N. Groth, T. Herrling, R. Milbradt, G. Zimmer, and L. Packer, J. Invest. Dermat. **98**, 713 (1992).
[233] C. Michel, N. Groth, T. Herrling, P. Rudolph, J. Fuchs, J. Kreuter, and H.-J. Fresleben, Internat. J. Pharm. **98**, 131 (1993).
[234] M. Sueki, R. W. Quine, S. S. Eaton, and G. R. Eaton, J. Chem. Soc. Faraday Soc. **86**, 3181 (1990).
[235] M. Sueki, S. S. Eaton, and G. R. Eaton, Anal. Chem. **63**, 883 (1991).
[236] K. Nakagawa, G. R. Eaton, and S. S. Eaton, J. Appl. Polymer Sci: Appl. Polymer Symp. **52**, 235 (1993).
[237] E. V. Galtseva, O. Ya. Grinberg, A. A. Dubinskiy, G. G. Lazarev, Ya. S. Lebedev, O. G. Poluektov, A. I. Smirnov, and O. E. Yakimchenko, Proceedings of 24th Colloque Ampere (Elsevier, Amsterdam, 1989), p. 461.
[238] Ya. S. Lebedev and O. Ye. Yakimchenko, in *EPR Imaging and in vivo EPR*, edited by G. R. Eaton, S. S. Eaton, and K. Ohno (CRC Press, Boca Raton, FL, 1991), Ch. 4.
[239] M. Sueki, S. S. Eaton, and G. R. Eaton, J. Magn. Reson. **A105**, 25 (1993).
[240] S. A. Marshall, B. H. Suits, M. T. Umlor, and Y. N. Zhang, J. Magn. Reson. **76**, 494 (1988).
[241] M. Sueki, G. A. Rinard, S. S. Eaton, and G. R. Eaton, J. Magn. Reson. **A103**, 208 (1993).
[242] T. Miki, Appl. Radiat. Isotop. **40**, 1243 (1989).
[243] Y. Morita, K. Ohno, K. Ohashi, and J. Sohma, Appl. Radiat. Isotop. **40**, 1237 (1989).
[244] K. Ohno, Y. Morita, S. Sawamura, and K. Ohashi, Physica Medica **5**, 277 (1989).
[245] K. Ohno, in *EPR Imaging and in vivo EPR*, edited by G. R. Eaton, S. S. Eaton, and K. Ohno (CRC Press, Boca Raton, FL, 1991), Ch. 18.
[246] K. Ohno, Y. Morita, S. Sawamura, and K. Ohashi, Physica Medica **5**, 277 (1989).
[247] K. Ohno, Japan. J. Appl. Phys. **20**, L179 (1981).
[248] K. Ohno, J. Yonezawa, and Y. Morita, J. Phys, Chem. **93**, 1657 (1989).
[249] M. Sueki, S. S. Eaton, and G. R. Eaton, Appl. Radiat. Isotop. **44**, 377 (1993).
[250] M. Sueki, W. R. Austin, L. Zhang, D. B. Kerwin, R. G. Leisure, G. R. Eaton, and S. S. Eaton, J. Appl. Phys. **77**, 790 (1995).
[251] M. Sueki, S. S. Eaton, and G. R. Eaton, Appl. Radiat. Isotop. **47**, 1595 (1996).
[252] M. Furusawa, M. Kasuya, S. Ikeda, and M. Ikeya, Nucl. Tracks Radiat. Meas. **18**, 185 (1991).
[253] H. Ishii, M. Ikeya, M. Kasuya, and M. Furusawa, Nucl. Tracks Radiat. Meas. **18**, 189 (1991).
[254] S. Ikeda, D. Neil, M. Ikeya, A. Kai, and T. Miki, Japan. J. Appl. Phys. **31**, L1644 (1992).
[255] M. Furusawa and M. Ikeya, J. Phys. Soc. Japan. **59**, 2340 (1990).
[256] M. Kasuya, M. Furusawa, and M. Ikeya, Nucl. Tracks Radiat. Meas. **17**, 563 (1990).
[257] K. Ohno, T. Yokono, J. Yamada, Y. Sanada, Carbon **24**, 517 (1986).
[258] Y. Xu, M. Furusawa, M. Ikeya, Y. Kera, and K. Kuwata, Chem. Lett. 293 (1991).
[259] I. N. Kartsivadze, O. E. Yakimchenko, A. I. Alekandrov, B. G. Ershov, and Ya. S. Lebedev, Dokl. Akad. Nauk SSSR **280**, 924 (1985) (p. 109, in translation).
[260] R. G. Yakhin, O. V. Trofanchuk, Yu. G. Nazarov, and B. M. Odintsov, J. Appl. Spectros. **61**, 628 (1994).
[261] G. Kordas and Y.-H. Kang, J. Amer. Ceram. Soc. **74**, 709 (1991).
[262] M. Ikeya and M. Yamamoto, Japan. J. Appl. Phys. **33**, L1087 (1994).
[263] J. Dahm, V. Macho, G. G. Maresch, and H. W. Spiess, Molecular Phys. **74**, 591 (1991).
[264] O. E. Yakimchenko, E. V. Gal'tseva, E. N. Degtyarev, and Y. S. Lebedev, Vysokomol. Soedin., Ser. A **28**, 1880 (1986).

[265] V. A. Silaev, K. M. Mysin, O. I. Gnezdilov, E. S. Nefed'ev, L. A. Abdrakhmanova, and V. G. Khozin, Bull. Magn. Reson. **15**, 133 (1993).
[266] E. N. Degtyarev, S. G. Kiryushkin, A. I. Smirnov, and O. E Yakimchenko, High Energy Chem. **27**, 274 (1993).
[267] O. E. Yakimchenko, E. N. Degtyarev, E. A. Ermakova, and Ya. S. Lebedev, High Energy Chem. **27**, 279 (1993).
[268] E. N. Degtyarev, T. V. Pokholok, G. B. Pariiskii, and O. E. Yakimchenko, Russian J. Phys. Chem. **68**, 414 (1994).
[269] K. Ohno, J. Magn. Reson, **49**, 56 (1982).
[270] K. Ohno, in *EPR Imaging and in vivo EPR*, edited by G. R. Eaton, S. S. Eaton, and K. Ohno (CRC Press, Boca Raton, FL, 1991), Ch. 17.
[271] K. Ohno, in *EPR Imaging and in vivo EPR*, edited by G. R. Eaton, S. S. Eaton, and K. Ohno (CRC Press, Boca Raton, FL, 1991), Ch. 21.
[272] K. Ohno, Chem. Lett. **17** (1986).

CHAPTER XI

HIGH MAGNETIC FIELDS AND HIGH FREQUENCIES

Sandra S. Eaton and Gareth R. Eaton

Department of Chemistry and Biochemistry, University of Denver, Denver, Colorado 80208, USA

A literature survey is given of instrumentation systems and application of ESR measurement carried out at very high magnetic field and frequencies.

Some comment is needed about what could be meant by "high-field" ESR. In colloquial use scientists usually mean high-frequency when the say high-field, and refer to ESR frequency greater than that commonly used or commercially available. In this sense, high-field ESR has meant the use of frequencies above 35 GHz (called Q-band[a] by most ESR spectroscopists). Now that 95 GHz spectrometers are commercially available from Bruker Instruments, and frequencies of 95, 150, and 250 GHz are offered by Resonance Technologies, this common definition moves the lower bound to higher frequencies. Nevertheless, until these commercially available frequencies become commonly used, most spectroscopists will continue to consider them "high" frequencies. The assumption is that we also refer to the magnetic field for $g = 2$, so high frequency implies high magnetic field. In this review frequencies above 90 GHz are defined as "high." To achieve resonance at $g = 2$ at these frequencies requires magnetic fields greater than 3.2 T.

There are numerous incentives to use high microwave frequencies for ESR spectroscopy. The following is a list of those for which examples are included in the tables:

(1) At a given temperature, increasing magnetic field increases the population difference between the upper and lower Zeeman levels. Thus, for the same number of unpaired electrons the ESR signal increases, which implies increased sensitivity at higher field.
(2) For a given g anisotropy higher magnetic field gives greater dispersion of spectra, so greater spectral resolution is achieved at higher magnetic field, provided the

[a] There are a variety of one- and two-letter abbreviations that have been used to specify microwave frequency bands, especially at lower frequencies. Tables of these abbreviations can be found in refs. 1 and 2. Since "standard" notation has changed with time, we prefer to specify actual frequency rather than band abbreviations in describing instruments and experiments.

linewidths do not increase as rapidly as the signal dispersion. For example, at 250 GHz we can read the three g-values for an immobilized nitroxyl radical directly from the spectrum. At X-band the three nitroxyl g-values can be determined only by careful computer simulation.

(3) As a result of the increased dispersion, ESR signals become sensitive to faster motions as magnetic field strength increases. For example, at 250 GHz nitroxyl linewidths are sensitive to motions about a thousand times faster than linewidths at 9 GHz.

(4) Increased dispersion can provide increased orientation selection in polarization and excitation of spins. For example, ENDOR spectra of powder samples at higher frequencies can be much more orientation selective than at X-band.

(5) In electron–electron spin-coupled systems if the exchange interaction, J, is greater than the energy separation between the resonances for the two centers, an average signal is observed and it is difficult to determine the value of J. Increased dispersion at higher frequency increases the energy separation, which converts signals that are "averaged" at X-band to AB splitting patterns, and permits determination of values of J.

(6) In complicated spin systems, simulation of spectra at several microwave frequencies permits selection of Hamiltonian parameters with less ambiguities than can be achieved at a single frequency. For example, small hyperfine splittings may be better resolved at lower frequencies and g-value difference may be better defined at higher frequency.

(7) The external magnetic field may become larger than the internal field due to spin–spin interactions. For example, for paramagnetic centers with $S > \frac{1}{2}$, transitions are not observable if the zero field splitting (ZFS) is greater than the ESR quantum ($h\nu$). At higher magnetic field the Zeeman interaction becomes larger relative to the ZFS and $h\nu$ is larger so additional transitions become observable, which permits determination of larger values of ZFS than can be measured at lower field/frequency.

(8) At high microwave frequencies and low temperature we can easily achieve $h\nu > k_B T$ so spin polarization is increased, i.e., almost all of the spins are in the Zeeman ground state. The changing populations of levels as a function of temperature and resulting changes in relative intensity of transitions permit determination of the sign of the ZFS.

(9) In many antiferromagnets the zero-field transition is in the far-infrared region, so high frequencies are needed to detect the resonance.

(10) In some materials new phases are induced by high magnetic fields.

(11) At high magnetic fields magnetic anisotropy may be sufficient to partially orient microcrystalline solid samples, samples with large anisotropy, or small molecules oriented in liquid crystals. For example, a high magnetic field is commonly employed to orient superconducting grains.

There are additional characteristics of high-frequency ESR that pertain to pulsed experiments.

(1) For the same resonator Q the frequency bandwidth is greater at higher frequency, which makes it easier to excite a broad spectrum in a pulsed experiment.
(2) For the same Q ringdown times also are faster at higher frequency which makes the deadtime shorter.
(3) At high field ESEEM (see Chapter V) from strongly coupled nuclei can be observed when it is not observed at lower fields.
(4) Measurements of electron spin relaxation times at higher microwave frequencies extend the range of available frequencies providing more stringent tests of models of relaxation mechanisms. Some of the higher magnetic fields overlap those employed for NMR experiments which permits determination of relaxation times under conditions relevant for interpretation of NMR dynamic nuclear polarization experiments.

There are several technical challenges in designing high-field ESR instrumentation and experiments.

(1) Although an increase in sensitivity due to the increasing difference between Boltzmann populations of energy levels with increasing magnetic field is observed, to some extent and for some spectrometers, several factors can diminish the gains in observed sensitivity relative to that which is expected.
   (a) Increased dispersion with increasing magnetic field leads to a smaller number of spins per Gauss at higher fields.
   (b) Increases in sample lossiness and decreases in dimensions of resonators may require smaller sample sizes at higher frequency.
   (c) Filling factors are typically lower in the Fabry–Perot resonators used at higher frequencies than in X-band resonators. At the highest frequencies experiments typically are performed without a resonator, which greatly decreases sensitivity.
   (d) The detectors used at higher frequencies have substantially poorer performance than X-band detectors.
(2) The highest magnetic fields that can be achieved with traditional iron-core magnets are about 2.5 T, which limits ESR experiments at $g = 2$ to less than 70 GHz. Experiments at higher frequencies require superconducting magnets, Bitter magnets (or other hybrids), or pulsed magnets. The highest magnetic fields are produced in pulsed mode and provide much lower magnetic field homogeneity than is routinely available at lower magnetic field. These magnets are useful for some types of experiments, but not for high resolution.
(3) The types of microwave sources and resonators that are used at lower frequencies are not practical at higher frequencies where the instrumentation takes on characteristics of optics or quasioptics.

There is a small, but growing, number of laboratories that have capabilities for ESR at frequencies about 90 GHz. Table 1 lists papers that describe technique and instrument development. Table 2 lists applications of high-frequency ESR. To provide the reader with a sense of the wide variety of areas in which high-frequency ESR has

(continued on page 362)

TABLE 1. Technique development and use of ESR resonance to demonstrate achievement of high magnetic fields.

Frequency (GHz)	Max. field (T)	Location	Instrumentation/technique	Ref.
9 to 115	8	Tokyo Institute of Technology	Seven klystron sources, superconducting magnet, reflection cavities at 9 to 32 GHz, transmission at 50 to 115 GHz	29
50 to 150	2.8	Bell Labs	Spectrometer with backward wave oscillator and Varian magnet, designed to measure ZFS	30
50 to 250	30	Nijmegen	Use vector network analyzer as source and detector, oversized cylindrical resonator	31
60 to 383	30, pulsed	Kobe	Gunn oscillator or BWO source, pulsed magnetic field is generated by discharge of capacitor bank	32
65 to 135	5	Fukui	Spectrometer based on tunable gyrotron with resonances about 10 GHz apart	33
79 to 535	1	Kazan	Spectrometer with backward wave oscillators and quasioptics lens system	34
94	5.87	Berlin	Spectrometer with klystron, superconducting magnet, and Fabry–Perot resonator	35
94	5.87	Berlin	Circuit for magnet stabilization	36
95	6	Berlin	ESR/ENDOR spectrometer with Gunn diode or klystron source and Fabry–Perot or cylindrical resonator	37
95		Berlin	Fabry–Perot resonator with two-axis goniometer	38
97	5.87	Berlin	ENDOR with Gunn oscillator or klystron, Fabry–Perot resonator, and superconducting magnet	39
95	3.5	Berlin	Pulsed ESR/ENDOR with klystron, heterodyne bridge, and cylindrical cavity.	40, 41
94	4.7	University of Illinois	Spectrometer with Gunn diode source, TE_{103} resonator, and superconducting magnet	9
95	6	University of Leiden	Pulsed spectrometer with Gunn diode, Fabry–Perot resonator and superconducting magnet	42–44
95		University of Leiden	Insert for ESE spectrometer that permits studies of photo-excited states in single crystals at liquid helium temperatures	45
95	6	University of Leiden	Pulsed ESR/ENDOR, cylindrical cavity gives factor of 10 better sensitivity and factor of 6 improvement in time resolution compared to Fabry–Perot resonator	46
95 to 3000	17	NHMFL, Tallahassee	95 to 550 GHz with solid state Gunn diodes, 160 to 2000 GHz with far-infrared laser, Fabry–Perot resonator under development	47
140	6.5	Ukraine	Spectrometer with klystron, open resonator with spherical mirrors, and superconducting magnet	48

140	7	Moscow	Spectrometer with klystron, cylindrical reflection resonator, and superconducting magnet	3, 5, 49
140	5.5	Moscow	Pulsed spectrometer with IMPATT diode source, cylindrical resonator	50
140	5	Moscow	ESR imaging, ferromagnetic wedges produce larger gradient when static field is larger	51, 52
140	5	MIT	Pulsed spectrometer with extended interaction klystron, Fabry–Perot resonator, and superconducting magnet	53
	5	MIT	Precision field-sweep system for superconducting solenoids for ESR spectroscopy	54
140	5	MIT	Spectrometer for ESR and DNP, Gunn diode for low power and gyrotron for high power, cylindrical or Fabry–Perot resonators, can do CW, ESE, and FT ESR	55
140 to 3109	to 550, pulsed	University of Tokyo	Produce megagauss fields by flux compression or single-coil techniques	56
150		Russia	Simulation of slow-tumbling nitroxide spectra	57
	7.5	Russia	Superconducting solenoid for ESR	58
150 to 600	5.6 to 21.6	Grenoble	Pulsed spectrometer with pick-up coil to measure relaxation times	59, 60
158 to 525	20	Grenoble	Spectrometer with far-infrared laser to operate at 158 to 525 GHz, Bitter or superconducting magnet	8, 61–63
239	20	Grenoble	Laser source, ENDOR, lack of resonant cavity makes it difficult to saturate spins	64
159 to 161	6.5	Orsay	Transmission spectrometer with carcinotron source and superconducting magnet	65
170	9.2	Cornell	Broad-banded (100–300 GHz) spectrometer, reflection-mode bridge with polarized coding, Fabry–Perot resonator, and superconducting magnet	19, 66
220 to 3109	17, 30 pulsed	Kobe	BWO or far-infrared laser as source, pulsed magnetic field is generated by discharge of capacitor bank	67, 68
250	9.2	Cornell	Spectrometer with phase-locked Gunn, Fabry–Perot resonator, and superconducting magnet	69
	20	Grenoble	Importance of two-phonon spin–lattice relaxation processes at high magnetic fields	70
604	21.5	Grenoble	2-pulse or 3-pulse ESE with two pulsed FIR lasers, Fabry–Perot resonator, and resistive magnet	71, 72
	30	Leuven	Design of magnet for pulsed fields	73, 74
	50	Moscow	Solenoid design	75
890	31.8	Osaka	Magnet for pulsed fields	76
891, 2567	50	Osaka	Spectrometer with pulsed fields	77
891, 2567	32, 91	Tokyo, Osaka	g-values for ruby are field independent, useful for calibration even at very high fields	14, 78
	100	Osaka	Design of magnet for pulsed fields	79, 80

TABLE 2. Applications of ESR at high field/frequency.

Frequency[a] (GHz)	Source[b]	Resonator[c]	Field (T)	Sample	Experiment	Ref.
				Organic and main-group radicals		
96	Gunn oscillator	Cylindrical	3.4	Coal	Measure g-values	81, 82
94	Gunn oscillator	Cylindrical	3.4	Coal	Sulfur-containing radicals	83, 84
94	Gunn oscillator	Cylindrical	3.4	Iodine-treated coal	Iodinated radicals	85
140	Klystron	Cylindrical	5	Coal	Effect of O_2 on g-value, linewidth, and power saturation	86
140	Klystron	Cylindrical	5	Coal	Fast passage dispersion, g-anisotropy	87
140	Klystron	Cylindrical	5	Coal	$g_{\|\|}$ correlates with degree of carbonization	88
245 to 428	FIR	None	to 17	Coal	g-anisotropy and spin exchange	89
135	Klystron	Cylindrical, TE_{011}	5	Nitroxide	Effect of substitution on g- and A-values	90
135	Klystron	Cylindrical	5	Fluorinated nitroxide	Analyze fluorine hyperfine tensor, determine geometry	91
140	Klystron	Cylindrical	5	Nitroxide	g- and A-values for deuterated and ^{15}N-derivatives	92
604	FIR	Fabry–Perot	to 21.5	Nitroxide in polystyrene	T_m at 4 K longer at 604 than 9.5 GHz, attributed to lower probability of flip-flop processes	71
94	Klystron	Fabry–Perot	to 5.87	Irradiated solid	Distinguish small g-value differences	35
140	Klystron	Cylindrical	5	Peroxy radicals	Irradiated polytetrafluoroethylene	93
140	Klystron	Cylindrical	5	Phenoxy radicals, including Coppinger's	Determine geometry of radical from g-values	94
891	HCN laser	Fabry–Perot	1.65	O_2 gas	Observe certain transitions	95
2.5×10^3	H_2O laser	None	to 2.5	CH radicals	Gas phase, rotational levels	96
95	Gunn oscillator	Fabry–Perot	to 3.5	$^3C_{60}$ in single crystal	ESE indicates triplet excitation is delocalized	97
66 to 160	Network analyzer	Oversized resonator	5.8	Er_m @ C_{82}	Assign two resonances to unpaired electron on cage and to Er^{3+}	31

94 to 428			to 15	*cis*- and *trans*-Polyacetylene	Distinguish localized and delocalized defects, frequency dependence of linewidths attributed to spin exchange	98
140	Klystron	Cylindrical	5	Nitroxide in polypyrrole	Probe distance to spinless charge carriers and size of carriers	99
140	Klystron	Cylindrical	5	Iodine-doped poly(tetrathiafulvalene)	CW saturation and saturation transfer, spin dynamics	100
140	Klystron	Cylindrical	5	Poly(*p*-phenylene)	Axial spectrum, defects on chains, exchange velocity $\sim 4 \times 10^{-7}$ s^{-1}	101, 102
140	Klystron	Cylindrical	5	*trans*-Polyacetylene	Temperature dependent anisotropic spin dynamics fit inter-soliton electron hopping model	103, 104
140	Klystron	Cylindrical	5	Polythiophene	*g*-values depend on anion, Dysonian lineshape, estimate conductivity	105
95	Gunn oscillator	TE_{103}	to 3.5	Aromatics on metal oxide	Resolve *g*-value differences	106
245, 337	FIR	Optical	to 12.5	GaAs, single crystal	Detect As antisite defect, $g = 2.04$, $A = 0.090$ cm^{-1}	107
245	FIR	None	to 9	*n*-type Si, doped with P	Observe linewidths ~ 5 G	108
428	FIR	None	15.3	*n*-type Si, doped with P and As	Detect *g*-value difference between P and As	61
349	FIR	None	12.5	Thermal donors in Si	Multiple sites, *g*-values consistent with ENDOR	109, 110
250	Phase-locked Gunn	Fabry–Perot	8.9	Cs^+ (18-crown-6) X, $X = e^-, Na^-, Cs^-$	Electrons trapped at anion vacancies	111
95	Gunn oscillator	Fabry–Perot	to 3.5	Self-trapped exciton in AgCl	Optically detected ESR, exchange splitting $= 5$ cm^{-1}	112
95	Gunn oscillator	Fabry–Perot or cylindrical	to 3.5	Self-trapped exciton in AgCl	Laser excitation, spin echo detection, S-T_o splitting $= 5.37$ cm^{-1}, $D < 0$ and small	113
95	Gunn oscillator	Cylindrical	to 3.5	Shallow electron center in AgCl	Ag and Cl pulsed ENDOR, defines wave function	114

TABLE 2 (*continued*)

Frequency[a] (GHz)	Source[b]	Resonator[c]	Field (T)	Sample	Experiment	Ref.
				Photosynthesis and related radicals		
135	Klystron	Cylindrical, TE_{011}	5	Radicals in photosynthetic systems	g-values for oxidized primary donor of *R. sphaeroides* and for tyrosyl radical in photosystem II	115
140	Gunn oscillator	Fabry–Perot	5	Primary donor, photosystem I	Determine g-tensor for preparations grown in D_2O	116
94	Gunn diode	TE_{013}	to 4.7	Photosynthetic reaction center	Measure g-values of special pair from *Rb. sphaeroides*	9
95	Gunn oscillator	Fabry–Perot	to 3.5	Photosynthetic reaction center, single crystal	g-tensor of primary donor of *Rb. sphaeroides*	38, 117
95	Gunn oscillator	Fabry–Perot or cylindrical	to 3.5	Photosynthetic reaction centers, frozen solution	Compare g-values for primary donor from several species	118
95	Gunn oscillator	Cylindrical	to 3.5	Photosynthetic reaction center, single crystal	Orientation of g-tensor of primary donor in mutant heterodimer	119
95	Klystron	Cylindrical	to 3.5	Photosynthetic reaction centers, Zn-substituted	Time-resolved, spin polarized spectra very sensitive to relative orientations of g-matrices	120
95	Klystron	Cylindrical	to 3.5	Photosynthetic reaction centers, Zn-substituted	Two-dimensional pulsed ESR, orientation dependence of T_2 due to anisotropic motion	121
250				Photosynthetic reaction centers, Zn-substituted	Calculations predict increased sensitivity of polarization to intermediate radicals at higher fields	122
94.5	Klystron	Fabry–Perot	to 8.8	Semiquinone anions	Anisotropic g- and A-values analyze with MO calculations	123
95	Klystron	Cylindrical TE_{011}	to 3.5	Plastoquinone anion radicals	Orientation-selected ENDOR	41
132.2	Klystron	Cylindrical, TE_{011}	5	Semiquinone of vitamin K and menadione	Anisotropic g-values change with protonation, substituents	124

135	Klystron	Cylindrical	5	Semiquinones	Adsorbed on zeolite and in frozen solution, hyperfine splitting and *g*-values	125
140	EIK	Fabry–Perot	5	Tyrosyl in *E. coli* ribonucleotide reductase	Resolved *g*- and *A*-values	126
140	EIK	Fabry–Perot	4.9 to 5.0	Radical in inhibited ribonucleotide reductase	Resolved *g*- and *A*-values, nitrogen centered radical is attached to a cysteine	127
245	FIR	None	8.7	Tyrosyl radical in photosystem II	Orientation of radical relative to Mn cluster and nonheme iron	128
245	FIR	None	8.7	Tyrosyl radicals in photosystem II and ribonucleotide reductase	Use MO calculations to analyze anisotropic *g*-values in terms of local hydrogen bonding	129
245	FIR	None	8.7	Tyrosyl radical in mutant photosystem II	Effect of mutations on hydrogen bonding and *g*-values	130
245	FIR	None	8.7	Tyrosyl radical in ribonucleotide reductase from *S. typhimurium*	*g*-anisotropy similar to *E. coli*, but conformation similar to photosystem II tyrosyl radicals	131
			Other biological samples			
94	Gunn oscillator	Cylindrical TE_{013}	to 3.5	Nitroxide in liposomes	Distinguish probe inside and outside liposome, rotational correlation times	132
250	Phase-locked Gunn	Fabry–Perot	8.9	Spin-labeled phosphatidylcholines	Polarity gradient along chain in phospholipid membranes	133
140	Klystron	Cylindrical	5	Spin-labeled albumin	Effect of polarity on *g*-values	134
140	Klystron	Cylindrical	5	Spin-labeled cotton	Temperature dependent shifts of g_z-components	135
140	Klystron	Cylindrical	5	Spin-labeled fatty acids	Mobility in liposome (180–260 K) or ethanol (110–220 K)	136
140	Klystron	Cylindrical	5	Spin-labeled lysozyme	Hydrogen bonding and mobility	137
94	Gunn oscillator	Cylindrical	3.4	Gd^{3+} and spin label	Effect of Gd^{3+} chelates on synthetic membranes	138
60 to 400	Far infrared	None	to 9	Fe^{2+} hemes	ZFS $D \sim 5\ cm^{-1}$ for deoxy	139
90 to 3000	Far infrared	None	to 5	Fe^{3+} hemes	Determined ZFS up to 33 cm^{-1}	140, 141
70 to 400	Klystron, BWO	None	to 5	Methemoglobin	Determined ZFS ($D = 10.7\ cm^{-1}$) from field dependence of $g_\perp$	142

TABLE 2 (*continued*)

Frequency[a] (GHz)	Source[b]	Resonator[c]	Field (T)	Sample	Experiment	Ref.
95	Gunn oscillator	Fabry–Perot	to 3.5	Blue Cu^{2+} in azurin, frozen solution, and single crystal	Field-swept ESE-detected, anisotropic *g*-values, orientation of *g*-axes	143
95	Gunn oscillator	Fabry–Perot	to 3.5	Blue Cu^{2+} in azurin, single crystal	Pulsed ENDOR, define nitrogen hyperfine and quadrupole tensors	144
				Motion/mobility		
94	Gunn oscillator	Cylindrical TE_{013}	to 3.5	Nitroxide	Use proton hyperfine from X-band to analyze 94 GHz spectra and study motion	145
140	Klystron	Cylindrical	5	Nitroxide	*g*-value differences in mixtures and glasses, sensitivity of faster motion	146, 147
140	Klystron	Cylindrical	5	Nitroxide	Effect of hydration of inverse micelle on mobility	148
140	Klystron	Cylindrical	5	Nitroxide	Rotational anisotropy	149
250	Phase-locked Gunn	Fabry–Perot	9.2	Nitroxide	Increased sensitivity to motion	69, 150
250	Phase-locked Gunn	Fabry–Perot	8.9	Perdeuterated tempone in toluene-d_8	Fully determined the diffusion tensor, motion ~ isotropic	151
250	Phase-locked Gunn	Fabry–Perot	8.9	Nitroxide in slow-motion regime, −70 to −136° C	Found free diffusion for small nitroxide and Brownian diffusion for larger probe	152
140	Klystron	Cylindrical	5	Perchlorotriphenylmethyl radical	Motional probe without resolved nuclear hyperfine	153
140	Klystron	Cylindrical	5	Cu (SQ) $(PPh_3)_2$, SQ = semiquinone	Obtained accurate *g*-values to interpret dynamic effects on X-band linewidths	154
140	Klystron	Cylindrical	5	Semiquinones	Correlation between rate of intramolecular rearrangement and mobility of medium	155
140	Klystron	Cylindrical	5	Oxy-phenoxy radical	Tumbling and exchange	156
140	Klystron	Cylindrical	5	Gd^{3+} complexes in aqueous solution	Analyze frequency dependence of linewidths to find ZFS	157

Magnetically dilute metal ions						
73 to 310	Klystron, BWO	None	to 5.8	V^{3+} in $V_{1-x}Nb_xO_2$, single crystal	Hamiltonian parameters	158
65 to 350	BWO	None	0.8	Cr^{3+} in Mg_2SiO_4, single crystal	Determined ZFS in two sites	159
65 to 300	BWO	None	0.8	Cr^{2+} in SrF_2, single crystal	Determined all Hamiltonian parameters	160
247 to 337	FIR	None	to 10	Cr^{2+} in GaAs, single crystal	ZFS $D = -1.86$ cm^{-1}	161
250	Phase-locked Gunn	Fabry–Perot	4.5 to 9.5	Mn^{2+} complexes	ZFS D between 0.124 and 0.775 cm^{-1} measured accurately	162
140	Klystron	Cylindrical	4.5	Mn^{2+} in β-Ca_2SiO_4, powder and single crystal	Simplification of spectrum at higher frequency	163
150 to 600	FIR	None	to 25	$Cd_{0.99}Mn_{0.01}Te$	Measured spin–lattice relaxation times with pick-up coil	60
250	FIR	None	9	$Cd_{0.995}Mn_{0.005}Te$	Spin–lattice relaxation times	59
95	Gunn oscillator	Fabry–Perot	to 3.5	Framework Fe^{3+} in T sites of zeolites	Pulsed ^{57}Fe ENDOR, observe isotropic coupling	164
150 to 1200	FIR	None	9	$FeSiF_6 \cdot 6H_2O$, $Fe(SPh)_4^{2-}$	Determined ZFS of 11.78 cm^{-1} and 5.98 cm^{-1}	165
891 to 2567	HCN and H_2O lasers	None	to 35, pulsed	Fe^{2+} in $CdPS_3$, single crystal	ZFS for $S = 1$, $D = -24$ cm^{-1}	14
96	Gunn oscillator	None	to 33, pulsed	Co^{2+} in CdS	$D = 0.64$ cm^{-1}, data suggest BS^3 term in Hamiltonian	166
158	FIR	None	to 5.8	Co^{2+} macrocycle	g-values	167
370	BWO	None	8 to 14	Cu^{2+} in $Ca(PO_3)_2$ glass	$g_{\|} = 2.44$, $g_{\perp} = 2.07$, linewidths weakly temperature dependent	168
110, 208, 370	Gunn or BWO	None	to 15, pulsed	Cu^{2+} in CaO–P_2O_5 glassy or polycrystalline	Below 5 K relaxation times, longer in glassy than polycrystalline sample	169
140	Klystron	Fabry–Perot	to 5	Gd^{3+} complexes in solution	Relaxivity of MRI contrast reagents	170
79 to 535	BWO	None	to 1	Dy^{3+} in CaF_2	ZFS $\Delta = 257$ GHz	34

TABLE 2 (*continued*)

Frequency[a] (GHz)	Source[b]	Resonator[c]	Field (T)	Sample	Experiment	Ref.
				Pairs and clusters		
135	Klystron	Cylindrical	5	Dimers of imidazoline nitroxides	Analyzed splittings to determine interspin distance and geometry	171
140	Klystron	Cylindrical	5	Nitroxide	Structures of biradicals	172
240	FIR	None	to 8.8	Nitroxide biradical	$D = -64$ G, $E = 1$ G	173
245, 294	FIR	None	to 11	Nitroxide biradical	$2J = -25$ K, $2D = -244$ G	174
245	FIR	None	to 10	Cu^{2+}-semiquinone	ZFS for $S = 1$ due to strong spin–spin interaction	175
158 to 526	FIR	None	5 to 20	S doped GaP	Heteropolar pairs of neutral and ionized donors	176
95	Gunn diode	Fabry–Perot	6	Porphyrin–quinone radical pairs	ESE of mechanically induced pairs, ZFS $D > 0$, relaxation times were orientation dependent	177
140	Klystron	Cylindrical	5	Porphyrin–quinone radical pairs	Spin polarization permitted determination of sign of D	178
140	Klystron	Cylindrical	5	Porphyrin–quinone radical pairs	Mechanically induced pairs, $D > 0$ although photoexcited pairs had $D < 0$, stable for days	179, 180
95, 140	Klystron	Cylindrical	to 5	Metal semiquinone complexes, M = Al, Ga, In	Chemically and mechanically prepared samples have same ESR with triplet or quartet ground states	181
159 to 294	FIR	None	to 9.2	Two polyoxovanadate clusters	Magnetic interaction between localized and delocalized V^{4+}	182
240	FIR	None	to 9.2	Polyoxovanadate cluster with 12 vanadiums	Four localized V^{4+} and one delocalized V^{4+}	183
65 to 300	BWO	None	0.8	Cr^{2+} in BaF_2, single crystal	Hamiltonian parameters for clusters	184
96 to 950	Gunn or laser	None	to 40, pulsed	$Zn_{0.96}Mn_{0.04}Se$, $Cd_{0.95}Mn_{0.05}Se$, $Zn_{0.95}Co_{0.05}S$	Identify satellite lines due to spin-coupled pairs	185
245	FIR	None	to 10	Mn_{12} cluster, polycrystalline	$S = 10$, $D = -0.60$ cm^{-1}, crystallites aligned in field	186, 187

245	FIR	None	to 10	Mn_{10} cluster, polycrystalline	$S = 12$, $D = -0.047\ cm^{-1}$, crystallites aligned in field	188
256	FIR	None	to 10	Fe_8 cluster, polycrystalline	$S = 10$, $D = -0.191\ cm^{-1}$, $E = 0.032\ cm^{-1}$	189
220 to 762	BWO or FIR	None	Pulsed	Co^{2+} in Rb_2MgF_4, single crystal	Smaller g-values for dimers and trimers attributed to mixing with excited states	67, 190
245	FIR	None	to 12	Cu^{2+}-polyorganosiloxanolate cluster, six Cu^{2+}	$S = 3$, $D = 0.30\ cm^{-1}$, Cu–O–Cu $J = -42\ cm^{-1}$	191
245	FIR	None	to 8.2	Polyolate-bridged Cu^{2+} clusters, 3 or 16 Cu	Increasing field relative to J caused resolution of signals	192
			Magnetically concentrated solids			
245	FIR	None	8.8	Zn^{2+}-nitronyl nitroxide complex, powder and single crystal	Angular dependence of line-width and lineshape indicated one-dimensional antiferromagnet	193
110, 232, 370	Gunn or BWO	None	to 30, pulsed	$SrCr_xGa_{12-x}O_{19}$, $x = 0.6$ to 8.78, powder	Temperature dependence of resonance	194, 195
521	FIR	None	to 30, pulsed	$Cd_{1-x}Mn_xSe$, $x = 0.1$, 0.3, 0.5; single crystal	20 to 300 K	196
55 to 135		None	to 6 T	$Al_2Mn_3Si_4O_{16}$, Pb_2MnFeF_9, $PbMnFeF_7$	Spin glasses, resonance field as function of temperature above and below transition temperature	197
891, 2567	HCN and H_2O lasers	None	to 36.3, pulsed	Mn^{2+} in FeF_2, single crystal	Antiferromagnet, magnon–phonon modes	198
1360	H_2O laser	None	to 9	FeF_2 and Mn^{2+} in FeF_2, single crystal	Antiferromagnet, polariton model	199
1360 to 1580	H_2O, D_2O or DCN laser	None	to 9	$Mn_xFe_{1-x}F_2$, $x < 10^{-2}$, single crystal	Antiferromagnet, magnon pair modes	200
50 to 310	BWO	None	to 6	$CsMnCl_3 \cdot 2H_2O$ $(CH_3)_2NH_2MnCl_3$ $(CH_3)_4NMnCl_3$, single crystals	One-dimensional antiferromagnets, frequency dependence of resonance, magnon energy	201
70 to 310	BWO	None	to 6	$(CH_3)_4NMnCl_3$, TMMC, single crystal	Temperature dependence, antiferromagnetic resonance	202
73 to 135	BWO	None	to 6	$(CH_3)_4NMnCl_3$, single crystal	Temperature dependence, antiferromagnetic resonance	203

TABLE 2 (*continued*)

Frequency[a] (GHz)	Source[b]	Resonator[c]	Field (T)	Sample	Experiment	Ref.
216 to 320	FIR	None	to 9	$(CH_3)_4NMnCl_3$, single crystal	Field along chain axis, $g = 2.17$	204
891	HCN laser	None	to 35, pulsed	$(CH_3)_4NMnCl_3$, single crystal	Temperature dependence of resonance	14, 205, 206
9 to 115			to 8	$(CH_3)_4MnBr_3$, single crystal	One-dimensional antiferromagnet, observed one-dimensional quantum effects	207
2526, 2913, 3108	FIR	None	to 17	$CsFeCl_3$, single crystal	Splitting of ground-state energy levels as function of field	208
200 to 384	BWO	None	to 10, pulsed	$CsFeCl_3$, single crystal	Low-lying energy levels, single-ion model not adequate	209
220 to 3109	BWO or FIR	None	to 17, pulsed	$CsFeCl_3$, $RbFeCl_3$, single crystal	Ground-state singlet systems, $D = 14$, 14.8 for Cs, Rb	210
56 to 580	FIR	None	1 to 6	$Fe_{1-x}Co_xCl_2$, single crystal	Mixture of two antiferromagnets, competition of anisotropies	211
90 to 370	Gunn or BWO	None	to 15, pulsed	Fe/MgO thin films	Orientation dependence of FMR, dependence on film thickness	212
891	HCN laser	Stripline	35, pulsed	Ni, Fe	Ferromagnetic resonance	213, 214
400 to 900	FIR	None	to 35, pulsed	Ni, Co, Fe	Ferromagnetic resonance	215
37 to 300	BWO	None	to 20	CoF_2, single crystal	Antiferromagnetic resonance	75
2913, 3108	FIR	None	to 17, pulsed	$CsCoCl_3$, single crystal	g-values 4.2 to 4.8, depending upon absorption line	216
220 to 762	BWO or FIR	None	to 12, pulsed	$Rb_2Co_xMg_{1-x}F_4$, $x = 0.3$, single crystal	g-value for paired spins due to mixing with excited state	217
220 to 762	BWO or FIR	None	to 15, pulsed	$CsCo_xMg_{1-x}Cl_3$, $x = 0.2$, 0.4, 0.6, single crystal	Determine intrachain and inter-chain exchange interaction	218
813	FIR	None	to 30, pulsed	$Cd_{1-x}Co_xSe$, $x = 0.3$, single crystal	20 to 80 K, $g = 2.30$	196
891, 2567	HCN and H_2O lasers	None	to 40, pulsed	$(NH_4)_2Co(SO_4)_2 \cdot 6H_2O$, single crystal	Field dependence of g-values, evidence for SH^3 term in Hamiltonian	77, 219

891 to 2567	HCN and H_2O lasers	None	to 35, pulsed	Rb_2NiF_4, single crystal	Antiferromagnetic resonance, temperature dependence	14
9 to 390	Klystron, Gunn, or BWO	Cavity or none	to 16	$CsNiBr_3$, $CsMnI_3$, single crystals	Antiferromagnetic resonance, diagonal interchain exchange	29, 220
90 to 383	Gunn or BWO	None	to 16, pulsed	$LiNiO_2$, powder	Temperature dependence of resonance	221
70 to 1000	IMPATT or FIR	None	to 18	$Ni(C_2H_8N_2)_2NO_2ClO_4$, NENP, single crystal	Haldane gap, observe singlet ground state and lowest triplet	222
160 to 1000	FIR	None	to 17	NENP, single crystal	Haldane spin system, observed transitions for $S = 1$	223
160 to 1042	FIR	None	to 21	NENP, single crystal	One-dimensional antiferromagnet, distinguished high- and low-frequency regimes	224, 225
35 to 131	IMPATT or Gunn diode	None	to 10	$CsCuCl_3$, single crystal	Ferromagnetic resonance, interchain interaction	226
90 to 380	Gunn or BWO	None	to 16, pulsed	$CsCuCl_3$, single crystal	Anomalies in g-values above H_c attributed to quantum effect	227, 228, 229
891	HCN laser	None	31.8, pulsed	$CuCl_2 \cdot H_2O$, single crystal	Spin exchange $J = 0.4$ K at 4.3 K, ferromagnetic	76, 77
891	HCN laser	None	31.8, pulsed	$(NH_4)_2CuCl_4 \cdot 2H_2O$, single crystal	Spin exchange $J = 0.25$ K, ferromagnetic, at 4.2 K	77
891	HCN laser	None	31.8, pulsed	$Cu(NH_3)_4SO_4 \cdot H_2O$, single crystal	Spin exchange $J = 0.24$ K, antiferromagnetic	77
891	HCN laser	None	50, pulsed	Cu^{2+} salts, single crystals	Ferromagnetic and antiferromagnetic exchange	230
220, 380, 585	BWO or FIR	None	to 30, pulsed	Bi_2CuO_4, single crystal	$g_{\|} = 2.26$, $g_{\perp} = 2.04$, antiferromagnetic resonance below T_N	68
220 to 649	BWO or FIR	None	to 15, pulsed	Li_2CuO_2, powder	$g_{\|} = 2.30$, $g_{\perp} = 2.01$, antiferromagnetic resonance below T_N	231
55 to 1627	Solid-state diode or laser	None	to 14	$CuGeO_3$, single crystal, one-dimensional antiferromagnet	Observed two types of transitions in dimer phase, $q = 0$ Peierls gap	232
90 to 340	Gunn or BWO	None	to 15, pulsed	$CuGeO_3$, single crystal	Different g-values in singlet and magnetic phases, excitation gap decreased with increasing B_0	233

TABLE 2 (*continued*)

Frequency[a] (GHz)	Source[b]	Resonator[c]	Field (T)	Sample	Experiment	Ref.
90 to 380	Gunn or BWO	None	to 16, pulsed	$CuGeO_3$, single crystal	Temperature dependence of signal intensity and linewidth, estimated spin-Peierls gap	234
90 to 370	Gunn or BWO	None	to 16, pulsed	$CuGeO_3$, single crystal	Observed different g-values in spin-Peierls and magnetic phases	235
160 to 1627	FIR	None	to 14	$CuGeO_3$, dimerized phase	Observed low- (<500 GHz) and high- (>1000 GHz) frequency lines, evidence for Zeeman splitting of spin-Peierls gap	236
245 to 525	FIR	None	to 17	$CuGeO_3$, single crystal	Distinguished uniform, dimerized, and incommensurate phases	237
90, 370	Gunn or BWO	None	10 to 14	$CuGeO_3$, single crystal	distinguished spin-Peierls and magnetic phases	238
525, 761	FIR	None	to 40, pulsed	$CuGeO_3$, single crystal	At 78 K, $g = 2.17$, 2.26, and 2.11, distinguished three phases	239
90 to 383	Gunn or BWO	None	to 16, pulsed	$Ba_2Cu_3O_4Cl_2$, single crystal	Temperature dependence of AFMR	240
525	FIR	None	to 40, pulsed	$Ba_2Cu_3O_4Cl_2$, single crystal	For $H\|\|a$, resonance changed drastically around 30 K, observed AFMR at 5 K	239
90 to 383	Gunn or BWO	None	to 16, pulsed	$R_2Cu_2O_5$ with R = Y, Lu, Sc, In	Temperature dependence, g-values and antiferromagnetic resonance	241, 242
90 to 383	Gunn or BWO	None	to 16, pulsed	Y_2BaCuO_5, powder	Antiferromagnetic ordering, transition ~ 15 K in powder	243
60 to 383	Gunn or BWO	None	to 30, pulsed	$Y_2Cu_2O_5$, single crystal	Antiferromagnetic resonance, $J_0 \sim 340$ K (dimer, $S = 1$)	32, 244
245	FIR	None	to 9	Gd^{3+} in $YBa_2(Cu_{0.97}Zn_{0.03})_3O_y$, $y = 7.0$ and 6.76	Superconducting and spin-fluctuation gaps suppressed by Zn substitution	245
60 to 535	BWO	None	0.8	Ho in KY_3F_{10}, single crystal	B_1 parallel to B_0, hyperfine structure, crystal field	246

70 to 600	Klystron, BWO	None	to 6	$LiTbF_4$, $LiHoF_4$, and $LiErF_4$, single crystals	g-factors, local fields, and temperature dependence	247
150 to 1600	FIR	None	to 7	$LiHoF_4$, single crystal	Determined g-values, ZFS, and hyperfine	248
580 to 900	FIR	None	to 4	$LiHoF_4$, single crystal	$g_{\|\|} = 14.1$	249
890	HCN laser	None	to 32, pulsed	$LiErF_4$, single crystal	Observed four doublets	249
891	HCN laser	None	35, pulsed	$LiErF_4$, single crystal	Resonant absorption	214, 250
90 to 240	Gunn or BWO	Stripline	to 16, pulsed	$Ce_xGd_{1-x}Co_2$, $0 \leq x \leq 1$	Concentration dependence, FMR except for $x = 0$	251
1360	H_2O laser	None	6.8	$DyPO_4$	Resolved hyperfine	252

[a]To facilitate comparison the microwave energy is designated in GHz. For systems that were described by the wavelength of the microwaves the conversion to GHz may be approximate.
[b]Abbreviations for sources: BWO, backward wave oscillator, EIK, extended interaction klystron; EIO, extended interaction oscillator; FIR, far-infrared laser.
[c]The designation "none" indicates that the sample was measured in transmission or reflection mode without a resonator.

(continued from page 347)

furnished insight that could not be obtained by ESR at lower frequency, we have endeavored to be comprehensive in the literature coverage. However, the coverage is not exhaustive. Much of the pioneering work in the laboratory of Professor Lebedev was initially published in Russian. Although English translations of some of the papers are not widely available, that work is described in detail in English-language reviews by Lebedev[3,4] and Krinichnyi.[5,6] In addition, as high-field ESR becomes more routine, the fact that it was used in a particular study may not be evident from the title of a paper. The literature coverage is through early 1996.

Reviews related to high-field/frequency ESR include overviews,[1,3,4,7–10] descriptions of pulsed electron spin echo (ESE) at 95 GHz,[11] ESR at 140 GHz,[6,12] ESR at 250 GHz,[13] ESR with pulsed magnetic fields,[14,15] high-field magnetism,[16] incentives for high-field ESR,[17,18] multifrequency ESR,[19] nitroxides at 9 and 95 GHz,[20] orientation selection,[12] quasioptical techniques,[21] and applications to bioorganic systems,[22] condensed matter physics,[23,24] conducting polymers at 140 GHz,[6] organic radicals at 140 GHz,[5,6,25,26] photosynthesis,[27] and transition metal clusters.[28]

ACKNOWLEDGMENTS

We are grateful to Professors Louis-Claude Brunel, Robert Clarkson, Jack Freed, and Hitoshi Ohta for providing us with preprints and reprints of their papers.

REFERENCES

[1] S. S. Eaton and G. R. Eaton, Magn. Reson. Rev. **16**, 157 (1993).

[2] S. Y. Liao, *Microwave Circuit Analysis and Amplifier Design* (Prentice-Hall, Englewood Cliffs, NJ, 1987), Ch. 1.

[3] Ya. S. Lebedev, in *Modern Pulsed and Continuous Wave Electron Spin Resonance*, edited by L. Kevan and M. K. Bowman (Wiley, New York, 1990), Ch. 8.

[4] Ya. S. Lebedev, Appl. Magn. Reson. **7**, 339 (1994).

[5] V. I. Krinichnyi, J. Biochem. Biophys. Meth. **23**, 1 (1991); Appl. Magn. Reson. **2**, 29 (1991); duplicate publications.

[6] V. I. Krinichnyi, *2-mm Wave Band EPR Spectroscopy of Condensed Systems* (CRC Press, Boca Raton, FL, 1995).

[7] R. L. Belford, R. B. Clarkson, J. B. Cornelius, K. S. Rothenberger, M. J. Nilges, and M. D. Timken, in *Electronic Magnetic Resonance of the Solid State*, edited by J. A. Weil, M. K. Bowman, J. R. Morton, and K. F. Preston (The Canadian Society for Chemistry, Ottawa, Canada, 1987), p. 21.

[8] L.-C. Brunel, Appl. Magn. Reson, **3**, 83 (1992).

[9] W. Wang, R. L. Belford, R. B. Clarkson, P. H. Davis, J. Forrer, M. J. Nilges, M. D. Timken, T. Walczak, M. C. Thurnauer, J. R. Norris, A. L. Morris, and Y. Zhang, Appl. Magn. Reson. **6**, 195 (1994).

[10] Ya. S. Lebedev, Mendeleev Chem. J. **30**, 99 (1985) (in transl.) (p. 187 in Russian).

[11] J. Allgeier, J. A. J. M. Disselhorst, R. T. Weber, W. Th. Wenckebach, and J. Schmidt, in *Modern Pulsed and Continuous-Wave Electron Spin Resonance*, edited by L. Kevan, and M. K. Bowman (Wiley-Interscience, New York, 1990), Ch. 6.

[12] Ya. S. Lebedev, in *Electron Magnetic Resonance of Disordered Systems*, edited by N. D. Yordanov (World Scientific, Singapore, 1991), p. 69.

[13] D. E. Budil, K. A. Earle, W. B. Lynch, and J. H. Freed, in *Advanced EPR: Applications in Biology and Biochemistry*, edited by A. J. Hoff (Elsevier, Amsterdam, 1989), Ch. 8.

[14] M. Motokawa, in *High Field Magnetism*, edited by M. Date (North-Holland, Amsterdam, 1983), p. 219.

[15] J. Witters and F. Herlach, Bull. Magn. Reson. **9**, 132 (1987).

[16] M. Date, Aust. J. Phys. **48**, 187 (1995).

[17] G. R. Eaton, in *Proceedings of National High Magnetic Field Laboratory Workshop on Nuclear Magnetic Resonance*, edited by E. R. Andrew, T. Mareci, and N. S. Sullivan (University of Florida, 1992), p. 57.

[18] J. Freed, in *Proceedings of National High Magnetic Field Laboratory Workshop on Nuclear Magnetic Resonance*, edited by E. R. Andrew, T. Mareci, and N. S. Sullivan (University of Florida, 1992), p. 47.

[19] R. L. Belford and R. B. Clarkson, in *Magnetic Resonance of Carbonaceous Solids*, edited by R. Botto and Y. Sanada, Adv. Chem. Ser. **229**, 107 (1993).

[20] A. I. Smirnov, R. L. Belford, and R. B. Clarkson, in *Spin Labelling: the Next Millenium*, edited by L. B. Berliner, Biol. Magn. Reson. **14**, 83 (1998).

[21] K. A. Earle, D. E. Budil, and J. H. Freed, in *Advances in Magnetic and Optical Resonance*, edited by W. Warren (Academic Press, New York, 1996), Vol. 19, Ch. 3.

[22] K. Möbius, Biol. Magn. Reson. **13**, 253 (1993).

[23] L. C. Brunel, A. L. Barra, and G. Martinez, Physica B **204**, 298 (1995).

[24] L. C. Brunel, Physica B **211**, 360 (1995).

[25] O. Ya. Grinberg, A. A. Dubinskii, and Ya. S. Lebedev, Russian Chem. Rev. **52**, 850 (1983) (in transl.) (p. 1490 in Russian).

[26] Yu. D. Tsvetkov, V. I. Gulin, S. A. Dikanov, and I. A. Grigor'ev, in *Electronic Magnetic Resonance of the Solid State*, edited by J. A. Weil, M. K. Bowman, J. R. Morton, and K. F. Preston (The Canadian Society for Chemistry, Ottawa, Canada, 1987), p. 45.

[27] K. Möbius, Appl. Magn. Reson. **9**, 389 (1995).

[28] D. Gatteschi, in *Molecular Magnetism: From Molecular Assemblies to the Devices*, edited by E. Coronado, P. Delhaès, D. Gatteschi, and J. S. Miller (Kluwer Academic, Amsterdam, 1996), p. 141.

[29] T. Kambe, H. Tanake, S. Kimura, H. Ohta, M. Motokawa, and K. Nagata, J. Phys. Soc. Japan. **65**, 1799 (1996).

[30] J. B. Mock, Rev. Sci. Instrum. **31**, 551 (1960).

[31] M. E. J. Boonman, P. H. M. van Loosdrecht, D. S. Bethune, I. Holleman, G. J. M. Meijer, and P. J. M van Bentum, Physica B **211**, 323 (1995).

[32] S. Kimura, H. Ohta, M. Motokawa, S. Mitsudo, W.-J. Jang, M. Hasegawa, and H. Takei, Internat. J. Infrared MM Wave **17**, 833 (1996).

[33] T. Tatsukawa, T. Maeda, H. Sasai, T. Idehara, M. Mekata, T. Saito, and T. Kanemaki, Internat. J. Infrared MM Wave **16**, 293 (1995).

[34] V. F. Tarasov and G. S. Shakurov, Appl. Magn. Res. **2**, 571 (1991).

[35] E. Haindl, K. Möbius, and H. Oloff, Z. Naturforsch. **40a**, 169 (1985).

[36] O. Burghaus, E. Haindl, M. Plato, and K. Möbius, J. Phys. E **18**, 294 (1985).

[37] O. Burghaus, M. Rohrer, T. Götzinger, M. Plato, and K. Möbius, Meas. Sci. Technol. **3**, 765 (1992).

[38] R. Klette, J. T. Törring, M. Plato, K. Möbius, B. Bönigk, and W. Lubitz, J. Phys. Chem. **97**, 2015 (1993).

[39] O. Burghaus, A. Toth-Kischkat, R. Klette, and K. Möbius, J. Magn. Reson. **80**, 383 (1988).

[40] T. F. Prisner, M. Rohrer, and K. Möbius, Appl. Magn. Reson. **7**, 167 (1994).

[41] M. Rohrer, M. Plato, F. MacMillan, Y. Grishin, W. Lubitz, and K. Möbius, J. Magn. Reson. **A116**, 59 (1995).

[42] J. Schmidt, Bull. Magn. Reson. **11**, 314 (1989).

[43] R. T. Weber, A. J. M. Disselhorst, L. J. Prevo, J. Schmidt, and W. Th. Wenckebach, in *Pulsed EPR: A New Field of Applications*, edited by C. P. Keijzers, E. J. Reijerse, and J. Schmidt (North-Holland, Amsterdam, 1989), p. 186.

[44] R. T. Weber, A. J. M. Disselhorst, L. J. Prevo, J. Schmidt, and W. Th. Wenckebach, J. Magn. Reson. **81**, 129 (1989).

[45] H. J. van der Meer, J. A. J. M. Disselhorst, J. Allgeier, J. Schmidt, and W. Th. Wenckebach, Meas. Sci. Technol. **1**, 396 (1990).

[46] J. A. J. M. Disselhorst, H. van der Meer, O. G. Poluektov, and J. Schmidt, J. Magn. Reson. **A115**, 183 (1995).
[47] Louis-Claude Brunel, personal communication, Nov. 1996.
[48] A. A. Galkin, G. G. Grinev, V. I. Kurochkin, and E. D. Nemchenko, Soviet J. Nondestructive Testing **12**, 443 (1977) (no. 4, p. 121 in Russian).
[49] A. A. Galkin, O. Ya. Grinberg, A. A. Dubinskii, N. N. Kabdin, V. N. Krymov, V. I. Kurochkin, Ya. S. Lebedev, L. F. Oranskii, and V. F. Shuvalov, Inst. Exper. Technol. **20**, 1229 (1977) (in transl.) (no. 4, p. 284 in Russian).
[50] A. Yu. Bresgunov, A. A. Dubinskii, V. N. Krimov, Yu. G. Petrov, O. G. Poluektov, and Ya. S. Lebedev, Appl. Magn. Reson. **2**, 715 (1991).
[51] O. G. Poluektov, A. I. Smirnov, and Ya. S. Lebedev, Russian J. Phys. Chem. **64**, 1549 (1990) (in transl.) (p. 2867 in Russian).
[52] A. I. Smirnov, O. G. Poluectov, and Ya. S. Lebedev, J. Magn. Reson. **97**, 1 (1992).
[53] T. F. Prisner, S. Un, and R. G. Griffin, Israel J. Chem. **32**, 357 (1992).
[54] S. Un, J. Bryant, and R. G. Griffin, J. Magn. Reson. **A101**, 92 (1993).
[55] L. R. Becerra, G. J. Gerfen, B. F. Bellew, J. A. Bryant, D. A. Hall, S. J. Inati, R. T. Weber, S. Un, T. F. Prisner, A. E. McDermott, K. W. Fishbein, K. E. Kreischer, R. J. Temkin, D. J. Singel, and R. G. Griffin, J. Magn. Reson. **A117**, 28 (1995).
[56] N. Miura, H. Nojiri, Y. Shimamoto, and Y. Imanaka, Physica B **211**, 23 (1995).
[57] L. I. Antsiferova and E. V. Lyubashevskaya (Academy of Sciences USSR, Institute of Chemical Physics, Chernogolovka, 1989). Preprint.
[58] V. I. Kurochkin, A. Ya. Laptienko, N. V. Taryanik, V. V. Sukhoi, and V. V. Shaparenko, Inst. Exper. Technol. **25**, 1485 (1982) (in transl.) (no. 6, p. 154 in Russian).
[59] T. Strutz, A. M. Witowski, R. E. M. de Bekker, and P. Wyder, Appl. Phys. Lett. **57**, 831 (1990).
[60] I. Strutz, A. M. Witowski, and P. Wyder, Rev. Sci. Instrum. **64**, 1853 (1993).
[61] A. L. Barra, L. C. Brunel, F. Muller, and J. B. Robert, Bull. Magn. Reson. **12**, 55 (1990).
[62] A. L. Barra, L. C. Brunel, and J. B. Robert, Chem. Phys. Lett. **165**, 107 (1990).
[63] F. Muller, M. A. Hopkins, N. Coron, M. Grynberg, L. C. Brunel, and G. Martinez, Rev. Sci. Instrum. **60**, 3681 (1989).
[64] L. Paschedag, J. van Tol, and P. Wyder, Rev. Sci. Instrum. **66**, 5098 (1995).
[65] M. Sauzade, J. Pontnau, and B. Girard, C. R. Acad. Sci. Paris **258**, 4458 (1964).
[66] K. A. Earle, D. S. Tipikin, and J. H. Freed, Rev. Sci. Instrum. **67**, 2502 (1996).
[67] M. Motokawa, H. Ohta, and N. Maki, Internat. J. Infrared MM Wave **12**, 149 (1991).
[68] H. Ohta, K. Yoshida, T. Matsuya, T. Nanba, M. Motokawa, K. Yamada, Y. Endoh, and S. Hosoya, J. Phys. Soc. Japan. **61**, 2921 (1992).
[69] B. Lynch, K. A. Earle, and J. H. Freed, Rev. Sci. Instrum. **59**, 1345 (1988).
[70] A. M. Witowski, Solid State Commun. **77**, 23 (1991).
[71] C. Kutter, H. P. Moll, J. van Tol, H. Zuckermann, J. C. Mann, and P. Wyder, Phys. Rev. Lett. **74**, 2925 (1995).
[72] C. Kutter, H. P. Moll, T. van Tol, H. Zuckermann, and P. Wyder, Infrared Phys. Technol. **36**, 245 (1995).
[73] F. Herlach, P. de Groot, P. Janssen, G. De Vos, and J. Witters, in *High Field Magnetism*, edited by M. Date (North-Holland, Amsterdam 1983), p. 229.
[74] F. Herlach, editor, *Strong and Ultrastrong Magnetic Fields and Their Applications* (Springer-Verlag, New York, 1985).
[75] V. I. Ozhogin, K. G. Gurtovoj, and A. S. Lagutin, in *High Field Magnetism*, edited by M. Date (North-Holland, Amsterdam, 1983), p. 267.
[76] M. Date, M. Motokawa, A. Seki, S. Kuroda, K. Matsui, H. Nakazato, and H. Mollymoto, J. Phys. Soc. Japan **39**, 898 (1975).
[77] M. Motokawa, S. Kuroda, and M. Date, J. Appl. Phys. **50**, 7762 (1979).
[78] G. Kido and N. Miura, Appl. Phys. Lett. **41**, 569 (1982).
[79] M. Date, J. Phys. Soc. Japan **39**, 892 (1975).

[80] M. Date and A. Yamagishi, IEEE Trans. Magnetics **MAG-23**, 3257 (1987).
[81] R. B. Clarkson, W. Wang, M. J. Nilges, and R. L. Belford, *in Processing and Utilization of High-Sulfur Coal*, edited by R. Markuszewski and T. D. Wheelock (Elsevier, New York, 1990), p. 67.
[82] R. B. Clarkson, W. Wang, D. R. Brown, H. C. Crookham, and R. L. Belford, Fuel **69**, 1405 (1990).
[83] R. B. Clarkson, W. Wang, D. L. Brown, H. C. Crookham, and R. L. Belford in *Magnetic Resonance of Carbonaceous Solids*, edited by R. Botto and Y. Sanada Adv. Chem. Ser. **229**, 507 (1993).
[84] R. B. Clarkson, R. L. Belford, P. Ceroke, W. Wang, W. Shi, K. Motsegood, A. Smirnov, T. Smirnova, and M. J. Nilges, Amer. Chem. Soc. Preprints Fuel Chem. **40**, 420 (1995).
[85] K. S. Rothenberger, R. F. Sprecher, H. L. Retcofsky, W. Wang, R. B. Clarkson, and R. L. Belford, Fuel **73**, 1600 (1994).
[86] A. Yu. Brezgunov, G. A. Vorob'ev, A. A. Dubinskii, O. G. Poluektov, and Ya. S. Lebedev, Russian J. Phys. Chem. **65**, 509 (1991) (in transl.) (p. 969 in Russian).
[87] A. Yu. Bresgunov, A. A. Dubinsky, O. G. Poluektov, G. A. Vorob'eva, and Ya. S. Lebedev, J. Chem. Soc. Faraday Trans. **86**, 3185 (1990).
[88] Yu. D. Tsvetkov, S. A. Dzuba, and V. I. Gulin, in *Magnetic Resonance of Carbonaceous Solids*, edited by R. Botto and Y. Sanada Adv. Chem. Ser. **229**, 443 (1993).
[89] A. Yu. Bresgunov, O. G. Poluektov, Ya. S. Lebedev, A. L. Barra, L. C. Brunel, and J. B. Robert, Chem. Phys. Lett. **175**, 621 (1990).
[90] S. A. Dikanov, V. I. Gulin, Y. D. Tsvetkov, and I. A. Grigor'ev, J. Chem. Soc. Faraday Trans. **86**, 3201 (1990).
[91] V. I. Gulin, S. A. Dikanov, and Yu. D. Tsvetkov, in *Electron Magnetic Resonance of Disordered Systems*, edited by N. D. Yordanov (World Scientific, Singapore, 1991), p. 92.
[92] M. A. Ondar, O. Ya. Grinberg, L. G. Oranskii, V. I. Kurochkin, and Ya. L. Lebedev, J. Struct. Chem. **22**, 626 (1981) (in transl.) (no. 4, p. 173 in Russian).
[93] O. Ya. Grinberg, A. A. Dubinskii, O. G. Poluektov, and Ya. S. Lebedev, Theor. Exp. Chem. **17**, 633 (1981) (in transl.) (p. 806 in Russian).
[94] A. Yu. Bresgunov, A. A. Dubinsky, O. G. Poluektov, Ya. S. Lebedev, and A. I. Prokof'ev, Molecular Phys. **75**, 1123 (1992).
[95] K. M. Evenson, H. P. Broida, J. S. Wells, R. J. Mahler, and M. Mizushima, Phys. Rev. Lett. **21**, 1038 (1968).
[96] K. M. Evenson, H. E. Radford, and M. M. Moran, Jr., Appl. Phys. Lett. **18**, 426 (1971).
[97] E. J. J. Groenen, O. G. Poluektov, M. Matsushita, J. Schmidt, J. H. van der Waals, and G. Meijer, Chem. Phys. Lett. **197**, 314 (1992).
[98] V. I. Krinichnyi, A. E. Pelekh, Ya. S. Lebedev, L. I. Tkachenko, G. I. Kozub, A. L. Barra, L. C. Brunel, and J. B. Robert, Appl. Magn. Reson. **7**, 459 (1994).
[99] A. E. Pelekh, L. M. Goldenberg, and V. I. Krinichnyi, Synth. Met. **44**, 205 (1991).
[100] V. I. Krinichnyi, A. E. Pelekh, H.-K. Roth, and K. Luders, Appl. Magn. Reson. **4**, 345 (1993).
[101] L. M. Goldenberg, A. E. Pelekh, V. I. Krinichnyi, O. S. Roshchupkina, A. F. Zueva, R. N. Lyubovskaya, and O. N. Efimov, Synth. Met. **36**, 217 (1990).
[102] L. M. Goldenberg, A. E. Pelekh, V. I. Krinichnyi, O. S. Roshchupkina, A. F. Zueva, R. N. Lyubovskaya, and O. N. Efimov, Synth. Met. **41–43**, 3071 (1991).
[103] V. I. Krinichnyi, A. E. Pelekh, L. I. Tkachenko, and G. I. Kozub, Synth. Met. **46**, 1 (1992).
[104] V. I. Krinichnyi, A. E. Pelekh, L. I. Tkachenko, and G. I. Kozub, Synth. Met. **46**, 13 (1992).
[105] V. I. Krinichnyi, O. Ya. Grinberg, I. B. Nazarova, G. I. Kozub, L. I. Tkachenko, M. L. Khidekel', and Ya. S. Lebedev, Akad. Nauk. SSSR Bull., Div. Chem. Sci. **34**, 425 (1985) (in transl.) (p. 467 in Russian).
[106] R. B. Clarkson, K. Mattson, W. Shi, W. Wang, and R. L. Belford, Molecular Engng. **4**, 89 (1994).
[107] R. J. Wagner, J. J. Krebs, and G. H. Strauss, Solid State Commun. **36**, 15 (1980); **88**, 887 (1993).
[108] F. Muller, S. Huant, K. Karrai, G. Dampne, M. Grynberg, G. Martinez, and L. C. Brunel, J. Phys. Colloq. **C7**, C7–717 (1987).
[109] T. Gregorkiewicz, H. H. P. Th. Bekman, C. A. J. Ammerlaan, W. Knap, L. C. Brunel, and G. Martinez, Phys. Rev. B **45**, 5873 (1992).
[110] T. Gregorkiewicz, W. Knap, H. H. P. Th. Bekman, L. C. Brunel, C. A. J. Ammerlaan, and G. Martinez, Physica B **177**, 476 (1992).

[111] D. H. Shin, J. L. Dye, D. E. Budil, K. A. Earle, and J. H. Freed, J. Phys. Chem. **97**, 1213 (1993).
[112] M. C. J. M. Donckers, O. G. Poluektov, J. Schmidt, and P. G. Baranov, Phys. Rev. B **45**, 13061 (1992).
[113] O. G. Poluektov, M. C. J. M. Donckers, P. G. Baranov, and J. Schmidt, Phys. Rev. B **47**, 10226 (1993).
[114] M. T. Bennebroek, O. G. Poluektov, A. J. Zakrzewski, P. G. Baranov, and J. Schmidt, Phys. Rev. Lett. **74**, 442 (1995).
[115] V. I. Gulin, S. A. Dikanov, Yu. D. Tsvetkov, R. G. Evelo, and A. J. Hoff, Pure Appl. Chem. **64**, 903 (1992).
[116] T. F. Prisner, A. E. McDermott, S. Un, J. R. Norris, M. C. Thurnauer, and R. G. Griffin, Proc. Nat. Acad. Sci. U.S.A. **90**, 9485 (1993).
[117] O. Burghaus, M. Plato, D. Bumann, B. Neumann, W. Lubitz, and K. Möbius, Chem. Phys. Lett. **185**, 381 (1991).
[118] M. Huber, J. T. Törring, M. Plato, U. Fink, W. Lubitz, R. Feick, C. C. Schenck, and K. Möbius, Solar Energy Mater. Solar Cells **38**, 119 (1995).
[119] M. Huber and J. T. Törring, Chem. Phys. **194**, 379 (1995).
[120] T. F. Prisner, A. van der Est, R. Bittl, W. Lubitz, D. Stehlik, and K. Möbius, Chem. Phys. **194**, 361 (1995).
[121] M. Rohrer, P. Gast, K. Möbius, and T. F. Prisner, Chem. Phys. Lett. **259**, 523 (1996).
[122] J. Tang, S. Bondeson, and M. C. Thurnauer, Chem. Phys. Lett. **253**, 293 (1996).
[123] O. Burghaus, M. Plato, M. Rohrer, K. Möbius, F. MacMillan, and W. Lubitz, J. Phys. Chem. **97**, 7639 (1993).
[124] M. K. Bowman, M. C. Thurnauer, J. R. Norris, S. A. Dikanov, V. I. Gulin, A. M. Tyryshkin, R. I. Samoilova, and Yu. D. Tsvetkov, Appl. Magn. Reson. **3**, 353 (1992).
[125] V. I. Gulin, R. I. Samailova, S. A. Dikanov, and Yu. D. Tsvetkov, Appl. Magn. Reson. **2**, 425 (1991).
[126] G. J. Gerfen, B. F. Bellew, S. Un, J. M. Bollinger, Jr., J. Stubbe, R. G. Griffin, and D. J. Singel, J. Amer. Chem. Soc. **115**, 6420 (1993).
[127] W. A. van der Donk, J. Stubbe, G. J. Gerfen, B. F. Bellew, and R. G. Griffin, J. Amer. Chem. Soc. **117**, 8908 (1995).
[128] S. Un, L.-C. Brunel, T. M. Brill, J.-L. Zimmermann, and A. W. Rutherford, Proc. Nat. Acad. Sci. U.S.A. **91**, 5262 (1994).
[129] S. Un, M. Atta, M. Fontecave, and A. W. Rutherford, J. Amer. Chem. Soc. **117**, 10713 (1995).
[130] S. Un, X.-S. Tang, and B. A. Diner, Biochem. **35**, 679 (1996).
[131] P. Allard, A. L. Barra, K. K. Andersson, P. P. Schmidt, M. Atta, and A. Gräslund, J. Amer. Chem. Soc. **118**, 895 (1996).
[132] A. I. Smirnov, T. I. Smirnova, and P. D. Morse, II, Biophys. J. **68**, 2350 (1995).
[133] K. A. Earle, J. K. Moscicki, M. Ge, D. E. Budil, and J. H. Freed, Biophys. J. **66**, 1213 (1994).
[134] V. I. Krinichnyi, O. Ya. Grinberg, V. R. Bogatyrenko, G. I. Likhtenshtein, and Ya. S. Lebedev, Biophys. **30**, 233 (1985) (in transl.) (p. 216 in Russian).
[135] V. I. Krinichnyi, O. Ya. Grinberg, I. Kh. Yusupov, R. M. Marupov, P. Kh. Bobodzhanov, G. I. Likhtenshtein, and Ya. S. Lebedev, Biophys. **31**, 526 (1986) (in transl.) (p. 482 in Russian).
[136] V. I. Krinichnyi, O. Ya. Grinberg, Ye. I. Yudanova, M. L. Borin, Ya. S. Lebedev, and G. I. Likhtenshtein, Biophys. **32**, 59 (1987) (in transl.) (p. 59 in Russian).
[137] V. I. Krinichnyi, O. Ya. Grinberg, Ye. I. Yudanova, Ye. V. Lyubashevskaya, L. I. Antsiferova, G. I. Likhtenstein, and Ya. S. Lebedev, Biophys. **32**, 229 (1987) (p. 215 in Russian).
[138] T. I. Smirnova, A. I. Smirnov, R. L. Belford, and R. B. Clarkson, Acta Chem. Scand. **51**, 562 (1997).
[139] P. M. Champion and A. J. Sievers, J. Chem. Phys. **72**, 1569 (1980).
[140] G. C. Brackett, P. L. Richards, and W. S. Caughey, J. Chem. Phys. **54**, 4383 (1971).
[141] P. L. Richards, W. S. Caughey, H. Eberspaecher, G. Feher, and M. Malley, J. Chem. Phys. **47**, 1187 (1967).
[142] Y. Alpert, Y. Couder, J. Tuchendler, and H. Thome, Biochim. Biophys. Acta **322**, 34 (1973).
[143] J. W. A. Coremans, O. G. Poluektov, E. J. J. Groenen, G. W. Canters, H. Nar, and A. Messerschmidt, J. Amer. Chem. Soc. **116**, 3097 (1994).

[144] J. W. A. Coremans, M. van Gastel, O. G. Poluektov, E. J. J. Groenen, T. den Blaauwen, G. van Pouderoyen, G. W. Canters, H. Nar, C. Hammann, and A. Messerschmidt, Chem. Phys. Lett. **235**, 202 (1995).
[145] T. I. Smirnova, A. I. Smirnov, R. B. Clarkson, and R. L. Belford, J. Phys. Chem. **99**, 9008 (1995).
[146] O. Ya. Grinberg, A. A. Dubinskii, V. F. Shuvalov, L. G. Oranskii, V. I. Kurochkin, and Ya. S. Lebedev, Dokl. Akad. Nauk SSR, **230**, 923 (1976) (in transl.) (p. 884 in Russian).
[147] O. Ya. Grinberg, A. A. Dadali, A. A. Dubinskii, A. M. Vasserman, A. L. Buchachenko, and Ya. S. Lebedev, Theor. Exper. Chem. **15**, 457 (1979) (in transl.) (p. 583 in Russian)
[148] V. I. Krinichnyi, L. I. Antsiferova, E. V. Lyubashevskaya, O. V. Belonogova, O. Ya. Grinberg, and G. I. Likhtenshtein, Russian J. Phys. Chem. **63**, 1653 (1989) (in transl.) (p. 3015 in Russian).
[149] A. A. Dubinskii, O. Ya. Grinberg, V. I. Kurochkin, L. G. Oranskii, O. G. Poluektov, and Ya. S. Lebedev, Theor. Exp. Chem. **17**, 180 (1981) (in transl.) (p. 231 in Russian).
[150] J. H. Freed, J. Chem. Soc. Faraday Trans. **86**, 3173 (1990).
[151] D. E. Budil, K. A. Earle, and J. H. Freed, J. Phys. Chem. **97**, 1294 (1993).
[152] K. A. Earle, D. E. Budil, and J. H. Freed, J. Phys. Chem. **97**, 13289 (1993).
[153] O. Ya. Grinberg, A. A. Dubinskii, O. G. Poluektov, and Ya. S. Lebedev, Russian J. Phys. Chem. **64**, 849 (1990) (in transl.) (p. 1586 in Russian).
[154] R. R. Rakhimov, N. P. Benetis, A. Lund, J. S. Hwang, A. I. Prokof'ev, and Ya. S. Lebedev, Chem. Phys. Lett. **255**, 156 (1996).
[155] R. R. Rakhimov, A. I. Prokof'ev, and Ya. S. Lebedev, Russian J. Phys. Chem. **67**, 979 (1993) (in transl.) (p. 1093 in Russian).
[156] Ya. S. Lebedev, R. R. Rakhimov, A. I. Prokof'ev, and A. Yu. Bresgunov, Pure Appl. Chem. **64**, 873 (1992).
[157] D. H. Powell, A. E. Merbach, G. Gonzáles, E. Brücher, K. Micskei, M. F. Ottaviani, K. Köhler, A. von Zelewsky, O. Ya. Grinberg, and Ya. S. Lebedev, Helv. Chim. Acta **76**, 2129 (1993).
[158] J. Magariño, J. Tuchendler, and J. P. D'Haenens, Phys. Rev. B **14**, 865 (1976).
[159] V. F. Tarasov, G. S. Shakurov, and A. N. Gavrilenko, Phys. Solid State **37**, 270 (1995).
[160] M. M. Zaripov, V. F. Tarasov, V. A. Ulanov, G. S. Shakurov, and M. L. Popov, Phys. Solid State **37**, 437 (1995).
[161] R. J. Wagner and A. M. White, Solid State Commun. **32**, 399 (1979).
[162] W. B. Lynch, R. S. Boorse, and J. H. Freed, J. Amer. Chem. Soc. **115**, 10909 (1993).
[163] R. Lück, R. Stösser, O. G. Poluektov, O. Ya. Grinberg, and Ya. S. Lebedev, Z. Anorg. Allg. Chem. **607**, 183 (1992).
[164] D. Goldfarb, K. G. Strohmaier, D. E. W. Vaughan, H. Thomann, O. G. Poluektov, and J. Schmidt, J. Amer. Chem. Soc. **118**, 4665 (1996).
[165] P. M. Champion and A. J. Sievers, J. Chem. Phys. **66**, 1819 (1977).
[166] V. Bindilatti, A. N. Anisimov, N. F. Oliveira, Jr., Y. Shapira, M. Goiran, F. Yang, S. Isber, M. Averous, and M. Demianiuk, Phys. Rev. B **50**, 16464 (1994).
[167] A. Ceulemans, R. Debuyst, F. Dejehet, P. Janssen, and L. G. Vanquickenborne, Inorg. Chem. **33**, 3188 (1994).
[168] N. Kitamura, H. Ohta, S. Imagawa, and M. Motokawa, J. Phys. Soc. Japan **63**, 781 (1994).
[169] N. Kitamura, H. Ohta, S. Imagawa, H. Ushiroyama, K. Kita, and M. Motakawa, Physica B **201**, 119 (1994).
[170] D. H. Powell, O. M. N. Dhubhghaill, D. Pubanz, L. Helm, Y. S. Lebedev, W. Schlaepfer, and A. E. Merbach, J. Amer. Chem. Soc. **118**, 9333 (1996).
[171] V. I. Gulin, S. A. Dikanov, and Yu. D. Tsvetkov, Chem. Phys. Lett. **170**, 211 (1990).
[172] M. A. Ondar, A. A. Dubinskii, O. Ya. Grinberg, I. A. Grigor'ev, L. B. Volodarskii, and Ya. S. Lebedev, J. Struct. Chem. **22**, 525 (1981) (in transl.) (no. 4, p. 59 in Russian).
[173] C. Fabre, S. Gambarelli, D. Jaouen, A. Rassat, L.-C. Brunel, and C. Chachaty, Magn. Reson. Chem. **33**, S129 (1995).
[174] S. Gambarelli, D. Jaouen, A. Rassat, L.-C. Brunel, and C. Chachaty, J. Phys. Chem. **100**, 9605 (1996).

[175] A. Dei, D. Gatteschi, L. Pardi, A. L. Barra, and L. C. Brunel, Chem. Phys. Lett. **175**, 589 (1990).
[176] F. Muller, L. C. Brunel, M. Grynberg, J. Blinowski, and G. Martinez, Europhys. Lett. **8**, 291 (1989).
[177] D. S. Tipikin, Ya. S. Lebedev, O. G. Poluektov, and J. Schmidt, Chem. Phys. Lett. **215**, 199 (1993).
[178] S. D. Chemerisov, D. S. Tipikin, O. Ya. Grinberg, and Ya. S. Lebedev, Russian J. Phys. Chem. **67**, 1999 (1993) (in transl.) (p. 2220 in Russian).
[179] S. D. Chemerisov, O. Ya. Grinberg, D. S. Tipikin, Ya. S. Lebedev, H. Kurreck, and K. Möbius, Chem. Phys. Lett. **218**, 353 (1994).
[180] G. D. Perekhodtsev, D. S. Tipikin, and Ya. S. Lebedev, Russian J. Phys. Chem. **68**, 1131 (1994) (in transl.) (p. 1252 in Russian).
[181] S. D. Chemerisov, G. D. Perekhodtsev, D. S. Tipikin, Ya. S. Lebedev, A. I. Prokof'ev, A. I. Aleksandrov, A. A. Dubinskii, K. Möbius, O. G. Poluektov, and J. Schmidt, J. Chem. Soc. Faraday Trans. **92**, 1959 (1996).
[182] D. Gatteschi, B. Tsukerblatt, A. L. Barra, L. C. Brunel, A. Müller, and J. Döring, Inorg. Chem. **32**, 2114 (1993).
[183] A. L. Barra, D. Gatteschi, B. S. Tsukerblatt, J. Döring, A. Müller, and L.-C. Brunel, Inorg. Chem. **31**, 5132 (1992).
[184] M. M. Zaripov, V. F. Tarasov, V. A. Ulanov, and G. S. Shakurov, Phys. Solid State **38**, 249 (1996).
[185] J. L. Martin, M. Goiran, Z. Golacki, J. Leotin, and S. Askenazy, Physica B **211**, 388 (1995).
[186] A. Caneschi, D. Gatteschi, R. Sessoli, A. L. Barra, L.-C. Brunel, and M. Guillot, J. Amer. Chem. Soc. **113**, 5873 (1991).
[187] R. Sessoli, H.-L. Tsai, A. R. Schake, S. Wang, J. B. Vincent, K. Folting, D. Gatteschi, G. Christou, and D. N. Hendrickson, J. Amer. Chem. Soc. **115**, 1804 (1993).
[188] A. L. Barra, A. Caneschi, D. Gatteschi, and R. Sessoli, J. Amer. Chem. Soc. **117**, 8855 (1995).
[189] A. L. Barra, P. Debrunner, D. Gatteschi, Ch. E. Schulz, and R. Sessoli, Europhys. Lett. **35**, 133 (1996).
[190] M. Motokawa, H. Ohta, K. Yoshida, and H. Ikeda, J. Magn. Magn. Mater. **104–107**, 947 (1992).
[191] E. Rentschler, D. Gatteschi, A. Cornia, A. C. Fabretti, A.-L. Barra, O. I. Shchegolikhina, and A. A. Zhdanov, Inorg. Chem. **35**, 4427 (1996).
[192] B. Pilawa and J. Schuhmacher, J. Phys.: Conden. Matter **8**, 1539 (1996).
[193] A. Caneschi, D. Gatteschi, R. Sessoli, C. I. Cabello, P. Rey, A. L. Barra, and L. C. Brunel, Inorg. Chem. **31**, 1882 (1991).
[194] H. Ohta, M. Sumikawa, M. Motokawa, H. Kikuchi, and H. Nagasawa, J. Phys. Soc. Japan. **65**, 848 (1996).
[195] M. Sumikawa, H. Ohta, M. Motokawa, H. Kikuchi, and H. Nagasawa, Physica B **201**, 123 (1994).
[196] N. Adachi, G. Kido, Y. Nakagawa, Y. Oka, and J. R. Anderson, J. Magn. Magn. Mater. **90/91**, 778 (1990).
[197] J. Tuchendler and J. P. Renard, Solid State Commun. **71**, 1073 (1989).
[198] M. Motokawa, C. Uyeda, A. Otsuka, M. Date, and R. M. Toussaint, J. Phys. Soc. Japan. **54**, 2317 (1985).
[199] R. W. Sanders, R. M. Belanger, M. Motokawa, V. Jaccarino, and S. M. Rezende, Phys. Rev. **B23**, 1190 (1981).
[200] R. M. Belanger, D. Hone, and M. Motokawa, Phys. Rev. **B25**, 3186 (1982).
[201] J. Tuchendler, J. Magariño, and J. P. Renard, Phys. Rev. **B24**, 5363 (1981).
[202] J. Magariño, J. Tuchendler, and J. P. Renard, Solid State Commun. **26**, 721 (1978).
[203] J. Tuchendler, J. Magariño, and J. P. Renard, Phys. Rev. **B20**, 2637 (1979).
[204] M. Hagiwara, K. Katsumata, and J. Tuchendler, Physica B **211**, 190 (1995).
[205] P. de Groot, P. Janssen, F. Herlach, G. De Vos, and J. Witters, J. Magn. Magn. Mater. **31–34**, 637 (1983).
[206] M. Motokawa, A. Otsuka, C. Uyeda, and M. Date, J. Phys. Soc. Japan. **53**, 1857 (1984).
[207] S. Teraoka, T. Kambe, N. Koido, S. Hirai, and K. Nagata, J. Magn. Magn. Mater. **140–144**, 1659 (1995).
[208] M. Motokawa, H. Ohta, K. Fukuda, N. Kitamura, T. Hirano, and K. Yamazaki, Physica B **155**, 340 (1989).

[209] H. Ohta, N. Makita, K. Yoshida, T. Nanba, and M. Motokawa, Internat. J. Infrared MM Wave **13**, 457 (1992).

[210] H. Ohta and M. Motokawa, in *Recent Advances in Magnetism of Transition Metal Compounds*, edited by A. Kotani and N. Suzuki (World Scientific, Singapore, 1993), p. 316.

[211] K. Katsumata and J. Tuchendler, in *High Field Magnetism*, edited by M. Date (North-Holland, Amsterdam, 1983), p. 237.

[212] H. Ohta, S. Imagawa, M. Motokawa, and E. Kita, J. Phys. Soc. Japan. **62**, 4467 (1993).

[213] P. de Groot, P. Janssen, F. Herlach, J. Witters, and G. De Vos, in *High Field Magnetism*, edited by M. Date (North-Holland, Amsterdam, 1983), p. 245.

[214] P. de Groot, P. Janssen, F. Herlach, G. De Vos, and J. Witters, Bull. Magn. Reson. **5**, 187 (1983).

[215] L. van Bockstal and F. Herlach, J. Phys.: Conden Matter **2**, 7187 (1990).

[216] H. Ohta, K. Fukuda, N. Kitamura, and M. Motokawa, J. Phys. Colloque **C8**, 1457 (1988).

[217] M. Motokawa, H. Ohta, N. Makita, and H. Ikeda, J. Phys. Soc. Japan. **61**, 322 (1992).

[218] H. Ohta, S. Imagawa, M. Motokawa, and H. Ikeda, J. Phys. Soc. Japan. **62**, 2481 (1993).

[219] S. Kuroda, M. Motokawa, and M. Date, J. Phys. Soc. Japan. **44**, 1797 (1978).

[220] T. Kambe, H. Tanake, S. Kimura, H. Ohta, M. Motokawa, and K. Nagata, J. Magn. Magn. Mater. **140–144**, 1967 (1995).

[221] H. Ohta, Y. Yamamoto, M. Sumikawa, M. Motokawa, A. Hirano, and R. Kanno, Physica B **211**, 217 (1995).

[222] W. Lu, J. Tuchendler, M. von Ortenberg, and J. P. Renard, Phys. Rev. Lett. **67**, 3716 (1991).

[223] L. C. Brunel, T. M. Brill, I. Zaliznyak, J. P. Boucher, and J. P. Renard, Phys. Rev. Lett. **69**, 1699 (1992).

[224] T. M. Brill, J. P. Boucher, L. C. Brunel, J. P. Renard, and M. Verdaguer, J. Magn. Magn. Mater. **140–144**, 1655 (1995).

[225] T. M. Brill, J. P. Boucher, L. C. Brunel, J. P. Renard, and M. Verdaguer, Physica B **204**, 303 (1995).

[226] W. Palme, F. Mertens, O. Born, B. Lüthi, and U. Schotte, Solid State Commun. **76**, 873 (1990).

[227] H. Ohta, S. Imagawa, M. Motokawa, and H. Tanaka, Phys. Soc. Japan. **62**, 3011 (1993).

[228] H. Ohta, S. Imagawa, M. Motokawa, and H. Tanaka, Physica B **201**, 208 (1994).

[229] M. Motokawa, M. Arai, H. Ohta, M. Mino, H. Tanaka, and K. Ubukata, Physica B **211**, 199 (1995).

[230] S. Kuroda, M. Motokawa, and M. Date, Phys. Soc. Japan. **61**, 1036 (1992).

[231] H. Ohta, N. Yamauchi, T. Nanba, M. Motokawa, S. Kawamata, and K. Okuda, J. Phys. Soc. Japan. **62**, 785 (1993).

[232] T. M. Brill, J. P. Boucher, J. Voiron, W. Palme, B. Lüthi, G. Dhalenne, and A. Revcolevschi, J. Magn. Magn. Mater. **140–144**, 1683 (1995).

[233] H. Ohta, S. Imagawa, H. Ushiroyama, M. Motokawa, O. Fujita, and J. Akimitsu, J. Phys. Soc. Japan. **63**, 2870 (1994).

[234] S. Imagawa, H. Ohta, M. Motokawa, O. Fujita, and J. Akimitsu, Physica B **201**, 178 (1994).

[235] H. Ohta, Y. Yamamoto, S. Imagawa, T. Nanba, K. Watanabe, M. Arai, M. Motokawa, O. Fujita, and J. Akimitsu, Internat. J. Infrared MM Wave **16**, 501 (1995).

[236] T. M. Brill, J. P. Boucher, J. Voiron, G. Dhalenne, A. Revcolevschi, and J. P. Renard, Phys. Rev. Lett. **73**, 1545 (1994).

[237] W. Palme, G. Ambert, J. P. Boucher, G. Dhalenne, and A. Revcolevschi, J. Appl. Phys. **79**, 5384 (1996).

[238] H. Ohta, S. Imagawa, Y. Yamamoto, M. Motokawa, O. Fujita, and J. Akimitsu, J. Magn. Magn. Mater. **140–144**, 1685 (1995).

[239] N. Adachi, T. Hamamoto, G. Kido, M. Hase, Y. Sasago, K. Uchinokura, S. Noro, and T. Yamadaya, Physica B **201**, 174 (1994).

[240] H. Ohta, M. Sumikawa, M. Motokawa, S. Noro, and T. Yamadaya, J. Phys. Soc. Japan. **64**, 1759 (1995).

[241] S. Kimura, H. Ohto, and M. Motokawa, J. Phys. Soc. Japan. **65**, 297 (1996).

[242] S. Kimura, K. Kaneko, H. Ohta, and M. Motokawa, Physica B **201**, 115 (1994).

[243] H. Ohta, S. Kimura, and M. Motokawa, J. Phys. Soc. Japan. **64**, 3934 (1995).

[244] S. Kimura, H. Ohta, S. Mitsudo, M. Motokawa, W.-J. Jang, M. Hasegawa, and H. Takei, J. Phys.: Conden. Matter **8**, 5461 (1996).

[245] A. Jánossy, J. R. Cooper, L.-C. Brunel, and A. Carrington, Phys. Rev. B **50**, 3442 (1994).
[246] B. Z. Malkin, V. F. Tarasov, and G. S. Shakurov, JETP Lett. **62**, 811 (1995).
[247] J. Magariño, J. Tuchendler, P. Beauvillain, and I. Laursen, Phys. Rev. B **21**, 18 (1980).
[248] P. Janssen, I. DeWolf, and I. Laursen, J. Phys. Chem. Solids **46**, 1387 (1985).
[249] P. Janssen, P. de Groot, G. De Vos, F. Herlach, and J. Witters, in *High Field Magnetism*, edited by M. Date (North-Holland, Amsterdam 1983), p. 241.
[250] P. de Groot, P. Janssen, F. Herlach, G. De Vos, and J. Witters, Internat. J. Infrared MM Waves **5**, 135 (1984).
[251] H. Ohta, M. Sumikawa, K. Kita, M. Motokawa, T. Seixas, and J. M. Machado da Silva, Physica B **216**, 341 (1996).
[252] R. J. Wagner and G. A. Prinz, Appl. Optics **10**, 2060 (1971).

INDEX